Complexes of Differential Operators

Mathematics and Its Applications

Volume 340

Complexes of Differential Operators

by

Nikolai N. Tarkhanov
Institute of Physics,
Siberian Academy of Sciences,
Krasnoyarsk, Russia

SPRINGER SCIENCE+BUSINESS MEDIA, B.V.

Library of Congress Cataloging-in-Publication Data

Tarkhanov, N. N. (Nikolaĭ Nikolaevich)
[Metod parametriksa v teorii different͡sial'nykh kompleksov. English]
Complexes of differential operators / by Nikolai N. Tarkhanov.
p. cm. -- (Mathematics and its applications ; v. 340)
Includes bibliographical references and indexes.
ISBN 978-94-010-4144-7 ISBN 978-94-011-0327-5 (eBook)
DOI 10.1007/978-94-011-0327-5
1. Differential operators. 2. Complexes. I. Title. II. Series: Mathematics and its applications (Kluwer Academic Publishers) ; v. 340.
QA329.4.T3713 1995
514'.72--dc20 95-31287

ISBN 978-94-010-4144-7

Printed on acid-free paper

This is a completely revised and updated translation from the Russian original work *Parametrix Method in the Theory of Differential Complexes,* Novobirsk, Nauka © 1990.

Translation by
P.M. Gauthier, Université de Montréal, Mathématiques,
Montréal, Québec, Canada.

To the shining memory of my mother
Alexandra Fedorovna Tarkhanova

Contents

Preface to the English Translation

This English translation of my book "Parametrix Method in the Theory of Differential Complexes" differs from the Russian edition to a marked degree.

There are considerably enlarged Sections: 9, 13, 16 and 18. Sections 3, 14, 15 and 20 have been revised. A new chapter, "The Atiyah-Bott-Lefschetz theorem on fixed points for elliptic complexes", has been added. The number of examples has been increased in order to help in understanding the general theory. Finally, many exercises have been inserted. Most of them are pedagogical. However some exercises reflect questions which are still open.

I would like to thank Kluwer Academic Publishers for making the English translation of my book available. Professor P.M. Gauthier has kindly agreed to make corrections on my original translation and has also suggested several improvements. O.Karepov and A.Shlapunov rendered invaluable assistance in organizing the material for the manuscript.

The new research contained herein was carried out during my stay at the *Max-Planck-Arbeitsgruppe "Partielle Differentialgleichungen und Komplexe Analysis"* at the *Universität Potsdam* for which I would like to express my profound gratitude to the Alexander von Humboldt Foundation and to Professor B.-W. Schulze for the kind invitation.

Finally, I am greatly indebted to Dr. T. Hirschmann, who acquainted me with EMTEX, in which the book was written.

Potsdam
October, 1993

N. Tarkhanov

Preface to the Russian Edition

This book gives a systematic account of the facts concerning complexes of differential operators on differentiable manifolds.

The central place is occupied by the study of general complexes of differential operators between sections of vector bundles. Although the global situation often contains nothing new as compared with the local one (that is, complexes of partial differential operators on an open subset of $\mathbb{R}^n$), the invariant language allows one to simplify the notation and to distinguish better the algebraic nature of some questions.

In the last 2 decades within the general theory of complexes of differential operators, the following directions were delineated: 1) the formal theory; 2) the existence theory; 3) the problem of global solvability; 4) overdetermined boundary problems; 5) the generalized Lefschetz theory of fixed points, and 6) the qualitative theory of solutions of overdetermined systems. All of these problems are reflected in this book to some degree. It is superfluous to say that different directions sometimes whimsically intersect.

Considerable attention is given to connections and parallels with the theory of functions of several complex variables.

One of the reproaches avowed beforehand by the author consists of the shortage of examples. The framework of the book has not permitted their number to be increased significantly.

Certain parts of the book consist of results obtained by the author in 1977-1986. They have been presented in seminars in Krasnoyarsk, Moscow, Ekaterinburg, and Novosibirsk.

A great part of the material was presented as a special course for the students of the Department of Mathematics of Krasnoyarsk University in 1981-1986.

This book has been written at the suggestion of Professor L.A. Aizenberg. I am grateful to him for his constant help.

Consultations with Professors M.A. Shubin and V.P. Palamodov were very valuable for me.

Let us make some remarks on the use of this book.

It is assumed that the reader is acquainted with the main theorems of linear functional analysis and the theory of distributions (Edwards' book [62] is more than sufficient). A knowledge of the main notions connected with manifolds and vector bundles, sheaf theory, pseudo-differential operators, and elliptic equations (for example, in the first four chapters of Wells' book [254]) is also assumed.

The book is divided into 6 chapters, the numeration of the sections is throughout. The formula (a.b) is located in section "a" and has number "b" there. The symbol □ signals the end of a proof.

Krasnoyarsk
February, 1989

N. Tarkhanov

Introduction

0.0.1 *Timeliness*

Loosely speaking, the universe around us seems to be overdetermined. That is why an adequate picture of it, especially with respect to cause-effect relations, can be written only in the language of overdetermined systems of differential operators.

A heuristic definition of an overdetermined differential operator P_0 is that there exists a differential operator $P_1 \neq 0$ for which the composition $P_1 P_0$ (is defined and) $= 0$. Although the main equations of mathematical physics are not overdetermined, the situation here is similar to that in the problem of finding a conditional extremum, where "superfluous" variables have already been excluded with the help of side conditions.

Complexes of differential operators or, as we shall say, differential complexes, naturally appear as resolutions of overdetermined differential operators (see Spencer's survey [208])

$$\mathfrak{C}(E^{\cdot}) : \; 0 \longrightarrow \mathfrak{C}(E^0) \xrightarrow{P} \mathfrak{C}(E^1) \xrightarrow{P} \ldots \xrightarrow{P} \mathfrak{C}(E^N) \longrightarrow 0. \qquad (0.0.1)$$

Here $\mathfrak{C}(E^i)$ $(i = 0 \ldots N)$ are the spaces of sections of a class $\mathfrak{C}$ of differentiable (vector) bundles E^i over a differentiable manifold X, $P_i \in \mathrm{do}_{p_i}(E^i \to E^{i+1})$ are differential operators satisfying $P_{i+1} P_i = 0$ and P is the graduated operator defined by $Pf = P_i f$ for $f \in \mathfrak{C}(E^i)$.

By the Abstract de Rham Theorem the cohomologies of the sheaf of solutions of a homogeneous overdetermined system of differential equations are isomorphic to the cohomologies of a resolution of the differential operator which defines this system. This suggests that any serious investigation of overdetermined systems must include the study of corresponding differential complexes. It is known, say, that a domain $\mathcal{O} \subset \mathbb{C}^n$ is the domain of existence of some holomorphic function if and only if the Dolbeault complex over this domain is acyclic in positive degrees.

Generally, complex analysis and the theory of differential complexes are two closely related subjects. On the one hand the study of the Dolbeault complex is essential to understanding holomorphic functions of several complex variables. This complex is a particular example of a complex of differential operators and thus many results in complex analysis could be considered as particular instances of the more general theory of differential complexes. Our knowledge of the Dolbeault complex is a powerful source of intuition to forecast the behaviour of general differential complexes. Moreover, holomorphic functions play a very central role because they are also an essential tool for the study of differential operators. The relationship of these two fields has been well observed in the book of Palamodov [175] and the lectures of Nacinovich [166].

It is clear from what has been said that the theory of overdetermined systems of differential equations is in essence the theory of complexes of differential operators. Spencer [208] defined the notion of an overdetermined system in this manner.

Complexes are a part of homological algebra and therefore its language and methods are naturally used in the study of differential complexes. They have turned out to be especially useful in investigation of so-called Hilbert complexes in $\mathbb{R}^n$ which appear as resolutions of overdetermined systems of differential operators with constant coefficients. A symbiosis of Fourier analysis, algebraic geometry and $\overline{\partial}$-cohomologies with estimates had been expressed in the Ehrenpreis-Malgrange-Palamodov theorem on the exponential representation of solutions of systems of differential equations with constant coefficients. On this basis it has proved to be possible to construct a unified theory of Hilbert complexes (see Palamodov [175]).

The so-called elliptic complexes, which are defined by the condition of exactness of the sequence (0.0.1) on the symbolic level, have the strongest resemblance by their properties to complexes of differential operators with constant coefficients. This notion, naturally generalizing the definition of an elliptic differential operator, was introduced by Atiyah and Bott [19] who had developed ideas of Spencer [205]. Classical examples of elliptic complexes are the de Rham complex and the Dolbeault complex which play a great role in differential geometry. The class of elliptic complexes is all the more significant in that the second Spencer resolution of an overdetermined differential operator P_0 with injective symbol is an elliptic complex (see Spencer [208]).

In connection with an increased interest in overdetermined systems of differential equations of general form, a further study of differential complexes on manifolds and an elaboration of new constructive approaches to them seems to be timely.

0.0.2 *Directions*

Over the past sixty years the following directions within the theory of general complexes of differential operators emerged: 1) the formal theory; 2) the existence theory; 3) the problem of global solvability; 4) overdetermined boundary problems; 5) the generalized Lefschetz theory of fixed points, and 6) the qualitative theory of solutions of overdetermined systems.

The formal theory is concerned with the construction of compatibility conditions for overdetermined non-homogeneous systems of differential equations. This means conditions under which there is a formal power series with center in a given point satisfying the system. The formal theory has taken a satisfactory appearance mainly thanks to the fundamental works of Quillen [186] and Goldschmidt [74, 75] based on the earlier geometrical approach of Spencer [203, 204, 207]. For applications the aspect of this theory connected with systems of non-linear differential equations is especially important (see Pommaret [182] and Vinogradov, Krasil'shik, Lychagin [244]).

The more difficult existence theory is the second step in the study of overdetermined systems which consists in the search for an acceptable condition for the local exactness in the C^∞_{loc} sense of a given formally exact complex. By the efforts of Sweeney [210, 211, 212], MacKichan [154, 155] and others a homological condition for the local exactness in the C^∞_{loc} sense, the so-called δ-estimate, has been found. Andreotti and Nacinovich [14] provided some further conditions for the validity of the analytic and C^∞_{loc} Poincaré lemma. But the existence theory is far from complete. For example, it is not known whether a formally exact elliptic complex is exact on the C^∞_{loc} sheaves level.

The problem of global solvability in the widest sense consists of the study of cohomologies in some class of sections of a given differential complex on a manifold. This problem has been solved only in its classical variant connected with elliptic complexes on compact manifolds where the answer is given by the Hodge theory (see Wells [254, Ch.IV]. In one of its aspects the problem of global solvability comes into contact with non-coercive boundary problems. Namely, in connection with a study of the solvability in the C^∞_{loc} sense of overdetermined non-homogeneous systems of differential equations, Spencer [205] suggested an approach which in certain cases reduced this question to the solvability of some boundary problem for an elliptic system, the so-called Neumann problem. But the boundary conditions turned out to be non-coercive in general. The principal difficulty arose in proving the smoothness of a solution up to the boundary. Kohn and Nirenberg [131] showed that in the presence of a subelliptic estimate the Neumann problem is solvable and the solution is regular up to the boundary. Hörmander [111] described the widest class of boundary conditions for elliptic systems such that for solutions of corresponding boundary problems there is a subelliptic estimate with a loss of $\frac{1}{2}$ in the order of the estimated derivatives. But in the application to the Neumann problem his results do not shed much light. Nacinovich [167] extended the method of L^2 estimates with weights, originally introduced for the $\bar{\partial}$-operator by Hörmander [109], to obtain global solvability conditions for general overdetermined non-homogeneous systems. By this approach it is possible to avoid the question of boundary regularity.

The theory of overdetermined boundary problems, or the theory of boundary differential complexes, is on the whole parallel to the theory of overdetermined systems of differential equations. The formal theory of such problems was constructed by Samborskii [191]. Principles of the theory of elliptic complexes of pseudo-differential operators on a manifold with boundary have been set forth in the papers of Dynin [60], Pillat and Schulze [181] Dudnikov and Samborskii [57] and others. The boundary differential complexes related to the overdetermined Cauchy problem were investigated in the fundamental work of Andreotti, Hill, Lojasiewicz and MacKichan [11].

The generalized Lefschetz theory of fixed points involves first of all the Atiyah-Singer index theorem and its generalization to group actions (see Gilkey's survey [70]). Later on, Atiyah and Bott [19, 20] showed that the classical Lefschetz theory of fixed points also made sense in the context of general elliptic complexes on compact manifolds. Brenner and Shubin [32] obtained an analogous result for elliptic complexes of pseudo-differential operators on a manifold with boundary. Toledo

[238] extended the Atiyah-Bott theorem to the case of isolated fixed points. But nothing is known, say, for subelliptic complexes.

Finally, the qualitative theory deals with specific properties of solutions of overdetermined systems of differential equations such as the continuous dependence of solutions on their jets on sets of low dimension, a forced continuation of solutions to a neighborhood of a compact set of singularities, and so on (see Palamodov [175, Ch.8]). But for the most part these results relate to solutions of overdetermined systems of differential equations with constant coefficients.

0.0.3 *Purpose*

This book has been planned as an attempt to shed light on and to systematize some of the numerous facts concerning complexes of differential operators of general form on manifolds which could lay down a basis for more specialized investigations of both theoretical and applied character.

0.0.4 *Methods*

For elliptic complexes on manifolds there is a classical approach due to Levi [149] and Hilbert [106] and developed by de Rham [50] in his theory of currents on a differentiable manifold. It is based on the notion of a parametrix.

The term "parametrix" in the theory of second-order elliptic equations was introduced by Hilbert and meant something which was sort of a rough approximation to some fundamental solution. In the case of homogeneous equations with constant coefficients a fundamental solution is found without difficulty. For equations with variable coefficients a fundamental solution of the principal part with "frozen" coefficients gives a parametrix. Then the finding of a fundamental solution is reduced to the solving of some integral equation of the second-kind with a weak singularity of the kernel.

This is the idea of the "parametrix method" of Levi [149] and Hilbert [106].

The construction of paramertrices for general elliptic equations has been known at least since the middle of the 50's. As a matter of fact, out of it arose the theory of pseudo-differential operators, now called *classical* pseudo-differential operators. At the same time the notion of parametrix itself had been transformed: it came to mean an operator Π satisfying $\Pi P = P\Pi = 1$ up to a smoothing operator (see Hörmander [111]). By contrast a parametrix in the previous sense is called a rough one. A true parametrix is obtained from the rough one by means of an iterative procedure often called the Levi method.

The construction of parametrices for parabolic or, more generally, for hypoelliptic equations led Hörmander [110] to an investigation of a wider class of pseudo-differential operators than the classical ones. The construction of parametrices for (the Cauchy problem for) hyperbolic equations takes us out of this framework and requires the utilization of Fourier integral operators (see Hörmander [112]).

Systematic applications of parametrices to scalar differential operators can be found in the books of John [119], Bers, John and Schechter [22], Shilov [200], Friedman [68], Hörmander [114] and others.

Qualitatively new obstacles appear in the consideration of overdetermined systems. The point is that an overdetermined differential operator P_i can not have a right fundamental solution because for the solvability of the equation $P_i u = f$ it is necessary that $P_{i+1} f = 0$. That is why the notion of the right parametrix loses its sense. The same may be said concerning the left parametrix provided the differential operator P_i is underdetermined (that is the differential operator P_i' is overdetermined). Having joined the topological idea of homotopy with the analytical idea of current, de Rham [50] guessed that a substitute for the equations $\Pi P = 1$ and $P\Pi = 1$ was the homotopy equation $\Pi P + P\Pi = 1$. The regularization (or parametrix) method formed the principal analytic contents of de Rham's treatise [50].

The notion of a parametrix for arbitrary elliptic complexes on compact manifolds attracted attention after the deep paper of Atiyah and Bott [19] devoted to a generalization of the Lefschetz theory. Following that, a series of papers appeared in which parametrices were applied for the calculation of the generalized Euler characteristics for elliptic complexes (see Kotake [137], Toledo [238], Toledo and Tong [239], Gilkey [70] and others).

Together with the fundamental discovery of Atiyah and Bott [19], at quite the same time in complex analysis there was an almost imperceptible find of Koppelman [136]. These two discoveries cried out to one another. Koppelman had constructed a convolution type fundamental solution for the Dolbeault complex in $\mathbb{C}^n$. However the general concept of fundamental solution for the Dolbeault complex appeared only after 12 years in the papers of Harvey and Polking [100, 101]. Roughly speaking, the parametrix of a differential complex is a cochain homotopy between the identity endomorphism of the complex and a smoothing one while the fundamental solution is a cochain homotopy between the identity endomorphism and the null one. Later on, Koppelman's work [136] caused a stream of papers devoted to solutions with estimates of the $\bar{\partial}$-problem. Apparently the parametrix method in complex analysis, traditionally called the method of integral representations, took a finished shape with the discovery of the Koppelman formula. It is given in detail in the book of Aizenberg and Yuzhakov [6] and in the survey of Henkin [104].

On compact manifolds the smoothing operators are compact in the Sobolev spaces. This has served as a basis for the abstract notion of parametrix in the general theory of Fredholm operators (left and right regularizations!) or complexes. By a parametrix is understood a cochain homotopy between the identity endomorphism of the complex and a compact one. For complexes of Banach spaces the existence of a parametrix is equivalent to the complex being Fredholm, in other words its cohomology spaces are finite-dimensional (see Atkinson [21], Rempel and Shulze [188]). These general definitions are finding a favourable ground in the theory of elliptic complexes of pseudo-differential operators on manifolds with boundaries (see Dynin [60], Pillat and Schulze [181], Dudnikov and Samborskii [57] and others).

Elliptic complexes on non-compact manifolds are usually considered in classes of locally convex (non-normed!) spaces of sections. And here there is no connection between the existence of a parametrix and the complex being Fredholm. For example, the complex $C^{\infty}_{loc}(E^{\cdot})$ always has a parametrix in the above indicated (abstract) sense but is not a Fredholm complex in general. Hence in that case the parametrix does not answer the principal question concerning the complex (0.0.1), that is, the question on the solvability of the equation $Pu = f$. This fact evoked a certain scepticism in the appraisal of a role for parametrices for elliptic complexes on non-compact manifolds manifested by a limited interest in them.

On the other hand the de Rham theory and the method of integral representations in complex analysis pointed out the rich possibilities of joining the parametrix method with other means: 1) the theory of distributions with vector values of Schwartz, in particular, the apparatus of topological tensor products; 2) homological algebra and sheaf theory; 3) main theorems of linear functional analysis; 4) the modern theory of pseudo-differential operators and others.

The construction of parametrices for elliptic complexes is one of the simplest applications of the calculus of pseudo-differential operators. Thus, both the method of exponential representation and the parametrix method have the same foundations because the Fourier transform lies at their base. Moreover, the parametrix method is equally well applicable to both complexes of differential operators with constant coefficients in $\mathbb{R}^n$ and elliptic complexes on differentiable manifolds.

0.0.5 *Approach*

In conclusion we shall explain in a few words the essence of our approach. Let us concentrate upon the i-th fragment of the complex (0.0.1) which will be written schematically in the form $L^{i-1} \xrightarrow{P} L^i \xrightarrow{P} L^{i+1}$, where L^j are some spaces of distributions with vector values. Consider the equation $Pf = 0$ where $f \in L^i$. The set of solutions of this equation is $Z^i(L^{\cdot}) = \ker P_i$. It is natural to consider the solutions of the form $f = Pu$ $(u \in L^{i-1})$ to be trivial. The set of all trivial solutions is denoted by $B^i(L^{\cdot}) = \operatorname{im} P_{i-1}$. Then the quotient space $H^i(L^{\cdot}) = Z^i(L^{\cdot})/B^i(L^{\cdot})$ reflects the availability of non-trivial solutions of the equation $Pf = 0$, and one can interpret its elements as "generalized " solutions. It is clear that the trivial solutions may have very bad local properties. But it turns out that, if the operators P_{i-1} and P_i are correctly selected, in each cohomology class from $H^i(L^{\cdot})$ there is a representative with sufficiently good regularity properties. An important aspect of the parametrix is that it allows one to pick out such a distinguished representative. A distinguished representative of a class in $H^i(L^{\cdot})$ of a solution of $Pf = 0$ contains essential information about this solution that is transparent.

0.0.6 *Results*

In Chapter 1 we present a sketch of the formal theory for overdetermined systems of differential equations both with constant coefficients and of general type.

In Chapter 2 the applications of parametrices and fundamental solutions to complexes of differential operators are elaborated.

Chapter 3 contains a systematic account of boundary properties of integrals of double layer potential type, associated with a differential complex.

In Chapter 4 a constructive approach to the overdetermined Cauchy problem for cohomologies of differential complexes is given.

In Chapter 5 we present a constructive duality theory for cohomologies of differential complexes.

Finally Chapter 6 is devoted to a generalization of the Atiyah-Bott-Lefschetz theorem to the case of an arbitrary set of fixed points.

0.0.7 *Authorship*

It is very difficult to give the correct references in a book of such a type. Many ideas and methods were slowly being developed over a long time. Besides there is a boundless number of recent results concerning complexes of differential operators of particular forms. At the same time the questions of proper credits are usually perceived too sharply. However I have decided in each section to give a few references to corresponding papers. Sooner or later such an attempt had to be made. I hope these references are not too inadequate. I present my sincere apologies to these authors whose important papers have not been mentioned.

List of Main Notations

$\mathbb{N}$	the set of positive integers (i.e., natural numbers);
$\mathbb{Z}_+$	the set of nonnegative integers;
$\mathbb{Z}_+^n$	the Cartesian product of n copies of $\mathbb{Z}_+$, whose elements are called multi-indexes and are usually denoted by α, β, γ etc.;
1_j	the multi-index from $\mathbb{Z}_+^n$ which is 1 in the j-th place and 0 in each other one. The length n will always be clear from the context;
$\|\alpha\|$	$= \alpha_1 + \ldots + \alpha_n$ for $\alpha = (\alpha_1, \ldots, \alpha_n) \in \mathbb{Z}_+^n$;
$\alpha!$	$= \alpha_1! \times \ldots \times \alpha_n!$ for $\alpha = (\alpha_1, \ldots, \alpha_n) \in \mathbb{Z}_+^n$;
$\mathbb{R}$	the set of real numbers;
$\mathbb{R}^n$	the n-dimensional real space of points with coordinates $x = (x_1, \ldots, x_n)$ endowed with its Euclidean structure;
dx	$= dx_1 \wedge \ldots \wedge dx_n$ is the Lebesgue measure in $\mathbb{R}^n$;
D	$= (D_1, \ldots, D_n)$ is the vector of the differentiations $D_j = \frac{1}{\sqrt{-1}} \partial / \partial x_j$;
D^α	$= D_1^{\alpha_1} \ldots D_n^{\alpha_n}$ for $\alpha = (\alpha_1, \ldots, \alpha_n) \in \mathbb{Z}_+^n$;
$\mathbb{C}$	the set of complex numbers;
$\mathbb{C}^n$	the n-dimensional complex space of points with coordinates $z = (z_1, \ldots, z_n)$ endowed with its Hermitian structure;
z^α	$z_1^{\alpha_1} \ldots z_n^{\alpha_n}$ for $z = (z_1, \ldots, z_n) \in \mathbb{C}^n$ and $\alpha = (\alpha_1, \ldots, \alpha_n) \in \mathbb{Z}_+^n$;
L^k	$L \times \ldots \times L$ (k copies) for a vector space L;
M^t	the transposed matrix of a matrix M;
M^*	the conjugate matrix of a matrix M of complex numbers;
I_k	the identity $(k \times k)$-matrix;
L'	the dual space of a topologocal vector space L endowed with the strong topology;
$\mathrm{tr} = \langle \cdot, \cdot \rangle$	is a duality pairing $L' \otimes L \to \mathbb{C}$, for a topological vector space L;
$\sigma_1 \subset\subset \sigma_2$	for subsets σ_ν ($\nu = 1, 2$) of a manifold X means that $\overline{\sigma_1}$ (the closure of σ_1) is compact and $\overline{\sigma_1} \subset \sigma_2$.

Chapter 1

Resolution of Differential Operators

1.1 Differential Complexes and Their Cohomology

1.1.1 *Manifolds countable at infinity*

Let X be a differentiable (that is of class C^{∞}_{loc}) manifold of dimensions n that is countable at infinity. Later on the following two properties of such manifolds will be used.

Proposition 1.1.1 *On any manifold X countable at infinity there is a sequence $\{U_\nu\}$ of relatively compact open sets forming a covering of X and such that $\overline{U_\nu} \subset U_{\nu+1}$ for all ν.*

Proof. See Bourbaki [27, Ch.I, §9]

□

A covering $\{U_\alpha\}_{\alpha \in A}$ is said to be a refinement of a covering $\{U'_\beta\}_{\beta \in B}$ if there exists a mapping $M \cdot A \to B$ such that $U_\alpha \subset U'_{M(\alpha)}$ for all $\alpha \in A$.

Proposition 1.1.2 *For any open covering U of a manifold X countable at infinity there is a countable locally finite open covering U' of X that is a refinement of U and consists of relatively compact sets.*

Proof. See Bourbaki [27, Ch.I, §9].

□

In particular each manifold X countable at infinity is paracompact or, what is equivalent, metrisable.

Proposition 1.1.3 *For any manifold X, the following 2 conditions are equivalent: 1) X is countable at infinity, and 2) the topology of X has a countable base.*

Proof. It follows from the Corollary of Bourbaki [28, Ch.IX, §2,n.9] because, for manifolds, both the condition of paracompactness and the condition of metrisability coincide (see Spivak [209, p. A-1]).

□

Corollary 1.1.4 *Any open subset of a manifold X countable at infinity is itself countable at infinity.*

Proof. Let $\mathcal{O}$ be an open subset of X. It follows from Proposition 1.1.3 that the topology of X has a countable base. Then the same is true concerning the induced topology of $\mathcal{O}$. Since $\mathcal{O}$ is a manifold itself, it follows from Proposition 1.1.3 that $\mathcal{O}$ is countable at infinity.

□

1.1.2 *Partitions of unity*

A family $\{\sigma_\alpha\}$ of subsets X is said to be locally finite if any point $x \in X$ has a neighborhood $U = U(x)$ such that $U \cap \sigma_\alpha \neq \emptyset$ for only a finite number of indices α.

The following theorem represents a fundamental result of analysis on differentiable manifolds. De Rham [50, §2] refers to Dieudonné (1937) in connection with this theorem.

Theorem 1.1.5 *For any open covering $\{U_\alpha\}_{\alpha\in A}$ of a paracompact manifold X, there is a countable family $\{\psi_\nu\} \subset C^\infty_{comp}(X)$ of non-negative functions such that:*

1) $\operatorname{supp}\psi_\nu \subset\subset U_{M(\nu)}$ *for some index* $M(\nu) \in A$*;*

2) $\{\operatorname{supp}\psi_\nu\}$ *is locally finite;*

3) $\sum \psi_\nu \equiv 1$ *on* X.

Proof. See de Rham [50, §2].

□

Because of condition 3) the family $\{\psi_\nu\}$ is called a *partition of unity* on X. Condition 1) is often formulated by saying that $\{\psi_\nu\}$ is subordinate to $\{U_\alpha\}_{\alpha\in A}$. Condition 2) is often expressed by saying that $\{\psi_\nu\}$ is locally finite.

Corollary 1.1.6 *For any open covering $\{U_\alpha\}_{\alpha\in A}$ of a paracompact manifold X, there is a family $\{\varphi_\alpha\}_{\alpha\in A} \subset C^\infty_{loc}(X)$ of non-negative functions such that*

1) $\operatorname{supp}\varphi_\alpha \subset U_\alpha$, *i.e.*, $\varphi_\alpha \equiv 0$ *in a neighborhood of* $X \setminus U_\alpha$*;*

2) $\{\operatorname{supp}\varphi_\alpha\}$ *is locally finite;*

3) $\sum \varphi_\alpha \equiv 1$ *on* X.

Proof. Let $\{\psi_\nu\}$ be a partition of unity on X subordinate to $\{U_\alpha\}$ that has been constructed in Theorem 1.1.5. Now set $\varphi_\alpha = \sum_{M(\nu)=\alpha}\psi_\nu$ and 1) - 3) follow.

□

We emphasize that the sets of indices for the covering $\{U_\alpha\}$ and the partition $\{\varphi_\alpha\}$ are the same.

1.1.3 *Sections of vector bundles*

Let E be a differentiable $\mathbb{C}$-vector bundle of rank k over X, or, more concisely, $\pi : E \to X$. By $\sigma(E)$ we denote the vector space of all sections of E, that is, all mappings $f : X \to E$ such that the composition πf gives the identity mapping of X. In other words f maps a point $x \in X$ into the fiber E_x of the bundle E over this point. Locally each section of E is represented by a vector-function on an open subset of the space $\mathbb{R}^n$ with values in $\mathbb{C}^k$. How does this happen? We suppose that U is some coordinate neighborhood on X over which E is trivial, and let $t : E|_U \to U \times \mathbb{C}^k$ be this trivialization. For $f \in \sigma(E)$, the composition $U \xrightarrow{f} E|_U \xrightarrow{t} U \times \mathbb{C}^k \xrightarrow{proj} \mathbb{C}^k$ defines a vector-function $f_U : U \to \mathbb{C}^k$ that is called the representation of f in U. If U_1 and U_2 are some coordinate neighborhoods on X with $U_1 \cap U_2 \neq \emptyset$ over which the bundle E is trivial with trivializations t_1 and t_2 respectively, then the representations f_1 and f_2 of the section f in U_1 and U_2 are connected in the intersection $U_1 \cap U_2$ by means of the equality $f_2 = t_{21} f_1$ where $t_{21} : U_1 \cap U_2 \to GL(k, \mathbb{C})$ are the transition matrices for the bundle E induced by the mapping $t_2 t_1^{-1} : (U_1 \cap U_2) \times \mathbb{C}^k \to (U_1 \cap U_2) \times \mathbb{C}^k$. The smoothness of the transition matrices is determined by the smoothness of the bundle E so an analysis of differentiability properties of sections of vector bundles is reduced to an investigation of such properties for their local representations, where in each neighborhood it is sufficient to limit oneself to some one representation. The use of local representations also gives a way to define a topology in spaces of differentiable sections of E. We shall illustrate this by the examples of the smooth, Hölder, Sobolev and Besov sections of E.

1.1.4 *Spaces of smooth sections*

For $s \in \mathbb{Z}_+$ we denote by $C^s_{loc}(E)$ the vector space of all s times continuously differentiable sections of E. The topology on $C^s_{loc}(E)$ is introduced in the following way. We fix some locally finite coordinate covering $U = \{U_\nu\}_{\nu \in \mathbb{N}}$ of the manifold X, over which the maps of the bundle E are trivial, and compact sets $K_\nu \subset U_\nu$ such that $\{K_\nu\}$ still cover X. Let f_ν be the representation of f in U_ν. Then the topology in $C^s_{loc}(E)$ may be given with the help of the family of seminorms $\{p^s_{(\nu)}\}_{\nu \in \mathbb{N}}$, where

$$p^s_{(\nu)}(f) = \sup_{\substack{x \in K_\nu \\ |\alpha| \le s}} |D^\alpha f_\nu(x)| \quad (f \in C^s_{loc}(E)). \tag{1.1.1}$$

It is easy to see that this topology does not depend on the choice of the covering of X and the local trivializations of E, and turns $C^s_{loc}(E)$ into a Fréchet space.

For a (closed) subset σ of X, we denote by $C^s_\sigma(E)$ the subspace of $C^s_{loc}(E)$ consisting of all sections with supports in σ. It is closed in $C^s_{loc}(E)$ so with the induced topology is a Fréchet space itself. If σ is compact, $C^s_\sigma(E)$ is a Banach space.

On the space $C^s_{comp}(E)$ of all sections from $C^s_{loc}(E)$ with compact supports, we introduce the inductive limit topology induced by the family of Banach spaces

$C^s_K(E)$, where $K \subset X$ is compact. By definition, this topology is the strongest one of all topologies on $C^s_{comp}(E)$ under which all embedding maps $i_K : C^s_K(E) \to C^s_{comp}(E)$ are continuous (see Palamodov [175, Pt 2, §1]). With this topology $C^s_{comp}(E)$ is complete.

We now define the spaces $C^s_\cdot(E|_\Sigma)$, where Σ is a closed subset of X, and the point denotes one of the symbols "*loc*", "σ" or "*comp*" (otherwise, they admit non-unique interpretations). By $C^s_\cdot(E|_\Sigma)$ we shall always understand the quotient space of $C^s_\cdot(E)$ over the subspace of all sections that are s-flat on Σ, in other words, of all $f \in C^s_\cdot(E)$ such that $D^\alpha f|_\Sigma = 0$ for all $|\alpha| \leq s$. Thus the elements of $C^s_\cdot(E|_\Sigma)$ are s-jets on Σ of sections of E of corresponding class in a neighborhood of Σ.

The quotient topology on $C^s_\cdot(E|_\Sigma)$ is separable and complete. In particular, $C^s_{loc}(E|_\Sigma)$ is a Fréchet space and even a Banach space if $\Sigma = K$ is compact. In this case the topological vector spaces $C^s_{loc}(E|_K)$ and $C^s_{comp}(E|_K)$ coincide, and we shall write the common notation $C^s(E|_K)$ for them.

It is convenient to also permit the case $s = \infty$. For $s_1 \leq s_2$, we have continuous embedding maps $C^{s_2}_{loc}(E) \hookrightarrow C^{s_1}_{loc}(E)$, and we define $C^\infty_{loc}(E)$ to be the projective limit of the spaces $C^s_{loc}(E)$ for $s \to \infty$. This means that $C^\infty_{loc}(E) = \cap_s C^s_{loc}(E)$, and the topology on $C^\infty_{loc}(E)$ is the weakest of all topologies on this space under which each embedding map $C^\infty_{loc}(E) \hookrightarrow C^s_{loc}(E)$ is continuous (see Palamodov [175, Pt 2, §1]). Now we can introduce the spaces $C^\infty_\sigma(E)$, $C^\infty_{comp}(E)$ and $C^\infty_\cdot(E|_\Sigma)$ as above. In particular, $C^\infty_\cdot(E|_\Sigma)$ is the quotient space of $C^\infty_\cdot(E)$ over the subspace of sections which are flat (i.e. ∞-flat) on the set $\Sigma \subset X$.

The following property of the spaces $C^\infty_\sigma(E|_\Sigma)$ will often be used later on. It is well-known.

Proposition 1.1.7 *For any closed subsets σ and Σ of X, $C^\infty_\sigma(E|_\Sigma)$ is a Fréchet-Schwartz space.*

Proof. With the help of Proposition 1.1.1 it is easy to represent $C^\infty_{loc}(E)$ as the projective limit of the sequence of Banach spaces $C^\nu(E|_{\overline{U_\nu}})$ with compact embedding maps $C^{\nu+1}(E|_{\overline{U_{\nu+1}}}) \hookrightarrow C^\nu(E|_{\overline{U_\nu}})$ having dense images. By Proposition 5 of Palamodov [175, Pt 2, §1] this implies that $C^\infty_{loc}(E)$ is a Fréchet-Schwartz space. We can now use Proposition 6 in the same source and conclude that $C^\infty_\sigma(E|_\Sigma)$ is a Fréchet-Schwartz space.

□

For the spaces $C^\infty_{loc}(E|_\Sigma)$ and $C^\infty_{comp}(E|_\Sigma)$ we have the generally accepted notations $\mathcal{E}(E|_\Sigma)$ and $\mathcal{D}(E|_\Sigma)$ due to Schwartz. We note that $\mathcal{D}(E)$ is a Montel space (see Edwards [62, 8.4.7]).

1.1.5 *Hölder spaces*

For $s \in \mathbb{Z}_+$ and $0 < \lambda \leq 1$ we denote by $C^{s,\lambda}_{loc}(E)$ the vector space of all sections $f \in C^s_{loc}(E)$ such that, for any coordinate neighborhood U on X over which the bundle E is trivial, a representation of f in U together with the derivatives up to order s satisfy a uniform Hölder condition of order λ on compact subsets of U.

The topology in $C^{s,\lambda}_{loc}(E)$ is introduced by the family of seminorms $\{p^{s,\lambda}_{(\nu)}\}_{\nu \in \mathbb{N}}$ where

$$p^{s,\lambda}_{(\nu)}(f) = p^{s}_{(\nu)}(f) + \sup_{\substack{x,y \in K_\nu \\ |\alpha|=s}} \frac{|D^\alpha f_\nu(x) - D^\alpha f_\nu(y)|}{|x-y|^\lambda} \qquad (f \in C^{s,\lambda}_{loc}(E)). \tag{1.1.2}$$

As before, this topology doesn't depend on the choice of the covering $\{U_\nu\}$ of X and the local trivializations of E, and turns $C^{s,\lambda}_{loc}(E)$ into a Fréchet space.

Now we can define the spaces $C^{s,\lambda}_{\sigma}(E)$ $C^{s,\lambda}_{comp}(E)$ and $C^{s,\lambda}_{\cdot}(E|_\Sigma)$ where the point denotes one of the symbols "*loc*", "σ" or "*comp.*" In particular, $C^{s,\lambda}_{\cdot}(E|_\Sigma)$ is the quotient space of $C^{s,\lambda}_{\cdot}(E)$ over the subspace of sections which are s-flat on the set $\Sigma \subset X$, endowed with the quotient topology.

If $\sigma = K$ is a compact subset of X, the topological vector spaces $C^{s,\lambda}_{loc}(E|_K)$ and $C^{s,\lambda}_{comp}(E|_K)$ coincide, and for this Banach space we use the notation $C^{s,\lambda}(E|_K)$.

1.1.6 *Sobolev spaces*

For $s \in \mathbb{Z}_+$ and $1 \le q \le \infty$, we denote by $W^{s,q}_{loc}(E)$ the vector space of all sections $f \in \sigma(E)$ such that, for any coordinate neighborhood U on X over which the bundle E is trivial, a local representation of f in U together with the (generalized) derivatives up to order s are measurable and their q-th powers are summable on compact subsets of U.

This space is a Fréchet space with the topology defined by the family of seminorms $\{p^{s,q}_{(\nu)}\}_{\nu \in \mathbb{N}}$ where

$$p^{s,q}_{(\nu)}(f) = \left(\sum_{|\alpha| \le s} \int_{K_\nu} |D^\alpha f_\nu(x)|^q dx \right)^{1/q} \qquad (f \in W^{s,q}_{loc}(E)). \tag{1.1.3}$$

Again this topology doesn't depend on the choice of the covering $\{U_\nu\}$ of X and the local trivializations of E.

If σ is a closed subset of X, $W^{s,q}_{\sigma}(E)$ means the subspace of $W^{s,q}_{loc}(E)$ consisting of all sections with supports in σ. This subspace is closed in $W^{s,q}_{loc}(E)$ so with the induced topology it is a Fréchet space and even a Banach space provided σ is compact.

On the space $W^{s,q}_{comp}(E)$ of all sections from $W^{s,q}_{loc}(E)$ with compact supports we introduce the inductive limit topology induced by the family of Banach spaces $W^{s,q}_{K}(E)$, where $K \subset X$ is compact.

It is useful to introduce the spaces $W^{s,q}_{\cdot}(E|_\Sigma)$, where the point denotes one of the symbols "*loc*", "σ" or "*comp*" and Σ is a closed subset of X. This denotes the quotient space of $W^{s,q}_{\cdot}(E)$ by the (certainly closed) subspace of those sections $f \in W^{s,q}_{\cdot}(E)$ such that $D^\alpha f = 0$ a.e. on Σ for all $|\alpha| \le s$. Thus the elements of $W^{s,q}_{\cdot}(E|_\Sigma)$ are s-jets on Σ of sections of E of corresponding class in a neighborhood of Σ. Of course, these spaces are non-trivial only if the measure of Σ is positive.

In particular, $W^{s,q}_{loc}(E|_\Sigma)$ is a Fréchet space and even a Banach space if $\Sigma = K$ is compact. In the last case the topological vector spaces $W^{s,q}_{loc}(E|_\Sigma)$ and $W^{s,q}_{comp}(E|_\Sigma)$ coincide, and for these Banach space the notation $W^{s,q}(E|_K)$ is used.

For $1 < q < \infty$ the spaces $W^{s,q}_{loc}(E)$ are reflexive which gives the possibility to extend their scale to (integral) negative indices s. Namely, for an integer $s < 0$ one can set by definition $W^{s,q}_{loc}(E) = (W^{-s,q'}_{comp}(E'))'$, E' being the dual vector bundle of E, where $1/q + 1/q' = 1$ and the space of functionals on the right-hand side is endowed with the strong topology.

We can then define the spaces $W^{s,q}_{\sigma}(E)$ and $W^{s,q}_{comp}(E)$ as above. We then have $W^{s,q}_{comp}(E) = (W^{-s,q'}_{loc}(E'))'$ for all integers $s < 0$. Thus for all integral indices s and $1 < q < \infty$ the spaces $W^{s,q}_{loc}(E)$ and $W^{-s,q'}_{comp}(E')$ are reciprocally dual.

For these spaces the Sobolev Embedding Theorems remain valid in full generality.

Theorem 1.1.8 *If $1 < q < \infty$ and $\lambda > 0$, for $1/q \le (s_1 - s_2 - \lambda)/n$ there are continuous embedding maps $W^{s_1,q}_{loc}(E) \hookrightarrow C^{s_2,\lambda}_{loc}(E)$.*

Proof. See Hörmander [113, 4.5.13].

□

In order to continue the scale of the spaces $W^{s,q}_{loc}(E)$ to the case of non-integral indices $s \in \mathbb{R}$, there are mainly 2 possibilities. Either for such s they are defined to be locally spaces of (vector-valued) Bessel potentials, which is done in a number of books (see Taylor [236] and others), or as Besov spaces (see Rempel and Schulze [188] and others). These approaches coincide if $q = 2$.

1.1.7 *Besov spaces*

We shall require the Besov spaces $B^{s,q}_{\cdot}(E)$ for non-integral real indices s. For this reason we limit ourselves to the case $s = [s] + \lambda$ where $[s]$ is the integral part of s and $0 < \lambda < 1$.

For $s > 0$ and $1 \le q \le \infty$, we denote by $B^{s,q}_{loc}(E)$ the vector space of those sections $f \in W^{[s],q}_{loc}(E)$ such that, for any coordinate neighborhood U on X over which the bundle E is trivial, a local representation of f in U together with its (generalized) derivatives up to order $[s]$ satisfy a Hölder condition of order λ in the L^q metric on compact subsets of U.

The topology in $B^{s,q}_{loc}(E)$ is induced by the family of seminorms $\{p^{s,q}_{(\nu)}\}_{\nu\in\mathbb{N}}$ where

$$p^{s,q}_{(\nu)}(f) = p^{[s],q}_{(\nu)}(f) + \left(\sum_{|\alpha|=s} \int_{K_\nu} \int_{K_\nu} \frac{|D^\alpha f_\nu(x) - D^\alpha f_\nu(y)|^q}{|x-y|^{n+\lambda q}} dx dy \right)^{1/q}$$
$$(f \in B^{s,q}_{loc}(E)). \tag{1.1.4}$$

This topology proves to be independent of the choice of the covering of X and the local trivializations of E, and turns $B^{s,q}_{loc}(E)$ into a Fréchet space.

For a closed subset σ of X, $B^{s,q}_{\sigma}(E)$ denotes the subspace of $B^{s,q}_{loc}(E)$ consisting of all sections with supports in σ. It is closed in $B^{s,q}_{loc}(E)$ and so with the induced topology it is a Fréchet space and even a Banach space if σ is compact.

We endow the space $B^{s,q}_{comp}(E)$ of those sections from $B^{s,q}_{loc}(E)$ with compact supports with the inductive limit topology induced by the family of Banach spaces $B^{s,q}_{K}(E)$, where $K \subset X$ is compact.

Moreover we stipulate the definition of the spaces $B^{s,q}_{\cdot}(E|_{\Sigma})$, where Σ is a closed subset of X, because they admit non-equivalent interpretations. This denotes the quotient space of $B^{s,q}_{\cdot}(E)$ by the closed subspace of those sections $f \in B^{s,q}_{\cdot}(E)$ such, that $D^{\alpha}f = 0$ a.e. on Σ for all $|\alpha| \le [s]$. Of course, these spaces are non-trivial only if the measure of Σ is positive. In particular, $B^{s,q}_{loc}(E|_{\Sigma})$ is a Fréchet space and even a Banach space provided $\Sigma = K$ is compact. In this case the topological vector spaces $B^{s,q}_{loc}(E|_{\Sigma})$ and $B^{s,q}_{comp}(E|_{\Sigma})$ coincide, and for this Banach space we use the notation $B^{s,q}(E|_{K})$

For $1 < q < \infty$ the spaces $B^{s,q}_{loc}(E)$ are reflexive which gives the possibility of extending their scale to (non-integral) negative indices s. Namely, for a non-integral index $s < 0$ we set by definition $B^{s,q}_{loc}(E) = (B^{-s,q'}_{comp}(E'))'$, where $1/q + 1/q' = 1$ and the space of functionals on the right side is endowed with the strong topology.

We can now define the spaces $B^{s,q}_{\sigma}(E)$ and $B^{s,q}_{comp}(E)$ as above. Thus we have $B^{s,q}_{comp}(E) = (B^{-s,q'}_{loc}(E'))'$ for all non-integers $s < 0$. Hence for all non-integral indices s and $1 < q < \infty$ the spaces $B^{s,q}_{loc}(E)$ and $B^{-s,q'}_{comp}(E')$ are mutually dual.

We shall often use the notation $B^{s,q}_{\cdot}(E|_{S})$ where S is a smooth hypersurface in X. Since the measure of S is equal to zero, the space $B^{s,q}_{\cdot}(E|_{S})$ must be defined internally only, that is, using the induced bundle $E|_{S}$ over the "initial" manifold S. Moreover, the Trace Theorem remains valid.

Theorem 1.1.9 *Let $\mathcal{O}$ be an open subset of X whose boundary S is of class C^{s+1}_{loc}. For $1 < q < \infty$ and $s - 1/q > 0$, the trace on S of each section $f \in W^{s,q}_{loc}(E|_{\mathcal{O}\cup S})$ belongs to the space $B^{s-1/q,q}_{loc}(E|_{S})$, and the corresponding embedding map is surjective, continuous and has a continuous inverse branch.*

Proof. See Triebel [241, 4.7.1].

□

1.1.8 *Spaces of generalized sections*

It is convenient to define the spaces of generalized sections of E in an invariant way.

Let Ω denote the bundle of volumes of the manifold X. This is the line bundle over X associated with the tangent bundle $T(X)$ of X by means of the representation $M \to |\det M|$ of the group $GL(n, \mathbb{R})$. The fiber Ω_x of the bundle Ω over the point $x \in X$ is the one-dimensional space of all Haar measures on $T_x(X)$.

The differentiable sections of Ω are the measures on X that can be represented in any coordinate neighborhood U in the form $\omega(x)dx$ where dx is the Lebesgue measure in $\mathbb{R}^n$ and ω is an (infinitely) differentiable function in U. By a change of variables $x = x(y)$ the representation is transformed by the rule $\omega(x)dx \to$

$\omega(x(y))|\det\frac{\partial x}{\partial y}|dy$. Hence, for any section $\omega \in L^1_{comp}(\Omega)$, one can define the integral $\int_X \omega$ in an invariant manner.

We note that, if X is oriented, Ω can be naturally identified with the bundle of exterior forms of degree n on X denoted by Λ^n.

Now, for the bundle E, we introduce the dual bundle $E' = \Omega \otimes E^* \cong \mathrm{Hom}\,(E \to \Omega)$, where $E^* = \mathrm{Hom}\,(E \to \mathbb{C})$ is the (algebraic) conjugate bundle for E and the tensor product is taken over the field $\mathbb{C}$. Then there exists a natural pairing $\langle , \rangle_x$ (or tr) : $E'_x \otimes E_x \to \Omega_x$ and for $g \in \mathcal{E}(E')$, $f \in \mathcal{E}(E)$ we can define an element $\omega \in \mathcal{E}(\Omega)$ by $\omega(x) = \langle g(x), f(x)\rangle_x$. By integrating we obtain the separately continuous bilinear pairing

$$\langle , \rangle_X \ (\text{or tr}) : \ \mathcal{D}(E') \otimes \mathcal{E}(E) \to \mathbb{C}.$$

Hence we have a continuous linear map $\mathcal{E}(E) \to (\mathcal{D}(E'))'$ where, for a topological vector space L, L' denotes the topological dual space with the strong topology. It is easy to see that this map is an injection with dense image.

The elements of $(\mathcal{D}(E'))'$ are called the distribution sections of the bundle E. Following Schwartz, we shall denote the space $(\mathcal{D}(E'))'$ by $\mathcal{D}'(E)$.

Thus we obtain vector spaces of distribution sections $\mathcal{D}'(E)$ and distribution sections with compact supports $\mathcal{E}'(E) = (\mathcal{E}(E'))'$. We always consider them to be endowed with the strong topologies of dual spaces.

The properties of the topological vector spaces $\mathcal{D}'(E)$ and $\mathcal{E}'(E)$ are like the properties of the usual spaces of distributions on open subsets of $\mathbb{R}^n$. In particular, these are Montel spaces (see Edwards [62, 8.4.7]).

In conclusion we note that all spaces of sections $\mathfrak{C}(E|_\Sigma)$ which we have considered are defined so that, if the bundle E is trivial, say, $E = X \times \mathbb{C}^k$, we have $\mathfrak{C}(E|_\Sigma) \cong [\mathfrak{C}(\Sigma)]^k$ where on the right side stands the (topological) product of k copies of the function space $\mathfrak{C}(\Sigma)$.

1.1.9 *Differential operators*

Let F be an other differentiable $\mathbb{C}$-vector bundle of rank l over X.

A differential operator P of type $E \to F$ is defined to be any $\mathbb{C}$-linear mapping $P: \ \mathcal{E}(E) \to \mathcal{E}(F)$ which is a local operator in the sense that $\operatorname{supp} Pf \subset \operatorname{supp} f$ for all $f \in \mathcal{E}(E)$.

The localness property implies that each differential operator P of type $E \to F$ induces a differential operator P of type $E|_U \to F|_U$ for any open set $U \subset X$.

The order of P at the point $x \in X$ will be denoted by $p(x)$. It is the largest integer p such that $P(\varphi^p f)(x) \neq 0$ for some smooth function φ which vanishes at x and some local section f near x. The order of a differential operator P is defined as $\sup_{x\in X} p(x)$.

We denote the vector space of all differential operators of type $E \to F$ and order $\leq p$ on X by $\mathrm{do}_p(E \to F)$.

The following result sheds light upon the local structure of differential operators.

Theorem 1.1.10 *If P is a differential operator of type $E \longrightarrow F$, then, for any choice of a coordinate neighborhood U on X, over which E and F are trivial, any local trivializations of these bundles and any open set $V \subset\subset U$, the mapping P_V : $[\mathcal{E}(V)]^k \longrightarrow [\mathcal{E}(V)]^l$ defined by the commutative diagram*

$$\begin{array}{ccc} \mathcal{E}(E)|_V & \xrightarrow{P} & \mathcal{E}(F)|_V \\ \downarrow \subset & & \downarrow \subset \\ \mathcal{E}(E|_V) & \longrightarrow & \mathcal{E}(F|_V) \\ \downarrow \cong & & \downarrow \cong \\ [\mathcal{E}(V)]^k & \xrightarrow{P_V} & [\mathcal{E}(V)]^l \end{array}$$

can be expressed by a matrix $P(x, D) = \sum_{|\alpha| \leq p} P_\alpha(x) D^\alpha$ where $P_\alpha(x)$ are $(l \times k)$-matrices of differentiable functions on V.

Proof. See Peetre [177].

□

We note that the relative compactness condition of V is used here in order for p to be finite.

It follows from Peetre's Theorem and Proposition 1.1.2, that for each differential operator P of type $E \longrightarrow F$ there exists a locally finite coordinate covering $\{U_\nu\}_{\nu \in \mathbb{N}}$ of the manifold X such that E and F are trivial over $\{U_\nu\}$, and in each neighborhood U_ν the differential operator P is represented by an $(l \times k)$-matrix $P^{(\nu)}$ of scalar partial differential operators.

Since the representations f_ν of the section f of the bundle E (or F) in U_ν are related by transition matrices of E (or F) in intersections $U_\mu \cap U_\nu$, there are also relations between the representations $P^{(\nu)}$. Namely, let $x^{(\nu)} : U_\nu \longrightarrow \mathbb{R}^n$ be local coordinates in U_ν, and $t^{(E)}_{\nu\mu} : U_\mu \cap U_\nu \longrightarrow GL(k, \mathbb{C})$ and $t^{(F)}_{\nu\mu} : U_\mu \cap U_\nu \longrightarrow GL(l, \mathbb{C})$ be transition matrices for the bundles E and F respectively. Then the condition $P : \mathcal{E}(E) \longrightarrow \mathcal{E}(F)$ means that for any vector function $f \in [\mathcal{E}(x^{(\mu)}(U_\mu \cap U_\nu))]^k$ the following relations are fulfilled:

$$P^{(\nu)}\left(\left(t^{(E)}_{\nu\mu} \circ x^{(\nu)^{-1}}\right) f(x^{(\mu)} \circ x^{(\nu)^{-1}})\right) = \left(t^{(F)}_{\nu\mu} \circ x^{(\nu)^{-1}}\right)\left(P^{(\mu)} f\right)\left(x^{(\mu)} \circ x^{(\nu)^{-1}}\right) \quad \text{in } x^{(\nu)}(U_\mu \cap U_\nu). \tag{1.1.5}$$

So, the local analysis of differential operators between sections of vector bundles is reduced to the analysis of partial differential operators on open subsets of $\mathbb{R}^n$.

Remark 1.1.11 *The question of when it is possible to choose the local representations $P^{(\nu)}$ with constant coefficients, is connected with an analysis of the relations (1.1.5). It leads to the notion of a so-called Γ-manifold where Γ is a transitive pseudo-group of local diffeomorphisms of the space $\mathbb{R}^n$ (see Spencer* [206]*). A similar question has been discussed by Guillemin and Singer* [95].

Let $\pi : T^*(X) \to X$ be the (real) cotangent bundle of the manifold X, and π^*E and π^*F be the induced bundles over $T^*(X)$ (see Wells [254, Ch.1]. The fiber of π^*E over a point $(x,z) \in T^*(X)$ coincides with E_x, and for π^*F by analogy.

The symbol (or, more precisely, *principal symbol*) $\sigma(P) \in \mathrm{Hom}(\pi^*E \to \pi^*F)$ of a differential operator $P \in \mathrm{do}_p(E \to F)$ for a fixed point $(x,z) \in T^*(X)$ represents the linear mapping $E_x \to F_x$ defined in the following way. Let $e \in E_x$. There exist a function $\varphi \in \mathcal{E}(X)$ such that $\varphi(x) = 0$ and $d\varphi(x) = z$, and a section $f \in \mathcal{E}(E)$ such that $f(x) = e$. Then we set $\sigma(P)(x,z)e = \frac{(\sqrt{-1})^p}{p!} P(\varphi^p f)(x) \in F_x$.

It is easily seen that this definition does not depend on the choice of φ and f with the required properties.

If in some coordinate neighborhood $U \subset X$ the differential operator P is given by the matrix $P(x,D) = \sum_{|\alpha|\le p} P_\alpha(x) D^\alpha$ and $z = \sum_{j=1}^n z_j dx_j$, then

$$\sigma(P)(x,z) = \sum_{|\alpha|=p} P_\alpha(x) z^\alpha.$$

As we have said above, each differential operator P of type $E \to F$ defines a continuous linear operator $P : \mathcal{D}(E) \to \mathcal{D}(F)$. Let $P' : (\mathcal{D}(F))' \to (\mathcal{D}(E))'$ be the transposed mapping. By identifying E'' with E we have the injection with dense image $\mathcal{E}(F') \hookrightarrow (\mathcal{D}(F))'$. So P' also acts on $\mathcal{E}(F')$, and standard considerations show that $P' : \mathcal{E}(F)' \to \mathcal{E}(E)'$ is a differential operator of type $F' \to E'$. It is called the *transposed operator* for P.

If $P(x,D) = \sum_{|\alpha|\le p} P_\alpha(x) D^\alpha$ is a local representation of the differential operator P, then the differential operator P' has the local representation

$$P'(x,D) = \sum_{|\alpha|\le p} (-1)^{|\alpha|} D^\alpha (P_\alpha(x)^t \times \cdot).$$

It is easy to check that $\sigma(P') = (-1)^p \sigma(P)'$ where $\sigma(P)' : F'_x \to E'_x$ is the transposed mapping for $\sigma(P) : E_x \to F_x$ by the identifications $E^*_x \cong E'_x$ and $F^*_x \cong F'_x$.

Each differential operator P of type $E \to F$ extends to a continuous linear mapping $P : \mathcal{D}'(E) \to \mathcal{D}'(F)$ as the transposed mapping of $P' : \mathcal{D}(F') \to \mathcal{D}(E')$.

1.1.10 *Information from homological algebra*

Before passing to complexes of differential operators we make some remarks on the language of homological algebra which is very useful for the study of complexes (see Dold [56, Ch.2]).

By a (cochain) complex of topological vector spaces $L^\cdot$ we understand any sequence of topological vector spaces L^i $(i \in \mathbb{Z})$ together with continuous linear mappings $d_i : L^i \to L^{i+1}$ satisfying $d_{i+1} \circ d_i = 0$ for all i.

With each such complex we associate the graduated topological vector space $L^\cdot = \{L^i\}$ endowed with the product topology, and the differential $d : L^\cdot \to L^\cdot$ of degree 1 defined by means of the equality $dx = d_i x$ for $x \in L^i$.

The spaces of cocycles, coboundaries and cohomologies of the complex $L^\cdot$ at degree i are denoted by $Z^i(L^\cdot)$, $B^i(L^\cdot)$ and $H^i(L^\cdot)$ respectively. We always consider the spaces $Z^i(L^\cdot)$ and $B^i(L^\cdot)$ with the induced topologies, and the space $H^i(L^\cdot)$ with the quotient topology.

Proposition 1.1.12 *For the space $H^i(L^\cdot)$ to be separated it is necessary and sufficient that $B^i(L^\cdot)$ should be closed in L^i.*

Proof. See Edwards [62, 1.8.5].

□

Let $K^\cdot$ and $L^\cdot$ be two (cochain) complexes of topological vector spaces with differentials d_K and d_L respectively. When discussing mappings $M : L \to K^\cdot$ one usually demands that they be compatible with the graduations of $L^\cdot$ and $K^\cdot$. Namely, by a mapping $M : L^\cdot \to K^\cdot$ of degree r is understood any sequence of mappings $M_i : L^i \to K^{i+r}$ $(i \in \mathbb{Z})$. We shall write $Mx = M_i x$ for $x \in L^i$.

It is well-known that the continuity of $M : L^\cdot \to K^\cdot$ is equivalent to the continuity of each of the mappings M_i (see Bourbaki [27, Ch.1, §4]).

Let $\mathcal{L}(L^\cdot \to K^\cdot)^r$ denote the vector space of all continuous linear mappings $M : L^\cdot \to K^\cdot$ of degree r endowed with the topology of uniform convergence on bounded subsets of $L^\cdot$. These topological vector spaces are assembled into a new complex $\mathcal{L}(L^\cdot \to K^\cdot)^\cdot$ with the differential d of degree 1 defined by means of the equality $dM = \{(-1)^{r+1}M_{i+1}d_L + d_K M_i\}$ for $M \in \mathcal{L}(L^\cdot \to K^\cdot)^r$. This complex is called the *complex of homomorphisms.*

Cocycles of the complex $\mathcal{L}(L^\cdot \to K^\cdot)^\cdot$ at degree r carry the name of cochain mappings $L^\cdot \to K^\cdot$ of degree r. They preserve the spaces of cocycles and coboundaries, so passing to cohomologies they induce a mapping $HM : H^\cdot(L^\cdot) \to H^\cdot(K^\cdot)$ of degree r which is continuous along with M (see Edwards [62, 0.2.16]).

Cochain mapping $M^1, M^2 \in \mathcal{L}(L^\cdot \to K^\cdot)^r$ are said to be homotopic (written $M^1 \simeq M^2$) if they lie in the same cohomology class of $H^r(\mathcal{L}(L^\cdot \to K^\cdot)^\cdot)$. In this case the mapping $h \in \mathcal{L}(L^\cdot \to K^\cdot)^{r-1}$ such, that $dh = M^2 - M^1$, is called a *cochain homotopy* between M^1 and M^2.

Proposition 1.1.13 *If cochain mappings $M^1, M^2 \in \mathcal{L}(L^\cdot \to K^\cdot)^r$ are homotopic, we have $HM^1 = HM^2$.*

Proof. See Dold [56, Ch.2, 3.4].

□

In applications Proposition 1.1.13 is usually used for $r = 0$.

A cochain mapping $M \in \mathcal{L}(L^\cdot \to K^\cdot)^0$ is said to be a homotopic equivalence if there exists a cochain mapping $M^{-1}\mathcal{L}(K^\cdot \to L^\cdot)^0$ such that $M^{-1}M \simeq 1_L$ and $MM^{-1} \simeq 1_K$ where 1_L and 1_K are the indentity endomorphisms of $L^\cdot$ and $K^\cdot$ respectively. We then say that the complexes $K^\cdot$ and $L^\cdot$ are *homotopically equivalent,* and write $K^\cdot \simeq L^\cdot$.

Corollary 1.1.14 *If $M : L^\cdot \to K^\cdot$ is a homotopic equivalence, then $HM : H^\cdot(L^\cdot) \to H^\cdot(K^\cdot)$ is a topological isomorphism.*

Proof. In fact, since $M^{-1}M \simeq 1_L$, Proposition 1.1.13 implies that

$$(HM^{-1})(HM) = H(M^{-1}M) = H(1_L) = 1.$$

Moreover, since $MM^{-1} \simeq 1_K$, Proposition 1.1.13 implies exactly $(HM)(HM^{-1}) = 1$. Thus the continuous linear mapping $HM: H^{\cdot}(L^{\cdot}) \to H^{\cdot}(K^{\cdot})$ has the continuous inverse mapping $HM^{-1}: H^{\cdot}(K^{\cdot}) \to H^{\cdot}(L^{\cdot})$ which completes the proof.

□

In particular, if the zero and identity endomorphisms of the complex $L^{\cdot}$ are homotopic (such complexes $L^{\cdot}$ are said to be *contractible* ones), then $H^{\cdot}(L^{\cdot}) = 0$.

It is useful to compare the following result with its algebraic variant (see Dold [56, Ch.2, 3.6]).

Proposition 1.1.15 *For a complex of Fréchet spaces $L^{\cdot}$ to be contractible it is necessary and sufficient that $H^{\cdot}(L^{\cdot}) = 0$ and for all i the $Z^i(L^{\cdot})$ be a direct summand of L^i.*

Proof. *Necessity.* Let $h \in \mathcal{L}(L^{\cdot} \to L^{\cdot})^{-1}$ be a cochain homotopy between the zero and identity endomorphism of the complex $L^{\cdot}$. We have $hd + dh = 1$ on L^i and $dh = 1$ on $Z^i(L^{\cdot})$ for all $i \in \mathbb{Z}$. This implies that $Z^i(L^{\cdot}) = B^i(L^{\cdot})$, and the operator $dh: L^i \to Z^i(L^{\cdot})$ is a continuous projector on $Z^i(L^{\cdot})$. Hence the kernel of this operator gives a topological complement of the subspace $Z^i(L^{\cdot})$ in L^i.

Sufficiency. It follows from the condition that for all $i \in \mathbb{Z}$ there are decompositions as topological direct sums $L^i = C^i \overset{top}{\oplus} B^i(L^{\cdot})$ where C^i and $B^i(L^{\cdot})$ are closed subspaces of L^i, and the projection operators on the summands are continuous. Denote these projectors by c_i and b_i respectively.

Each operator d_i defines by restriction a continuous linear operator $\delta_i: C^i \to B^{i+1}(L^{\cdot})$ which is, obviously, injective and surjective. Moreover, by the Open Mapping Theorem (see Edwards [62, 8.9.3.(1)]) the inverse operator δ_i^{-1} is continuous also.

We now set $h_i = \delta_{i-1}^{-1} \circ (1 - \pi_i)$ $(i \in \mathbb{Z})$. Then $h = \{h_i\}$ defines a continuous linear mapping $L^{\cdot} \to L^{\cdot}$ of degree (-1). For all elements $x \in L^i$ we have

$$\begin{aligned}
(hd + dh)\, x &= h_{i+1}d_i x + d_{i-1}h_i x = \\
&= h_{i+1}d_i\, (\pi_i x + (1 - \pi_i)x) + d_{i-1}h_i\, (\pi_i x + (1 - \pi_i)x) = \\
&= h_{i+1}d_i\, (\pi_i x) + d_{i-1}h_i\, ((1 - \pi_i)x) = \\
&= \delta_i^{-1}(1 - \pi_{i+1})\, d_i\, (\pi_i x) + d_{i-1}\delta_{i-1}^{-1}(1 - \pi_i)\, ((1 - \pi_i)x) = \\
&= \pi_i x + (1 - \pi_i)x = \\
&= x.
\end{aligned}$$

Hence h is the desired cochain homotopy between the zero and identity endomorphisms of the complex $L^{\cdot}$. This proves the proposition.

□

Besides cochain complexes we shall deal with chain complexes $L.$ which differ from the cochain ones only in that the differential ∂ of $L.$ has degree (-1). In general chain complexes appear as transposed complexes to cochain ones. The definitions of this subsection carry over to chain complexes without any difficulty. In particular the spaces of cycles, boundaries and homologies of the complex L in dimension i are traditionally denoted by $Z_i(L)$, $B_i(L)$ and $H_i(L)$.

1.1.11 *Differential complexes*

Definition 1.1.16 *A (cochain) differential complex E on a manifold X is a sequence of differentiable $\mathbb{C}$-vector bundles E^i ($i \in \mathbb{Z}$) over X together with differential operators P_i of respective types $E^i \to E^{i+1}$ satisfying $P_{i+1} \circ P_i = 0$ for all i.*

Unless otherwise stated, we always assume that the complex E is supported in the non-negative degrees and has a finite length N, that is $E^i \neq 0$ for $0 \le i \le N$ only.

For each of the classes $\mathfrak{C} = \mathcal{E}, \mathcal{D}, \mathcal{D}', \mathcal{E}'$ and many others, a (cochain) differential complex E gives rise to a (cochain) complex of topological vector spaces $L = \mathfrak{C}(E)$ with the differential $d = P$ defined by $Pf = P_i f$ for $f \in \mathfrak{C}(E)$. Namely

$$\mathfrak{C}(E): \ 0 \longrightarrow \mathfrak{C}(E^0) \xrightarrow{P} \mathfrak{C}(E^1) \xrightarrow{P} \dots \xrightarrow{P} \mathfrak{C}(E^N) \longrightarrow 0. \qquad (1.1.6)$$

The spaces of cocycles, coboundaries and cohomologies of complex (1.1.6) in degree i, endowed with the natural topologies, will be denoted respectively by

$$\begin{aligned} Z^i(\mathfrak{C}(E)) &= \ker P: \ \mathfrak{C}(E^i) \to \mathfrak{C}(E^{i+1}), \\ B^i(\mathfrak{C}(E)) &= \operatorname{im} P: \ \mathfrak{C}(E^{i-1}) \to \mathfrak{C}(E^i) \end{aligned}$$

and

$$H^i(\mathfrak{C}(E)) = Z^i(\mathfrak{C}(E))/B^i(\mathfrak{C}(E)).$$

Sometimes the notations become more simple if we write $\mathfrak{C}^i(\sigma)$ instead of $\mathfrak{C}(E^i|_\sigma)$. In particular this defines, for each differential complex E, the sheaves $\mathcal{E}^i$ and $\mathcal{D}'^i$, and cosheaves $\mathcal{D}^i$ and $\mathcal{E}'^i$ of germs of sections of the bundles E^i over X.

A particularity of differential complexes is that the differential P of (1.1.6) may be lowered to a differential of the sheaf (or cosheaf) complex $\mathfrak{C}$.

The principal question concerning differential complexes E is the investigation of the cohomologies of the complexes $\mathfrak{C}(E)$ (and $\mathfrak{C}$). From this point of view it is convenient to introduce an equivalence relation on the set of all differential complexes such that complexes with the same cohomologies are identified.

Definition 1.1.17 *(Cochain) differential complexes $E = \{E^i, d_i^{(E)}\}$ and $F = \{E^i, d_i^{(E)}\}$ on X are said to be homotopically equivalent if there exist sequences of differential operators $\{M_i\}$ of types $F^i \to E^i$, $\{M_i^{-1}\}$ of types $E^i \to F^i$, $\{h_i^{(F)}\}$ of types $F^i \to F^{i-1}$ and $\{h_i^{(E)}\}$ of types $E^i \to E^{i-1}$ such that for all $i \in \mathbb{Z}$ we have*

$$\begin{aligned} 1)\quad & M_{i+1}d_i^{(F)} - d_i^{(E)}M_i = 0, \\ & M_{i+1}^{-1}d_i^{(E)} - d_i^{(F)}M_i^{-1} = 0; \end{aligned}$$

and

$$\begin{aligned} 2)\quad & h_{i+1}^{(F)}d_i^{(F)} + d_{i-1}^{(F)}h_i^{(F)} = 1 - M_i^{-1}M_i, \\ & h_{i+1}^{(E)}d_i^{(E)} + d_{i-1}^{(E)}h_i^{(E)} = 1 - M_iM_i^{-1}. \end{aligned}$$

The importance of Definition 1.1.17 will become clear from the following simple observation.

Proposition 1.1.18 *If (cochain) differential complexes $E^\cdot$ and $F^\cdot$ are homotopically equivalent, then for each of the classes $\mathfrak{C} = \mathcal{E}, \mathcal{D}, \mathcal{D}', \mathcal{E}'$ and many others, the complexes of topological vector spaces $\mathfrak{C}(E^\cdot)$ and $\mathfrak{C}(F^\cdot)$ are homotopically equivalent.*

Proof. Set $K^\cdot = \mathfrak{C}(E^\cdot)$ and $L^\cdot = \mathfrak{C}(F^\cdot)$. The mapping $M = \{M_i\}$, where the M_i are given in Definition 1.1.17, is a continuous linear mapping $L^\cdot \to K^\cdot$ of degree 0. Then the first of the conditions in 1) means that M is a cochain mapping. Analogously, $M^{-1} = \{M_i^{-1}\}$ is a cochain mapping $K^\cdot \to L^\cdot$ of degree 0. The conditions in 2) mean that $M^{-1}M \simeq 1_L$ and $MM^{-1} \simeq 1_K$ which finishes the proof. □

1.1.12 *Elliptic complexes*

Let the orders of the differential operators P_i in a (cochain) differential complex $E^\cdot$ be stipulated so that the symbols $\sigma(P_i)$ may be calculated.

Since $P_{i+1}P_i = 0$, we have $\sigma(P_{i+1})\sigma(P_i) = 0$ for all i. So with the complex $E^\cdot$ it is possible to associate the symbolic complex

$$\pi^*E^\cdot : 0 \longrightarrow \pi^*E^0 \xrightarrow{\sigma(P_0)} \pi^*E^1 \xrightarrow{\sigma(P_1)} \dots \xrightarrow{\sigma(P_{N-1})} \pi^*E^N \longrightarrow 0. \qquad (1.1.7)$$

Definition 1.1.19 *A cotangent vector $z \in T_x^*(X)$ is said to be characteristic for the complex $E^\cdot$ at a point $x \in X$ if the symbolic complex (1.1.7) is non-exact over the point $(x, z) \in T^*(X)$. Otherwise the vector $z \in T_x^*(X)$ is said to be non-characteristic for the $E^\cdot$ at x.*

The notion of elliptic complex, due to Atiyah and Bott [19], is a natural generalization of the notion of elliptic differential operator. However variants of this notion had already occured in Spencer's works [205, 206].

Definition 1.1.20 *A differential complex $E^\cdot$ is said to be elliptic if, for any point $x \in X$, no cotanqent vector $z \in T_x^*(X) \setminus \{0\}$ is characteristic for $E^\cdot$ at x.*

In other words a complex $E^\cdot$ is elliptic if its symbolic complex $\pi^*E^\cdot$ is exact off the zero section of $T^*(X)$.

Example 1.1.21 Let $P \in \mathrm{do}_p\ (E \to F)$ be some differential operator on X. Consider a (cochain) differential complex E of length $N = 1$ on X by setting $E^0 = E$, $E^1 = F$ and $E^i = 0$ for $i \neq 0, 1$, and $P_0 = P$. Ellipticity of such a complex means that, for all $x \in X$ and $z \in T^*_x(X)\setminus\{0\}$, the symbolic mapping $\sigma(P)(x,z) : E_x \to F_x$ is an isomorphism. By definition this is equivalent to ellipticity of the differential operator P in the classical sense (see Wells [254, Ch.4]).

□

It is convenient to distinguish the properties of the symbol of an elliptic differential operator in the following way.

Definition 1.1.22 *A differential operator $P \in \mathrm{do}_p\ (E \to F)$ is said to be a differential operator with injective (respectively, surjective) symbol if the symbolic mapping $\sigma(P)(x,z) : E_x \to F_x$ is an injective (respectively, surjective) one for all points $(x, z) \in T^*(X)$ with $z \neq 0$.*

Obviously, in an elliptic (cochain) complex E of length N the differential operator P_0 is a differential operator with injective symbol, and the differential operator P_{n-1} is a differential operator with surjective symbol.

We note that there is a more general definition of elliptic complexes due to Andreotti and Nacinovich [13] (see Nacinovich [166] also). This definition is connected with a suitable choice of the principal parts of the differential operators P_i, and for differential complexes E of length 1 it coincides with the requirement of ellipticity of the differential operator P_0 in the Douglis-Nirenberg sense. Such complexes will therefore be said to be elliptic in the Douglis-Nirenberg sense.

Generally, the ellipticity property of a differential complex is only a property of the way in which the complex is written, and it may appear or disappear by replacing the complex by a homotopically equivalent complex. The cohomologies of a complex $\mathfrak{C}(E)$ are homotopically invariant but it may be necessary to change the spaces of sections for the calculation of cohomologies.

1.1.13 *Transposed complex*

In many problems chain differential complexes appear along with cochain ones.

Definition 1.1.23 *A chain differential complex E on a manifold X is a sequence of differentiable $\mathbb{C}$-vector bundles E_i $(i \in \mathbb{Z})$ over X together with differential operators P_i of types $E_i \to E_{i-1}$ satisfying $P_{i-1} \circ P_i = 0$ for all i.*

All that has been said on cochain differential complexes may be repeated for chain complexes with slight modifications. In particular, for each of the classes $\mathfrak{C} = \mathcal{E}, \mathcal{D}, \mathcal{D}', \mathcal{E}'$ and many others, a chain differential complex E defines a chain complex of topological vector spaces $L = \mathfrak{C}(E)$ with the differential $\partial = P$ given by $Pf = P_i f$ for $f \in \mathfrak{C}(E_i)$. More precisely,

$$\mathfrak{C}(E) :\ 0 \longleftarrow \mathfrak{C}(E_0) \xleftarrow{P} \mathfrak{C}(E_1) \xleftarrow{P} \dots \xleftarrow{P} \mathfrak{C}(E_N) \longleftarrow 0. \tag{1.1.8}$$

The spaces of cycles, boundaries and homologies of the complex (1.1.8) in dimension i, endowed with the natural topologies, will be denoted respectively by

$$\begin{aligned} Z_i(\mathfrak{C}(E.)) &= \ker P: \mathfrak{C}(E_i) \to \mathfrak{C}(E_{i-1}), \\ B_i(\mathfrak{C}(E.)) &= \operatorname{im} P: \mathfrak{C}(E_{i+1}) \to \mathfrak{C}(E_i) \end{aligned}$$

and

$$H_i(\mathfrak{C}(E.)) = Z_i(\mathfrak{C}(E.))/B_i(\mathfrak{C}(E.)).$$

Chain differential complexes appear naturally as transposed complexes to cochain complexes.

Definition 1.1.24 *Let $E^\cdot = \{E^i, P_i\}$ be a (cochain) differential complex on X. The transposed complex to $E^\cdot$ is a chain differential complex $E^{\cdot\prime}$ consisting of the dual vector bundles $E^{i\prime}$ and the transposed differential operators P_i'.*

It is easy to carry this definition over to the case of chain differential complexes. In particular, we have $(E^{\cdot\prime})' = E^\cdot$.

The following simple statement justifies the term "transpose differential complex."

Proposition 1.1.25 *The complexes of topological vector spaces $\mathcal{E}'(E^{\cdot\prime})$, $\mathcal{D}'(E^{\cdot\prime})$, $\mathcal{D}(E^{\cdot\prime})$, and $\mathcal{E}(E^{\cdot\prime})$ are transposed (in the topological sense) to the complexes of topological vector spaces $\mathcal{E}(E^\cdot)$, $\mathcal{D}(E)$, $\mathcal{D}'(E^\cdot)$ and $\mathcal{E}'(E^\cdot)$ respectively.*

Proof. This follows from the reflexivity property of Montel spaces and the definition of differential operators on distribution sections.

□

The notion of ellipticity carries over to chain differential complexes in the obvious way. It is easy to see that the transposed complex to an elliptic one is again an elliptic complex.

1.2 The Hilbert Resolution of a Differential Operator with Constant Coefficients

1.2.1 *Compatibility problem for overdetermined systems of differential equations*

Let P be a differential operator of type $E \to F$ on a manifold X.

Definition 1.2.1 *A differential operator P is said to be overdetermined (respectively, underdetermined) if there exists a differential operator $P_1 \neq 0$ (respectively, $P_{-1} \neq 0$) such that the composition $P_1 P$ (respectively, PP_{-1}) is defined and equal to zero.*

Obviously, a differential operator P is overdetermined iff the transposed operator P' is underdetermined.

If P is an overdetermined differential operator on X, then, for the non-homogeneous system $Pu = f$ to be solvable, it is necessary that f satisfy $P_1 f = 0$.

In the study of overdetermined systems the initial object of investigation is usually a differential operator $P_0 = P$ only. The problem consists of finding a full set of solvability conditions for $Pu = f$ having the form of a system of differential equations.

The following definition gives an exact sense to the previous word "full."

Definition 1.2.2 *A differential operator P_1 of type $F \to E^2$ is said to be a compatibility operator for P if $P_1 P = 0$ and, for any differential operator $\tilde{P}_1$ of type $F \to \tilde{E}^2$ satisfying $\tilde{P}_1 P = 0$, there exists a differential operator T of type $E^2 \to \tilde{E}^2$ such that $\tilde{P}_1 = T P_1$.*

Let us consider an important classical example (see Cartan [36, Ch.3, §6]).

Example 1.2.3 Let X be an open subset of $\mathbb{R}^n$, $E = X \times \mathbb{C}^k$, and $F = \oplus E$ (n copies) be trivial vector bundles over X. Let T_i $(i = 1, \ldots, n)$ be $(k \times k)$-matrices of differentiable functions on X, and $P \in \mathrm{do}_1\,(E \to F)$ be the differential operator given by $Pf = (P^{(i)})$ where $P^{(i)} = D_i + T_i$ $(i = 1, \ldots, n)$. The Frobenius conditions $(D_j T_i - D_i T_j) + (T_j T_i - T_i T_j) = 0$ $(i < j)$ are assumed to be satisfied which is equivalent to the assertion that the differential operators $P^{(i)}$ commute in pairs. Now we set $E^2 = \otimes_{i<j} E$ and consider the differential operator $P_1 \in \mathrm{do}_1(F \to E^2)$ given by $P_1 f = (P^{(j)} f_i - P^{(i)} f_j)_{i<j}$ where $f = \oplus f_i$ is a decomposition of f in accordance with the decomposition of F It follows from the Frobenius Theorem that P_1 is a compatibility operator for P.

□

Definition 1.2.4 *A (cochain) differential complex E on X is said to be a compatibility complex for the differential operator $P = P_0$ if, for each integer $i \geq 0$, the differential operator P_{i+1} is a compatibility operator for the differential operator P_i.*

We wish to introduce an equivalence relation on the set of all differential operators P in such a way that, for equivalent differential operators, compatibility complexes exist simultaneously and are homotopically equivalent.

Definition 1.2.5 *Two differential operators $d_0^{(E)}$ of type $E^0 \to E^1$ and $d_0^{(F)}$ of type $F^0 \to F^1$ are said to be equivalent if there exist differential operators M_i $(i = 0, 1)$ of type $F^i \to E^i$, M_i^{-1} $(i = 0, 1)$ of type $E^i \to F^i$, $h_1^{(F)}$ of type $F^1 \to F^0$ and $h_1^{(E)}$ of type $E^1 \to E^0$ such that the following conditions are fulfilled:*

1) $M_1 d_0^{(F)} - d_0^{(E)} M_0 = 0,$
$M_1^{-1} d_0^{(E)} - d_0^{(F)} M_0^{-1} = 0;$

and

2) $h_1^{(F)} d_0^{(F)} = 1 - M_0^{-1} M_0,$
$h_1^{(E)} d_0^{(E)} = 1 - M_0 M_0^{-1}.$

It is easy to see that, if two differential operators $d_0^{(E)}$ and $d_0^{(F)}$ are equivalent, then there is a (topological) isomorphism between the sheaves (or cosheaves) of topological vector spaces of solutions of classes $\mathfrak{C} = \mathcal{E}, \mathcal{D}, \mathcal{D}', \mathcal{E}'$ and many others of the systems $d_0^{(E)} f = 0$ and $d_0^{(F)} f = 0$. This isomorphism is defined by the differential operators M_0 and M_0^{-1} from Definition 1.2.5.

Example 1.2.6 Let P be a differential operator of type $E \to F$ on X, and D be a differential operator of type $F \to G$. Then the differential operators P and $(1 \oplus D)P$ are equivalent. In fact, we have $E^0 = E$, $E^1 = F$, $F^0 = E$ and $F^1 = F \oplus G$ here. So it is possible to take $M_0 = M_0^{-1} = 1$, $M_1 f = f_1$ where $f = f_1 \oplus f_2$ is a decomposition of f in accordance with the decomposition of F^1 $M_1^{-1} = 1 \oplus D$ and $h_1^{(F)} = h_1^{(E)} = 0$. The conditions 1) and 2) of Definition 1.2.5 are fulfilled.

□

By this example we can see that equivalent differential operators are in fact different manifestations of a mathematical object describing the same physical (or some other) process.

Proposition 1.2.7 *Let $d_0^{(E)}$ and $d_0^{(F)}$ be equivalent differential operators on X. If for $d_0^{(E)}$ there exists a compatibility complex, then for $d_0^{(F)}$ there also exists a compatibility complex.*

Proof. Let E be a compatibility complex for $d_0^{(E)}$ with differential $d^{(E)}$.

We would like to construct a compatibility complex F for $d_0^{(F)}$ with a differential $d^{(F)}$. Let us begin with the construction of a differential operator $d_1^{(F)}$.

We fix some differential operators M_i $(i = 0, 1)$, M_i^{-1} $(i = 0, 1)$, $h_1^{(F)}$ and $h_1^{(E)}$ on X given by Definition 1.2.5. We set $F^2 = E^2 \oplus F^1$ and consider the differential operator $d_1^{(F)}$ of type $F^1 \to F^2$ given by

$$d_1^{(F)} = (d_1^{(E)} M_1) \oplus (1 - M_1^{-1} M_1 - d_0^{(F)} h_1^{(F)}).$$

We claim that $d_1^{(F)}$ is a compatibility operator for $d_0^{(F)}$.

In fact, by Definition 1.2.5 we obtain

$$d_1^{(F)} d_0^{(F)} = (d_1^{(E)} d_0^{(E)} M_0) \oplus d_0^{(F)} (1 - M_0^{-1} M_0 - h_1^{(F)} d_0^{(F)}) = 0.$$

On the other hand, let $\tilde{d}_1^{(F)}$ be a differential operator of type $F^1 \to \tilde{F}^2$ satisfying $\tilde{d}_1^{(F)} d_0^{(F)} = 0$. It follows from Definition 1.2.5 that $(\tilde{d}_1^{(F)} M_1^{-1}) d_0^{(E)} = 0$. Since $d_1^{(E)}$ is a compatibility operator for $d_0^{(E)}$, there exists a differential operator T_2 of type $E^2 \to \tilde{F}^2$ such that $\tilde{d}_1^{(F)} M_1^{-1} = T_2 d_1^{(E)}$. Hence

$$\begin{aligned}
\tilde{d}_1^{(F)} &= \tilde{d}_1^{(F)} (M_1^{-1} M_1) + \tilde{d}_1^{(F)} (1 - M_1^{-1} M_1) = \\
&= T_2 (d_1^{(E)} M_1) + \tilde{d}_1^{(F)} (1 - M_1^{-1} M_1 - d_0^{(F)} h_1^{(F)}) = \\
&= T \, d_1^{(F)}.
\end{aligned}$$

So $d_1^{(F)}$ is a compatibility operator for the differential operator $d_0^{(F)}$.

For the next step we require additional constructions.

Since $(d_1^{(E)}\, M_1) d_0^{(F)} = 0$, there exists a differential operator M_2 of type $F^2 \to E^2$ such that $(d_1^{(E)}\, M_1) = M_2\, d_1^{(F)}$. By analogy, since $(d_1^{(F)}\, M_1^{-1}) d_0^{(E)} = 0$, there exists a differential operator M_2^{-1} of type $E^2 \to F^2$ such that $(d_1^{(F)}\, M_1^{-1}) = M_2^{-1}\, d_1^{(E)}$. Further we consider a differential operator $\tilde{d}_1^{(F)}$ of type $F^1 \to F^1$ given by $\tilde{d}_1^{(F)} = 1 - M_1^{-1}\, M_1 - d_0^{(F)}\, h_1^{(F)}$. It follows from Definition 1.2.5 that $\tilde{d}_1^{(F)}\, d_0^{(F)} = 0$. So there exists a differential operator $h_2^{(F)}$ of type $F^2 \to F^1$ such that $\tilde{d}_1^{(F)} = h_2^{(F)}\, d_1^{(F)}$. By analogy there exists a differential operator $h_2^{(E)}$ of type $E^2 \to E^1$ such that $1 - M_1\, M_1^{-1} - d_0^{(E)}\, h_1^{(E)} = h_2^{(E)}\, d_1^{(E)}$. Thus we have

1) $M_2 d_1^{(F)} - d_1^{(E)} M_1 = 0,$
$M_2^{-1} d_1^{(E)} - d_1^{(F)} M_1^{-1} = 0;$

and

2) $h_2^{(F)} d_1^{(F)} + d_0^{(F)} h_1^{(F)} = 1 - M_1^{-1} M_1,$
$h_2^{(E)} d_1^{(E)} + d_0^{(E)} h_1^{(E)} = 1 - M_1 M_1^{-1}.$

Let us suppose that a compatibility operator $d_i^{(F)}$ for the differential operator $d_{i-1}^{(F)}$ $(i > 1)$ and also differential operators M_j $(j = i, i+1)$ of type $F^j \to E^j$, M_j^{-1} $(j = i, i+1)$ of type $E^j \to F^j$, $h_{i+1}^{(F)}$ of type $F^{i+1} \to F^i$ and $h_{i+1}^{(E)}$ of type $E^{i+1} \to E^i$ have already been constructed, so that the following relationships hold:

1) $M_{i+1} d_i^{(F)} - d_i^{(E)} M_i = 0,$
$M_{i+1}^{-1} d_i^{(E)} - d_i^{(F)} M_i^{-1} = 0;$

and

2) $h_{i+1}^{(F)} d_i^{(F)} + d_{i-1}^{(F)} h_i^{(F)} = 1 - M_i^{-1} M_i,$
$h_{i+1}^{(E)} d_i^{(E)} + d_{i-1}^{(E)} h_i^{(E)} = 1 - M_i M_i^{-1}.$

Then we can construct a compatibility operator $d_{i+1}^{(F)}$ for the differential operator $d_i^{(F)}$ by setting $F^{i+2} = E^{i+2} \oplus F^{i+1}$ and

$$d_{i+1}^{(F)} = (d_{i+1}^{(E)}\, M_{i+1}) \oplus (1 - M_{i+1}^{-1}\, M_{i+1} - d_i^{(F)} h_{i+1}^{(F)}).$$

Now an application of the Mathematical Induction Principle finishes the proof.

□

The following useful result complements Proposition 1.2.7.

Proposition 1.2.8 *Let E and F be compatibility complexes for differential operators $d_0^{(E)}$ and $d_0^{(F)}$ respectively. Then, if the differential operators $d_0^{(E)}$ and $d_0^{(F)}$ are equivalent, the complexes $E^{\cdot}$ and F are homotopically equivalent.*

Proof. See the proof of Proposition 1.2.7.

□

In this section the problem of constructing a compatibility operator for a differential operator P with constant coefficients in $\mathbb{R}^n$ will be considered. It follows from Remark 1.1.11 that this problem can also be interpreted as one of constructing a compatibility operator for an arbitrary differential operator P but on a Γ-manifold X associated with the differential operator P only.

1.2.2 *Formal theory in the case of constant coefficients (Hilbert complexes)*

Let P be a differential operator of type $E \to F$ with constant coefficients on an open set $X \subset \mathbb{R}^n$. This means that the bundles E and F are trivial, namely $E = X \times \mathbb{C}^k$ and $F = X \times \mathbb{C}^k$, and P is given by a $(l \times k)$-matrix of scalar partial differential operators with constant coefficients on X, namely $P(D) = (p_{ij}(D))$.

We denote by $\mathcal{P}$ the ring of all polynomials (with complex coefficients) in $z = (z_1, \ldots, z_n) \in \mathbb{C}^n$. Let us compare with the differential operator P the matrix $P(z) = (p_{ij}(z))$ with elements of $\mathcal{P}$ which is obtained from $P(D)$ by means of the formal change of " variables " $\frac{1}{\sqrt{-1}}\partial/\partial x_j \to z_j$ $(j = 1, \ldots, n)$. This correspondence between (matrix) differential operators with constant coefficients and matrices of polynomials is one-to-one.

The notation $\mathcal{P}^k$ is used for the $\mathcal{P}$-module, which is the direct sum of k copies of the ring $\mathcal{P}$. We interpret elements of $\mathcal{P}^k$ as (k)-columns of polynomials. Then the matrix $P(z)^t$ determines a $\mathcal{P}$-mapping $\mathcal{P}^l \to \mathcal{P}^k$ given by means of multiplication by this matrix from the left.

We consider the $\mathcal{P}$-module $M = \operatorname{coker} P(z)^t$, that is, $M = \mathcal{P}^k / P(z)^t \mathcal{P}^l$. Since the ring $\mathcal{P}$ is Noetherian, the $\mathcal{P}$-module M is finitely generated.

Proposition 1.2.9 *For $\mathcal{P}$-matrices $d_0^{(E)}(z)$ and $d_0^{(F)}(z)$ the $\mathcal{P}$-modules* $\operatorname{coker} d_0^{(E)}(z)^t$ *and* $\operatorname{coker} d_0^{(F)}(z)^t$ *coincide up to a $\mathcal{P}$-isomorphism if and only if the differential operators $d_0^{(E)}(D)$ and $d_0^{(F)}(D)$ are equivalent in the sense of Definition 1.2.5 where the differential operators M_i $(i = 0,1)$, M_i^{-1} $(i = 0,1)$, $h_1^{(F)}$ and $h_1^{(E)}$ have constant coefficients.*

Proof. This is an easy exercise in homological algebra.

□

Thus the $\mathcal{P}$-module $M = \operatorname{coker} P(z)^t$ does not depend on a choice of the differential operator $P(D)$ in the class of equivalent differential operators.

The next result is usually called the Hilbert Syzygies Theorem.

Theorem 1.2.10 *For each finitely generated $\mathcal{P}$-module M, there exists a free resolution of length $N \le n$ of the form*

$$M.: \; 0 \longleftarrow M \longleftarrow \mathcal{P}^{k_0} \overset{P_0^t}{\longleftarrow} \mathcal{P}^{k_1} \overset{P_1^t}{\longleftarrow} \ldots \overset{P_{N-1}^t}{\longleftarrow} \mathcal{P}^{k_N} \longleftarrow 0. \tag{1.2.1}$$

Proof. See Bourbaki [30, §8].

□

Applying this theorem to the $\mathcal{P}$-module $M = \operatorname{coker} P(z)^t$ we get an exact sequence of the form (1.2.1).

Let us denote by $P_i(D)$ the $(k_{i+1} \times k_i)$-matrix differential operator with constant coefficients corresponding to the matrix $P_i(z)$.

Definition 1.2.11 *A (cochain) differential complex $E^{\cdot}$ on X, obtained in this way, consisting of (trivial) bundles $E^i = X \times \mathbb{C}^{k_i}$ and differential operators $P_i(D)$ is said to be a Hilbert complex associated with the $\mathcal{P}$-module M.*

We turn our attention to a few formal properties of Hilbert complexes.

Proposition 1.2.12 *If $E^{\cdot}$ is a Hilbert complex associated with a $\mathcal{P}$-module $M = \operatorname{coker} P(z)^t$, the differential operators $P(D)$ and $P_0(D)$ are equivalent in the sense of Proposition 1.2.9*

Proof. This follows from Proposition 1.2.9 since the exactness of (1.2.1) implies that M and $\operatorname{coker} P_0(z)^t$ coincide up to a $\mathcal{P}$-isomorphism.

□

The property of being a compatibility complex is a characteristic one for Hilbert complexes.

Proposition 1.2.13 *For a (cochain) complex $E^{\cdot}$ of differential operators with constant coefficients on X to be a compatibility complex for the differential operator $P_0(D)$ it is necessary and sufficient that $E^{\cdot}$ should be a Hilbert complex associated with the $\mathcal{P}$-module $M = \operatorname{coker} P_0(z)^t$.*

Proof. *Necessity.* Let $E^{\cdot}$ be a compatibility complex for P_0 where $E^i = X \times \mathbb{C}^{k_i}$ are non-zero for $i = 0, \dots, N$ only, and $P_i = P_i(D)$. We must show that the complex (1.2.1) is exact.

The exactness at the terms M and $\mathcal{P}^{k_0}$ is obvious.

Let $p \in \mathcal{P}^{k_i}$ satisfy $P_{i-1}(z)^t p = 0$ where $i \geq 1$. Then $p^t P_{i-1}(z) = 0$ and so $p(D)^t P_{i-1}(D) = 0$. Since $P_i(D)$ is a compatibility operator for $P_{i-1}(D)$, there exists a column $q \in \mathcal{P}^{k_{i+1}}$ such that $P(D)^t = q(D)^t P_i(D)$. Hence we have $p = P_i(z)^t q$ which proves necessity.

Sufficiency. We must prove that, for a Hilbert complex $E^{\cdot}$ on X, the differential operator $P_i(D)$ is a compatibility operator for $P_{i-1}(D)$ for every $i \geq 1$.

It is obvious that $P_i(D)P_{i-1}(D) = 0$. Let $\tilde{P}_i(D)$ be some $(\tilde{k}_{i+1} \times k_i)$-matrix differential operator with constant coefficients satisfying $\tilde{P}_i(D)P_{i-1}(D) = 0$. Then $P_{i-1}(z)^t \tilde{P}_i(z)^t = 0$. Since the sequence (1.2.1) is exact at the term $\mathcal{P}^{k_i}$, there exists a $(\tilde{k}_{i+1} \times k_{i+1})$-matrix $T(z)$ with elements in $\mathcal{P}$ such that $\tilde{P}_i(z)^t = P_i(z)^t T(z)^t$. Hence we have $\tilde{P}_i(D) = T(D)P_i(D)$ which finishes the proof.

□

We have seen in particular that each differential operator P with constant coefficients has a compatibility complex.

Proposition 1.2.14 *Let $E^{\cdot}$ and $F^{\cdot}$ be Hilbert complexes associated with $\mathcal{P}$-modules $M^{(E)}$ and $M^{(F)}$ respectively. If $M^{(E)}$ and $M^{(F)}$ coincide up to a $\mathcal{P}$-isomorphism, then $E^{\cdot}$ and $F^{\cdot}$ are homotopically equivalent in the sense of Definition 1.1.17 where the differential operators $\{M_i\}$, $\{M_i^{-1}\}$, $\{h_i^{(F)}\}$ and $\{h_i^{(E)}\}$ have constant coefficients.*

Proof. It is enough to invoke Propositions 1.2.9 and 1.2.8.

□

This result together with Corollary 1.1.14 show that, for a Hilbert complex $E^{\cdot}$ on an open set $X \subset \mathbb{R}^n$, the cohomologies of the complex $\mathfrak{C}(E^{\cdot})$ depend on the $\mathcal{P}$-module $M = \operatorname{coker} P_0(z)^t$ and the (stable under differentiations) class $\mathfrak{C}$ only.

Remark 1.2.15 *In fact by one of the possible definitions of the functors Tor and Ext (see Bourbaki [30]), for each $i \in \mathbb{Z}$ and $x^0 \in \mathbb{R}^n$, we have*

$$\begin{array}{rcl} H^i(\mathfrak{C}(E^\cdot)) & = & \mathrm{Ext}^i(M, \mathfrak{C}(x^0 + X)), \\ H_i(\mathfrak{C}(E^{\cdot\prime})) & = & \mathrm{Tor}_i(M, \mathfrak{C}(x^0 - X)). \end{array} \qquad (1.2.2)$$

Thus the compatibility problem for a non-homogeneous system of differential equations with constant coefficients is completely solved by the *Hilbert Syzygies Theorem.*

1.2.3 *Elliptic Hilbert complexes*

Let $E^\cdot$ be a Hilbert complex on an open set $X \subset \mathbb{R}^n$ corresponding to some free resolution of a finitely generated $\mathcal{P}$-module M of the form (1.2.1).

For Hilbert complexes the sequence of full symbols behaves in a predictable way but not the sequence of principal symbols. For a fixed point $z^0 \in \mathbb{C}^n$,

$$\mathbb{C}^k : 0 \longrightarrow \mathbb{C}^{k_0} \stackrel{P_0(z^0)}{\longrightarrow} \mathbb{C}^{k_1} \stackrel{P_1(z^0)}{\longrightarrow} \ldots \stackrel{P_{N-1}(z^0)}{\longrightarrow} \mathbb{C}^{k_N} \longrightarrow 0. \qquad (1.2.3)$$

Denote by $\mathcal{R}$ the field of all rational functions of n variables $z = (z_1, \ldots, z_n) \in \mathbb{C}^n$.

Lemma 1.2.16 *Let $M(z)$ be a matrix with entries in $\mathcal{P}$. For every $z^0 \in \mathbb{C}^n$ we have* $\mathrm{rank}_{\mathbb{C}} M(z^0) \leq \mathrm{rank}_{\mathcal{R}} M(z)$.

Proof. Indeed, if a minor determinant of rank r of the matrix $M(z)$ is not equal to zero for $z = z^0$, it is in particular a polynomial not equal to zero.

□

The following technical result is due to Andreotti and Nacinovich [14].

Proposition 1.2.17 *For a point $z^0 \in \mathbb{C}^n$, the sequence (1.2.3) is exact at the positive degrees if and only if* $\mathrm{rank}_{\mathbb{C}} P_0(z^0) = \mathrm{rank}_{\mathcal{R}} P_0(z)$.

Proof. *Necessity.* Since $\mathcal{R}$ is flat over $\mathcal{P}$, we deduce from the exact sequence (1.2.1), an exact sequence

$$\mathcal{R}^{k_0} \stackrel{P_0^t}{\longleftarrow} \mathcal{R}^{k_1} \stackrel{P_1^t}{\longleftarrow} \ldots \stackrel{P_{N-1}^t}{\longleftarrow} \mathcal{R}^{k_N} \longleftarrow 0.$$

Therefore, if we set $r_i = \mathrm{rank}_{\mathcal{R}} P_i(z)$, we obtain that $k_i = r_{i-1} + r_i$ for $i = 1, \ldots, N$.

Assume that (1.2.3) is an exact sequence at degree 1, that is, at $\mathbb{C}^{k_1}$. Then we have $k_1 = \mathrm{rank}_{\mathbb{C}} P_0(z^0) + \mathrm{rank}_{\mathbb{C}} P_1(z^0)$. Because of the previous lemma it follows that we must have $\mathrm{rank}_{\mathbb{C}} P_i(z^0) = r_i$ for $i = 0, 1$.

This is more than enough to infer necessity.

Sufficiency. Conversely, let us assume that $\mathrm{rank}_{\mathbb{C}} P_0(z^0) = \mathrm{rank}_{\mathcal{R}} P_0(z)$.

By renumbering the components of $\mathcal{P}^{k_0}$ and $\mathcal{P}^{k_1}$ we may assume that in the matrix $P_0(z)$ the determinant $D(z)$ of the first r_0 rows and columns is different from zero for $z = z^0$.

Let $P_0(z) = (p_{ij}(z))$ where $1 \le i \le k_1$ and $1 \le j \le k_0$. For $j = 1, \dots, r_0$ all minor determinants of order (r_0+1) of the matrix

$$\begin{pmatrix} p_{11}(z) & \dots & p_{1r_0}(z) & p_{1j}(z) \\ & \dots & & \\ p_{r_01}(z) & \dots & p_{r_0r_0}(z) & p_{r_0j}(z) \\ & \dots & & \\ p_{k_11}(z) & \dots & p_{k_1r_0}(z) & p_{k_1j}(z) \end{pmatrix}$$

are zero because two columns of that matrix are equal. In particular all minor determinants obtained by augmenting the first r_o rows and columns by another row and the last column are zero.

Since the last $(k_0 - r_0)$ columns of the matrix $P_0(z)$ may be expressed by the first r_0 ones linearly over $\mathcal{R}$, this will give $k_1 - r_0 = r_1$ vectors in $\ker P_0(z)^t$ of the form

$$v_1(z) = \begin{pmatrix} c_{11}(z) \\ \dots \\ c_{r_01}(z) \\ D(z) \\ 0 \\ \dots \\ 0 \end{pmatrix}, \quad v_2(z) = \begin{pmatrix} c_{12}(z) \\ \dots \\ c_{r_02}(z) \\ 0 \\ D(z) \\ \dots \\ 0 \end{pmatrix}, \quad \dots, \quad v_{r_1}(z) = \begin{pmatrix} c_{1r_1}(z) \\ \dots \\ c_{r_0r_1}(z) \\ 0 \\ 0 \\ \dots \\ D(z) \end{pmatrix}$$

It follows from the exactness of the sequence (1.2.1) at $\mathcal{P}^{k_1}$ that $v_1(z), v_2(z), \dots, v_{r_1}(z)$ belong to $\operatorname{im} P_1(z)^t$ and therefore, in particular, $v_1(z^0), v_2(z^0), \dots, v_{r_1}(z^0)$ are in the image of $P_1(z^0)^t$.

Now, since $D(z^0) \ne 0$, these last r_1 vectors are linearly independent over $\mathbb{C}$ and therefore, $\operatorname{rank}_{\mathbb{C}} P_1(z^0) \ge r_1$. By the preceding lemma we deduce then that $\operatorname{rank}_{\mathbb{C}} P_1(z^0) = r_1$. This proves the exactness of (1.2.3) at $\mathbb{C}^{k_1}$.

We now have an exact sequence

$$0 \longleftarrow \operatorname{coker} P_1(z)^t \longleftarrow \mathcal{P}^{k_1} \xleftarrow{P_1^t} \mathcal{P}^{k_2} \xleftarrow{P_2^t} \dots \xleftarrow{P_{N-1}^t} \mathcal{P}^{k_N} \longleftarrow 0$$

with $\operatorname{rank}_{\mathbb{C}} P_1(z^0) = \operatorname{rank}_{\mathcal{R}} P_1(z)$, and arguing as before we deduce the exactness of the sequence (1.2.3) at $\mathbb{C}^{k_2}$ and also that $\operatorname{rank}_{\mathbb{C}} P_2(z^0) = \operatorname{rank}_{\mathcal{R}} P_2(z)$.

In this way, iterating the argument, we obtain the desired conclusion.

□

This proposition allows us in certain cases to give an effective description of elliptic Hilbert complexes.

A partial differential operator P on an open set $X \subset \mathbb{R}^n$ is said to be homogeneous if P coincides with its principal part, that is, $P = \sigma(P)(x, D)$.

Corollary 1.2.18 *For a Hilbert complex $E^\cdot$ of homogeneous differential operators to be elliptic, it is necessary and sufficient that the symbol of the differential operator P_0 be injective.*

Proof. The necessity is obvious. Let us prove the sufficiency.

The fiber of the bundle $E^i = X \times \mathbb{C}^{k_i}$ over a point $x \in X$ is $\mathbb{C}^{k_i}$, and we have $\sigma(P_i)(x,z) = P_i(z)$ by homogeneity. So we are to prove the exactness of the sequence (1.2.3) for $z^0 \in \mathbb{R}^n \setminus \{0\}$. But the exactness of (1.2.3) at $\mathbb{C}^{k_0}$ follows from the fact that the differential operator P_0 has an injective symbol. The exactness of (1.2.3) at other terms $\mathbb{C}^{k_i}$ $(i > 0)$ follows from Proposition 1.2.17 since we have $\operatorname{rank}_{\mathbb{C}} P_0(z^0) = \operatorname{rank}_{\mathcal{R}} P_0(z)$ for all $z^0 \in \mathbb{R}^n \setminus \{0\}$. In fact

$$k_0 = \operatorname{rank}_{\mathbb{C}} P_0(z^0) \leq \operatorname{rank}_{\mathcal{R}} P_0(z) \leq k_0.$$

This finishes the proof.

□

For arbitrary Hilbert complexes $E^{\cdot}$, Corollary 1.2.18 is false.

Example 1.2.19 We fix some polynomials $P^{(i)}(z)$ $(i = 1,2)$ with the property that the dimension of the algebraic variety $\mathcal{N}(M)$ in $\mathbb{C}^n$ formed by their common zeros is not more than $(n-2)$. Let us consider a (cochain) differential complex $E^{\cdot}$ of length 2 on X consisting of the (trivial) vector bundles $E^0 = X \times \mathbb{C}^1$, $E^1 = X \times \mathbb{C}^2$ and $E^2 = E^0$, and the differential operators $d_0^{(E)} = \begin{pmatrix} P^{(1)}(D) \\ P^{(2)}(D) \end{pmatrix}$ and $d_1^{(E)} = (P^{(2)}(D), -P^{(1)}(D))$. It follows from Proposition 2 (see Palamodov [175, Ch.VIII, §13]) that $E^{\cdot}$ is a Hilbert complex corresponding to some free resolution of the $\mathcal{P}$-module $M = \operatorname{coker}\left(P^{(1)}(z), P^{(2)}(z)\right)$. Moreover the polynomials $P^{(i)}(z)$ $(i = 1,2)$ may be chosen so that $d_0^{(E)}$ will be a differential operator with injective symbol. Further let D be a differential operator of type $E^1 \to G$ with constant coefficients on X such that the composition $Dd_0^{(E)}$ has an injective symbol. For example we may choose as D the differential operator which is the formal adjoint for $d_0^{(E)}$ (see Section 2.1). We consider a (cochain) differential complex F of length 2 on X consisting of the (trivial) vector bundles $F^0 = E^0, F^1 = E^1 \oplus G$ and $F^2 = E^2 \oplus G$, and the differential operators $d_0^{(F)} = (1 \oplus D) d_0^{(E)}$ and $d_1^{(F)} = \begin{pmatrix} d_1^{(E)} & 0 \\ -D & 1 \end{pmatrix}$. It is easy to see from Example 1.2.6 and the proof of Proposition 1.2.7 that $F^{\cdot}$ is a compatibility complex for the differential operator $d_0^{(F)}$. Hence by Proposition 1.2.12 $F^{\cdot}$ is a Hilbert complex. By its construction, the differential operator $d_0^{(F)}$ has injective symbol. But, if the order of the differential operator D is more than that of the differential operator $d_0^{(E)}$, the complex F is not elliptic. Moreover any complex beginning with $d_0^{(F)}$ is not elliptic.

□

At the same time the complex $F^{\cdot}$ in Example 1.2.19 is homotopically equivalent to the elliptic Hilbert complex $E^{\cdot}$. Perhaps this fact reflects a general conformity to natural laws but it is purely an algebraic question.

Problem 1.2.20 *To show or to refute that any Hilbert complex beginning with a differential operator with injective symbol is homotopically equivalent to an elliptic Hilbert complex.*

1.2.4 *Existence theory in the case of constant coefficients*

In this subsection we sketch briefly the existence theory for overdetermined systems of differential equations with constant coefficients.

Apparently, the principal results in this direction are due to Malgrange ([156, 157, 158]). Malgrange [158] was the first to suggest the language of homological algebra in this subject.

Suppose that we are given a Hilbert complex E on an open set $X \subset \mathbb{R}^n$ corresponding to some free resolution of a $\mathcal{P}$- module M of the form (1.2.1).

Let us recall that a mapping of topological vector spaces is said to be a (topological) homomorphism if it is linear, continuous and open (see Edwards [62, 1.9]).

One of the main results concerning Hilbert complexes is the following.

Theorem 1.2.21 *For a convex open set $X \subset \mathbb{R}^n$, the complex $\mathcal{E}(E)$ is exact at positive degrees, and its differential P is a (topological) homomorphism.*

Proof. See Palamodov [175, Ch. VII, §8].

□

For "quadratic" differential operators P_0 (= Hilbert complexes E of length 1) Theorem 1.2.21 was proved by Ehrenpreis [64] in the case $X = \mathbb{R}^n$ and in Malgrange's thesis [156] for any convex X. In the general case Theorem 1.2.21 was proved by Malgrange ([157, 158]) and independently by Palamodov [173].

As Malgrange [158] observed, the convexity of X is also necessary for $P\mathcal{E}(X) = \mathcal{E}(X)$ to be fulfilled for each scalar differential operator $P \neq 0$ with constant coefficients.

Corollary 1.2.22 *For a convex open set $X \subset \mathbb{R}^n$, the complex $\mathcal{E}'(E')$ is exact at positive dimensions and its differential P' is a (topological) homomorphism.*

Proof. Indeed, duality considerations (see Subsection 5.1.1) imply that the statements of Theorem 1.2.21 and Corollary 1.2.22 are equivalent.

□

The general problem of solvability of non-homogeneous systems of differential equations with constant coefficients in classes of distributions arose from the classical problem of constructing a global fundamental solution. In the papers of Fredholm [67], Zeylon [260], Herglotz [105] and Petrowsky [178] fundamental solutions for special classes of differential operators with constant coefficients had been constructed. The existence of a fundamental solution for any scalar differential operator $P \neq 0$ with constant coefficients established by Ehrenpreis [64] and Malgrange [156] was one of the first successes of the general theory of differential operators with constant coefficients. This result was obtained as a consequence of more general theorems on solvability of the equation $Pu = f$ for any $f \in \mathcal{D}'^F(X)$ obtained by Ehrenpreis [64] in the case $X = \mathbb{R}^n$ and in Malgrange's thesis [156] for arbitrary convex open sets X. Later on, Ehrenpreis [64, pt.3] and Malgrange [157] succeeded in proving an analogous theorem on solvability in $\mathcal{D}'(X)$.

Theorem 1.2.23 *For a complex open set $X \subset \mathbb{R}^n$, the complex $\mathcal{D}'(E^{\cdot})$ is exact at positive degrees, and its differential P is a (topological) homomorphism.*

Proof. See Palamodov [175, Ch. VII, §8].

□

This theorem was proved in full generality by Palamodov [173].

Corollary 1.2.24 *For a convex open set $X \subset \mathbb{R}^n$, the complex $\mathcal{D}(E^{\cdot\,\prime})$ is exact at positive dimensions, and its differential P' is a (topological) homomorphism.*

Proof. It follows from duality considerations (see Subsection 5.1.1) that the statements of Theorem 1.2.23 and Corollary 1.2.24 are equivalent.

□

The problem of solvability of non-homogeneous systems of differential equations with constant coefficients in the class of real analytic functions originated from the following two very pertinent remarks.

Example 1.2.25 Let $P = \Delta$ be the Laplace operator in the real plane $\mathbb{R}^2$. De Giorgi [47] showed that there exists some f real-analytic in $\mathbb{R}^3$ such that the equation $Pu = f$ has no real analytic solutions u defined on $\mathbb{R}^3$. Piccinini [180] obtained an analogous fact for the Cauchy-Riemann operator $P = \partial/\partial\bar{z}$ in the plane.

□

The second observation is the following.

Remark 1.2.26 *Let P be a non-zero scalar differential operator with constant coefficients in $\mathbb{R}^2$. De Giorgi and Cattabriga [49, 48] proved that for every f real-analytic in $\mathbb{R}^2$ the equation $Pu = f$ always admits a real analytic solution u in $\mathbb{R}^2$.*

Thus to encounter examples as in the first remark the number of variables needs to be ≥ 3.

Nevertheless there are no obstacles to the exactness of Hilbert complexes in the class of real-analytic functions at the sheaves level Apparently, the following result was well-known at mathematical socials in the 60-s.

Theorem 1.2.27 *The complex $\mathcal{A}^{\cdot}$ of sheaves of germs of real-analytic sections of the bundles $E^{\cdot}$ endowed with the differential P is exact at positive degrees.*

Proof. See, for example, Palamodov [176] and Andreotti and Nacinovich [14].

□

For real-analytic objects X and E, we shall denote by $\mathcal{A}(E)$ the space of real-analytic sections of E over X.

It follows from Theorems 1.2.21 and 1.2.27 and the Abstract de Rham Theorem (see 5.2) that, if the initial differential operator P_0 in the complex $E^{\cdot}$ is analytically hypoelliptic, we have an isomorphism of vector spaces (without topologies) $H^{\cdot}(\mathcal{E}(E^{\cdot})) \cong H^{\cdot}(\mathcal{A}(E^{\cdot}))$. In this case the complex $\mathcal{A}(E^{\cdot})$ is exact at positive degrees for any convex open set $X \subset \mathbb{R}^n$. But in other cases, for a convex X, the cohomologies of the complex $\mathcal{A}(E^{\cdot})$ may be supported at the degrees 0 and 1 only.

Theorem 1.2.28 *For a convex open set $X \subset \mathbb{R}^n$, the complex $\mathcal{A}(E)$ is exact at all degrees ≥ 2.*

Proof. See Andreotti and Nacinovich [14, Theorem 4]

□

Apparently, Andreotti and Nacinovich [14] were the first to prove this theorem. Moreover they showed that, if X is open and convex, $H^1(\mathcal{A}(E))$ is either zero or infinite-dimensional (see ibid, Theorem 6). Later on, Andreotti and Nacinovich [15] gave, for X convex, necessary and sufficient conditions for the vanishing of the group $H^1(\mathcal{A}(E))$ in terms of a generalized Phragmén-Lindelöf principle.

Finally, for real-analytic objects X and E, we shall denote by $\mathcal{H}(E)$ the space of hyperfunction sections of E over X.

Theorem 1.2.29 *Let X be open and convex. Then the complex $\mathcal{H}(E)$ is exact at positive degrees.*

Proof. See Komatsu [135] or Schapira [192, Ch. III, §1].

□

Theorem 1.2.29 was first proved by Komatsu [135].

As Nacinovich [166] has observed, each of the properties listed in Theorems 1.2.21, 1.2.23 or 1.2.27 is characteristic for Hilbert complexes.

The following definition due to Malgrange [158] plays a key role in the problem of solvability of general systems of differential equations with constant coefficients.

Definition 1.2.30 *Let $\mathfrak{C}(X)$ be a space of functions on X stable under differentiations. Then $\mathfrak{C}(X)$ is said to be M-convex if the complex $\mathfrak{C}(E)$ is exact at positive degrees and its differential P is a (topological) homomorphism.*

Many algebraic and geometric conditions of M-convexity are presented in Palamodov's book [175].

We shall now consider examples of Hilbert complexes.

1.2.5 *Recovering a solution of an elliptic system by means of a "scalar" component*

Let P be a differential operator with constant coefficients on an open set $X \subset \mathbb{R}^n$ given by a $(k \times k)$-matrix of scalar differential operators, $P(D) = (p_{ij}(D))_{\substack{1 \leq i \leq k \\ 1 \leq j \leq k}}$. We suppose that the differential operator $\det P(D)$ is elliptic. Let us consider the system of differential equations for a vector-function $f \in [\mathcal{D}'(X)]^k$ on X,

$$Pf = 0 \tag{1.2.4}$$

It is clear that the differential operator P is analytically hypoelliptic and so all generalized solutions of the system (1.2.4) are in fact real-analytic (see the commentary of Volevich and Oleinik [179]).

We shall interpret a solution f of (1.2.4) as a column of real-analytic functions on X, i.e. $f = \begin{pmatrix} f_1 \\ \dots \\ f_k \end{pmatrix}$.

The determination of a solution $f \in [\mathcal{D}'(X)]^k$ of the system (1.2.4) by means of its "scalar" component $f_1 \in \mathcal{D}'(X)$ is a problem going back to the classical result of the reconstruction of a holomorphic function from its real part.

Problem 1.2.31 *Let $f_1^0 \in \mathcal{D}'(X)$ be a given function on X. Find a solution f of (1.2.4) on X such that $f_1 = f_1^0$.*

We shall give a necessary condition for the solvability of Problem 1.2.31. Denote by $f^{(2)} = \begin{pmatrix} f_2 \\ \dots \\ f_k \end{pmatrix}$ the truncated column, and consider the differential operators

$$P_1 = (p_{i1})_{1 \le i \le k} \quad \text{and} \quad P_2 = (p_{ij})_{\substack{1 \le i \le k, \\ 2 \le j \le k}}$$

so that $Pf = P_1 f_1 + P_2 f^{(2)}$.

We now rewrite Problem 1.2.31 in the following form.

Problem 1.2.32 *Given $f_1^0 \in \mathcal{D}'(X)$, find $f^{(2)} \in [\mathcal{D}'(X)]^{k-1}$ such that $P_2 f^{(2)} = -P_1 f_1^0$.*

For $1 \le i \le k$, let m_i be the greatest common divisor of the polynomial det P and the polynomials in the i-th row of the matrix *adj* P, and let the polynomials Δ_i be defined by $\Delta_i = \frac{\det P}{m_i}$.

Lemma 1.2.33 *For Problem 1.2.32 to be solvable it is necessary that f_1^0 satisfy*

$$\Delta_1(D) f_1^0 = 0 \quad on \quad X. \tag{1.2.5}$$

Proof. Set $M = \text{diag}\,(\frac{1}{m_1}, \dots, \frac{1}{m_k}) \cdot adj\ P$. It is easy to see that

$$MP = \text{diag}\,(\Delta_1, \dots, \Delta_k). \tag{1.2.6}$$

We write $M = (m_{ij})_{\substack{1 \le i \le k \\ 1 \le j \le k}}$ and by analogy with the decomposition of P define

$$M_1 = (m_{1j})_{1 \le j \le k} \quad \text{and} \quad M_2 = (m_{ij})_{\substack{2 \le i \le k. \\ 1 \le j \le k}}$$

It follows from (1.2.6) that

$$\begin{pmatrix} M_1 P_1 & M_1 P_2 \\ M_2 P_1 & M_2 P_2 \end{pmatrix} = \text{diag}\,(\Delta_1, \dots, \Delta_k),$$

or

$$\begin{aligned} M_1 P_1 &= \Delta_1, & (1.2.7)\\ M_1 P_2 &= 0, & (1.2.8)\\ M_2 P_1 &= 0, & (1.2.9)\\ M_2 P_2 &= \operatorname{diag}(\Delta_2, \dots, \Delta_k) & (1.2.10) \end{aligned}$$

Now we can easily conclude from (1.2.7) and (1.2.8) that for Problem 1.2.32 to be solvable it is necessary that f_1^0 satisfy the equations (1.2.5).

This completes the proof.

□

Since the differential operator Δ_1 is obviously elliptic, all generalized solutions of the equation (1.2.5) are real-analytic in X.

We are ready to formulate the principal result of this subsection obtained in the author's note [235].

Theorem 1.2.34 *Problem 1.2.31 is solvable for all functions $f_1^0 \in \mathcal{D}'(X)$ satisfying (1.2.5) iff the space $\mathcal{E}(X)$ is* (coker ${P_2}^t$)*-convex.*

Proof. *Sufficiency.* Since Problems 1.2.31 and 1.2.32 are equivalent, we need only prove that the following sequence is a free resolution of the $\mathcal{P}$-module coker ${P_2}^t$:

$$0 \longleftarrow \operatorname{coker} {P_2}^t \longleftarrow \mathcal{P}^{k-1} \xleftarrow{{P_2}^t} \mathcal{P}^k \xleftarrow{{M_1}^t} \mathcal{P} \longleftarrow 0. \qquad (1.2.11)$$

In view of (1.2.7) the column ${M_1}^t$ is non-zero because the differential operator Δ_1 is elliptic. Hence we must verify the exactness of (1.2.11) only at the term $\mathcal{P}^k$.

Suppose that $p = \begin{pmatrix} p_1 \\ \dots \\ p_k \end{pmatrix} \in \mathcal{P}^k$ satisfies ${P_2}^t p = 0$. We have $PM = P_1 M_1 + P_2 M_2$, whence it follows that

$$(PM)^t p = {M_1}^t {P_1}^t p + {M_2}^t {P_2}^t p = {M_1}^t {P_1}^t p. \qquad (1.2.12)$$

According to (1.2.8) we have ${P_2}^t {M_1}^t = 0$ so (1.2.12) implies ${P_2}^t (PM)^t p = 0$, and the repeated application of (1.2.12) gives for all integers $\imath \geq 0$

$${M_1}^t {P_1}^t [(PM)^t]^\imath p = [(PM)^t]^{\imath+1} p. \qquad (1.2.13)$$

Invoking (1.2.6) we get

$$\begin{aligned} [(PM)^t]^{\imath+1} &= [(PM)^{\imath+1}]^t = \\ &= [P \operatorname{diag}(\Delta_1^\imath, \dots, \Delta_k^\imath) M]^t = \\ &= M^t \operatorname{diag}(\Delta_1^\imath, \dots, \Delta_k^\imath) P^t, \end{aligned}$$

so that (1.2.13) can be rewritten in the form

$${M_1}^t {P_1}^t M^t \operatorname{diag}(\Delta_1^{\imath-1}, \dots, \Delta_k^{\imath-1}) P^t p = M^t \operatorname{diag}(\Delta_1^\imath, \dots, \Delta_k^\imath) P^t p.$$

It follows from (1.2.7) and (1.2.9) that

$$\begin{aligned} {P_1}^t M^t &= {P_1}^t\,({M_1}^t, {M_2}^t) = \\ &= ({P_1}^t {M_1}^t, {P_1}^t {M_2}^t) \\ &= (\Delta_1, \underbrace{0, \ldots, 0}_{(k-1)\ times}), \end{aligned}$$

and hence

$$\begin{aligned} {M_1}^t \Delta_1^i {P_1}^t p &= M^t \operatorname{diag}(\Delta_1^i, \ldots, \Delta_k^i) P^t p = \\ &= (P^t)^{-1} \operatorname{diag}(\Delta_1^{i+1}, \ldots, \Delta_k^{i+1}) P^t p. \end{aligned} \tag{1.2.14}$$

Since i is arbitrary, the right side of (1.2.14) is equal to

$$(P^t)^{-1} \operatorname{diag}(\Delta_1^{i+1}, \ldots, \Delta_k^{i+1}) P^t p = \Delta_1^{i+1} p,$$

so we have from (1.2.14) that ${M_1}^t {P_1}^t p = \Delta_1 p$, that is

$$\begin{cases} m_{11}\ {P_1}^t p &= \Delta_1 p_1, \\ \quad \ldots & \\ m_{1k}\ {P_1}^t p &= \Delta_1 p_k. \end{cases} \tag{1.2.15}$$

We can conclude from (1.2.15) that ${P_1}^t p$ is divisible by Δ_1 because the greatest common divisor of the polynomials Δ_1 and $m_{11}, \ldots, m_{1k}$ is equal to 1. Hence the polynomial $\tilde{p} = {P_1}^t p / \Delta_1$ satisfies the equation ${M_1}^t \tilde{p} = p$.

The exactness of the sequence (1.2.11) has been established.

Necessity. Since $\mathcal{E}(X)$ is a Fréchet-Schwartz space, by the Open Mapping Theorem (see Edwards [62, 8.9.3. (1)]) it is sufficient to prove the algebraic exactness of the sequence

$$[\mathcal{E}(X)]^{k-1} \xrightarrow{P_2} [\mathcal{E}(X)]^k \xrightarrow{M_1} \mathcal{E}(X) \longrightarrow 0. \tag{1.2.16}$$

Let us verify the exactness of (1.2.16) at $\mathcal{E}(X)$. We fix some function $f \in \mathcal{E}(X)$. As the differential operator Δ_1 is elliptic, by the well-known theorem of Malgrange [156, p.341] there exists a function $v \in \mathcal{E}(X)$ such that $\Delta_1 v = f$. If we set $u = P_1 v$, we obtain $u \in [\mathcal{E}(X)]^k$ and $M_1 u = f$ due to (1.2.7).

There remains to check the exactness of the sequence (1.2.16) at $[\mathcal{E}(X)]^k$. Let $f \in [\mathcal{E}(X)]^k$ satisfy $M_1 f = 0$. Since the differential operator $\det P$ is elliptic, by the same theorem of Malgrange [156] there exists a vector-function $v \in [\mathcal{E}(X)]^k$ such that $(\det P)v = f$. By setting $w = (adj\ P)v$ we have that $w \in [\mathcal{E}(X)]^k$ and $Pw = f$. Let us write $w = \begin{pmatrix} w_1 \\ w^{(2)} \end{pmatrix}$ where $w^{(2)} \in [\mathcal{E}(X)]^{k-1}$. Then we get $P_1 w_1 + P_2 w^{(2)} = f$ so that we have $\Delta_1 w_1 = M_1 P_1 w_1 = 0$ in view of (1.2.7) and (1.2.8). By the condition of the theorem we can find a vector-function $f^{(2)} \in [\mathcal{E}(X)]^{k-1}$ such that $P_2 f^{(2)} = -P_1 w_1$. Consequently, the vector-function $u = w^{(2)} - f^{(2)}$ belongs to $[\mathcal{E}(X)]^{k-1}$ and satisfies $P_2 u = f$ in X. This finishes the proof.

□

Obviously a solution of Problem 1.2.32 is determined uniquely up to a solution of the homogeneous equation $P_2 f^{(2)} = 0$.

1.2.6 The de Rham complex

Let X be a differentiable manifold of dimension n. We denote by $T^*_{\mathbb{C}}(X)$ the complexified cotangent bundle of X, i.e. $T^*_{\mathbb{C}}(X) = \mathbb{C} \otimes_{\mathbb{R}} T^*(X)$, and by $\Lambda^q = \Lambda^q T^*_{\mathbb{C}}(X)$ the bundle of complex-valued exterior forms of degree q over X. These bundles are non-zero only for $0 \le q \le n$.

In each coordinate neighborhood U on X with coordinates $x = x(p)$ the bundles Λ^q have the natural trivializations obtained by choosing the repères $\{dx_{\mathcal{I}}\}_{\#\mathcal{I}=q}$ for $\Lambda^q|_U$. Here $dx_{\mathcal{I}} = dx_{\imath_1} \wedge \ldots \wedge dx_{\imath_q}$ where $\{dx_\imath\}_{1\le\imath\le n}$ is the basis of $T^*_{\mathbb{C},p}(X)$ dual to the basis $\{\frac{\partial}{\partial x_\imath}\}_{1\le\imath\le n}$ of $T_{\mathbb{C},p}(X)$, and $\mathcal{I}$ varies over all increasing multi-indices of integers $\{1,\ldots,n\}$ and of power $\#\mathcal{I} = q$, i.e. $\mathcal{I} = (\imath_1,\ldots,\imath_q)$ with $1 \le \imath_1 < \ldots < \imath_q \le n$.

A section $f \in \mathfrak{C}(\Lambda^q)$ is said to be a differential form of degree q and class $\mathfrak{C}$ on X.

For a coordinate neighborhood U on X, any differential form $f \in \mathfrak{C}(\Lambda^q|_U)$ has a local representation

$$f_U(x) = \sum\nolimits'_{\#\mathcal{I}=q} f_{\mathcal{I}}(x) dx_{\mathcal{I}}$$

where the prime on the summation symbol means that the sum is over increasing multi-indices $\mathcal{I}$, and $f_{\mathcal{I}} \in \mathfrak{C}(U)$.

We denote by $d_q \in \mathrm{do}_1(\Lambda^q \to \Lambda^{q+1})$ the operator of exterior derivation acting on differential forms of degree q. The subscript q for d_q is traditionally omitted. Locally this operator is given by

$$df_U(x) = \sum\nolimits'_{\#\mathcal{I}=q} \sum_{\imath=1}^{n} \frac{\partial f_{\mathcal{I}}(x)}{\partial x_\imath} dx_\imath \wedge dx_{\mathcal{I}}. \qquad (1.2.17)$$

Proposition 1.2.35 $d^2 = 0$.

Proof. See Cartan [36, Ch.3, §2.5].

□

The (cochain) differential complex $\Lambda^\cdot = \Lambda^\cdot T^*_{\mathbb{C}}(X)$ on X consisting of the vector bundles Λ^q and the differential operators d_q is called the de Rham complex.

Proposition 1.2.36 *The de Rham complex is an elliptic complex.*

Proof. Indeed, for a point $(x,z) \in T^*(X)$, the symbol mapping $\sigma(d_q)(x.z) : \Lambda^q_x \to \Lambda^{q+1}_x$ consists in the exterior multiplication by the cotangent vector $z \in T^*_x(X)$. This proves the theorem.

□

If we write the coefficients of the differential forms f_U and df_U one after another, we obtain vector-functions $f_U^{(v)}$ and $(df_U)^{(v)}$ with $\begin{pmatrix} n \\ q \end{pmatrix}$ and $\begin{pmatrix} n \\ q+1 \end{pmatrix}$ components respectively. It is easy to see that the action of the differential operator d_q on the form f_U given in (1.2.17) corresponds to the action of some differential operator with constant coefficients $d_q^{(v)} \in \mathrm{do}_1(\mathbb{R}^n \times \mathbb{C}^{k_q} \to \mathbb{R}^n \times \mathbb{C}^{k_{q+1}})$ on the vector-function $f_U^{(v)}$ where $k_q = \begin{pmatrix} n \\ q \end{pmatrix}$.

Let U_1 and U_2 be some coordinate neighborhoods on X with $U_1 \cap U_2 \neq \emptyset$ and local coordinates $x^{(\nu)} : U_\nu \to \mathbb{R}^n$ ($\nu = 1, 2$) respectively. If f_ν is a local representation of the differential form $f \in \mathfrak{C}(\Lambda^q)$ in U_ν, we have $f_2 = m^\sharp f_1$ where $m^\sharp$ is the so-called "pull-back" operator on forms induced by the mapping $m : x^{(2)}(U_1 \cap U_2) \to x^{(1)}(U_1 \cap U_2)$ given via $m = x^{(1)} \circ x^{(2)^{-1}}$. By passing to vector-functions we obtain $f_2^{(v)} = \bigwedge^q(\frac{\partial m}{\partial x})' \cdot m^* f_1^{(v)}$ where $\bigwedge^q(\frac{\partial m}{\partial x})'$ is the q-th exterior power of the transpose of the tangent mapping $\frac{\partial m}{\partial x} : T_x(x^{(2)}(U_1 \cap U_2)) \to T_{m(x)}(x^{(1)}(U_1 \cap U_2))$, and $(m^* f_1^{(v)})(x) = f_1^{(v)}(m(x))$.

Hence (1.2.17) means that for each vector-function $f \in [\mathcal{E}(x^{(2)}(U_1 \cap U_2))]^{k_q}$ we have

$$d_q^{(v)} \left(\bigwedge^q \left(\frac{\partial m}{\partial x} \right)' \cdot m^* f \right) = \bigwedge^{q+1} \left(\frac{\partial m}{\partial x} \right)' \cdot m^* \left(d_q^{(v)} f \right) \quad \text{in} \quad x^{(2)}(U_1 \cap U_2). \tag{1.2.18}$$

Remark 1.2.37 *The relations (1.2.18) are a particular case of (1.1.5). The pseudo-group Γ of local diffeomorphisms m of the space $\mathbb{R}^n$ satisfying (1.2.18) consists of all local diffeomorphisms. This property allows one to lift the differential operators $d_q^{(v)}$ on any differentiable manifold X up to the global differential operators d_q.*

The relations (1.2.18) express the well-known property of invariance of the exterior derivative under a differentiable change of variables.

Proposition 1.2.38 *For any mapping $m : X \to Y$ of class C^1_{loc} between differentiable manifolds, we have $d \circ m^\sharp = m^\sharp \circ d$.*

Proof. See Cartan [36, Ch. 3, § 2.9]

□

Thus an investigation of local problems connected with the de Rham complex on X is reduced to a study of the differential complex $\Lambda^\cdot T^*_{\mathbb{C}}(\mathbb{R}^n)$.

Proposition 1.2.39 *The de Rham complex on $\mathbb{R}^n$ is a Hilbert complex.*

Proof. This follows immediately from Proposition 2 of Palamodov [175, Ch. VIII, § 13].

□

Proposition 1.2.40 *If X is orientable, the bundle homomorphism $\Lambda^{n-q} \to (\Lambda^q)'$ induced by exterior multiplication defines an isomorphism of complexes $\Lambda^{\cdot\prime} \cong \Lambda^\cdot$.*

Proof. Indeed, there is a natural pairing $\langle , \rangle_x$ (or tr) $: \Lambda^{n-q} \otimes \Lambda^q \to \Lambda^n$ given by exterior multiplication in each fiber, i.e. $\langle g, f \rangle_x = f \wedge g$. Obviously it is non-degenerate in both variables. If X is oriented, the bundle of volumes Ω is identified with Λ^n so Λ^{n-q} can be identified with the dual bundle for Λ^q, i.e. $(\Lambda^q)' \cong \Lambda^{n-q}$. Here the transposed operator to the exterior derivative d_q is given by $d_q' = (-1)^{q+1} d_{n-q-1}$. This completes the proof.

□

It follows from Theorem 1.2.21 that for any point $x^0 \in X$ there exists a neighborhood U such that the complex $\mathcal{E}(\Lambda\,|_U)$ is exact at positive degrees. However this result can also be obtained from the very simple Poincaré Theorem (see Cartan [36, Ch.3, 2.12]).

Proposition 1.2.41 *If the identity mapping of X and a mapping with image in a q-dimensional submanifold of X are homotopic, the complex $\mathcal{E}(\Lambda\,)$ is exact at degrees $> q$.*

Proof. Let $m_\imath : X \to Y$ $(\imath = 0, 1)$ be differentiable mappings between differentiable manifolds. They are assumed to be homotopic, i.e. there exists a differentiable mapping $m_t(x)$ of $X \times [0,1]$ into Y which coincides with m_0 for $t = 0$, and with m_1 for $t = 1$. We consider a homotopy operator $h : C^s_{loc}(\Lambda\, T^*_{\mathbb{C}}(Y)) \to C^s_{loc}(\Lambda\, T^*_{\mathbb{C}}(X))$ of degree (-1) given by

$$hf = \int_0^1 \left(dt \rfloor m_t^\sharp f\right) dt \qquad \text{for } f \in C_{loc}\left(\Lambda^q T^*_{\mathbb{C}}(Y)\right), \tag{1.2.19}$$

where $\rfloor$ means the operator of convolution, or inner product (see de Rham [50, § 26]). It is easy to verify that $hd + dh = m_1^\sharp - m_0^\sharp$, so h is a cochain homotopy between the cochain mappings $m_0^\sharp$ and $m_1^\sharp$ of $\mathcal{E}(\Lambda\, T^*_{\mathbb{C}}(Y))$ into $\mathcal{E}(\Lambda\, T^*_{\mathbb{C}}(X))$. If the image of m_0 lies in some q-dimensional submanifold of Y, the mapping $m_0^\sharp$ acts non-trivially at degrees $\le q$ only. Hence the desired result follows immediately.

□

In particular, if X is an open subset of $\mathbb{R}^n$ starlike with respect to a point $x^0 \in X$, we can take

$$m_t(x) = x^0 + t(x - x^0).$$

In this case (1.2.19) gives the Poincaré-Cartan homotopy operator (see Cartan [36, Ch.3, 2.13]). For the general case we refer to Gugenheim and Spencer [93].

The cohomology spaces of the complex of topological vector spaces $\mathfrak{C}(\Lambda\,)$ are said to be the de Rham cohomologies of X in the class $\mathfrak{C}$.

Proposition 1.2.42 (De Rham Theorem) *For all $0 \le q \le n$, it follows that*

$$H^q(\mathcal{E}(\Lambda\,)) \cong H^q(X, \mathbb{C}).$$

Proof. Indeed, as we have said just before Proposition 1.2.41, the complex of sheaves $\mathcal{E}$ associated with the de Rham complex gives a fine resolution of the sheaf of germs of locally constant functions on X. Now it is enough to use the Abstract de Rham Theorem (see 5.2).

□

Sections $f \in \mathcal{D}'(\Lambda^q) = (\mathcal{D}(\Lambda^{n-q}))'$ were called currents of dimension $(n - q)$ on X by de Rham [50, § 8].

1.2.7 *The Dolbeault complex*

Let X be a complex manifold of (complex) dimension n. We denote by $\Lambda^{p,q} = \Lambda^{p,q}T^*_{\mathbb{C}}(X))$ the bundle of complex-valued exterior forms of bidegree (p,q) over X. They are non-zero only $0 \le p, q \le n$, and besides

$$\Lambda^d = \bigoplus_{p+q=d} \Lambda^{p,q}.$$

In each coordinate neighborhood U on X with (holomorphic) coordinates $z = z(p)$ the bundles $\Lambda^{p,q}$ have natural trivializations. Exactly, if we set $z_\imath = x_\imath + \sqrt{-1}x_{n+\imath}$ for $\imath = 1,\dots,n$, the complexified tangent space $T_{\mathbb{C},p}(X)$ at any point $p \in U$ has the basis $\{\frac{\partial}{\partial z_\jmath}, \frac{\partial}{\partial \bar{z}_\imath}\}_{\substack{1 \le \jmath \le n \\ 1 \le \imath \le n}}$ where $\frac{\partial}{\partial z_\jmath} = \frac{1}{2}(\frac{\partial}{\partial x_\jmath} - \sqrt{-1}\frac{\partial}{\partial x_{n+\jmath}})$ and $\frac{\partial}{\partial \bar{z}_\imath} = \frac{1}{2}(\frac{\partial}{\partial x_\imath} + \sqrt{-1}\frac{\partial}{\partial x_{n+\imath}})$. Then the bundles $\Lambda^{p,q}|_U$ may be trivialized by choosing the repères $\{dz_{\mathcal{J}} \wedge d\bar{z}_{\mathcal{I}}\}_{\substack{\#\mathcal{J}=p \\ \#\mathcal{I}=q}}$ where $\{dz_\jmath, d\bar{z}_\imath\}_{\substack{1 \le \jmath \le n \\ 1 \le \imath \le n}}$ is the basis of $T^*_{\mathbb{C},p}(X)$ dual to the chosen basis of $T_{\mathbb{C},p}(X)$, and $\mathcal{J}$ and $\mathcal{I}$ vary over all increasing multi-indices of integers $\{1,\dots,n\}$.

Sections $f \in \mathfrak{C}(\Lambda^{p,q})$ are said to be differential forms of bidegree (p,q) and class $\mathfrak{C}$ on X. For a coordinate neighborhood U on X, any differential form $f \in \mathfrak{C}(\Lambda^{p,q}|_U)$ has a local representation

$$f_U(z) = \sum\nolimits'_{\substack{\#\mathcal{J}=p \\ \#\mathcal{I}=q}} f_{\mathcal{J}\mathcal{I}}(z)\, dz_{\mathcal{J}} \wedge d\bar{z}_{\mathcal{I}}$$

where $f_{\mathcal{J}\mathcal{I}} \in \mathfrak{C}(U)$.

We denote by $\bar{\partial}^{(p)}_q \in \mathrm{do}_1(\Lambda^{p,q} \to \Lambda^{p,q+1})$ the Cauchy-Riemann operator extended to (p,q)-forms. The indexes p and q at $\bar{\partial}^{(p)}_q$ are usually omitted. Locally this operator is given by

$$\bar{\partial} f_U(z) = \sum\nolimits'_{\substack{\#\mathcal{J}=p \\ \#\mathcal{I}=q}} \sum_{\imath=1}^{n} \frac{\partial f_{\mathcal{J}\mathcal{I}}(z)}{\partial \bar{z}_\imath} d\bar{z}_\imath \wedge dz_{\mathcal{J}} \wedge d\bar{z}_{\mathcal{I}}. \tag{1.2.20}$$

Proposition 1.2.43 $\bar{\partial}^2 = 0$.

Proof. This easily follows from Proposition 1.2.35 because $d = \partial + \bar{\partial}$. □

Since $\Lambda^{p,q} = \Lambda^{p,0} \otimes \Lambda^{0,q}$ and $\bar{\partial}^{(p)}_q = 1 \otimes \bar{\partial}^{(0)}_q$, the (cochain) differential complex $\Lambda^{p,\cdot} = \Lambda^{p,\cdot}T^*_{\mathbb{C}}(X)$ on X consisting of the vector bundles $\Lambda^{p,q}$ and the differential operators $\bar{\partial}^{(p)}_q$ is called the Dolbeault complex with coefficients in the (holomorphic!) vector bundle $\Lambda^{p,0}$. As we have seen, $\Lambda^{p,\cdot} = \Lambda^{p,0} \otimes \Lambda^{0,\cdot}$ so the main object is the complex $\Lambda^{0,\cdot}$. We shall refer to it as (if it were) the Dolbeault complex of X.

Proposition 1.2.44 *For any holomorphic vector bundle E over X, the complex $E \otimes \Lambda^{0,\cdot}$ is elliptic.*

Proof. Indeed, for a point $(x,z) \in T^*(X)$, the symbol mapping $\sigma(\bar{\partial}^{(E)}_q)(x,z) : E_x \otimes \Lambda^{0,q}_x \to E_x \otimes \Lambda^{0,q+1}_x$ consists in exterior multiplication by the $(0,1)$-component of the cotangent vector $z \in T^*_x(X)$. This implies the desired statement.

□

As in 1.2.4, the action of the differential operator $\overline{\partial}_q^{(p)}$ on the differential form f_U given by (1.2.20) corresponds to the action of some differential operator with constant coefficients ${\overline{\partial}_q^{(p)}}^{(v)} \in \mathrm{do}_1(\mathbb{R}^n \times \mathbb{C}^{k_q} \to \mathbb{R}^n \times \mathbb{C}^{k_{q+1}})$ on the vector-function $f_U^{(v)}$ constructed from the components of f_U, where $k_q = \begin{pmatrix} n \\ p \end{pmatrix}\begin{pmatrix} n \\ q \end{pmatrix}$.

Let U_1 and U_2 be coordinate neighborhoods on X with $U_1 \cap U_2 = \emptyset$ and local coordinates $z^{(\nu)} : U_\nu \to \mathbb{C}^n$ $(\nu = 1, 2)$ respectively. If f_ν is a local representation of the differential form $f \in \mathfrak{C}(\Lambda^{p,q})$ in U_ν, we have $f_2 = m^\sharp f_1$ where $m : z^{(2)}(U_1 \cap U_2) \to z^{(1)}(U_1 \cap U_2)$ is the transition mapping given by $m = z^{(1)} \circ {z^{(2)}}^{-1}$. By passing to vector-functions we obtain $f_2^{(v)} = \left(\Lambda^p \left(\frac{\partial m}{\partial z}\right)' \otimes \Lambda^q \left(\overline{\frac{\partial m}{\partial z}}\right)'\right) m^* f_1^{(v)}$ where $\frac{\partial m}{\partial z}$ is the complex tangent mapping for m of the complex (i.e. generated by the basis $\left\{\frac{\partial}{\partial z_j}\right\}_{1 \le j \le n}$) tangent spaces.

Thus (1.2.20) means that for any vector-function $f \in [\mathcal{E}(z^{(2)}(U_1 \cap U_2))]^{k_q}$

$$
{\overline{\partial}_q^{(p)}}^{(v)}\left(\left(\Lambda^p\left(\frac{\partial m}{\partial z}\right)' \otimes \Lambda^q\left(\overline{\frac{\partial m}{\partial z}}\right)'\right) \cdot m^* f\right) = \left(\Lambda^p\left(\frac{\partial m}{\partial z}\right)' \otimes \Lambda^{q+1}\left(\overline{\frac{\partial m}{\partial z}}\right)'\right) \cdot m^* \left({\overline{\partial}_q^{(p)}}^{(v)} f\right) \quad \text{in } z^{(2)}(U_1 \cap U_2). \tag{1.2.21}
$$

Remark 1.2.45 *The relationships (1.2.21) are a particular case of those in (1.1.5). The pseudo-group Γ of local diffeomorphisms m of the space $\mathbb{C}^n$ satisfying (1.2.21) consists of all the holomorphic ones. This property allows one to lift the differential operators ${\overline{\partial}_q^{(p)}}^{(v)}$ on any complex manifold X up to the global differential operators $\overline{\partial}_q^{(p)}$.*

Equality (1.2.21) expresses the well-known property of invariance of the $\overline{\partial}$-derivative under a holomorphic change of variables

Proposition 1.2.46 *For any holomorphic mapping $m : X \to Y$ of complex manifolds, we have $\overline{\partial} \circ m^\sharp = m^\sharp \circ \overline{\partial}$.*

Proof. This follows from Proposition 1.2.38 because $d = \partial + \overline{\partial}$ and $m^\sharp$ preserves the bidegree of a form.

□

Thus an investigation of local problems connected with the complex $\Lambda^{p,\cdot} T^*_{\mathbb{C}}(X)$ is reduced to a study of the Dolbeault complex of $\mathbb{C}^n$.

Proposition 1.2.47 *The complex $\Lambda^{p,\cdot} T^*_{\mathbb{C}}(\mathbb{C}^n)$ is a Hilbert complex.*

Proof. This follows from Proposition 2 of Palamodov [175, Ch.VIII, § 13]

□

Proposition 1.2.48 *The bundle homomorphism $\Lambda^{n-p,n-q} \to (\Lambda^{p,q})'$ given by exterior multiplication defines an isomorphism of complexes $(\Lambda^{p,\cdot})' \cong \Lambda^{n-p,\cdot}$.*

Proof. In fact, there is a natural pairing $\langle , \rangle_x$ (or $tr: \ \Lambda^{n-p,n-q} \otimes \Lambda^{p,q} \longrightarrow \Lambda^{n,n}$ given by the exterior multiplication in fibers, i.e. $\langle g, f\rangle_x = f \wedge g$. It is well-known that this pairing is non-degenerate in both variables. Since $\Lambda^{n,n} \cong \Omega$, the dual bundle for $\Lambda^{p,q}$ can be identified with $\Lambda^{n-p,n-q}$, i.e. $(\Lambda^{p,q})' \cong \Lambda^{n-p,n-q}$. Moreover the transposed operator for $\overline{\partial}_q^{(p)}$ is equal to $(\overline{\partial}_q^{(p)})' = (-1)^{p+q+1}\overline{\partial}_{n-q-1}^{(n-p)}$, which finishes the proof.

□

It follows from Theorem 1.2.21 that for any point $z^0 \in X$ there exists a neighborhood U such that the complex $\mathcal{E}(\Lambda^{p,\cdot}|_U)$ is exact at the positive degrees. However this result can also be obtained from the much more elementary Dolbeault Lemma (see his paper [55], or Gunning and Rossi [96, Ch.1, D3]). But no simple homotopy operator like (1.2.19) is known for the Dolbeault complex.

The cohomology spaces of the complex of topological vector spaces $\mathfrak{C}(\Lambda^{0,\cdot})$ are called the Dolbeault cohomologies of X in the class $\mathfrak{C}$.

Let us denote by $\mathcal{O}^{(p)}$ the sheaf of germs of holomorphic p-forms over X, or the sheaf of germs of holomorphic sections of $\Lambda^{p,0}$.

Proposition 1.2.49 (Dolbeault Theorem) *For all $0 \le q \le n$, it follows that*

$$H^q(\mathcal{E}(\Lambda^{p,\cdot})) \cong H^q(X, \mathcal{O}^{(p)}).$$

Proof. Indeed, as we have said, the complex of sheaves $\mathcal{E}^\cdot$ associated with the complex $\Lambda^{p,\cdot}$ gives a fine resolution of the sheaf of germs of $\overline{\partial}$-closed $(p,0)$-forms over X. Now an application of the Abstract de Rham Theorem finishes the proof because this sheaf is equal to $\mathcal{O}^{(p)}$.

□

1.2.8 *The Koszul complexes*

Let $P = \|P^{(i)}\|_{1\le i\le N}$ be a column of scalar differential operators $P^{(i)}$ on an open set $X \subset \mathbb{R}^n$. The differential operators $P^{(i)}$ are assumed to commute, i.e. the commutators $[P^{(i)}, P^{(j)}] = P^{(i)}\, P^{(j)} - P^{(j)}\, P^{(i)}$ are zero for $1 \le i < j \le N$.

We consider the (trivial) vector bundles $E^q = X \times \mathbb{C}^{k_q}$ where $k_q = \begin{pmatrix} N \\ q \end{pmatrix}$, and the differential operators P_q of type $E^q \to E^{q+1}$ defined in the following way. A section of E^q is interpreted as a family $f = (f_\mathcal{I})_{\#\mathcal{I}=q}$ of functions on X defined for all multi-indices $\mathcal{I}$ of integers $\{1, \ldots, N\}$ in such a way that $f_\mathcal{I}$ is a skew-symmetric function of $\mathcal{I}$. Then

$$(P_q f)_\mathcal{I} = \sum_{k=1}^{q+1} (-1)^{k-1}\ P^{(i_k)}\ f_{i_1 \ldots \widehat{i_k} \ldots i_{q+1}} \quad \text{for} \quad \#\mathcal{I} = q+1 \tag{1.2.22}$$

where the sign "$\wedge$" over i_k means that this index has been omitted. In other words P_q is obtained from the exterior derivative operator d_q corresponding to the space $\mathbb{R}^N_y$ by the formal substitution $\frac{\partial}{\partial y_i} \to P^{(i)}(x, D)$ $(i = 1, \ldots, N)$.

Thus we have $P_0 = P$ and $P_q \ne 0$ only for $0 \le q \le N-1$. The index q at P_q will be omitted.

Proposition 1.2.50 $P^2 = 0$.

Proof. This is obtained by direct calculations.

□

The (cochain) differential complex E on X consisting of the bundles $E^q = X \times \mathbb{C}^{k_q}$ and the differential operators P_q given by (1.2.22) is said to be a Koszul complex (associated with the column P).

For the Koszul complexes the ellipticity condition is easily checked.

Proposition 1.2.51 *For a Koszul complex E to be elliptic it is necessary and sufficient that the differential operator P_0 have an injective symbol.*

Proof. The necessity is obvious; let us prove the sufficiency. We fix a point $(x, z) \in T^*(X)$ with $z \neq 0$. Since the symbol of the differential operator P_0 is injective, the vector $\zeta = \sigma(P_0)(x, z)$ is in $\mathbb{C}^N \setminus \{0\}$. Moreover, the symbol complex for E over the point (x, z) is the same as the one for the de Rham complex $\Lambda\, T^*_{\mathbb{C}}(\mathbb{R}^N)$ over the point $(y, \sum_{i=1}^{N} \zeta_i dy_i) \in T^*_{\mathbb{C}}(\mathbb{R}^N)$ up to a non-essential multiplier $\sqrt{-1}$. Hence the symbolic complex for E over (x, z) is exact because the symbolic complex for $\Lambda\, T^*_{\mathbb{C}}(\mathbb{R}^N)$ is exact not only over all points $(y, \zeta) \in T^*(\mathbb{R}^N)$ but also over all points $(y, \zeta) \in T^*_{\mathbb{C}}(\mathbb{R}^N)$ with $\zeta \neq 0$.

□

The concept of a Koszul complex has often been interpreted in a more limited sense, as in the following proposition.

Proposition 1.2.52 *If $P(D)$ is a column of scalar differential operators with constant coefficients on X such that the dimension of the algebraic variety $\mathcal{N}(M) = \{z \in \mathbb{C}^n : P(z) = 0\}$ is no more than $(n - N)$, the Koszul complex E associated with $P(D)$ is a Hilbert complex.*

Proof. This follows from Proposition 2 of Palamodov [175, Ch. VIII, § 13].

□

The column $P(D)$ satisfies the condition of Proposition 1.2.52, for example, if the differential operators $P^{(i)}(D)$ $(i = 1, \ldots, N)$ are acting on different groups of variables. The de Rham complex in $\mathbb{R}^n$ $(P(D) = \left(\frac{\partial}{\partial x_i}\right)_{1 \leq i \leq n})$ and the Dolbeault complex in $\mathbb{C}^n$ $(P(D) = \left(\frac{\partial}{\partial \bar{z}_i}\right)_{1 \leq i \leq n})$ have such a property so they are Koszul complexes in the limited sense.

Some results on global solvability for general Koszul complexes of first order differential operators can be obtained from papers of Hörmander [108], [111, Theorem 1.27]. We remark in passing that Koszul complexes are not in general compatibility complexes.

1.2.9 *The Asada complexes*

These complexes are intermediate between the general Koszul complexes and the de Rham and Dolbeault complexes.

Let M be a $(N\times n)$-matrix of complex numbers such that the following condition is fulfilled:

$$\begin{array}{l}\text{for some} \quad (N\times n)-\text{matrix} \quad \widetilde{M}, \quad \text{the} \\ \text{matrix } (\widetilde{M}^*M - I_n) \text{ is skew-symmetric.}\end{array} \tag{1.2.23}$$

The Koszul complex associated with the column $P(D) = M\cdot D$ is called an *Asada complex*.

Proposition 1.2.53 *The Asada complex is an elliptic complex.*

Proof. Indeed, by setting $\widetilde{P} = \widetilde{M}D$ we obtain

$$\widetilde{P}^*P = D^*D + D^*(\widetilde{M}^*M - I_n)D = D^*D = -\Delta$$

where Δ is the Laplace operator in $\mathbb{R}^n$. Thus the differential operator P has an injective symbol which completes the proof.

□

Apparently Asada [1] was the first who dealt with the analysis of such special overdetermined systems of differential equations $(MD)f = 0$.

In particular, the de Rham complex in $\mathbb{R}^n$ ($M = \widetilde{M} = \sqrt{-1}\, I_n$) and the Dolbeault complex in $\mathbb{C}^n$ ($M = \frac{1}{4}\widetilde{M} = \frac{\sqrt{-1}}{2}\, \|I_n, \sqrt{-1}\, I_n\|$) are Asada complexes.

1.3 The Spencer Resolution of a Formally Integrable Differential Operator

Let X be a differentiable manifold of dimension n, and $P \in \mathrm{do}_p(E\to F)$ be some overdetermined differential operator on X.

Unlike the case of constant coefficients, constructing a compatibility operator for P is a substantial problem in the general case. Progress here has been attained mainly thanks to works of American mathematicians close to Spencer, namely Quillen, Sternberg, Guillemin, Goldschmidt, Cenkl, Sweeney, MacKichan and others, who developed Spencer's earlier geometrical approach (see Spencer [203, 204, 207]).

We shall restrict ourselves to formulating the main results (see the survey of Spencer [208] and the book of Pommaret [182, Ch. 5]).

1.3.1 *Geometrical viewpoint for differential operators*

For $s \in \mathbb{Z}_+$, we denote by $\mathcal{J}^s(E)$ the bundle of s-jets of sections of a bundle E over X (see Bourbaki [29, 12.6]). The fiber of the bundle $\mathcal{J}^s(E)$ over a point $x\in X$ is the quotient of the space of germs of differentiable sections of E at x by the subspace of germs vanishing at x together with all derivatives up to order s.

If U is a coordinate neighborhood on X such that E is trivial over U, then $\mathcal{J}^s(E)|_U \cong U\times \left(\otimes_{j\le s}\mathcal{P}_j{}^k\right)$ where $\mathcal{P}_j$ is the space of all homogeneous polynomials of

n variables of degree j with complex coefficients. An arbitrary section $\sigma \in \sigma(\mathcal{J}^s(E))$ has in U the local representation

$$\sigma_U(x) = \sum_{|\alpha|\leq s} \sigma_\alpha(x)\frac{(y-x)^\alpha}{\alpha!} \quad (x \in U), \tag{1.3.1}$$

where σ_α ($|\alpha| \leq s$) are suitable vector-valued functions on U of smoothness corresponding to the smoothness of σ.

We identify $\mathcal{J}^0(E)$ and E, and denote by π^{s_1,s_2} the natural projection $\mathcal{J}^{s_1}(E) \to \mathcal{J}^{s_2}(E)$ where $s_1 \geq s_2$. In local coordinates of U

$$\pi^{s_1,s_2}\left(x, \sum_{|\alpha|\leq s_1} \sigma_\alpha(x)\frac{(y-x)^\alpha}{\alpha!}\right) = \left(x, \sum_{|\alpha|\leq s_2} \sigma_\alpha(x)\frac{(y-x)^\alpha}{\alpha!}\right).$$

The operator j^s that associates the section $f \in \mathcal{E}(E)$ with its s-jet, is a canonical differential operator of type $E \to \mathcal{J}^s(E)$ and order s on X. In the coordinate neighborhood U on X we have

$$(j^s f)(x) = \sum_{|\alpha|\leq s} \partial^\alpha f(x)\frac{(y-x)^\alpha}{\alpha!} \quad (x \in U).$$

Proposition 1.3.1 *For each differential operator $P \in \mathrm{do}_p(E \to F)$ there exists a bundle homomorphism $\eta(P) : \mathcal{J}^p(E) \to F$ such that $P = \eta(P) \circ j^p$, and this correspondence between differential operators of type $E \to F$ and order $\leq p$ and* $\mathrm{Hom}\,(\mathcal{J}^p(E) \to F)$ *is one-to-one.*

Proof. In a coordinate neighborhood U on X where the differential operator P has the representation $P(x,D) = \sum_{|\alpha|\leq p} \tilde{P}_\alpha(x)\ \partial^\alpha$, and $\sigma \in \sigma(\mathcal{J}^p(E))$ has the representation (1.3.1), we may set

$$(\eta(P)\sigma)(x) = \sum_{|\alpha|\leq p} \tilde{P}_\alpha(x)\ \sigma_\alpha(x) \quad (x \in U). \tag{1.3.2}$$

It follows from the bundle structure of $\mathcal{J}^p(E)$ that (1.3.2) defines in fact a global bundle homomorphism $\eta(P) : \mathcal{J}^p(E) \to F$ with $P = \eta(P) \circ j^p$.

In other respects the proposition is obvious which finishes the proof.

□

In the future the following statement will be useful.

Corollary 1.3.2 *For differential operators $P \in \mathrm{do}_p(E \to F)$ and $D \in \mathrm{do}_d(F \to G)$ on X, we have*

$$\eta(j^s D \circ P) = \eta(j^s D) \circ \eta(j^{s+d} P) \quad (s \geq 0).$$

Proof. This is an easy exercise (see, for example, Pommaret [182, Ch.2, 1.15]). □

Proposition 1.3.1 gives the possibility of outlining a slightly different approach to the notion of symbol of a differential operator P. Namely, let $S^rT^*(X)$ be the bundle of symmetric r-forms over X where $r \in \mathbb{Z}_+$, and let $\varepsilon^r : E \otimes S^rT^*(X) \to \mathcal{J}^r(E)$ be the embedding mapping. Then the principal symbol of the differential operator P is the homomorphism $\sigma(P) \in \mathrm{Hom}\,(E \otimes S^pT^*(X) \to F)$ given by $\sigma(P) = \eta(P)\varepsilon^p$.

Definition 1.3.3 *The differential operator $j^s \circ P \in \mathrm{do}_{p+s}(E \to \mathcal{J}^s(F))$ is said to be the s-th prolongation of P.*

Prolongations of a differential operators P bring information on all possible differential consequences of the system $Pf = 0$. For this reason the following technical result is especially important.

Proposition 1.3.4 *For any $s \geq 0$, the differential operators P and $j^s \circ P$ are equivalent.*

Proof. This follows from Example 1.2.6 because $j^s \circ P = (1 \otimes D) \circ P$ for some differential operator D.

□

1.3.2 *Regularity conditions*

Not every overdetermined differential operator with variable coefficients has a compatibility operator.

Example 1.3.5 Choose some smooth function $P_0(X)$ on $\mathbb{R}^1$ such that $P_0(x) = 0$ for $x \leq 0$ and $P_0(x) > 0$ for $x > 0$. Then the differential operator P given by $(Pf)(x) = P_0(x)f(x)$ is obviously overdetermined. However P has no compatibility operator. Indeed, if P_1 is some compatibility operator for P, we can conclude from equality $P_1P = 0$ that the coefficients of P_1 must vanish for $x > 0$. Hence the differential operator $\widetilde{P}_1 = \frac{1}{x}P_1$ has smooth coefficients and satisfies $\widetilde{P}_1P_1 = 0$. On the other hand it is easy to see that $\widetilde{P}_1$ can not be represented in the form $\widetilde{P}_1 = TP_1$ where T is some differential operator with smooth coefficients on $\mathbb{R}^1$.

□

In this example non-existence of a compatibility operator is a consequence of strong "degeneration" of coefficients of P. However, if the differential operator P satisfies some conditions of non-degeneracy of coefficients (regularity), a compatibility operator always exists for P.

For $s \geq p$, we consider a family of vector spaces

$$\mathfrak{R}^s(x) = \ker\{\eta(j^{s-p}P) : \mathcal{J}^s(E)_x \to \mathcal{J}^{s-p}(F)_x\} \quad (x \in X).$$

Lemma 1.3.6 *Given $p \le s_2 \le s_1$, we have*

$$\pi^{s_1-p,s_2-p}\eta(\jmath^{s_1-p}P) = \eta(\jmath^{s_2-p}P)\pi^{s_1,s_2}.$$

Proof. See Pommaret [182, Ch.2, 1.18].

□

This lemma implies that, for $p \le s_2 \le s_1$, the restriction of π^{s_1,s_2} on $\mathfrak{R}^{s_1}(x)$ takes its values in $\mathfrak{R}^{s_2}(x)$.

The following definition is of fundamental importance in Spencer's theory.

Definition 1.3.7 *The differential operator P is said to be sufficiently regular if: 1) the dimensions of the subspaces $\mathfrak{R}^s(x)$ $(s \ge p)$ do not depend on $x \in X$, and 2) the mappings $\pi^{s_1,s_2} : \mathfrak{R}^{s_1}(x) \to \mathfrak{R}^{s_2}(x)$ have constant rank for all $p \le s_2 \le s_1$.*

The condition 1) in the definition means that, for all $s \ge p$, the family $\mathfrak{R}^s = \cup_{x\in X}\mathfrak{R}^s(x)$ is a vector bundle over X (regularity). The condition 2) is more subtle and means that $\pi^{s_1,s_2}(\mathfrak{R}^{s_1})$ is a vector subbundle of $\mathcal{J}^{s_2}(E)$ for all $p \le s_2 \le s_1$.

Example 1.3.8 Each differential operator P with constant coefficients on an open set $X \subset \mathbb{R}^n$ is sufficiently regular.

□

Example 1.3.9 Let f_0 be a smooth differential form of degree 1 on an open set $X = \mathbb{R}^3$, and let the differential operator $P \in \mathrm{do}_1(\Lambda^1 T^*_{\mathbb{C}}(X) \to \Lambda^2 T^*_{\mathbb{C}}(X))$ be given by $Pf = df + f_0 \wedge f$. It is easily verified that the dimensions of the subspaces $\mathfrak{R}^1(x)$ and $\mathfrak{R}^2(x)$ do not depend on $x \in X$, and the range of $\pi^{2,1} : \mathfrak{R}^2(x) \to \mathfrak{R}^1(x)$ consists of all 1-jets $\left(x, \sum_{|\alpha|\le 1}\sigma_\alpha(x)\frac{(y-x)^\alpha}{\alpha!}\right) \in \mathfrak{R}^1(x)$ such that $df_0(x) \wedge \sigma_0(x) = 0$. Hence the projection $\pi^{2,1} : \mathfrak{R}^2(x) \to \mathfrak{R}^1(x)$ can have constant rank only in the following two cases: 1) $df_0(x) = 0$ for all $x \in X$; 2) $df_0(x) \ne 0$ for all $x \in X$. One can show that the operator P is sufficiently regular in each of these cases.

□

1.3.3 *Formally exact differential complexes*

Suppose we are given a short complex $\mathcal{E}(E') \xrightarrow{P'} \mathcal{E}(E'') \xrightarrow{P''} \mathcal{E}(E''')$ where the orders of the differential operators P' and P'' are equal to p' and p'' respectively. By Corollary 1.3.2, for every $s \ge 0$, the following sequence of bundle homomorphisms is also a complex:

$$\mathcal{J}^{s+p'+p''}(E') \xrightarrow{\eta(\jmath^{s+p''}P')} \mathcal{J}^{s+p''}(E'') \xrightarrow{\eta(\jmath^{s}P'')} \mathcal{J}^{s}(E'''). \tag{1.3.3}$$

Definition 1.3.10 *A short complex $\mathcal{E}(E') \xrightarrow{P'} \mathcal{E}(E'') \xrightarrow{P''} \mathcal{E}(E''')$ is said to be formally exact if the complex (1.3.3) is exact for every $s \ge 0$.*

One notices that, over a fixed point $x \in X$, the complex (1.3.3) is a complex of linear mappings of finite-dimensional vector spaces.

For a long differential complex E on X, the formal exactness means formal exactness of any short complex $\mathcal{E}(E^i) \xrightarrow{P_i} \mathcal{E}(E^{i+1}) \xrightarrow{P_{i+1}} \mathcal{E}(E^{i+2})$ for each $i \ge 0$.

Proposition 1.3.11 *Each formally exact differential complex $E^{\cdot}$ is a compatibility complex for the initial differential operator P_0.*

Proof. Via the proof we would like to illustrate a technique for applying the preceding notions.

Let $\widetilde{P}_{i+1}$ be a differential operator of type $E^{i+1} \to \widetilde{E}^{i+1}$ and order $\widetilde{p}_{i+1}$ on X such that $\widetilde{P}_{i+1}P_i = 0$ where $i \geq 0$. Choose an integer s such that $s + \widetilde{p}_{i+1} \geq p_{i+1}$. Since $\widetilde{P}_{i+1}P_i = 0$, the following sequence is a complex:

$$\mathcal{J}^{s+p_i+\widetilde{p}_{i+1}}(E^i) \overset{\eta(j^{s+\widetilde{p}_{i+1}}P_i)}{\longrightarrow} \mathcal{J}^{s+\widetilde{p}_{i+1}}(E^{i+1}) \overset{\eta(j^s\widetilde{P}_{i+1})}{\longrightarrow} \mathcal{J}^s(\widetilde{E}^{i+2})$$

Now we can conclude from the formal exactness of $E^{\cdot}$ that

$$\ker \eta\,(j^{s+\widetilde{p}_{i+1}-p_{i+1}}P_{i+1}) = \operatorname{im}\eta\,(j^{s+\widetilde{p}_{i+1}}P_i) \subset \ker \eta\,(j^s\widetilde{P}_{i+1}).$$

Thus there exist a bundle homomorphism

$$h: \mathcal{J}^{s+\widetilde{p}_{i+1}-p_{i+1}}(E^{i+2}) \to \mathcal{J}^s(\widetilde{E}^{i+2})$$

such that $\eta\,(j^s\widetilde{P}_{i+1}) = h \circ \eta\,(j^{s+\widetilde{p}_{i+1}-p_{i+1}}P_{i+1})$. By Proposition 1.3.1

$$j^s\widetilde{P}_{i+1} = \left(h \circ \eta\,\left(j^{s+\widetilde{p}_{i+1}-p_{i+1}}P_{i+1}\right)\right)j^{s+\widetilde{p}_{i+1}} = \left(h \circ j^{s+\widetilde{p}_{i+1}-p_{i+1}}\right)P_{i+1} = TP_{i+1}$$

where T is a differential operator of type $E^{i+2} \to \mathcal{J}^s(\widetilde{E}^{i+2})$.

This immediately implies the desired statement.

□

In particular, this proposition allows one "to paste together" a global compatibility operator for P on X from local ones.

Corollary 1.3.12 *If, for some open covering $\{U_\alpha\}_{\alpha\in A}$ of X, the restrictions of the differential operator P on U_α have compatibility operators, then P has a global compatibility operator on X.*

Proof. See for example Dudnikov and Samborskii [58].

□

We conclude this subsection with an important example of a formally exact complex.

Example 1.3.13 (see Spencer [203]) We take the canonical operator $j^s \in \mathrm{do}_s(E \to \mathcal{J}^s(E))$ as P, where $s \in \mathbb{Z}_+$. For each pair of non-integers s and q, there exists a first order differential operator D of type $\mathcal{J}^s(E) \otimes \Lambda^q T^*(X) \to \mathcal{J}^{s-1}(E) \otimes \Lambda^{q+1}T^*(X)$ which is defined locally in the following way. Each section $\sigma \in \mathcal{E}(\mathcal{J}^s(E) \otimes \Lambda^q T^*(X))$ has in U a representation of the form

$$\sigma_U(x) = {\sum}'_{\substack{\#I=q\\|\alpha|\leq s}} \sigma_{I,\alpha}(x)\frac{(y-x)^\alpha}{\alpha!}\,dx_I \qquad (x \in U),$$

where $\sigma_{I,\alpha} \in [\mathcal{E}(U)]^k$. Then $D\sigma$ is given in U by the equality

$$\begin{aligned} D\sigma(x) &= \pi^{s,s-1} d_x \sigma_U(x) \\ &= \sum\nolimits'_{\substack{\#I=q \\ |\alpha|\le s-1}} \sum_{\iota=1}^{n} \left(\frac{\partial \sigma_{I,\alpha}(x)}{\partial x_\iota} - \sigma_{I,\alpha+1_\iota}(x) \right) \frac{(y-x)^\alpha}{\alpha!}\, dx_\iota \wedge dx_I \\ & (x \in U). \end{aligned} \tag{1.3.4}$$

It is easy to verify that $D^2 = 0$. Hence, assuming $s - n \geq 0$, we obtain a complex

$$\begin{aligned} 0 \longrightarrow \mathcal{E}(E) \xrightarrow{j^s} \mathcal{E}(\mathcal{J}^s(E)) &\xrightarrow{D} \mathcal{E}(\mathcal{J}^{s-1}(E) \otimes \Lambda^1 T^*(X)) \xrightarrow{D} \dots \\ &\xrightarrow{D} \mathcal{E}(\mathcal{J}^{s-n}(E) \otimes \Lambda^n T^*(X)) \longrightarrow 0. \end{aligned}$$

This complex is known to be exact but not formally exact. Now we shall construct a second complex which has better formal properties. The second term on the right-hand side of (1.3.4) gives the so-called operator of formal exterior differentiation $\delta \in \mathrm{Hom}(E \otimes S^s T^*(X) \otimes \Lambda^q T^*(X) \to E \otimes S^{s-1} T^*(X) \otimes \Lambda^{q+1} T^*(X))$. Clearly, $\delta^2 = 0$. Thus we have a complex of bundle homomorphisms

$$\begin{aligned} 0 \longrightarrow E \otimes S^s T^*(X) &\xrightarrow{\delta} E \otimes S^{s-1} T^*(X) \otimes \Lambda^1 T^*(X) \xrightarrow{\delta} \dots \\ &\xrightarrow{\delta} E \otimes S^{s-n} T^*(X) \otimes \Lambda^n T^*(X) \longrightarrow 0. \end{aligned}$$

This complex is also verified to be exact. Hence it follows that, for each $s \in \mathbb{Z}_+$, we may define the following quotient bundles:

$$S^q = S^q(\mathcal{J}^s(E)) = \begin{cases} \mathcal{J}^s(E) & \text{for } q = 0, \\ \frac{\mathcal{J}^s(E) \otimes \Lambda^q T^*(X)}{\delta(E \otimes S^{s+1} T^*(X) \otimes \Lambda^{q-1} T^*(X))} & \text{for } q > 0. \end{cases}$$

The operator D gives rise to a first order differential operator $D^{(s)}$ of the type $S^q \to S^{q+1}$ with $(D^{(s)})^2 = 0$, and we have arrived at the desired complex

$$\begin{aligned} 0 \longrightarrow \mathcal{E}(E) \xrightarrow{j^s} \mathcal{E}(S^0(\mathcal{J}^s(E))) &\xrightarrow{D^{(s)}} \mathcal{E}(S^1(\mathcal{J}^s(E))) \xrightarrow{D^{(s)}} \dots \\ &\xrightarrow{D^{(s)}} \mathcal{E}(S^n(\mathcal{J}^{s-n}(E))) \longrightarrow 0. \end{aligned} \tag{1.3.5}$$

The complex (1.3.5) is known to be not only exact but also formally exact.

□

Remark 1.3.14 *In Example 1.3.13, we constructed the main Spencer sequences for the trivial operator $E \to 0$ (see below).*

1.3.4 *Formal integrability*

Definition 1.3.15 *A differential operator P (of order p) is said to be formally integrable if: 1) P is sufficiently regular, and moreover 2) the mappings $\pi^{s+1,s}$: $\mathfrak{R}^{s+1} \to \mathfrak{R}^s$ are surjective for all $s \geq p$.*

Formal integrability of the operator P (of order p) means that, for any $s > p$, all differential consequences of order s of the system $Pf = 0$ (i.e. consequences extracted by means of differentiations of any orders, equating mixed derivatives, and application of linear algebra for each $x \in X$) may be obtained by way of differentiations of order no more than $(s - p)$, and application of linear algebra.

Example 1.3.16 Each homogeneous differential operator P with constant coefficients on an open set $X \subset \mathbb{R}^n$ is formally integrable.

□

Example 1.3.17 As in Example 1.3.9, let f_0 be a smooth differential form of degree 1 on an open set $X \subset \mathbb{R}^3$, and the differential operator $P \in \mathrm{do}_1(\Lambda^1 T^*_{\mathbb{C}}(X) \to \Lambda^2 T^*_{\mathbb{C}}(X))$ be given by $Pf = df + f_0 \wedge f$. Formal integrability of the differential operator P is equivalent to the equality $df_0(x) = 0$ for all $x \in X$. If $df_0(x) \neq 0$ for all $x \in X$ (this condition provides sufficient regularity of P), then the differential operator P is equivalent to the differential operator $\widetilde{P}f = Pf \oplus (df_0 \wedge f) \oplus d^*(df_0 \wedge f)$ which is already formally integrable.

□

Example 1.3.18 Let P be a differential operator on an open set $X \subset \mathbb{R}^n$ (so that $E = X \times \mathbb{C}^k$ and $F = X \times \mathbb{C}^l$). We suppose that P has a non-characteristic covector, i.e. a covector $z \in \mathbb{R}^n$ such that the symbol mapping $\sigma(P)(x, z) : \mathbb{C}^k \to \mathbb{C}^l$ is an isomorphism for each $x \in X$. For example, P may be elliptic or hyperbolic. Then P is known to be formally integrable.

□

Each sufficiently regular differential operator is equivalent to some formally integrable differential operator. More precisely, the following result is true.

Proposition 1.3.19 *Let $P \in \mathrm{do}_p(E \to F)$ be a sufficiently regular operator. Then there is a differential operator D, which can be constructed in a finite number of steps, such that: 1) $\widetilde{P} = DP$ is formally integrable; 2) $\ker \widetilde{P} = \ker P$, and 3) the operators P and $\widetilde{P}$ are equivalent.*

Proof. As in Example 1.3.17, the operator $\widetilde{P}$ can be constructed from P by completely writing differential consequences of the correlation $Pf = 0$. Hence it is possible to consider without loss of generality that $\widetilde{P} = (1 \oplus D)P$ for some differential operator D. Then 2) is obvious, and 3) follows from Example 1.2 6 which completes the proof.

□

1.3.5 *Involutive differential operators*

For $s \geq p$, we denote by $\sigma^{(s)}$ the kernel of the bundle homomorphism $\pi^{s,s-1} : \mathfrak{R}^s \to \mathfrak{R}^{s-1}$. If the differential operator P is sufficiently regular (and $s > p$), $\sigma^{(s)}$ is a vector bundle over X.

Often $\sigma^{(s)}$ has been called the symbolic bundle of $(s-p)$-th prolongation of P because it may be identified with the kernel of $\sigma(\jmath^{s-p}P): \; E \otimes S^sT^*(X) \to F \otimes S^{s-p}T^*(X)$.

One can easily verify that the restriction of the formal exterior differentiation operator δ on $\sigma^{(s-q)} \otimes \Lambda^q T^*(X)$ takes its values in $\sigma^{(s-q-1)} \otimes \Lambda^{q+1}T^*(X)$ for any s and q with $s-q-1 \geq p$. Thus the following complex of bundle homomorphisms arises:

$$\begin{array}{llllll} 0 \longrightarrow \sigma^{(s)} & \xrightarrow{\delta} & \sigma^{(s-1)} \otimes \Lambda^1 T^*(X) & \xrightarrow{\delta} & \ldots & \\ & \xrightarrow{\delta} & \sigma^{(p)} \otimes \Lambda^{s-p}T^*(X) & \longrightarrow & \delta(\sigma^{(p)} \otimes \Lambda^{s-p}T^*(X)) & \longrightarrow 0. \end{array} \tag{1.3.6}$$

This complex (called the Spencer δ-complex) is not necessarily exact at all degrees however it is always exact at the first two degrees.

Definition 1.3.20 *The differential operator P (of order p) is said to be involutive if the complex (1.3.6) is exact for all $s \geq p$.*

E. Cartan formulated the property of involutivity in the language of Pfaffian systems. Moreover he imposed that the operation of prolongation of systems within a rather wide class of systems lead to an involutive system. There is also a definition of involutivity for first order systems due to Kuranishi. The equivalence of E. Cartan's and Kuranishi's definitions was proved by Kuranishi (1967). Definition 1.3.20 is due to Spencer [206]. Its equivalence to Kuranishi's definition was proved by Serre.

The following result is called the σ-Poincaré Lemma, by analogy with the Poincaré Lemma for the de Rham complex.

Theorem 1.3.21 (Quillen [186]) *Let P be a sufficiently regular differential operator (of order p). There exists an integer $s_0 \geq p$, depending on n, k and p, such that the complex (1.3.6) is exact for all $s \geq s_0$.*

Proof. See Pommaret [182, Ch.3, 4.1].

□

It follows that for each sufficiently regular differential operator P, its prolongation $\jmath^{s-p} \circ P$ with $s \geq s_0$ is an involutive operator. We note that, if P is a formally integrable operator, then $\jmath^{s-p} \circ P$ $(s \geq p)$ is also a formally integrable operator. Since the operators P and $\jmath^{s-p} \circ P$ are equivalent by Proposition 1.3.4, Theorem 1.3.21 together with Proposition 1.3.19 show that each sufficiently regular differential operator P can be replaced by an equivalent involutive and formally integrable one.

Remark 1.3.22 *Upper estimates for the number s_0 in terms of n, k and p from Theorem 1.3.21 are known. They may be obtained from the following correlations:*

$$\begin{aligned} s_o\,(0,k,1) &= 0, \\ s_0\,(n,k,1) &= k\begin{pmatrix} n+s_0(n-1,k,1) \\ n-1 \end{pmatrix} + s_0(n-1,k,1)+1, \\ s_0\,(n,k,p) &= s_0\left(n, \sum_{\jmath=0}^{p} \begin{pmatrix} n+\jmath-1 \\ n-1 \end{pmatrix} k, 1\right). \end{aligned}$$

1.3.6 *Normalized differential operators*

It follows from Corollary 1.3.12 that by constructing compatibility operators one concentrates on local considerations. Then we may write the operator P locally in a suitable form. It is convenient to reduce the investigation to the study of first order differential operators.

Definition 1.3.23 *An operator $P \in \mathrm{do}_p(E \to F)$ is said to be normalized if: 1) the order p of P equals 1; 2) P is formally integrable; 3) P is involutive, and 4) the symbol mapping $\sigma(P) : E \otimes T^*(X) \to F$ is surjective.*

The first three conditions have already been discussed. As to condition 4), it means that among the correlations $Pf = 0$ there are (evidently or non-evidently) no purely algebraic (i.e. non-differential) correlations between the functions f_j $(j = 1, \dots, k)$. If such correlations occur, one can exclude them by cutting down the number of functions $\{f_j\}$. Here, the transformed operator is equivalent to the initial one.

Theorem 1.3.24 (Samborskii [191]) *Each sufficiently regular differential operator P may be transformed by a finite number of steps (within the framework of differentiations and linear algebra at fibers of the bundles) into an equivalent normalized differential operator.*

Proof. The transformation is constructed in the following successive steps.

1) Transition from the differential operator $P^{(0)} = P$, to an equivalent formally integrable operator $P^{(1)}$ by writing out the missing differential consequences of $P^{(0)}f = 0$, as in Proposition 1.3.19.

2) Transition from the differential operator $P^{(1)}$ to an equivalent involutive formally integrable operator $P^{(2)}$ by taking the prolongation of a sufficiently high order, as in Theorem 1.3.21.

3) Transition from the differential operator $P^{(2)}$ to an equivalent involutive formally integrable operator $P^{(3)}$ of the first order in the following way.

Let P be a differential operator of type $E \to F$ and order p on X. We choose some operator $P' \in \mathrm{do}_1(\mathcal{J}^{p-1}(E) \to F)$ with the property that $P = P' \circ j^{p-1}$. Then we define the operator $\tilde{P} \in \mathrm{do}_1(\mathcal{J}^{p-1}(E) \to F \oplus S^1(E))$ by means of $\tilde{P}\sigma = P'\sigma \oplus D^{(p-1)}\sigma$ where $D^{(p-1)}$ is the operator from Example 1.3.13. This operator $\tilde{P}$ can be shown to be equivalent to P and to inherit its formal properties.

4) Transition from the differential operator $P^{(3)}$ to an equivalent normalized differential operator $P^{(4)}$, as explained before formulating Theorem 1.3.24.

□

Consider an easy example.

Example 1.3.25 Let $P = \Delta$ be the Laplace operator on an open set $X \subset \mathbb{R}^2$. According to Example 1.3.16, Δ is a formally integrable operator. Easy calculations show that

$$\sigma^{(s)} = \left\{ \sum_{|\alpha|=s} \sigma_\alpha(x) \frac{(y-x)^\alpha}{\alpha!} \ : \ \sigma_{\alpha+2\cdot 1_1}(x) + \sigma_{\alpha+2\cdot 1_2}(x) = 0 \quad (|\alpha| = s-2) \right\}$$

are two-dimensional for all $s \geq 2$. From here one can conclude that the following complexes are exact:

$$\begin{array}{ccccccccc}
0 & \to & \sigma^{(2)} & \to & \delta(\sigma^{(2)}) & \to & 0, & & \\
0 & \to & \sigma^{(3)} & \xrightarrow{\delta} & \sigma^{(2)} \otimes \Lambda^1 T^*(X) & \to & \delta\left(\sigma^{(2)} \otimes \Lambda^1 T^*(X)\right) & \to & 0, \\
0 & \to & \sigma^{(s)} & \xrightarrow{\delta} & \sigma^{(s-1)} \otimes \Lambda^1 T^*(X) & \xrightarrow{\delta} & \sigma^{(s-2)} \otimes \Lambda^2 T^*(X) & \to & 0 \quad (s > 3).
\end{array}$$

Thus Δ is an involutive operator. If we use the natural trivializations $\mathcal{J}^1(X \times \mathbb{C}^1) \cong X \times \mathbb{C}^3$ and $S^1(X \times \mathbb{C}^1) \cong X \times \mathbb{C}^3$, the operators $P' \in \mathrm{do}_1(X \times \mathbb{C}^3 \to X \times \mathbb{C}^1)$ such that $P'(j^1 f) = \Delta f$, and $D_0^{(1)} \in \mathrm{do}_1(X \times \mathbb{C}^3 \to X \times \mathbb{C}^3)$ will be given by means of

$$P' \begin{pmatrix} \sigma_0 \\ \sigma_1 \\ \sigma_2 \end{pmatrix} = \partial_1 \sigma_1 + \partial_2 \sigma_2, \text{ and } D_0^{(1)} \begin{pmatrix} \sigma_0 \\ \sigma_1 \\ \sigma_2 \end{pmatrix} = \begin{pmatrix} \partial_1 \sigma_0 - \sigma_1 \\ \partial_2 \sigma_0 - \sigma_2 \\ \partial_2 \sigma_1 - \partial_1 \sigma_2 \end{pmatrix}.$$

Note that in the equality $D_0^{(1)} \sigma = 0$ the first two equations correspond to introducing new unknown functions in the standard way of passing to a first order differential operator, and the third equation expresses the symmetry property of the mixed derivatives. Hence the operator $\widetilde{P} \in \mathrm{do}_1(X \times \mathbb{C}^3 \to X \times \mathbb{C}^4)$, which is given by

$$\widetilde{P} \begin{pmatrix} \sigma_0 \\ \sigma_1 \\ \sigma_2 \end{pmatrix} = \begin{pmatrix} \partial_1 \sigma_1 + \partial_2 \sigma_2 \\ \partial_1 \sigma_0 - \sigma_1 \\ \partial_2 \sigma_0 - \sigma_2 \\ \partial_2 \sigma_1 - \partial_1 \sigma_2 \end{pmatrix},$$

is normalized and equivalent to Δ. It is pertinent to note that the operator $\widetilde{P}$ has injective symbol.

□

As in this example, if a covector $z \in T_x^*(X)$ is non-characteristic for the differential operator P, then it is also non-characteristic for the equivalent normalized differential operator obtained by Theorem 1.3.24.

1.3.7 *Commutativity relations*

Let P be a sufficiently regular differential operator of type $E \to F$ and order 1 on X.

We take a coordinate neighborhood U on X, over which the bundles E and F are trivial, with coordinates $x = (x_1, \ldots, x_n)$. The coordinate x_n is assumed to be so chosen that the derivative D_n appears in the local expression of the differential operator P. Then one can decompose the fibers of $E|_U$ and $F|_U$ into direct sums $\mathbb{C}^k = \mathbb{C}^{k_1} \oplus \mathbb{C}^{k_2}$ and $\mathbb{C}^l = \mathbb{C}^{l_1} \oplus \mathbb{C}^{l_2}$ in such a way that $k_1 = l_2$ and after a suitable isomorphism between $\mathbb{C}^{k_1}$ and $\mathbb{C}^{l_2}$ the operator P is written in the form

$$Pu = \begin{pmatrix} Mu & = & M^{(1)} u^{(1)} + M^{(2)} u^{(2)} \\ & & D_n u^{(1)} + T u \end{pmatrix}, \qquad (1.3.7)$$

where the differential operators M and T do not contain the derivatives $D_n u$ and $D_n u^{(1)}$ respectively.

The most important fact of the formal theory is that in each equivalence class one can choose a representation whose local form (1.3.7) in a suitable coordinate system has the property that the kernel of M is mapped by the operator $u \to D_n u^{(1)} + Tu$ into the kernel of $M^{(1)}$. Such a "breaking up" of the operator P allows one easily to "disentangle" a complicated system of non-evident dependencies between the functions and their partial derivatives.

More precisely, Example 1.2.3 has the following generalization which is of key importance in the local construction of a compatibility operator.

Definition 1.3.26 *Commutativity relations are said to hold in (1.3.7) if, for some differential operator $T^{(1)}$ in U which contains only differentiations in $x_1, \ldots, x_{n-1}$, we have*

$$M^{(1)} \left(D_n u^{(1)} + Tu\right) = \left(D_n + T^{(1)}\right) Mu \qquad (u \in [\mathcal{E}(U)]^k). \tag{1.3.8}$$

The importance of commutativity relations was understood for the first time by Guillemin [94]. Samborskii [191] used them for a simple procedure of constructing a compatibility operator.

Apparently, if commutativity relations hold in (1.3.7), for solvability of the system $Pu = f$ it is necessary that f satisfy $(D_n + T^{(1)})f^{(1)} - M^{(1)}f^{(2)} = 0$ where $f = f^{(1)} \oplus f^{(2)}$ in accordance with the decomposition of F.

Proposition 1.3.27 *Let commutativity relations hold in (1.3.7), and M_1 be a compatibility operator for M. Then a compatibility operator for P is given by*

$$P_1 f = \begin{pmatrix} M_1 f^{(1)} \\ D_n f^{(1)} + (T^{(1)} f^{(1)} - M^{(1)} f^{(2)}) \end{pmatrix} \qquad (f \in [\mathcal{E}(U)]^l).$$

Proof. We refer the reader to Dudnikov and Samborskii [57, pp.15-16] where the complex $[\mathcal{E}(U)]^k \xrightarrow{P} [\mathcal{E}(U)]^l \xrightarrow{P_1} [\mathcal{E}(U)]^{k_2}$ is shown to be formally exact. Thus we may invoke Proposition 1.3.11.

□

We note that the variable x_n enters into the operator M only as a parameter. So application of Proposition 1.3.27 in the problem of constructing a compatibility operator decreases the number of variables (and the number of equations in the system at the same time). Therefore it is possible (if commutativity relations hold at each step) to decrease the dimension of x so that after no more than $(n-1)$ steps the problem will be reduced to constructing a compatibility operator for a differential operator of the form $f \to D_1 f^{(1)} + Tf$ where T does not contain the derivative $D_1 f^{(1)}$. Such a compatibility operator is seen to be zero.

Let us consider conditions on P under which commutativity relations hold for this differential operator.

Definition 1.3.28 *A covector $z \in T_x^*(X)$ is said to be quasiregular for a differential operator P at a point $x \in X$ if*

$$\dim \ker \sigma(P)(x,z) = \min_{\xi \neq 0} \dim \ker \sigma(P)(x,\xi).$$

For instance, each non-characteristic covector $z \in T_x^*(X)$ for a differential operator P (i.e., $\ker \sigma(P)(x,z) = \{0\}$) is a *quasiregular covector.*

Proposition 1.3.29 *Let P be an involutive formally integrable differential operator of the first order, and $x = (x_1, \ldots, x_n)$ be a coordinate system in U such that the covector dx_n is quasiregular for P at each point $x \in U$. Then commutativity relations hold in (1.3.7).*

Proof. Firstly, formal integrability is the property which allows one to reduce the question of commutativity relations (1.3.8) to an analogous question for symbols of operators only. This is an important step that translates the question from complicated differential algebra to much more simple linear algebra. Secondly, it turns out that, if the differential operator P is involutive and the covector dx_1 is quasiregular for P at each point $x \in U$, then commutativity relations (1.3.8) hold on the level of symbols. See Samborskii [191, pp. 8-10]

□

Assuming the coefficients of the operator $T^{(1)}$ to be undetermined, we obtain from (1.3.8) a system of linear algebraic equations for the coefficients.

Remark 1.3.30 *As we shall see from Proposition 1.3.31 for "almost all" coordinate systems $x = (x_1, \ldots, x_n)$ in U the covector dx_n is quasiregular for P at each point $x \in U$.*

1.3.8 *Compatibility complexes for normalized operators*

In this subsection we shall present an inductive procedure for constructing a compatibility operator.

Observe that as a topology on the set of all coordinate systems in a neighborhood $U \subset X$ one can take the topology of $[C^\infty_{loc}(U)]^n$.

Proposition 1.3.31 *Let P be a normalized differential operator of type $E \to F$ on X, and $U \subset X$ be a coordinate neighbourhood over which the bundles E and F are trivial. Then, for an everywhere dense open set of coordinate systems $x = (x_1, \ldots, x_n)$ in U:*

1) the bundles $E|_U$ and $F|_U$ may be decomposed into direct sums

$$E|_U = E^{(1)} \oplus \ldots \oplus E^{(n+1)},$$
$$F|_U = F^{(1)} \oplus \ldots \oplus F^{(n)}$$

in such a way that $P = P^{(1)} \oplus \ldots \oplus P^{(n)}$ in U where

$$P^{(j)}u = D_j \left(u^{(1)} \oplus \ldots \oplus u^{(j)}\right) + T_j^{(1)}(x, D_1, \ldots, D_{j-1}) \left(u^{(1)} \oplus \ldots \oplus u^{(j)}\right)$$
$$+ T_j^{(2)}(x, D_1, \ldots, D_j) \left(u^{(j+1)} \oplus \ldots \oplus u^{(n+1)}\right);$$

2) for every $1 \le j \le n$, the operator $P^{(1)} \oplus \ldots \oplus P^{(j)}$ (which contains the variables $(x_{j+1}, \ldots, x_n)$ as parameters) is normalized, and the covector dx_j is quasiregular for it at each point $x \in U$.

Proof. See Dudnikov and Samborskii [57, 58].

□

If in the fiber of $E|_U$ (i.e. in $\mathbb{C}^k$) a basis $(e_1, \ldots, e_k)$ has been fixed beforehand, it may be renumbered so that the decomposition $E|_U = E^{(1)} \oplus \ldots \oplus E^{(n+1)}$ corresponds to the decomposition

$$(e_1, \ldots, e_{k_1}), \quad (e_{k_1+1}, \ldots, e_{k_1+k_2}), \quad \ldots, \quad (e_{k_1+\ldots+k_n+1}, \ldots, e_{k_1+\ldots+k_{n+1}})$$

of the basis. For an everywhere dense set of bases this numeration may be arbitrary. So in the situation of "general position" neither change of variables nor change of searched functions are needed.

Definition 1.3.32 *The representation of a normalized operator P, as in point 1), 2) of Proposition 1.3.31, is called the (E.) Cartan normal form.*

It is convenient to use the term "regular" for a corresponding coordinate system.

Proposition 1.3.33 *Let P be a normalized differential operator written in the Cartan normal form in U, and the operators $M_j^{(1)}$, $M_j^{(2)}$ $(j = 1, \ldots, n-1)$ be defined from the correlations*

$$(P^{(1)} \oplus \ldots \oplus P^{(j)})\, u = M_j^{(1)} (u^{(1)} \oplus \ldots \oplus u^{(j+1)}) + M_j^{(2)} (u^{(j+2)} \oplus \ldots \oplus u^{(n+1)}).$$

Then there are differential operators $\widetilde{P}^{(j)}$ $(j = 2, \ldots, n)$ of the first order in U such that

$$M_{j-1}^{(1)}\, P^{(j)} = \widetilde{P}^{(j)} \left(P^{(1)} \oplus \ldots \oplus P^{(j-1)}\right), \qquad j = 2, \ldots, n.$$

Proof. This follows from Propositions 1.3.31 and 1.3.29

□

The operators $\widetilde{P}^{(j)}$ have the forms $\widetilde{P}^{(j)} = D_j + T_j$ where differential operators T_j do not contain the derivatives D_j. They may be found from the commutativity relationships above, for example, by the method of undefined coefficients.

Proposition 1.3.34 *If a normalized differential operator P is written in the Cartan normal form in U, then a compatibility operator*

$$P_1 \in \mathrm{do}_1(F|_U \to \oplus_{j=0}^{n-1}(F^{(1)} \oplus \ldots \oplus F^{(j)}))$$

for P in U is given by

$$P_1 f = \begin{pmatrix} \widetilde{P}^{(2)} f^{(1)} & - & M_1^{(1)} f^{(2)} \\ \widetilde{P}^{(3)} (f^{(1)} \oplus f^{(2)}) & - & M_2^{(1)} f^{(3)} \\ & \ldots & \\ \widetilde{P}^{(n)} (f^{(1)} \oplus \ldots \oplus f^{(n-1)}) & - & M_{n-1}^{(1)} f^{(n)} \end{pmatrix}, \qquad f \in \mathcal{E}(F|_U).$$

Proof. It is sufficient to use Propositions 1.3.33 and 1.3.27.

□

The compatibility operator P_1 for P, that is calculated by Proposition 1.3.34, is at once written in the Cartan normal form. This allows one to write a compatibility operator P_2 for P_1, and so on. So we come to a compatibility complex for the differential operator $P_0 = P$ in U:

$$\mathcal{E}(E|_U) \xrightarrow{P} \mathcal{E}(F|_U) \xrightarrow{P_1} \dots \xrightarrow{P_{N-1}} \mathcal{E}(E^N|_U) \longrightarrow 0. \qquad (1.3.9)$$

Definition 1.3.35 *A differential operator P is said to be an operator with constant defect if the number* $\dim\ker\sigma(P)(x,z)$ *does not depend on* $(x,z) \in T^*(X)$ $(z \neq 0)$.

For an operator P with constant defect each covector $z \in T_x^*(X)\setminus\{0\}$ is quasiregular.

Remark 1.3.36 *If a (normalized) differential operator P has constant defect, the length N of the complex (1.3.9) is equal to the number of non-zero operators among the $P^{(1)}, \dots, P^{(n)}$ in the Cartan normal form of P. In particular, we have $N \leq n$.*

Local compatibility operators for P, calculated by Proposition 1.3.34, may be "pasted together" to define a global compatibility operator P_1 on X because in intersections of coordinate neighborhoods they coincide up to bundle isomorphisms. Thus we get the following result.

Theorem 1.3.37 (see Samborskii [191]) *For each normalized differential operator P on X, one can construct, by a finite number of steps, a formally exact complex of normalized differential operators $E^\cdot$ on X such that $P_0 = P$.*

As Quillen [186] showed, if a complex $E^\cdot$ consists of first order differential operators with surjective mappings $\sigma(P_i) : E^i \otimes T^*(X) \to E^{i+1}$ and is formally exact, then for $E^\cdot$ to be elliptic it is necessary and sufficient that the differential operator P_0 have injective symbol. But one can say more of the compatibility complex for P obtained by Proposition 1.3.34.

Proposition 1.3.38 (see Samborskii [191]) *Let $E^\cdot$ be the compatibility complex for a normalized differential operator P given by Theorem 1.3.37. Then, if a covector $z \in T_x^*(X)$ is quasiregular for P at x, the following complex is exact:*

$$E_x^0 \xrightarrow{\sigma(P_0)(x,z)} E_x^1 \xrightarrow{\sigma(P_1)(x,z)} \dots \xrightarrow{\sigma(P_{N-1})(x,z)} E_x^N \longrightarrow 0.$$

Proof. This follows from obvious expressions for the compatibility operators given in Proposition 1.3.34 if the regular coordinate system $x = (x_1, \dots, x_n)$ on a neighbourhood of x has been chosen so that $z = dx_n$.

□

1.3.9 *Compatibility complexes for arbitrary regular operators*

If P is an arbitrary sufficiently regular differential operator, then by a finite number of steps one can construct a normalized differential operator $\tilde{P}$ that is equivalent to P. Using the compatibility complex for $\tilde{P}$ given by Theorem 1.3.37 and the obvious procedure of Proposition 1.2.7, we may obtain by a finite number of steps a compatibility complex for P.

We note that formal integrability and involutivity are very essential properties of the way we write $\tilde{P}$, and even the orders of the operators in the compatibility complex for P depend on the number of differentiations in passing from P to $\tilde{P}$. Similarly the requirements of first order and surjectivity of the symbol mapping in the definition of a normalized operator are of a purely technical character and have been accepted for simplification.

Thus we have arrived at a result published by Goldschmidt [73]. However it had earlier been obtained by Quillen [186] in his unpublished thesis.

Theorem 1.3.39 *For each sufficiently regular differential operator P on X, one can by finite number of steps construct a differential complex $E^{\cdot}$ on X with $P_0 = P$ such that :*

1) $E^{\cdot}$ is formally exact, and

2) $E^{\cdot}$ is minimal in the sense that, if $\tilde{P}_{i+1} \in \mathrm{do}_{\tilde{p}_{i+1}}(E^{i+1} \to \tilde{E}^{i+2})$, $i \geq 0$, is another operator for which the sequence $\mathcal{E}(E^i) \xrightarrow{P_i} \mathcal{E}(E^{i+1}) \xrightarrow{\tilde{P}_{i+1}} \mathcal{E}(\tilde{E}^{i+2})$ is formally exact, then $\tilde{p}_{i+1} \geq p_{i+1}$ and there is an operator $T \in \mathrm{do}_{\tilde{p}_{i+1}-p_{i+1}}(E^{i+2} \to \tilde{E}^{i+2})$ such that $\tilde{P}_{i+1} = TP_{i+1}$.

Proof. Use Theorems 1.3.24 and 1.3.37, and Proposition 1.2.7.

□

In particular, 1) means that $E^{\cdot}$ is a compatibility complex for the differential operator P.

1.3.10 *Existence theory in the real analytic case*

Suppose that we are in the category of real analytic manifolds and mappings.

It is easy to see that the formal theory holds in the real analytic case too. In particular, each real analytic sufficiently regular differential operator P has a real analytic compatibility complex $E^{\cdot}$. So, having denoted by $\mathcal{A}^i$ ($i \in \mathbb{Z}$) the sheaf of germs of real analytic sections of E^i, we obtain a complex of sheaves $\mathcal{A}^{\cdot}$ with differential P.

Theorem 1.3.40 (see Spencer [203]) *Let $E^{\cdot}$ be a compatibility complex for a sufficiently regular differential operator P in the real analytic category. Then the corresponding complex of sheaves $\mathcal{A}^{\cdot}$ is exact at positive degrees.*

Proof. We shall present a proof which is easily obtained by induction over n ($= \dim X$) and application of commutativity relationships (1.3.8).

In view of Proposition 1.2.8 one can consider without loss of generality that $E^\cdot$ is the compatibility complex for a normalized differential operator P given by Theorem 1.3.37. Moreover, it is sufficient to verify only the exactness of the complex $\mathcal{A}^0 \xrightarrow{P} \mathcal{A}^1 \xrightarrow{P_1} \mathcal{A}^2$.

Thus we should show that for each section $f \in \mathcal{A}(E|_U)$, satisfying $P_1 f = 0$ on an open set $U \subset X$, there is an open set $U' \subset U$ and a section $u \in \mathcal{A}(E|_{U'})$, such that $Pu = f$ in U'. One can assume that U is a coordinate neighbourhood on X with regular coordinate system $x = (x_1, \ldots, x_n)$, that the operator P is written in the form (1.3.7) in U, and that the compatibility operator P_1 is given by Proposition 1.3.27.

If $n = 1$, then we have $P_1 = 0$, and the desired result follows directly from the Cauchy-Kovalevskaya Theorem.

For $n > 1$, since the variable x_n enters into the operators M and M_1 only as a parameter, the inductive hypothesis allows us to consider the complex $[\mathcal{A}]^k \xrightarrow{M} [\mathcal{A}]^{l_1} \xrightarrow{M_1} [\mathcal{A}]^m$ to be exact where $\mathcal{A}$ is the sheaf of germs of real analytic functions in U.

Let $f \in [\mathcal{A}(U)]^l$ satisfy $P_1 f = 0$ in U. Writing $f = f^{(1)} \oplus f^{(2)}$ with $f^{(1)} \in [\mathcal{A}(U)]^{l_1}$, we obtain $M_1 f^{(1)} = 0$ in U.

Since the complex $[\mathcal{A}]^k \xrightarrow{M} [\mathcal{A}]^{l_1} \xrightarrow{M_1} [\mathcal{A}]^m$ is exact, there is an open set $U' \subset U$ and a section $v \in [\mathcal{A}(U')]^k$ such that $Mv = f^{(1)}$ in U'. We write $v = v^{(1)} \oplus v^{(2)}$ in accordance with the decomposition $\mathbb{C}^k = \mathbb{C}^{k_1} \oplus \mathbb{C}^{k_2}$.

Fix some point $x^0 = (x_1^0, \ldots, x_n^0)$ in U'. By the Cauchy-Kovalevskaya Theorem, we can find a connected neighbourhood $U'' \subset U'$ of x^0 and a section $u \in [\mathcal{A}(U'')]^k$ such that $u^{(2)} \equiv v^{(2)}$ in U'' and also

$$\begin{cases} D_n u^{(1)} + Tu = f^{(2)} & \text{in} \quad U'', \\ u^{(1)} = v^{(1)} & \text{for} \quad x_n = x_n^0. \end{cases}$$

We wish to show that $Mu = f^{(1)}$ in U''. With this aim let us consider the difference $\delta = Mu - f^{(1)} \in [\mathcal{A}(U'')]^{l_1}$. In view of the commutativity relationships (1.3.8) and the equality $P_1 f = 0$ we have in U'':

$$\begin{aligned} (D_n + T^{(1)})\delta &= (D_n + T^{(1)})Mu - (D_n + T^{(1)})f^{(1)} \\ &= M^{(1)}(D_n u^{(1)} + Tu) - M^{(1)} f^{(2)} \\ &= 0. \end{aligned}$$

Moreover, since the operator M does not contain the derivative D_n, we get

$$\begin{aligned} \delta \Big|_{x_n = x_n^0} &= \left(M^{(1)}(u^{(1)} - v^{(1)})\right)\Big|_{x_n = x_n^0} \\ &= M^{(1)}\left((u^{(1)} - v^{(1)})\Big|_{x_n = x_n^0}\right) \\ &= 0. \end{aligned}$$

By the uniqueness of analytic solutions of the Cauchy problem, $\delta \equiv 0$ in U''. Therefore, $Pu = f$ in U'' which completes the proof.

□

Following (E.) Cartan, we give a description of the set of all analytic solutions of the equation $Pu = f$. Such descriptions may be obtained for normalized differential operators P by using their Cartan normal forms.

Consider a normalized differential operator P of type $E \to F$ on X which is written in some regular coordinate system $x : U \to \mathbb{R}^n$ in the Cartan normal form, as in Proposition 1.3.31.

Theorem 1.3.41 (the Cartan-Kähler Theorem) *Let $x^0 \in U$, and $u_0^{(j)} = u_0^{(j)}(x_1, \dots, x_{j-1})$ $(j = 1, \dots n+1)$ be given real analytic functions in U with values in $E^{(j)}$. Then, for each section $f \in \mathcal{A}(F|_U)$ satisfying $P_1 f = 0$, there is a unique section u which is real analytic and satisfies $Pu = f$ and $u^{(j)}(x_1, \dots, x_{j-1}, x_j^0, \dots, x_n^0) = u_0^{(j)}$ $(j = 1, \dots n+1)$ in some neighbourhood of x^o.*

Proof. The proof may be easily obtained by inductive application of the commutativity relationships (1.3.8), as in the proof of Theorem 1.3.40. See also Pommaret [182, Ch.4, 4.3].

□

1.3.11 *The first Spencer sequence*

If we are not dealing with the real analytic category but only with the C^∞_{loc} category, the solvability questions get more complicated.

Example 1.3.42 Let $P \in \mathrm{do}_1(\mathbb{R}^3 \times \mathbb{C}^1 \to \mathbb{R}^3 \times \mathbb{C}^1)$ be "the Hans Lewy operator" given by $Pu = \frac{\partial u}{\partial \bar{z}} + iz\frac{\partial u}{\partial x_3}$ where $z = x_1 + \sqrt{-1}x_2$. This operator is known to be normalized, and its compatibility operator is $P_1 \equiv 0$. However, as Lewy [150] noted, the complex of sheaves $\mathcal{E} \xrightarrow{P} \mathcal{E} \to 0$ over $\mathbb{R}^3$ is not exact. See also Pommaret [182, Ch. 5, §7].

□

To study the local C^∞_{loc}-cohomologies of compatibility complexes for sufficiently regular differential operators P, the so-called first and second Spencer sequences (see Spencer [203, 204, 207]) have been used. In the Spencer sequences, information from the initial differential operator P has been "ciphered" in the structures of the bundles while the differential operators have been completely unified. When localizing, "simple" differential operators are obtained on a set of functions connected by non-differential linear correlations.

For a bundle B over X, it will be convenient to denote by $\widetilde{B}$ the sheaf of germs of differentiable sections of B.

Later on, we shall suppose that $P \in \mathrm{do}_p(E \to F)$ is some formally integrable operator. If P is only sufficiently regular, we should replace P by an equivalent formally integrable operator.

In the following proposition, D is the operator defined in Example 1.3.13.

Proposition 1.3.43 *For any integers s and q with $s-(q+1)-p \geq 0$, the following diagram is commutative:*

$$\begin{array}{ccc}
\mathcal{J}^{s-q}\widetilde{(E)\otimes \Lambda^q} T^*(X) & \xrightarrow{D} & \mathcal{J}^{s-q-1}\widetilde{(E)\otimes \Lambda^{q+1}} T^*(X) \\
\Big\downarrow {\scriptstyle \eta(\jmath^{s-q-p}\circ P)\otimes 1} & & \Big\downarrow {\scriptstyle \eta(\jmath^{s-q-1-p}\circ P)\otimes 1} \\
\mathcal{J}^{s-q-p}\widetilde{(F)\otimes \Lambda^q} T^*(X) & \xrightarrow{D} & \mathcal{J}^{s-q-1-p}\widetilde{(F)\otimes \Lambda^{q+1}} T^*(X)
\end{array}$$

Proof. See Pommaret [182, Ch.5, 3.3].

□

It follows from Proposition 1.3.43 that, if $s - \underbrace{q-1}\geq p$, the restriction of the operator D on $\mathfrak{R}^{(s-q)}\widetilde{\otimes \Lambda^q} T^*(X)$ acts to $\mathfrak{R}^{(s-q-1)}\widetilde{\otimes \Lambda^{q+1}} T^*(X)$. So, denoting by $\mathcal{E}_P$ the sheaf of germs of differentiable solutions of the system $Pf=0$ over X and assuming $s-n\geq p$, we get the following complex of sheaves:

$$0 \longrightarrow \mathcal{E}_P \xrightarrow{\jmath^s} \mathfrak{R}^{(s)} \xrightarrow{D} \mathfrak{R}^{(s-1)}\widetilde{\otimes \Lambda^1} T^*(X) \xrightarrow{D} \ldots \xrightarrow{D} \mathfrak{R}^{(s-n)}\widetilde{\otimes \Lambda^n} T^*(X) \longrightarrow 0. \tag{1.3.10}$$

Definition 1.3.44 *The differential complex (1.3.10) is called the first Spencer sequence for P (or, what is better, for the corresponding "equation" $\mathfrak{R}^{(s)}$).*

Spencer [203, 204, 207] observed the following interesting property of the complex (1.3.10).

Proposition 1.3.45 *The cohomologies of (1.3.10) do not depend on s provided $s\geq s_0+(n-1)$, where s_0 is the number from Theorem 1.3.21.*

Proof. This is easily obtained by diagram chasing.

□

The cohomologies of (1.3.10) are said to be *stable* if $s\geq s_0+(n-1)$.

On the whole, the importance of the first Spencer sequence for a differential operator P is determined by the following result.

Theorem 1.3.46 (Quillen [186]) *Let $E^\cdot$ be some formally exact compatibility complex for a formally integrable differential operator P. Then the cohomologies of the complex $0 \to \mathcal{E}_P \xrightarrow{\imath} \widetilde{E}^\cdot$ coincide with the stable cohomologies of the first Spencer sequence for P.*

Proof. The relationship between the complexes $0 \to \mathcal{E}_P \to \widetilde{E}$ and (1.3.10) is expressed by the commutative diagram

$$
\begin{array}{ccccccccc}
 & & 0 & & 0 & & 0 & & \\
 & & \downarrow & & \downarrow & & \downarrow & & \\
0 & \longrightarrow & \mathcal{E}_P & \xrightarrow{j^s} & \widetilde{\mathfrak{R}^{(s)}} & \xrightarrow{D} & \widetilde{\mathfrak{R}^{(s-1)} \otimes \Lambda^1 T^*(X)} & \xrightarrow{D} & \dots \\
 & & \downarrow \subset & & \downarrow & & \downarrow & & \\
0 & \longrightarrow & \widetilde{E^0} & \longrightarrow & \widetilde{\mathcal{J}^{(s)}(E^0)} & \longrightarrow & \widetilde{\mathcal{J}^{(s-1)}(E^0) \otimes \Lambda^1 T^*(X)} & \longrightarrow & \dots \\
 & & \downarrow P & & \downarrow & & \downarrow & & \\
0 & \longrightarrow & \widetilde{E^1} & \longrightarrow & \widetilde{\mathcal{J}^{(s-p_0)}(E^1)} & \longrightarrow & \widetilde{\mathcal{J}^{(s-p_0-1)}(E^1) \otimes \Lambda^1 T^*(X)} & \longrightarrow & \dots \\
 & & \downarrow P & & \downarrow & & \downarrow & & \\
 & & \dots & & \dots & & \dots & &
\end{array}
\tag{1.3.11}
$$

where s is large. Since E is formally exact and the first Spencer sequence for the trivial operator is exact, the diagram is exact except possibly for the first row and first column. Thus by diagram chasing the cohomologies of the first column are the same as the stable cohomologies of the first Spencer sequence.

□

1.3.12 *The second Spencer sequence*

In the previous subsection we saw how the exactness of the first Spencer sequence can be used to obtain existence theorems for the systems $Pu = f$. To this extent the theory generalizes the situation in complex analysis, where the Dolbeault sequence has proved to be an important tool for studying the Cauchy-Riemann equations. The problem of proving exactness is, of course, much more difficult in the general theory than in the case of the Dolbeault sequence. As Sweeney [211] observed, "in fact, it is more difficult than it should be."

To be specific, the sequence of symbol mappings associated with (1.3.10) is exact only in the most trivial cases. For instance, it is already non-exact when "$Pu = 0$" is the Cauchy-Riemann system. Thus the harmonic methods which establish exactness for the Dolbeault sequence (see Kohn [125, 126]) can not be applied to (1.3.10) even when "$Pu = 0$" is the Cauchy-Riemann system. That is why we are led to consider the second Spencer sequence which will be constructed from (1.3.10) by factoring out the degeneracy in the symbol sequence.

Supposing $s \geq p$, we set

$$
S^q = S^q(\mathfrak{R}^s) = \begin{cases} \mathfrak{R}^s(E) & \text{for} \quad q = 0, \\ \dfrac{\mathfrak{R}^s \otimes \Lambda^q T^*(X)}{\delta(\sigma^{(s+1)} \otimes \Lambda^{q-1} T^*(X))} & \text{for} \quad q > 0. \end{cases}
$$

Proposition 1.3.47 *Let $s \geq s_0$ (the number from Theorem 1.3.21). Then S^q $(d \in \mathbb{Z})$ are vector bundles.*

Proof. Indeed, from the exactness of the δ-sequence for $s \geq s_0$ we can conclude that $\delta(\sigma^{(s+1)} \otimes \Lambda^{q-1}T^*(X))$ $(q > 0)$ are vector subbundles of $\mathfrak{R}^{(s)} \otimes \Lambda^q T^*(X)$. □

Later on s will be chosen to satisfy $s \geq s_0$

Proposition 1.3.48 *The operator* $D' : \widetilde{\mathfrak{R}^{(s)} \otimes \Lambda^q T^*(X)} \to \widetilde{S^{q+1}}$ $(q \geq 0)$ *defined by the following commutative diagram factors through* $\widetilde{S^q}$:

$$\begin{array}{ccccccccc}
0 & \to & \widetilde{\sigma^{(s+1)} \otimes \Lambda^q T^*(X)} & \to & \widetilde{\mathfrak{R}^{(s+1)} \otimes \Lambda^q T^*(X)} & \to & \widetilde{\mathfrak{R}^{(s)} \otimes \Lambda^q T^*(X)} & \to & 0 \\
 & & \downarrow{\scriptstyle -\delta} & & \downarrow{\scriptstyle D} & & \downarrow{\scriptstyle D'} \searrow^{\varphi} \; \widetilde{S^q} \; \swarrow_{D^{(s)}} & & \\
0 & \to & \widetilde{\delta(\sigma^{(s+1)} \otimes \Lambda^q T^*(X))} & \to & \widetilde{\mathfrak{R}^{(s)} \otimes \Lambda^{q+1} T^*(X)} & \xrightarrow{\varphi} & \widetilde{S^{q+1}} & \to & 0
\end{array} \tag{1.3.12}$$

Proof. This is an easy exercise. See, for instance, Pommaret [182, Ch 5, 4.2] □

Accordingly, we obtain an operator $D_q^{(s)} : \widetilde{S^q} \to \widetilde{S^{q+1}}$ such that the operator D' from (1.3.11) is the composition of the quotient homomorphism $\varphi : \mathfrak{R}^s \otimes \Lambda^q T^*(X) \to S^q$ and $D_q^{(s)}$. It is easy to see from here that $D_q^{(s)}$ is a differential operator of type $S^q \to S^{q+1}$ and order 1 whose symbol mapping is the epimorphism $S^q \otimes T^*(X) \to S^{q+1}$ generated by the exterior multiplication mapping $\Lambda^q T^*(X) \otimes T^*(X) \to \Lambda^{q+1}T^*(X)$. Moreover, we have $D_{q+1}^{(s)} \circ D_q^{(s)} = 0$ for all q. So we have arrived at the following complex of sheaves over X:

$$0 \longrightarrow \mathcal{E}_p \xrightarrow{j^s} \widetilde{S^0} \xrightarrow{D^{(s)}} \widetilde{S^1} \xrightarrow{D^{(s)}} \ldots \xrightarrow{D^{(s)}} \widetilde{S^n} \longrightarrow 0. \tag{1.3.13}$$

Definition 1.3.49 *The differential complex* $S = S(\mathfrak{R}^{(s)})$ *is said to be the second Spencer sequence for* P *(or, what is better, for corresponding "equation"* $\mathfrak{R}^{(s)}$*).*

In those cases where confusion is excluded the index s for the differential $D^{(s)}$ will be omitted.

From the standpoint of the formal theory, the second Spencer sequence is better than the first one because it is formally exact while the first Spencer sequence is not formally exact even when P is the trivial operator $(E \to 0)$.

Theorem 1.3.50 (Goldschmidt [73]) *If* P *is a formally integrable differential operator on* X *and* $s \geq s_0$ *(the number from Theorem 1.3.21), then the complex* $S(\mathfrak{R}^{(s)})$ *is formally exact.*

Proof See Pommaret [182, Ch.5, 5.10]. □

This theorem allows us to apply the general result of Quillen [186] mentioned in Subsection 1.3.9 to the second Spencer sequence to obtain the following result

Theorem 1.3.51 (Quillen [186]) *A cotangent vector $z \in T_x^*(X)$ is non-characteristic for the complex $S^\cdot(\mathfrak{R}^{(s)})$ $(s \geq s_0)$ at a point $x \in X$ if and only if the mapping $\sigma(P)(x,z): E_x \to F_x$ is injective.*

It follows from Theorem 1.3.51 that the second Spencer sequence for a differential operator P is an elliptic complex if and only if P has injective symbol.

This result, together with the following theorem, removes the difficulty described at the beginning of this subsection.

Theorem 1.3.52 (Spencer [203]) *For $s \geq s_0$ the cohomologies of the second Spencer sequence (1.3.12) are the same as the stable cohomologies of the first Spencer sequence (1.3.10).*

Proof. Consider the commutative diagram (1.3.12). For each q we choose a mapping $M_q: S^q \to \mathfrak{R}^{(s)} \otimes \Lambda^q T^*(X)$ such that $\varphi_q \circ M_q = 1$ where $\varphi_q: \mathfrak{R}^{(s)} \otimes \Lambda^q T^*(X) \to S^q$ is the quotient homomorphism. We claim that $M = \{M_q\}$ and $\varphi = \{\varphi_q\}$ furnish the required isomorphism in cohomologies. Since $M_q \circ \varphi_q = 1$ modulo $\delta(\sigma^{(s+1)} \otimes \Lambda^{q-1}T^*(X))$ and the restriction of the operator D to $\sigma^{(s+1)} \otimes \Lambda^{q-1}T^*(X)$ coincides with $(-\delta)$, we need only show that φ maps cocycles to cocycles and coboundaries to coboundaries. The second assertion follows immediately from the diagram. To prove the first we let $\sigma \in \mathfrak{R}^{(s)} \otimes \Lambda^q T^*(X)$ satisfy $D\sigma = 0$. By chasing the diagram we can find $\tau \in \mathfrak{R}^{(s+1)} \otimes \Lambda^q T^*(X)$ such that $(\pi^{s+1,s} \otimes 1)\,\tau = \sigma$ and $D\tau = 0$. Hence it follows that

$$D^{(s)}\varphi\sigma = D^{(s)}\varphi(\pi^{s+1,s} \otimes 1)\,\tau = D'(\pi^{s+1,s} \otimes 1)\,\tau = \varphi D\tau = 0,$$

which completes the proof.

□

Let us mention two examples.

Example 1.3.53 When considering the system $df = 0$ for the complex-valued function f, the second Spencer sequence coincides with the de Rham sequence $\Lambda^\cdot T_{\mathbb{C}}^*(X)$ (see Spencer [206]).

□

Example 1.3.54 For the Cauchy-Riemann system $\overline{\partial} f = 0$ on a complex manifold X, the second Spencer sequence gives a resolution of the sheaf $\mathcal{O}^{(0)}$, different from the Dolbeault sequence $\Lambda^{0,\cdot}T_{\mathbb{C}}^*(X)$ (see Spencer [206]).

□

1.3.13 *Existence theory in the differentiable case*

The second, more difficult, step in studing overdetermined systems consists of searching for a satisfactory condition of exactness in the differentiable (i.e. C^∞_{loc}) sense for a given formally exact complex. The example of H. Lewy [150] shows that exactness does not always take place in the differentiable case. The condition that has attracted the most attention is the ellipticity of the complex, but even under this assumption exactness has not been established in the general case. As Spencer wrote [208], "it is possible, although this opinion is not wide-spread, that ellipticity is not a correct condition."

To establish exactness for the second Spencer sequence (1.3.12) in the elliptic case, Spencer [205, 206] formulated the so-called D-Neumann problem (see also Section 4.1 below). If this problem is locally solvable and the harmonic spaces are trivial at positive degrees, then the sequence (1.3.12) is exact.

However it is necessary to take into account the following remark of Sweeney [210].

Example 1.3.55 Let $x = (x_1, x_2, x_3)$ be the coordinates in $\mathbb{R}^3$. Set $z = x_1 + \sqrt{-1}x_2$, and consider the differential operators $P \in \mathrm{do}_1(\mathbb{R}^3 \times \mathbb{C}^2 \to \mathbb{R}^3 \times \mathbb{C}^4)$ and $P_1 \in \mathrm{do}_1(\mathbb{R}^3 \times \mathbb{C}^4 \to \mathbb{R}^3 \times \mathbb{C}^2)$ given by

$$P(D) = \begin{pmatrix} 0 & \partial/\partial\bar{z} \\ 0 & \partial/\partial x_3 \\ \partial/\partial\bar{z} & 0 \\ \partial/\partial x_3 & -\partial/\partial x_2 \end{pmatrix}, \quad P_1(D) = \begin{pmatrix} \partial/\partial x_3 & -\partial/\partial\bar{z} & 0 & 0 \\ -\partial/\partial x_2 & 0 & \partial/\partial x_3 & -\partial/\partial\bar{z} \end{pmatrix}$$

The complex $0 \to [\mathcal{E}]^2 \xrightarrow{P} [\mathcal{E}]^4 \xrightarrow{P_1} [\mathcal{E}]^2 \to 0$ is easily verified to be an elliptic Hilbert complex, so it is exact. However Sweeney [210] showed that the D-Neumann problem for P is not solvable on the unit ball of $\mathbb{R}^3$ in the L^2 sense (although it may be solvable in some other sense).

□

The main known results are connected with a cohomological condition, called the δ-estimate, on the δ-complex of P. As MacKichan showed [154, 155], this condition is a reasonable generalization to overdetermined systems of the notion of "diagonal" or "uncoupled" operators.

We assign metrics to E and $T^*(X)$, which then induce metrics on the $\sigma^{(s)}$'s.

Definition 1.3.56 *An operator $P \in \mathrm{do}_p(E \to F)$ is said to satisfy the δ-estimate if in the sequence*

$$0 \to \sigma^{(p+2)} \xrightarrow{\delta} \sigma^{(p+1)} \otimes \Lambda^1 T^*(X) \xrightarrow{\delta} \sigma^{(p)} \otimes \Lambda^2 T^*(X)$$

we have $\|\delta\sigma\|^2 \geq \frac{1}{2}(p+1)^2\|\sigma\|^2$ for all $\sigma \in (\sigma^{(p+1)} \otimes \Lambda^1 T^*(X)) \cap \ker \delta$.

If a differential operator P satisfies the δ-estimate, then its first prolongation $j^1 \circ P$ is an involutive operator.

Example 1.3.57 Let P be the gradient operator $d_0 \in \mathrm{do}_1(X \times \mathbb{C}^1 \to \Lambda^1 T^*_{\mathbb{C}}(X))$. Then $\sigma(d_0) : T^*_{\mathbb{C}}(X) \to T^*_{\mathbb{C}}(X)$ is the identity. Therefore $\sigma^{(1)} = 0$ and the δ-estimate holds vacuously.

□

Example 1.3.58 A first order differential operator ∇ of type $E \to E \otimes T^*(X)$ is said to be a covariant derivative if $\sigma(\nabla): E \otimes T^*(X) \to E \otimes T^*(X)$ is the identity. Again $\sigma^{(1)} = 0$ so the δ-estimate holds vacuously.

□

Example 1.3.59 The Cauchy-Riemann operator on a complex manifold X satisfies the δ-estimate (see MacKichan [154]).

□

Example 1.3.60 Sweeney's operator P from Example 1.3.55 does not satisfy the δ-estimate.

□

The δ-estimate, although it can be stated entirely in terms of the operator $D_0^{(s)}$, actually gives estimates for the other operators $D_q^{(s)}$ $(q \geq 1)$ in the second Spencer sequence $S^\cdot(\mathfrak{R}^{(s)})$. With symbol injectivity, it guarantees that the D-Neumann problem for P is solvable.

The δ-estimate was discovered in the early 60-s by I.M. Singer, who recognized its role in the proof of the main a priori estimate for the D-Neumann problem which is called the Kohn-Nirenberg estimate. He proved that, for domains satisfying necessary convexity conditions; the Kohn-Nirenberg estimate with $q = 1$ followed from the δ-estimate. MacKichan [154] strengthened this result by showing that the δ-estimate implied the Kohn-Nirenberg estimate for all degrees $q \geq 0$ on domains of the same type. Also Sweeney [212] proved that, if the operator P had injective symbol and satisfied the δ-estimate, then on sufficiently small coordinate balls the harmonic spaces in the D-Neumann problem were trivial at positive degrees. As we have already observed, this implies the following statement.

Theorem 1.3.61 (MacKichan [154]) *Let P be a formally integrable differential operator with injective symbol which satisfies the δ-estimate. Then the second Spencer sequence (1.3.12) for P is exact.*

We present some consequences.

Corollary 1.3.62 *Under the assumptions of Theorem 1.3.61, we have*

$$H^\cdot(\mathcal{E}(S^\cdot)) \cong H^\cdot(X, \mathcal{E}_P).$$

Proof. By Theorem 1.3.61, the complex (1.3.12) gives a fine resolution of the sheaf $\mathcal{E}_P$ over X. So an application of the Abstract de Rham Theorem leads to the desired result.

□

Spencer [206] called this statement the "*Theorem of de Rham for elliptic systems.*"

Corollary 1.3.63 *If P is a formally integrable differential operator with injective symbol which satisfies the δ-estimate, and $E^\cdot$ is a formally exact compatibility complex for P, then the complex of sheaves $0 \to \mathcal{E}_P \xrightarrow{\iota} \widetilde{E^\cdot}$ is exact.*

Proof. This follows from Theorems 1.3.46, 1.3.52 and 1.3.61.

□

1.4 Tensor products of differential complexes and Künneth's formula

1.4.1 *Topological tensor product of locally convex spaces*

An excellent introduction to the range of questions that will be considered in this section is given in Part III of Trèves [240].

In the following let K and L be locally convex (Hausdorff topological vector) spaces. In the space $\mathcal{L}(L \to K)$ of all continuous linear mappings $L \to K$ three important topologies are usually considered: 1) the topology of pointwise convergence; 2) the topology of uniform convergence on compact subsets of L, and 3) the topology of uniform convergence on bounded subsets of L. The corresponding topological vector spaces are denoted by $\mathcal{L}_s(L \to K)$, $\mathcal{L}_c(L \to K)$, and $\mathcal{L}_b(L \to K)$ respectively.

In particular, we have $L' = \mathcal{L}_b(L \to \mathbb{C})$. We set $L'_s = \mathcal{L}_s(L \to \mathbb{C})$, and let L'_τ denote the dual space $\mathcal{L}(L \to \mathbb{C})$ with Mackey's topology $\tau(L', L)$ (see Edwards [62, 8.3.3]). Finally, the vector space $\mathcal{L}(L'_\tau \to K)$ endowed with the topology of uniform convergence on equicontinuous sets of functionals in L'_τ is denoted by $\mathcal{L}_\varepsilon(L'_\tau \to K)$.

Schwartz [195] defines the ε-product $K \,\varepsilon\, L$ in a symmetric way (which allows a simple proof of the associativity of this product) and then proves that $K \,\varepsilon\, L = \mathcal{L}_\varepsilon(L'_\tau \to K)$. It plays a great role in his theory of vector-valued distributions (see Schwartz [195, 196])

We have $\mathcal{L}_\varepsilon(L'_\tau \to K) = \mathcal{B}_\varepsilon(K'_s, L'_s)$ where $\mathcal{B}_\varepsilon(K'_s, L'_s)$ is the space of all separately continuous bilinear mappings $K'_s \times L'_s \to \mathbb{C}$ with the topology of uniform convergence on sets of the form $\sigma_1 \times \sigma_2$ where $\sigma_1 \subset K'_s$ and $\sigma_2 \subset L'_s$ are equicontinuous (see Bourbaki [26, Proposition 7]). This isomorphism is given by mapping $M \in \mathcal{L}_\varepsilon(L'_\tau \to K)$ to the linear form $(k', l') \to \langle k', M(l')\rangle$ on $K'_s \times L'_s$.

There is a canonical mapping of the algebraic tensor product $K \otimes L$ to $\mathcal{B}_\varepsilon(K'_s, L'_s)$ which assigns to the element $k \otimes l$ the bilinear form $(k', l') \to \langle k', k\rangle\langle l', l\rangle$. The ε-topology on $K \otimes L$ is the topology induced by $\mathcal{B}_\varepsilon(K'_s, L'_s)$. The completion of $K \otimes L$ in the ε-topology is said to be the (completed) ε-tensor product of K and L. It is denoted by $K \otimes_\varepsilon L$.

The projective topology on $K \otimes L$ is the strongest topology which makes the canonical bilinear mapping $K \times L \to K \otimes L$ continuous. The complection of $K \otimes L$ in the projective topology is called the projective tensor product of K and L, and it is denoted by $K \hat{\otimes} L$.

One of the possible definition of a nuclear space L is the following. L is *nuclear* if for every locally convex topological vector space K, the projective topology on $K \otimes L$ coincides with the ε-topology, i.e. $K \hat{\otimes} L = K \otimes_\varepsilon L$ (see Grothendieck [92, Ch.2, Definition 4]).

Example 1.4.1 Let X and Y be differentiable manifolds which are countable at infinity and E and F be differentiable $\mathbb{C}$-vector bundles over X and Y respectively. Denote by $E \boxtimes F$ the external tensor product of the bundles E and F. This is the differentiable $\mathbb{C}$-vector bundle over $X \times Y$ whose fiber over a point (x,y) is $E_x \otimes F_y$, and whose transition matrices are tensor products of the transition matrices for E and F. If $X = Y$, the restriction of $E \boxtimes F$ to the diagonal gives the internal tensor product $E \otimes F$. This is a bundle on X. In the product $U \times V$ of coordinate neighbourhoods in X and Y, over which the bundles E and F are trivial, every section $K \in \sigma(E \boxtimes F)$ is represented by a $(k \times l)$-matrix of functions of a suitable class on $U \times V$ where k and l are the ranks of E and F respectively. It is well-known that each of the spaces $\mathcal{E}(E)$, $\mathcal{D}(E)$, $\mathcal{D}'(E)$ and $\mathcal{E}'(E)$ is nuclear (see Grothendieck [92, Ch.2, Theorem 10]). Moreover, for $\mathfrak{C} = \mathcal{E}$, $\mathcal{D}$, $\mathcal{D}'$ and $\mathcal{E}'$, we have $\mathfrak{C}(E)\widehat{\otimes}\mathfrak{C}(F) \cong \mathfrak{C}(E \boxtimes F)$ where the isomorphism is topological except in the case $\mathfrak{C} = \mathcal{D}$ when it is only algebraic (see Treves [240, pt.III]).

□

The following result was proved by Grothendieck [92, Ch.2, Theorem 6].

Theorem 1.4.2 *If K and L are complete locally convex topological vector spaces and one of them is nuclear, then the canonical mapping $K \otimes L \to \mathcal{L}_\varepsilon(L'_\tau \to K)$ may be extended to an isomorphism*

$$K\widehat{\otimes}L \overset{top}{\cong} \mathcal{L}_\varepsilon(L'_\tau \to K). \tag{1.4.1}$$

Later on, topological tensor products will be used, as a rule, in the situation where one of the factors is a nuclear space. That is why we shall consider projective tensor products only.

1.4.2 *Grothendieck's theorem on the topological tensor product of complexes*

Let $M_K : K^1 \to K^2$ and $M_L : L^1 \to L^2$ be continuous linear mappings of locally convex topological vector spaces. The algebraic tensor product $M_K \otimes M_L : K^1 \otimes L^1 \to K^2 \otimes L^2$ of the operators M_K and M_L is continuous provided the spaces $K^i \otimes L^i$ $(i = 1,2)$ have been endowed with the projective topologies. Therefore it may be extended in a unique way to a continuous linear mapping $M_K \widehat{\otimes} M_L : K^1\widehat{\otimes}L^1 \to K^2\widehat{\otimes}L^2$ (see Grothendieck [92, Ch.1, §14, n° 2]). This mapping $M_K \widehat{\otimes} M_L$ is called the topological tensor product of M_K and M_L.

Lemma 1.4.3 *The next equality holds provided all compositions therein are well-defined:*

$$(M''_K \widehat{\otimes} M''_L) \circ (M'_K \widehat{\otimes} M'_L) = (M''_K \circ M'_K)\widehat{\otimes}(M''_L \circ M'_L). \tag{1.4.2}$$

Proof. Equality (1.4.2) is obvious for the algebraic tensor products of operators. Then for the topological tensor products it may be obtained by passing to the limit.

□

Now we can define the topological tensor product of (cochain) complexes of locally convex topological vector spaces by analogy with such a product in homological algebra (see Dold [56, Ch.VI, 9.1]). Namely, let $K^\cdot$ and $L^\cdot$ be two such complexes endowed with differentials d_K and d_L respectively. We set $(K^\cdot\hat{\otimes}L^\cdot)^r = \oplus_{i+j=r} K^i \otimes L^j$ $(r \in \mathbb{Z})$, and provide these vector spaces with the sum topology as usual. These topological vector spaces are assembled into a new complex $(K^\cdot\hat{\otimes}L^\cdot)^\cdot$ with the differential d of degree 1 defined by means of the equality $d|_{K^i\hat{\otimes}L^j} = d_K\hat{\otimes}1_{L^j} + (-1)^i 1_{K^i}\hat{\otimes}d_L$. The equality $d^2 = 0$ is verified immediately by using Lemma 1.4.3. The complex $(K^\cdot\hat{\otimes}L^\cdot)^\cdot$ is called the topological tensor product of the complexes $K^\cdot$ and $L^\cdot$.

Formulas expressing the cohomologies of the complex $(K^\cdot\hat{\otimes}L^\cdot)^\cdot$ through the cohomologies of the complexes $K^\cdot$ and $L^\cdot$, are traditionally said to be Künneth formulas (see Dold, [56, Ch.VI, 9.1]).

There are natural linear mappings

$$\oplus_{i+j=r} H^i(K^\cdot) \otimes H^j(L^\cdot) \to H^r((K^\cdot\hat{\otimes}L^\cdot)^\cdot) \quad (r \in \mathbb{Z}),$$

and Grothendieck [90] found acceptable conditions under which these mappings may be extended to topological isomorphisms. Precisely, he proved the following result.

Theorem 1.4.4 (Künneth formula) *Let $K^\cdot$ and $L^\cdot$ be graded Fréchet spaces endowed with differentials d_K and d_L of degree 1, and let one of them be nuclear. If d_K and d_L are topological homomorphisms, then we have*

$$H^\cdot((K^\cdot\hat{\otimes}L^\cdot)^\cdot) = H^\cdot(K^\cdot)\hat{\otimes}H^\cdot(L^\cdot).$$

Proof. A full proof of this theorem may be found in the transactions of L. Schwartz's seminar devoted to studing the thesis of Grothendieck [92]. See Grothendieck [90].

□

Grosu and Vasilescu [88] obtained a Künneth formula for the topological tensor product of Fredholm complexes of Hilbert spaces. We mention that every nuclear normed space must be finite-dimensional.

1.4.3 *Tensor product of differential complexes. A Künneth formula*

We suppose that $E^\cdot = \{E^i, d_E\}_{i=0,\dots,l_E}$ and $F^\cdot = \{F^j, d_F\}_{j=0,\dots,l_F}$ are respectively (cochain) differential complexes on differentiable manifolds X and Y countable at infinity. For uniformity it is convenient here to write d_E instead of P and l_E instead of N.

To define a tensor product of these complexes we consider an example.

Example 1.4.5 Let P be a differential operator of type $E' \to E''$ on X, and F be a differentiable $\mathbb{C}$-vector bundle over Y. We consider P as a mapping $\mathcal{E}(E') \to \mathcal{E}(E'')$. In view of Example 1.4.1, the mapping $P\hat{\otimes}1_{\mathcal{E}(F)}$ may be identified with some mapping $\mathcal{E}(E' \boxtimes F) \to \mathcal{E}(E'' \boxtimes F)$ which is easily verified to be a differential operator of the type $E' \boxtimes F \to E'' \boxtimes F$ on $X \times Y$. Locally it may be identified with the operator P acting in the variable $x \in U \subset X$. That is why we shall use the notation $P(x)$ for $P\hat{\otimes}1_{\mathcal{E}(F)}$. Analogously, if E is a differentiable $\mathbb{C}$-vector bundle over X and Q is a differential operator of type $F' \to F''$ on Y, then the operator $1_{\mathcal{E}(E)}\hat{\otimes}Q$ defines a unique differential operator of type $E \boxtimes F' \to E \boxtimes F''$ on $X \times Y$ which can be identified with the operator Q acting in the variable $y \in V \subset Y$. We will refer to $1_{\mathcal{E}(E)}\hat{\otimes}Q$ as $Q(y)$.

□

Definition 1.4.6 *The tensor product of the complexes $E^{\cdot}$ and $F^{\cdot}$ is the differential complex $(E^{\cdot} \boxtimes F^{\cdot})$ on $X \times Y$ consisting of the bundles $(E^{\cdot} \boxtimes F^{\cdot})^r = \oplus_{i+j=r} E^i \boxtimes F^j$ $(r \in \mathbb{Z})$, and endowed with the differential $d_{E\boxtimes F}$ of degree 1 defined by*

$$d_{E\boxtimes F}|_{\mathcal{E}(E^i\boxtimes F^j)} = d_E(x) + (-1)^i d_F(y).$$

We note if $E^{\cdot}$ and $F^{\cdot}$ are elliptic complexes of differential operators of the same order, then $(E^{\cdot} \boxtimes F^{\cdot})$ is an elliptic complex too. For varying orders this fails, however it is true provided ellipticity is understood in some more wide sense.

For each of the classes $\mathfrak{C} = \mathcal{E}\ \mathcal{D}\ \mathcal{D}'\ \mathcal{E}'$ and many others, the complex $(E^{\cdot} \boxtimes F^{\cdot})$ defines a complex of topological vector space's

$$\mathfrak{C}((E^{\cdot} \boxtimes F^{\cdot})):$$
$$0 \longrightarrow \mathfrak{C}((E^{\cdot} \boxtimes F^{\cdot})^0) \xrightarrow{d_{E\boxtimes F}} \mathfrak{C}((E^{\cdot} \boxtimes F^{\cdot})^1) \xrightarrow{d_{E\boxtimes F}} \dots \xrightarrow{d_{E\boxtimes F}} \mathfrak{C}((E^{\cdot} \boxtimes F^{\cdot})^{l_E+l_F}) \longrightarrow 0.$$

Our definition of the tensor product of differential complexes justifies itself by the following result.

Lemma 1.4.7 *For $\mathfrak{C} = \mathcal{E}$, $\mathcal{D}$, $\mathcal{D}'$ and $\mathcal{E}'$, we have $\mathfrak{C}((E^{\cdot} \boxtimes F^{\cdot})) = \mathfrak{C}(E^{\cdot})\hat{\otimes}\mathfrak{C}(F^{\cdot})$ where the isomorphism is topological except in the case $\mathfrak{C} = \mathcal{D}$.*

Proof. Use the definitions and Example 1.4.1.

□

The next variant of Theorem 1.4.4 for differential complexes was presented in the author's paper [233].

Theorem 1.4.8 (Künneth formula) *Let $E^{\cdot}$ and $F^{\cdot}$ be (cochain) differential complexes on X and Y. If the cohomology spaces $H^{\cdot}(\mathcal{E}(E^{\cdot}))$ and $H^{\cdot}(\mathcal{E}(F^{\cdot}))$ are Hansdorff, then*

$$H^r(\mathcal{E}((E^{\cdot} \boxtimes F^{\cdot}))) \overset{top}{\cong} \oplus_{i+j=r} H^i(\mathcal{E}(E^{\cdot}))\hat{\otimes}H^j(\mathcal{E}(F^{\cdot})). \tag{1.4.3}$$

Proof. Since the space $H^{\cdot}(\mathcal{E}(E^{\cdot}))$ is separated, it follows from Proposition 7 of Bourbaki [27, Ch.1, §8] and Proposition 1.1.12 that the subspaces $B^i(\mathcal{E}(E^{\cdot}))$ are closed in $\mathcal{E}(E^i)$ for all $i \in \mathbb{Z}$. So by the Open Mapping Theorem the differential d_E is a topological homomorphism because $\mathcal{E}(E^i)$ are Fréchet spaces. The same is true also for the differential d_F of the complex $\mathcal{E}(F^{\cdot})$. Since $\mathcal{E}(E^{\cdot})$ and $\mathcal{E}(F^{\cdot})$ are nuclear (graded) spaces, Theorem 1.4.8 follows from Theorem 1.4.4 and Lemma 1.4.7 which completes the proof.

□

1.4.4 *Solvability of systems of differential equations with a parameter*

If a complex F has the length $l_F = 0$, i.e. it consists only of the non-zero vector bundle F over Y, then Theorem 1.4.8 asserts that the equality $H^q(\mathcal{E}(E^{\cdot})) = 0$ implies $H^q(\mathcal{E}(E^{\cdot} \boxtimes F)) = 0$ provided the space $H^{\cdot}(\mathcal{E}(E^{\cdot}))$ is separable. In other words, if the system $P_{q-1}u(x) = f(x)$ is solvable in $\mathcal{E}(E^{q-1})$ for each right-hand side $f \in \mathcal{E}_{P_q}(X)$, then the system $P_{q-1}(x)u(x,y) = f(x,y)$ is solvable in $\mathcal{E}(E^{q-1} \boxtimes F)$ for each right-hand side $f \in \mathcal{E}_{P_q(x)}(X \times Y)$. A more attentive analysis shows that the condition of separateness of the space $H^{\cdot}(\mathcal{E}(E^{\cdot}))$ is not needed here. But, if the right-hand side $f \in \mathcal{E}_{P_q(x)}(X \times Y)$ satisfies some additional differential conditions $Q(y)f(x,y) = 0$, can one guarantee that there exists a solution satisfying also these conditions? The answer to this question will be presented in Theorem 1.4.9 below.

For A differential operator Q of type $F \to G$ on Y, we denote by $\mathcal{E}^{i}_{Q}(X \times Y)$ the subspace of $\mathcal{E}(E^{i} \boxtimes F)$ consisting of all sections f which satisfy $Q(y)f = 0$. Since the differential $P(x)$ of the complex $\mathcal{E}(E^{\cdot} \boxtimes F)$ commutes with $Q(y)$, the graded space $\mathcal{E}^{\cdot}_{Q}(X \times Y)$ may be considered as a subcomplex of $\mathcal{E}(E^{\cdot} \boxtimes F)$

Theorem 1.4.9 *Let $E^{\cdot}$ be a (cochain) differential complex on X and Q be a differential operator of type $F \to G$ on Y. If the complex $\mathcal{E}^{\cdot}(X)$ is exact at degree q, then the complex $\mathcal{E}^{\cdot}_{Q}(X \times Y)$ is also exact at degree q.*

Proof. Consider the following complexes of topological vector spaces

$$0 \longrightarrow \mathcal{E}_{P_{q-1}}(X) \xrightarrow{\imath} \mathcal{E}^{q-1}(X) \xrightarrow{P_{q-1}} \mathcal{E}_{P_q}(X) \longrightarrow 0, \tag{1.4.4}$$

$$0 \longrightarrow \mathcal{E}_{Q}(Y) \longrightarrow 0, \tag{1.4.5}$$

where $\imath$ is the inclusion mapping.

Since the complex $\mathcal{E}^{\cdot}(X)$ is exact at degree q, the sequence (1.4.4) is exact at $\mathcal{E}_{P_q}(X)$. Exactness of this sequence at other degrees is obvious.

By the Open Mapping Theorem, P_{q-1} is a topological homomorphism. The same is clear concerning the other mappings in the sequence (1.4.4) and (1.4.5).

Since a subspace of a nuclear space is also nuclear, the graded topological vector spaces of (1.4.4) and (1.4.5) are nuclear. Thus, all conditions of Theorem 1 4.4 are fulfilled. By this Theorem, the topological tensor product of complexes (1.4.4) and (1.4.5) is exact, i.e.

$$\begin{array}{llll} 0 \longrightarrow \mathcal{E}_{P_{q-1}}(X)\hat{\otimes}\mathcal{E}_Q(Y) & \xrightarrow{\imath\hat{\otimes}1+1\hat{\otimes}0} & \mathcal{E}^{q-1}(X)\hat{\otimes}\mathcal{E}_Q(Y) & \xrightarrow{P_{q-1}\hat{\otimes}1-1\hat{\otimes}0} \\ \mathcal{E}_{P_q}(X)\hat{\otimes}\mathcal{E}_Q(Y) & \longrightarrow & 0. & \end{array} \tag{1.4.6}$$

According to Proposition 3 (3) of Grothendieck [92, Ch.1], we have

$$\begin{array}{rcl} \mathcal{E}^{q-1}(X)\hat{\otimes}\mathcal{E}_Q(Y) & = & \mathcal{E}^{q-1}_Q(X \times Y), \\ \mathcal{E}_{P_q}(X)\hat{\otimes}\mathcal{E}_Q(Y) & = & Z^q(\mathcal{E}^{\cdot}_Q(X \times Y)). \end{array}$$

So we obtain the following exact sequence from (1.4.6):

$$0 \longrightarrow Z^{q-1}(\mathcal{E}_Q(X\times Y)) \xrightarrow{\iota\hat{\otimes}1} \mathcal{E}_Q^{q-1}(X\times Y) \xrightarrow{P_{q-1}(X)} Z^q(\mathcal{E}_Q(X\times Y)) \longrightarrow 0.$$

Hence it follows that the complex $\mathcal{E}_Q(X\times Y)$ is exact at degree q which completes the proof.

□

It is useful to compare this result with Theorem 3 of Palamodov [175, Ch.VII, §9].

A particular case of Theorem 1.4.9 for the Dolbeault complex had been noted by Weinstock [253].

1.4.5 *Examples*

We shall limit ourselves to three well-known examples among numerous ones.

Example 1.4.10 Let $E = \Lambda\, T^*_{\mathbb{C}}(X)$ be the de Rham complex on a manifold X, and $F = \Lambda\, T^*_{\mathbb{C}}(Y)$ be the de Rham complex on a manifold Y. Then $(E \boxtimes F) = \Lambda\, T^*_{\mathbb{C}}(X\times Y)$ is the de Rham complex on the product $X\times Y$. Let us endow the spaces $H^{\cdot}(X,\mathbb{C})$ with the topologies induced from $H^{\cdot}(\mathcal{E}(\Lambda\, T^*_{\mathbb{C}}(X)))$ by the de Rham isomorphism of Proposition 1.2.42. Theorem 1.4.8 asserts that, if the cohomology spaces $H^{\cdot}(X,\mathbb{C})$ and $H^{\cdot}(Y,\mathbb{C})$ are Hausdorff, we have

$$H^{\cdot}(X\times Y,\mathbb{C}) = H^{\cdot}(X,\mathbb{C})\hat{\otimes} H^{\cdot}(Y,\mathbb{C}). \tag{1.4.7}$$

In particular, if the vector spaces $H^{\cdot}(X,\mathbb{C})$ and $H^{\cdot}(Y,\mathbb{C})$ are finite-dimensional, then $H^{\cdot}(X\times Y,\mathbb{C}) = H^{\cdot}(X,\mathbb{C})\hat{\otimes} H^{\cdot}(Y,\mathbb{C})$ (see Lemma 5.1.7). This is the classical Künneth formula for the cohomologies of products of manifolds (see Proposition 12.16 of Dold [56, Ch.VII]).

□

Example 1.4.11 Let X and Y be complex manifolds and $E = \Lambda^{0,\cdot}\, T^*_{\mathbb{C}}(X)$ and $F = \Lambda^{0,\cdot}\, T^*_{\mathbb{C}}(Y)$ be the Dolbeault complexes on X and Y respectively. Then $(E \boxtimes F) = \Lambda^{0,\cdot}\, T^*_{\mathbb{C}}(X\times Y)$ is the Dolbeault complex on the product $X\times Y$. Let us endow the space $H^{\cdot}(X,\mathcal{O})$ with the topologies induced from $H^{\cdot}(\mathcal{E}(\Lambda^{0,\cdot}\, T^*_{\mathbb{C}}(X)))$ by the Dolbeault isomorphism of Proposition 1.2.49. Theorem 1.4.9 asserts, that if the cohomology spaces $H^{\cdot}(X,\mathcal{O})$ and $H^{\cdot}(Y,\mathcal{O})$ are Hausdorff, then

$$H^{\cdot}(X\times Y,\mathcal{O}) = H^{\cdot}(X,\mathcal{O})\hat{\otimes} H^{\cdot}(Y,\mathcal{O}). \tag{1.4.8}$$

In particular, if the manifolds X and Y are compact, we have $H^{\cdot}(X\times Y,\mathcal{O}) = H^{\cdot}(X,\mathcal{O})\otimes H^{\cdot}(Y,\mathcal{O})$ (see Corollary 2.2.4 and Lemma 5.1.7).

□

Example 1.4.12 Let X and Y be compact differentiable manifolds and E and F be (cochain) elliptic complexes on X and Y respectively. According to Corollary

2.2.4, the cohomology spaces $H^{i}(\mathcal{E}(E))$ and $H^{j}(\mathcal{E}(F))$ are finite-dimensional. By Lemma 5.1.7, these spaces are Hausdorff. It follows now from Theorem 1.4.8 that

$$H\,(\mathcal{E}((E \boxtimes F))) = H\,(\mathcal{E}(E)) \otimes H\,(\mathcal{E}(F)). \tag{1.4.9}$$

This result was announced by Atiyah and Bott [19].

□

1.5 Cochain mappings of differential complexes

1.5.1 *A brief survey of operator kernels. Schwartz's kernel theorem*

We suppose that X and Y are differentiable manifolds countable at infnity and E and F are differentiable $\mathbb{C}$-vector bundles over X and Y respectively.

Sections $K(x,y) \in \mathcal{D}'(E \boxtimes F')$, where F' is the dual bundle for F, are said to be kernels of type $F \to E$ on $X \times Y$.

Let us agree to write $\mathfrak{C}'(E) = (\mathfrak{C}(E'))'$ when it can not evoke a misunderstanding. For example, we have $\mathcal{D}''(E) = (\mathcal{D}'(E'))' = (\mathcal{D}(E''))'' = \mathcal{D}(E)$ because the last space is reflexive.

We let $\mathfrak{C}$ be $\mathcal{E}$, $\mathcal{D}$, $\mathcal{D}'$ or $\mathcal{E}'$. Each kernel $K(x,y) \in \mathfrak{C}(E \boxtimes F')$ defines a continuous linear operator $K:\ \mathfrak{C}'(F) \to \mathfrak{C}(E)$ by the formula

$$\langle g, Kf\rangle_X = \langle K(x,y), g(x) \otimes f(y)\rangle_{X\times Y} \quad (g \in \mathfrak{C}'(E'), f \in \mathfrak{C}'(F)), \tag{1.5.1}$$

and the kernel theorem in its global formulation asserts that this correspondence is a topological isomorphism except in the case $\mathfrak{C} = \mathcal{D}$.

Theorem 1.5.1 (kernel theorem) *If* $\mathfrak{C} = \mathcal{E}, \mathcal{D}, \mathcal{D}'$ *or* $\mathcal{E}'$, *we have*

$$\mathfrak{C}(E \boxtimes F') \cong \mathcal{L}_b(\mathfrak{C}'(F) \to \mathfrak{C}(E)).$$

Let us emphasize that for $\mathfrak{C} = \mathcal{D}$ the isomorphism is only algebraic here.

Proof. We would like to deduce this theorem from Theorem 1.4.2 because all spaces dealt with are Montel and nuclear.

Set $K = \mathfrak{C}(E)$ and $L = \mathfrak{C}(F')$. It follows from Example 1.4.1 that $K \hat{\otimes} L = \mathfrak{C}(E \boxtimes F')$.

According to Theorem 8.4.8 of Edwards [62], we have $L'_\tau = L'$ because L has the Heine-Borel property.

According to Theorem 7.1.1 of Edwards [62], a set $\sigma \subset L'$ is equicontinuous if and only if it is bounded because L is barrelled. It means that $\mathcal{L}_\varepsilon(L' \to K) = \mathcal{L}_b(L' \to K)$.

Now we can apply Theorem 1.4.2 to get

$$\mathfrak{C}(E \boxtimes F') = K \hat{\otimes} L \overset{top}{\cong} \mathcal{L}_\varepsilon(L'_\tau \to K) \overset{top}{\cong} \mathcal{L}_b(\mathfrak{C}'(F) \to \mathfrak{C}(E))$$

which completes the proof.

□

The main (and classical) variant ($\mathfrak{C} = \mathcal{D}'$) of the kernel theorem was proved by Schwartz [194].

We need a more delicate characterization of kernels on $X \times Y$. It is based on the following result.

Proposition 1.5.2 *Let K be one of the spaces $\mathcal{E}(E)$, $\mathcal{D}(E)$, $\mathcal{D}'(E)$ or $\mathcal{E}'(E)$, and L be one of the spaces $\mathcal{E}(F)$, $\mathcal{D}(F)$, $\mathcal{D}'(F)$ or $\mathcal{E}'(F)$. Then we have $K\hat{\otimes}L' \overset{top}{\cong} \mathcal{L}_b(L \to K)$.*

Proof. This follows from Theorem 1.4.2 in the same way as Theorem 1.5.1.

□

The kernels $K(x,y) \in \mathcal{E}(E)\hat{\otimes}\mathcal{D}'(F')$ are said to be regular. They are characterized by the property that the corresponding operator K maps $\mathcal{D}(F)$ into $\mathcal{E}(E)$ continuously.

The kernels $K(x,y) \in \mathcal{D}'(E)\hat{\otimes}\mathcal{E}(F')$ are said to be continuable. Such kernels are characterized by the property that the corresponding operators $K : \mathcal{D}(F) \to \mathcal{D}'(E)$ may be extended to continuous linear operators from $\mathcal{E}'(F)$ to $\mathcal{D}'(E)$.

A kernel $K(x,y)$ ia said to be biregular if it is regular and continuable simultaneously.

Let us denote by $\Delta = \{(x,x) : x \in X\}$ the diagonal of the product $X \times X$.

An important property of biregular kernels is expressed in the following proposition. See Lemma 2.2 of Taylor [236, Ch.2] in connection with its local variant.

Proposition 1.5.3 *Let Y be a copy of X, and $K(x,y) \in \mathcal{D}'(E \boxtimes F')$ be a biregular kernel which is differentiable off the diagonal of $X \times X$. Then the corresponding operator $K : \mathcal{E}'(F) \to \mathcal{D}'(E)$ is pseudo-local, i.e.*

$$\operatorname{sing\,supp} Kf \subset \operatorname{sing\,supp} f \quad \textit{for all } f \in \mathcal{E}'(F).$$

Proof. We should prove that, if $f \in \mathcal{E}'(F)$ is differentiable in some neighbourhood U of a point $x^0 \in X$, then Kf is also differentiable in a (maybe smaller) neighbourhood of x^0. Choose $\varphi \in \mathcal{D}(U)$ such that $\varphi \equiv 1$ in a neighbourhood $U' \subset U$ of the point x^0. Write $f = \varphi f + (1-\varphi)f = f_1 + f_2$ where $f_1 \in \mathcal{D}(F)$, and $f_2 \in \mathcal{E}'(F)$ vanishes in U'. Since $K(x,y)$ is a regular kernel, we have $Kf_1 \in \mathcal{E}(E)$, and $Kf_2 \in \mathcal{D}'(E)$ is differentiable in U' because $K(x,y)$ is continuable and differentiable off the diagonal Δ. Hence Kf is differentiable in U'.

□

For an operator $K : \mathcal{D}(F) \to \mathcal{D}'(E)$, we define the transposed operator $K' : \mathcal{D}(E') \to \mathcal{D}'(F')$ by means of the equality

$$\langle g, Kf\rangle_X = \langle K'g, f\rangle_Y \quad \text{for all} \quad g \in \mathcal{D}(E'), f \in \mathcal{D}(F). \tag{1.5.2}$$

By identifying E'' with E it is easy to see that the kernel $K'(y,x)$ of the transposed operator is in $\mathcal{D}'(F' \boxtimes E)$ and its local representation is $K'(y,x) = (K(x,y))^t$. Moreover we have $(K')' = K$, and $(K_2 \circ K_1)' = K_1' \circ K_2'$.

Proposition 1.5.4 *A kernel $K(x,y)$ is regular if and only if the kernel $K'(y,x)$ is continuable and inversely.*

Proof. This follows immediately from the definition (1.5.2) of a transposed operator.

□

Definition 1.5.5 *A kernel $K(x,y) \in \mathcal{D}'(E \boxtimes F')$ is said to be properly supported if both projections $\pi_X : \operatorname{supp} K \to X$ and $\pi_Y : \operatorname{supp} K \to Y$ are proper mappings.*

Obviously, a kernel $K(x,y)$ is properly supported if and only if the kernel $K'(y,x)$ is properly supported.

Example 1.5.6 Let Y be a copy of X, and P be a differential operator of type $E \to F$ on X. It follows from the local property of P that the kernel $K_P(x,y) \in \mathcal{D}'(F \boxtimes E')$ of P (being of course biregular) is supported on the diagonal Δ. In particular, it is properly supported.

□

A role of properly supported kernels comes to light by using duality considerations.

Proposition 1.5.7 *1) If $K(x,y) \in \mathcal{D}'(E \boxtimes F')$ is a properly supported regular kernel, then the operator K determines a mapping $K : \mathcal{D}(F) \to \mathcal{D}(E)$ which may be extended to a continuous mapping $K : \mathcal{E}(F) \to \mathcal{E}(E)$.*

2) If $K(x,y) \in \mathcal{D}'(E \boxtimes F')$ is properly supported continuable kernel, then the operator K determines a mapping $K : \mathcal{E}'(F) \to \mathcal{E}'(E)$ which may be extended to a continuous mapping $K : \mathcal{D}'(F) \to \mathcal{D}'(E)$.

Proof. The proof follows the proof of Proposition 3.1 from Shubin's book [201]. For sets $\Sigma \subset X \times Y$ and $\sigma \subset Y$, we set

$$\Sigma \circ \sigma = \{x \in X : \text{ there exists } y \in \sigma \text{ such that } (x,y) \in \Sigma\}.$$

Let $K(x,y) \in \mathcal{D}'(E \boxtimes F')$ be a properly supported regular kernel. We claim that, for $f \in \mathcal{D}(F)$, the following inclusion takes place:

$$\operatorname{supp} Kf \subset (\operatorname{supp} K) \circ (\operatorname{supp} f). \tag{1.5.3}$$

Indeed, let us take a section $g \in \mathcal{D}(E')$ such that $(\operatorname{supp} g) \cap ((\operatorname{supp} K) \circ (\operatorname{supp} f)) = \emptyset$. Then we have $(\operatorname{supp} K) \cap (\operatorname{supp} g \otimes f) = \emptyset$ so $\langle g, Kf \rangle_X = 0$ in view of (1.5.1). This proves (1.5.3).

Further, it follows from the obvious formula

$$(\operatorname{supp} K) \circ (\operatorname{supp} f) = \pi_X((\operatorname{supp} K) \cap \pi_Y^{-1}(\operatorname{supp} f))$$

that the right-hand side of (1.5.3) is a compact set. Therefore K determines a continuous mapping $K : \mathcal{D}(F) \to \mathcal{D}(E)$ because $K(x,y)$ is regular.

If $K(x,y) \in \mathcal{D}'(E \boxtimes F')$ is a properly supported continuable kernel, the same reasoning shows that K determines a continuous mapping $K : \mathcal{E}'(F) \to \mathcal{E}'(E)$.

Let us apply these results to the transposed kernel $K'(y,x)$ and take into account Proposition 1.5.4. Then we get that in the case 1) the transposed operator K' determines a continuous mapping $K' : \mathcal{E}'(E') \to \mathcal{E}'(F')$, and in the case 2) it determines a continuous mapping $K' : \mathcal{D}(E') \to \mathcal{D}(F')$.

To complete the proof simply use formula (1.5.2).

□

Later on we will use the same terminology both for operators and their kernels. This will not lead to misunderstandings in view of Theorem 1.5.1.

Let us dwell upon two examples which are useful for the future.

1.5.2 *Kernel of the identity operator*

Let $M : X \to Y$ be a differentiable mapping, and $\Gamma_m = \{(x, m(x)) \in X \times Y\}$ be the graph of this mapping. If F is a differentiable $\mathbb{C}$-vector bundle over Y, then the induced bundle $E = m^*F$ is defined over X (see Wells [254]).

We denote by $\operatorname{tr}\Gamma_m^*$ the kernel in $\mathcal{D}'(E \boxtimes F')$ that is given by the equality

$$\langle \operatorname{tr}\Gamma_m^*, K\rangle_{X\times Y} = \int_X \operatorname{tr} K(x, m(x)) \quad \text{for} \quad K \in \mathcal{D}(E' \boxtimes F).$$

This kernel defines the continuous linear operator $f \to m^*f = f(m)$ from $\mathcal{E}(F)$ to $\mathcal{E}(E)$ which has traditionally been called the inverse image operator. Hence it follows that the kernel $\operatorname{tr}\Gamma_m^*$ is regular. It is properly supported if and only if the mapping m is proper.

In particular, if Y is a copy of X and $m(x) = x$ is the identity mapping, then $\Gamma_m = \Delta$ and the kernel $\operatorname{tr}\Delta^*$ defines the identity operator on $\mathcal{D}(F)$. It os clear that the kernel $\operatorname{tr}\Delta^*$ is biregular and supported on Δ.

For simplicity of notation, we do not indicate dependence of the kernel $\operatorname{tr}\Delta^*$ on the bundle F. It is always clear from context.

1.5.3 *Pseudo-differential operators*

Pseudo-differential operators form an important class of operators whose kernels satisfy the conditions of Proposition 1.5.3.

For the purposes of this book it will be enough to consider only the classical pseudo-differential operators first introduced by Kohn and Nirenberg [131]. Let us give a brief description of them following Seeley's [197] article.

Let E and F be differential $\mathbb{C}$-vector bundles over X of ranks k and l respectively. Pseudo-differential operators between sections of E and F are represented locally by $(l \times k)$-matrices of scalar pseudo-differential operators on an open set $U \subset \mathbb{R}^n$. That is why we shall start with a discussion of the latter.

For a real number p, we denote by $\tilde{S}^p(U)$ the space of all functions $P(x,z) \in \mathcal{E}(U \times \mathbb{R}^n)$ satisfying the following condition:

for each pair of multi − indices α, β *and compact set* $K \subset U$, *we have* $\sup_{\substack{x\in K\\ z\in\mathbb{R}^n}}(1+|z|)^{|\alpha|-p}|D_z^\alpha D_x^\beta P(x,z)| < \infty$.

We observe that, if $P(x,D) \in \mathrm{do}_p(U)$ where $\mathrm{do}_p(U)$ is a short notation for the space $\mathrm{do}_p(U\times\mathbb{C}\to U\times\mathbb{C})$, then the (full) symbol function $P(x,z)$ is in $\widetilde{S}^p(U)$.

Let us denote by $\widehat{f}$ the Fourier transformation of a distribution $f \in \mathcal{E}'(\mathbb{R}^n)$, i.e. $\widehat{f}(z) = \int e^{-\sqrt{-1}zy} f(y)\,dy$.

For $P \in \widetilde{S}^p(U)$, the canonical pseudo-differential operator of order $\le p$ on U is the mapping $P(x,D) : \mathcal{D}(U) \to \mathcal{E}(U)$ given by the formula

$$P(x,D)f = \frac{1}{(2\pi)^n}\int e^{\sqrt{-1}xz}P(x,z)\widehat{f}(z)\,dz. \tag{1.5.4}$$

(The function $P(x,z)$ is called the (complete) symbol of $P(x,D)$.) Notice that (1.5.4) agrees with a differential operator acting on functions if $P(x,z)$ is a polynomial in z.

If P is in some $\widetilde{S}^p(U)$ and $P_{p-j} \in \widetilde{S}^{p-j}(U)$ $(j = 0,1,\ldots)$, we write $P \sim \sum_{j=0}^{\infty} P_{p-j}$ if and only if $P - \sum_{j=0}^{r-1} P_{p-j}$ is in $\widetilde{S}^{p-r}(U)$ for all integer $r \ge 0$.

Let $S^p(U)$ be the subspace of $\widetilde{S}^p(U)$ consisting of functions $P \in \mathcal{E}(U\times\mathbb{R}^n)$ which have a homogeneous asymptotic expansion $P \sim \sum_{j=0}^{\infty} P_{p-j}$ where each $P_{p-j} \in \mathcal{E}(U\times\mathbb{R}^n)$ is (positively) homogeneous of degree $(p-j)$ in z off some neighbourhood of zero in $\mathbb{R}^n$, i.e. $P_{p-j}(x,tz) = t^{p-j}P_{p-j}(x,z)$ for $t \ge 1$ and $|z| \ge c$.

A linear mapping $P : \mathcal{D}(U) \to \mathcal{E}(U)$ is called a (classical) pseudo-differential operator of order $\le p$ in U if for each open set $V \subset\subset U$ there exists a function $P_V \in S^p(U)$ such that $Pf = P_V(x,D)$ for all $f \in \mathcal{D}(V)$. The space of such pseudo-differential operators is denoted by $\mathrm{pdo}_p(U)$. It is clear that $\mathrm{do}_p(U) \subset \mathrm{pdo}_p(U)$ $(p \in \mathbb{Z}_+)$.

It is well-known that the space $\mathrm{pdo}_p(U)$ is invariant under coordinate changes (see, for example, Theorem 4.2 of Shubin [201]). In view of the invariance it is clear how to globalize the definition of a pseudo-differential operator to sections of vector bundles over X.

Definition 1.5.8 *A linear mapping* $P : \mathcal{D}(E) \to \mathcal{E}(F)$ *is called a (classical) pseudo-differential operator of type* $E \to F$ *and order* $\le p$ *if, for any choice of a coordinate neighbourhood* U *on* X *over which* E *and* F *are trivial and local trivializations of these bundles, the mapping* $P_U : [\mathcal{D}(U)]^k \to [\mathcal{E}(U)]^l$ *defined by the commutative diagram*

$$\begin{array}{ccc}
\mathcal{D}(E) & \xrightarrow{P} & \mathcal{E}(F) \\
\uparrow\subset & & \downarrow\subset \\
\mathcal{D}(E|_U) & \longrightarrow & \mathcal{E}(F|_U) \\
\downarrow\cong & & \downarrow\cong \\
[\mathcal{D}(U)]^k & \xrightarrow{P_U} & [\mathcal{E}(U)]^l
\end{array}$$

can be expressed by an $(l \times k)$-matrix P_U of (scalar) pseudo-differential operators of order $\le p$ in U.

The space of (classical) pseudo-differential operators of type $E \to F$ and order $\le p$ is denoted by $\mathrm{pdo}_p(E \to F)$. By Peetre's theorem, $\mathrm{do}_p(E \to F) \subset \mathrm{pdo}_p(E \to F)$ $(p \in \mathbb{Z}_+)$.

Proposition 1.5.9 *If $P \in \mathrm{pdo}_p(E \to F)$ and $Q \in \mathrm{pdo}_q(F \to G)$ are pseudo-differential operators on X and one of them is properly supported, then $Q \circ P \in \mathrm{pdo}_{p+q}(E \to G)$.*

Proof. See Proposition 3.7 (b) of Shubin [201].

□

Proposition 1.5.10 *If $P \in \mathrm{pdo}_p(E \to F)$, then $P' \in \mathrm{pdo}_p(F' \to E')$ and $P^* \in \mathrm{pdo}_p(F \to E)$ where $P^* = *_E^{-1} P' *_F$ (see Section 2.1 below).*

It follows from Propositions 1.5.4 and 1.5.10 that pseudo-differential operators are biregular operators.

Define moreover the space $\mathrm{pdo}_p(E \to F)$ for $p = \infty$ by setting it equal to the intersection $\cap_{p<0} \mathrm{pdo}_p(E \to F)$. As we shall see in Subsection 1.5.4 operators $S \in \mathrm{pdo}_{-\infty}(E \to F)$ are smoothing. That is, they have kernels $S(x,y) \in \mathcal{E}(F \boxtimes E')$.

The main role of (classical) pseudo-differential operators is that they allow one to construct inverse operators for elliptic differential operators modulo smoothing operators.

Theorem 1.5.11 *Every elliptic operator $P \in \mathrm{do}_p(E \to F)$ has a properly supported parametrix $\Pi \in \mathrm{pdo}_{-p}(F \to E)$, i.e.*

$$\begin{aligned} \Pi P &= 1 \quad (\mathit{mod}\ \ \mathrm{pdo}_{-\infty}(E \to E)), \\ P\Pi &= 1 \quad (\mathit{mod}\ \ \mathrm{pdo}_{-\infty}(F \to F)). \end{aligned}$$

Proof. The proof of this well-known result may be found in most books on the theory of pseudo-differential operators. See Theorem 5.1 of Shubin [201] for example.

□

1.5.4 *Kernels of pseudo-differential operators. Seeley's theorem*

Let us clarify the local structure of kernels of pseudo-differential operators. For this it is sufficient to limit oneself to canonical pseudo-differential operators (1.5.4) on an open set $U \subset \mathbb{R}^n$.

A direct calculation shows that the kernel of the operator (1.5.4) is the distribution $K_P \in \mathcal{D}'(U \times U)$ defined by the oscillatory integral

$$\langle K_P, K\rangle_{U\times U} = \int\int \left(\frac{1}{(2\pi)^n}\int e^{\sqrt{-1}(x-y)\xi} P(x,\xi)d\xi\right) K(x,y)\,dxdy \qquad (K \in \mathcal{D}(U \times U))$$

(see Shubin [201, §1]. So, by a suitable interpretation of the integral in the sense of the theory of distributions, we have

$$K_P(x, x-y) = \frac{1}{(2\pi)^n}\int e^{\sqrt{-1}(x-y)\xi} P(x,\xi)d\xi \qquad ((x,y) \in U \times U). \tag{1.5.5}$$

The notation $K_P(x, x-y)$ instead of $K_P(x,y)$ here reflects the structure of the kernel.

For instance, choose integers r and s such that $r < n$ and $s - n > p$. Denoting by $\Delta = -\sum_{j=1}^n D_j^2$ the Laplace operator in $\mathbb{R}^n$, we may write for $(x,z) \in U \times \mathbb{R}^n$

$$K_P(x,z) = (1-\Delta_z)^{s/2}\left(\frac{1}{|z|^r}\frac{1}{(2\pi)^n}\int e^{\sqrt{-1}z\xi}(-\Delta_\xi)^{r/2}\left(\frac{P(x,\xi}{(1+|\xi|^2)^{s/2})}\right)d\xi\right). \tag{1.5.6}$$

The integral within the brackets here converges absolutely together with all derivatives up to order $(r+s-n-p)$ uniformly in x on compact subset of U and $z \in \mathbb{R}^n$. It is easy to see that K_P does not depend on the choice of r and s satisfying the above conditions.

Proposition 1.5.12 *Every operator $P \in \mathrm{pdo}_p(E \to F)$ on X is pseudo-local.*

Proof. If $(x,z) \in U \times (\mathbb{R}^n \setminus 0)$, formula (1.5.6) remains valid for all r. Hence it follows that the kernel $K_P(x, x-y)$ is (infinitely) differentiable in $(U \times U) \setminus \Delta$. To complete the proof simply invoke Proposotion 1.5.3.

□

To exhibit the singularity of the kernel $K_P(x,z)$ for $z = 0$ we need some additional definitions.

For real d ($d > -n$) let us introduce the class phf_d of pseudo-homogeneous functions of degree d. Namely, if $d \neq 0, 1, \ldots$, then $f \in \mathrm{phf}_d$ iff, outside the origin, f is a C^∞_{loc} function (positively) homogeneous of degree d. If $d = 0, 1, \ldots$, then $f \in \mathrm{phf}_d$ iff $f(z) = p(z)\log|z| + g(z)$, where $p(z)$ is a polynomial of degree d and, outside the origin, g is a C^∞_{loc} function (positively) homogeneous of degree d and satisfying $\int_{|z|=1} z^\alpha g(z)d\sigma = 0$ for all $|\alpha| = d$. Here $d\sigma$ is the standard volume form on the unit sphere of $\mathbb{R}^n$.

In the case of non-negative integers d, we have separated out the polynomial part of f and multiplied it by $\log|z|$. The Newtonian potential in $\mathbb{R}^2$, which is $\frac{1}{2\pi}\log|z|$, suggests that these "log" terms are necessary.

A more thorough discussion of homogeneous distributions is given by Hörmander [113, n. 3.2].

Definition 1.5.13 (See Seeley [197]) *Let $d > -n$. A kernel $K(x,z) \in \mathcal{E}(U \times (\mathbb{R}^n \setminus 0))$ is called a pseudo-homogeneous kernel of degree d if there are functions $K_{d+j}(x,z) \in \mathcal{E}(U \times (\mathbb{R}^n \setminus 0))$ $(j \in \mathbb{Z}_+)$, pseudo-homogeneous of degree $(d+j)$ in z, such that $K(x,z) - \sum_{j=0}^{r-1} K_{d+j}(x,z)$ is of the class $C^s_{loc}(U \times \mathbb{R}^n)$ for $s < d + r$.*

Notice that a pseudo-homogeneous function of degree $d > 0$ is a function of class $C^s_{loc}(\mathbb{R}^n)$ for all $s < d$, so the definition is reasonable.

The vector space of all pseudo-homogeneous kernels of degree d in U is denoted by $\mathrm{phk}_d(U)$.

The following remarkable result is due to Seeley [197].

Theorem 1.5.14 (Seeley [197]) *Let $p < 0$. Then $P : \mathcal{D}(U) \to \mathcal{E}(U)$ is a pseudo-differential operator of order p in U iff there exists a kernel $K_P(x,z) \in \mathrm{phk}_{-n-p}(U)$ such that for all $f \in \mathcal{D}(U)$*

$$Pf(x) = \int_U K_P(x, x-y) f(y) dy \quad (x \in U) \tag{1.5.7}$$

Proof. We shall give only a commentary to the proof. There are two simple general principles that underlie Theorem 1.5.14. Firstly, if K is homogeneous of degree d, then $\widehat{K}$ is homogeneous of degree $(-n-d)$. This is easy to check simply by writing $\widehat{K}(tz) = \int e^{-\sqrt{-1}tz\xi} K(\xi) d\xi$ and setting $t\xi = y$. However, the integral is not absolutely convergent for any value of d, and not even conditionally convergent for most values of s. This difficulty has to be circumvented in the proof. Secondly, the behavior of $\widehat{K}$ near ∞ is closely related to the behavior of K near zero. The proof exploits this to circumvent the convergence problem above, and relates the singularities of K near zero to the behavior of $\widehat{K}$ at ∞. See Seeley [197] for details. □

We could drop the restriction $p < 0$ by allowing distributional kernels. This would require a little more background on homogeneous distributions. See Kree [139].

As a consequence of Theorem 1.5.14, we deduce a relation between pseudo-differential operators and classical singular integral operators (*SIO's*, see Mikhlin [161, §5]. Namely if $P(x,z) \in S^0(U)$, then $P(x,D)$ is a classical *SIO* modulo $\mathrm{pdo}_{-1}(U)$. To prove this, let $Q(x,z) = P(x,z) \setminus (1 + |z|^2)$. Then by Theorem 1.5.14, $Q(x,D) \in \mathrm{pdo}_{-2}(U)$ has a kernel in $\mathrm{phk}_{-n+2}(U)$, and writing $P(x,D) = Q(x,D) \circ (1 - \Delta)$ we obtain $P(x,D)$ as a classical *SIO* modulo $\mathrm{pdo}_{-1}(U)$. Besides, Theorem 1.5.14 shows that (classical) pseudo-differential operators of order 0 are in the so-called class of Calderon-Zigmund operators (see Dyn'kin [61]).

Remark 1.5.15 *Theorem 1.5.14 allows one to prove boundedness theorems for pseudo-differential operators in Hölder and Sobolev spaces with classical tools. See Theorem 2.5 of Taylor [236, Ch. XI].*

Let us single out a particular case of Theorem 1.5.14 that will often be used.

Corollary 1.5.16 *Every operator $P \in \mathrm{pdo}_{-\infty}(E \to F)$ has a kernel $K_P(x,y) \in \mathcal{E}(F \boxtimes E')$.*

For a very useful discussion of symbols of pseudo-differential operators we refer to Shubin's [201, §3] book.

1.5.5 *The complex of homomorphisms*

Let $E^{\cdot} = \{E^i, d_E\}_{i=0,\dots,l_E}$ and $F^{\cdot} = \{F^j, d_F\}_{j=0,\dots,l_F}$ be (cochain) differential complexes on differential manifolds X and y respectively. We let $\mathfrak{C}$ be $\mathcal{E}$, $\mathcal{D}$, $\mathcal{D}'$ or $\mathcal{E}'$, and shall be interested in cochain mappings $M: \mathfrak{C}'(F^{\cdot}) \to \mathfrak{C}(E^{\cdot})$.

To deal with these mappings we should consider the complex of homomorphisms $\mathcal{L}(\mathfrak{C}'(F^{\cdot}) \to \mathfrak{C}(E^{\cdot}))^{\cdot}$ that is defined by the formulas

$$\begin{aligned} \mathcal{L}(\mathfrak{C}'(F^{\cdot}) \to \mathfrak{C}(E^{\cdot}))^r &= \textstyle\prod_{i\in\mathbb{Z}} \mathcal{L}_b(\mathfrak{C}'(F^i) \to \mathfrak{C}(E^{i+r})) \quad (z \in \mathbb{Z}), \\ d_r\{M_i\} &= \{(-1)^{r+1} M_{i+1} d_F + d_E M_i\} \end{aligned} \tag{1.5.8}$$

(see Subsection 1.1.11).

As usual, the spaces $\mathcal{L}(\mathfrak{C}'(F^{\cdot}) \to \mathfrak{C}(E^{\cdot}))^r$ $(r \in \mathbb{Z})$ are endowed with the product topologies. It is clear that each of the operators d_r is continuous in this topology.

Cocycles of the complex $\mathcal{L}(\mathfrak{C}'(F^{\cdot}) \to \mathfrak{C}(E^{\cdot}))^{\cdot}$ at degree r are just cochain mappings $M: \mathfrak{C}'(F^{\cdot}) \to \mathfrak{C}(E^{\cdot})$ of degree r.

Cohomologies of the complex $\mathcal{L}(\mathfrak{C}'(F^{\cdot}) \to \mathfrak{C}(E^{\cdot}))^{\cdot}$ at degree r are classes of homotopic cochain mappings $M: \mathfrak{C}'(F^{\cdot}) \to \mathfrak{C}(E^{\cdot})$ of degree r.

It is easy to see that there are natural linear mappings

$$H^r(\mathcal{L}(\mathfrak{C}'(F^{\cdot}) \to \mathfrak{C}(E^{\cdot}))^{\cdot}) \to \mathcal{L}(H^{\cdot}(\mathfrak{C}'(F^{\cdot})) \to H^{\cdot}(\mathfrak{C}(E^{\cdot})))^r \quad (r \in \mathbb{Z}),$$

and the question of when these mappings are topological isomorphisms is dual to the question of when there is a Künneth formula for the tensor product of complexes (see Subsection 1.4.2).

1.5.6 *A differential version of the complex of homomorphisms*

A differential version of the complex (1.5.8) can be easily constructed by using Theorem 1.5.1, according to which

$$\mathcal{L}(\mathfrak{C}'(F^{\cdot}) \to \mathfrak{C}(E^{\cdot}))^r \cong \mathfrak{C}\left(\oplus_{i\in\mathbb{Z}} E^{i+r} \boxtimes F^{i'}\right) \qquad (r \in \mathbb{Z}). \tag{1.5.9}$$

Let us define a new differential complex $(E^{\cdot} \boxtimes F^{\cdot\prime})^{\cdot}$ on $X \times Y$ in the following way. For $r \in \mathbb{Z}$, we set $(E^{\cdot} \boxtimes F^{\cdot\prime})^r = \oplus_{i\in\mathbb{Z}}(E^{i+r} \boxtimes F^{i'})$. Since the bundles E^i are non-zero only for $i = 0, \dots, l_E$ and the bundles F^j are non-zero only for $j = 0, \dots, l_F$, one may conclude that $(E^{\cdot} \boxtimes F^{\cdot\prime})^r$ are differentiable $\mathbb{C}$-vector bundles over $X \times Y$, and $(E^{\cdot} \boxtimes F^{\cdot\prime})^r \neq 0$ only for $-l_F \le r \le l_E$. We provide the graded bundle $(E^{\cdot} \boxtimes F^{\cdot\prime})^{\cdot}$ with a differential $d_{E\boxtimes F'}$ of degree 1 by setting $d_{E\boxtimes F'}|_{\mathcal{E}((E^{\cdot}\boxtimes F^{\cdot\prime})^r)} = d_E(x) + (-1)^{r+1} d_F{}'(y)$. An easy calculation shows that $(d_{E\boxtimes F'})^2 = 0$ so we get a (cochain) differential complex $(E^{\cdot} \boxtimes F^{\cdot\prime})^{\cdot}$ on $X \times Y$.

The complex $(E^{\cdot} \boxtimes F^{\cdot\prime})^{\cdot}$ is elliptic provided the complexes $E^{\cdot}$ and $F^{\cdot}$ are elliptic and consist of differential operators of the same order (being common for both complexes). For varying orders this fails, but is true if ellipticity is understood in some wider sense.

For each of the classes $\mathfrak{C} = \mathcal{E}, \mathcal{D}, \mathcal{D}', \mathcal{E}'$ and many others, the complex $(E \boxtimes F')$ defines a complex of topological vector spaces

$$\mathfrak{C}((E \boxtimes F')): \ 0 \longrightarrow \mathfrak{C}((E \boxtimes F')^{-l_F}) \xrightarrow{d_{E\boxtimes F'}} \mathfrak{C}((E \boxtimes F')^{-l_F+1}) \xrightarrow{d_{E\boxtimes F'}} \cdots \xrightarrow{d_{E\boxtimes F'}} \mathfrak{C}((E \boxtimes F')^{l_E}) \longrightarrow 0. \tag{1.5.10}$$

The following proposition shows that the differential $d_{E\boxtimes F'}$ corresponds to the differential of (1.5.8) via the isomorphism (1.5.9).

Proposition 1.5.17 *If $K(x,y) \in \mathcal{D}'((E \boxtimes F')^r)$, then*

$$(d_{E\boxtimes F'}K)\, f = (-1)^{r+1}\ K d_F f + d_E K f \quad \textit{for all} \quad f \in \mathcal{D}(F_\cdot).$$

Proof. Indeed, let $g \in \mathcal{D}(E^{\cdot+r+1'})$ be some test section. Using formula (1.5.1) and the definition of the transposed operator, we have

$$\begin{aligned}
&\langle g, (d_{E\boxtimes F'}K) f\rangle_X \\
&\quad = \langle d_{E\boxtimes F'}K(x,y), g(x) \otimes f(y)\rangle_{X\times Y} \\
&\quad = \left\langle (d_E(x) + (-1)^{r+1} d_F{}'(y))K(x,y), g(x) \otimes f(y)\right\rangle_{X\times Y} \\
&\quad = \langle K(x,y), (d_E{}'g)(x) \otimes f(y)\rangle_{X\times Y} + (-1)^{r+1} \langle K(x,y), g(x) \otimes (d_F f)(y)\rangle_{X\times Y} \\
&\quad = \langle d_E{}'g, Kf\rangle_X + (-1)^{r+1} \langle g, K d_F f\rangle_X \\
&\quad = \left\langle g, (-1)^{r+1} K d_F f + d_E K f\right\rangle_X ,
\end{aligned}$$

which completes the proof.

□

Corollary 1.5.18 *Let $\mathfrak{C}$ be $\mathcal{E}$, $\mathcal{D}$, $\mathcal{D}'$, or $\mathcal{E}'$. Then we have the following isomorphism which is topological with the exception of $\mathfrak{C} = \mathcal{D}$:*

$$\mathcal{L}(\mathfrak{C}'(F) \to \mathfrak{C}(E)) \cong \mathfrak{C}((E \boxtimes F')). \tag{1.5.11}$$

Proof. This follows from the isomorphism (1.5.9) and Proposition 1.5.17.

□

The complex $(E \boxtimes F')$ will mainly be used in the case where Y is a copy of X, and $F = E$.

Remark 1.5.19 *Sometimes the notations shorten if we denote by $\mathfrak{C}^r(X \times X)$ $(r \in \mathbb{Z})$ the space of sections of class $\mathfrak{C}$ of the bundle $(E \boxtimes F')^r$ over $X \times X$.*

1.5.7 *A Künneth formula*

Using the Künneth formula for the tensor products of complexes (1.4.3) and the Poincaré duality (5.1.2) (see below), we shall obtain a Künneth formula for the complex of homomorphisms.

Theorem 1.5.20 (Künneth formula) *Let E and F be (cochain) differential complexes on X and Y. If the spaces $H(\mathcal{E}(E))$ and $H(\mathcal{E}(E'))$ are Hansdorff, then*

$$H(\mathcal{L}(\mathcal{E}'(F) \to \mathcal{E}(E))) \overset{top}{\cong} \mathcal{L}(H(\mathcal{E}'(F)) \to H(\mathcal{E}(E))). \tag{1.5.12}$$

Proof. After the natural regrading $j \to -j$ the transposed complex F' may be considered as a cochain differential complex on Y and one can define the tensor product $(E \boxtimes F')$ of the complexes E and F' by formulas of Subsection 1.4.3. It is easy to see that the complex $(E \boxtimes F')$ so obtained coincides with the complex, constructed in Subsection 1.5.6, modulo the various sign conventions for the differentials.

Thus Corollary 1.5.18 together with Theorem 1.4.4 imply for all $r \in \mathbb{Z}$

$$\begin{aligned} H^r(\mathcal{L}(\mathcal{E}'(F) \to \mathcal{E}(E))) &\cong H^r(\mathcal{E}((E \boxtimes F'))) \\ &\cong \otimes_{i-j=r} H^i(\mathcal{E}(E)) \hat{\otimes} H_j(\mathcal{E}(F')). \end{aligned}$$

The spaces $H^i(\mathcal{E}(E))$ and $H_j(\mathcal{E}(F'))$ are nuclear Fréchet (-Schwartz) spaces because the passage from topological vector space to separated quotient space preserves the properties that the topological vector space be a Fréchet (-Schwartz) space or a nuclear one. Hence it folows by Theorem 1.4.2 that

$$H^i(\mathcal{E}(E)) \hat{\otimes} H_j(\mathcal{E}(F')) \cong \mathcal{L}_\varepsilon((H_j(\mathcal{E}(F')))'_\tau \to H^i(\mathcal{E}(E))).$$

Since the spaces $H_j(\mathcal{E}(F'))$ have the Heine-Borel property, the topologies in $(H_j(\mathcal{E}(F')))'_\tau$ coincide with the strong topologies of dual spaces. Hence, the Poincaré Duality Theorem 5.1.8 (see below) implies that $(H_j(\mathcal{E}(F')))'_\tau \cong H^j(\mathcal{E}'(F))$.

Thus

$$\begin{aligned} H^r(\mathcal{L}(\mathcal{E}'(F) \to \mathcal{E}(E))) &\cong \otimes_{i-j=r} \mathcal{L}_\varepsilon(H^j(\mathcal{E}'(F)) \to H^i(\mathcal{E}(E))) \\ &\cong \otimes_{i-j=r} \mathcal{L}_b(H^j(\mathcal{E}'(F)) \to H^i(\mathcal{E}(E))) \\ &= \mathcal{L}(H(\mathcal{E}'(F)) \to H(\mathcal{E}(E)))^r, \end{aligned}$$

where the second equality follows from Theorem 7.1.1 (b) of Edwards [62] because the spaces $H_j(\mathcal{E}(F'))$ are barelled.

This completes the proof of the theorem.

□

It is useful to compare this result with Theorem 10.11 of Dold [56, Ch.VI].

1.5.8 *Examples*

As in Subsection 1.4.5 we shall limit ourselves to three more or less known examples.

Example 1.5.21 Let $E^{\cdot} = \Lambda^{\cdot} T^{*}_{\mathbb{C}}(X)$ be the de Rham complex on a differentiable manifold X of dimension n, and $F^{\cdot} = \Lambda^{\cdot} T^{*}_{\mathbb{C}}(X)$ be the de Rham complex on a differentiable manifold Y of dimension m.

If Y is oriented and we identify the bundle of volumes of Y with $\Lambda^m T^{*}_{\mathbb{C}}(Y)$, then the usual wedge-product pairing $\Lambda^{m-j} T^{*}_{\mathbb{C}}(Y) \otimes \Lambda^{j} T^{*}_{\mathbb{C}}(Y) \to \Lambda^{m} T^{*}_{\mathbb{C}}(Y)$ allows one to identify the dual bundle for $\Lambda^{j} T^{*}_{\mathbb{C}}(Y)$ with $\Lambda^{m-j} T^{*}_{\mathbb{C}}(Y)$. Note that, if Y is not oriented, then $(\Lambda^{j} T^{*}_{\mathbb{C}}(Y))'$ is the bundle of twisted $(m-j)$-forms. From Stokes' Theorem we see that $(d_j)' = (-1)^{j+1} d_{m-j-1}$. So the transposed complex for the de Rham complex $\Lambda^{\cdot} T^{*}_{\mathbb{C}}(Y)$ may be identified with itself which is what has been observed in Proposition 1.2.40.

Let us denote by $\Lambda^{(i,j)} T^{*}_{\mathbb{C}}(X \times Y)$ the bundle of double exterior forms of degree i in coordinates x and degree j in coordinates y over $X \times Y$ (see de Rham [50, §7]. We have

$$\begin{aligned}(E^{\cdot} \boxtimes F^{\cdot\prime})^{r} &= \oplus_{i-j=r} \Lambda^{i} T^{*}_{\mathbb{C}}(X) \boxtimes (\Lambda^{j} T^{*}_{\mathbb{C}}(Y))' \\ &\cong \oplus_{i-j=r} \Lambda^{i} T^{*}_{\mathbb{C}}(X) \boxtimes \Lambda^{m-j} T^{*}_{\mathbb{C}}(Y) \\ &\cong \oplus_{j=0}^{m} \Lambda^{(j+r,m-j)} T^{*}_{\mathbb{C}}(X \times Y) \\ &\cong \Lambda^{m+r} T^{*}_{\mathbb{C}}(X \times Y),\end{aligned}$$

and

$$\begin{aligned}d_{E \boxtimes F'} \big|_{\mathcal{E}(\Lambda^{(j+r,m-j)} T^{*}_{\mathbb{C}}(X \times Y))} &= d(x) + (-1)^{r+1}\, d'(y) \\ &= d(x) + (-1)^{j+r}\, d(y) \\ &= d(x,y).\end{aligned}$$

Hence the complex $(E^{\cdot} \boxtimes F^{\cdot\prime})^{\cdot}$ is the de Rham complex on $X \times Y$ except that we use the grading from $(-m)$ to n instead of the usual grading from 0 to $(n+m)$.

Let us endow the cohomology space $H^{\cdot}_{comp}(Y, \mathbb{C})$ with the topology induced by the de Rham isomorphism $H^{\cdot}(\mathcal{E}'(\Lambda\, T^{*}_{\mathbb{C}}(Y))) \cong H^{\cdot}_{comp}(y, \mathbb{C})$. See Example 4.7.3 of Godement [72, Ch.II] concerning this isomorphism.

Now Theorem 1.5.20 together with Corollary 1.5.18 imply that, if the cohomology spaces $H^{\cdot}(X, \mathbb{C})$ and $H^{\cdot}(Y, \mathbb{C})$ are Hausdorff, then

$$H^{r}(X \times Y, \mathbb{C}) \overset{top.}{\cong} \mathcal{L}(H^{\cdot}_{comp}(Y, \mathbb{C}) \to H^{\cdot}(X, \mathbb{C}))^{r} \quad (r \in \mathbb{Z}). \tag{1.5.13}$$

□

Remark 1.5.22 *In connection with the general definition of cohomology spaces $H^{\cdot}_{\Phi}(X, S)$, called the cohomology spaces of X with values in a sheaf S and supports from a family Φ, we refer to Godement [72, Ch.II, 4.4].*

Another example of a great geometric interest is the following.

Example 1.5.23 Suppose $E^\cdot = E \otimes \Lambda^{0,\cdot} T^*_{\mathbb{C}}(X)$ is the Dolbeault complex with coefficients in a holomorphic vector bundle E over a complex manifold X of dimension n.

Further let $F^\cdot = F \otimes \Lambda^{0,\cdot} T^*_{\mathbb{C}}(Y)$ be the Dolbeault complex with coefficients in a holomorphic vector bundle F over a complex manifold Y of dimension m. We always identify the bundle of volumes of Y with $\Lambda^{m,m} T^*_{\mathbb{C}}(Y)$ using the natural orientation of Y. Again the wedge-product pairing $\Lambda^{m,m-j} T^*_{\mathbb{C}}(Y) \otimes \Lambda^{0,j} T^*_{\mathbb{C}}(Y) \to \Lambda^{m,m} T^*_{\mathbb{C}}(Y)$ allows one to identify the dual bundle for $F \otimes \Lambda^{0,j} T^*_{\mathbb{C}}(Y)$ with $F^* \otimes \Lambda^{m,m-j} T^*_{\mathbb{C}}(Y)$. From Stokes' Theorem we see that

$$\left(\overline{\partial}_j^{(F)}\right)' = (-1)^{j+1}\, \overline{\partial}_{m-j-1}^{(F^* \otimes \Lambda^{m,0} T^*_{\mathbb{C}}(Y))}.$$

So the transposed complex for $F^\cdot$ may be identified with the Dolbeault complex with coefficients in the bundle $F^* \otimes \Lambda^{m,0} T^*_{\mathbb{C}}(Y)$ over Y.

The complex $(E^\cdot \boxtimes F^{\cdot\prime})^\cdot$ is the Dolbeault complex with coefficients in the holomorphic vector bundle $E \boxtimes (F^* \otimes \Lambda^{m,0} T^*_{\mathbb{C}}(Y))$ over $X \times Y$ except that we use the grading from $(-m)$ to n instead of the usual grading from 0 to $(n+m)$.

For a holomorphic vector bundle B over X, let us denote by $\mathcal{O}^{(B)}$ the sheaf of germs of holomorphic sections of B over X. By a generalization of Dolbeault's theorem (see Proposition 1.2.49) due to Serre [198], we have

$$\begin{aligned} H^\cdot(\mathcal{E}(B \otimes \Lambda^{0,\cdot} T^*_{\mathbb{C}}(X))) &\cong H^\cdot(X, \mathcal{O}^{(B)}), \\ H^\cdot(\mathcal{E}'(B \otimes \Lambda^{0,\cdot} T^*_{\mathbb{C}}(X))) &\cong H^\cdot_{comp}(X, \mathcal{O}^{(B)}). \end{aligned}$$

Endow the right-hand sides of these equalities with the induced topologies.

Now Theorem 1.5.20 together with Corollary 1.5.18 imply that, if the cohomology spaces $H^\cdot(X, \mathcal{O}^{(E)})$ and $H^\cdot(Y, \mathcal{O}^{(F^* \otimes \Lambda^{m,0} T^*_{\mathbb{C}}(Y))})$ are Hausdorff, then

$$H^r\left(X \times Y, \mathcal{O}^{(E \boxtimes (F^* \otimes \Lambda^{m,0} T^*_{\mathbb{C}}(Y)))}\right) \overset{top.}{\cong} \mathcal{L}(H^\cdot_{comp}(Y, \mathcal{O}^{(F)}) \to H^\cdot(X, \mathcal{O}^{(E)}))^r \quad (r \in \mathbb{Z}) \tag{1.5.14}$$

□

Example 1.5.24 Let X an Y be compact differentiable manifolds, and $E^\cdot$ and $F^\cdot$ be (cochain) elliptic complexes on X and Y respectively. According to Corollary 2.2.4, the spaces $H^\cdot(\mathcal{E}(E^\cdot))$ and $H_\cdot(\mathcal{E}(F^{\cdot\prime}))$ are finite-dimensional. By Lemma 5.1.7, they are Hausdorff. Moreover, $H\,(\mathcal{E}'(F^\cdot)) \overset{top.}{\cong} H^\cdot(\mathcal{D}(F^\cdot))$ in view of Corollary 2.1.15. It follows now from Theorem 1.5.20 and Corollary 1.5.18 that

$$H^r(\mathcal{E}((E^\cdot \boxtimes F^{\cdot\prime})^\cdot) \overset{top.}{\cong} \mathcal{L}(H^\cdot(\mathcal{E}(F^\cdot)) \to H^\cdot(\mathcal{E}(E^\cdot)))^r \quad (r \in \mathbb{Z}). \tag{1.5.15}$$

□

Chapter 2

Parametrices and Fundamental Solutions of Differential Complexes

2.1 Parametrices of Differential Complexes

2.1.1 *Parametrices and the homotopy formula*

Assume that X is an n-dimensional differentiable manifold that is countable at infinity and E is a differential complex on X of length N, consisting of differentiable $\mathbb{C}$-vector bundles E^i and differential operators $P_i \in \mathrm{do}_{p_i}(E^i \to E^{i+1})$.

Using the formulas of Subsection 1.5.6, we construct the differential version $\mathcal{D}'((E \boxtimes E'))$ of the complex of homomorphisms $\mathcal{L}(\mathcal{D}(E) \to \mathcal{D}'(E))$, and denote by $\widehat{P}$ its differential $d_{E\boxtimes E'}$. This yields a complex of topological vector spaces and continuous linear mappings thereof

$$\mathcal{D}'((E \boxtimes E')): \quad 0 \longrightarrow \mathcal{D}'((E \boxtimes E')^{-N}) \xrightarrow{\widehat{P}} \mathcal{D}'((E \boxtimes E')^{-N+1}) \xrightarrow{\widehat{P}} \dots \xrightarrow{\widehat{P}} \mathcal{D}'((E \boxtimes E')^{N}) \longrightarrow 0. \tag{2.1.1}$$

The section $K_{(x,y)} \in \mathcal{D}'((E \boxtimes E')^r)$ will be called a kernel of type r on $X \times X$. Every such kernel determines a continuous linear operator $K: \mathcal{D}(E) \to \mathcal{D}'(E)$ of degree r by formula (1.5.1), and $\mathcal{D}'((E \boxtimes E')^r) \overset{top}{\cong} \mathcal{L}_b(\mathcal{D}(E) \to \mathcal{D}'(E))^r$.

The connection between the operators $\widehat{P}k$ and K is very clear.

Proposition 2.1.1 *Let $K \in \mathcal{D}'((E \boxtimes E')^r)$ be a kernel of type r, then for all $f \in \mathcal{D}(E^i)$.*

$$(\widehat{P}K)f = (-1)^{r+1}KPf + PKf. \tag{2.1.2}$$

This is a special case of Proposition 1.5.17.

Corollary 2.1.2 *For any kernels* $K^j \in \mathcal{D}'((E \boxtimes E')^{r_j})$ $(j = 1, 2)$ *whose composition is well-defined, we have*

$$\hat{P}(K^1 \circ K^2) = (\hat{P}K^1) \circ K^2 + (-1)^{r_1} K^1 \circ \hat{P}(K^2).$$

Proof. One must show that for all $f \in \mathcal{D}(E^{\cdot})$

$$(\hat{P}(K^1 \circ K^2))f = (\hat{P}K^1)K^2 f + (-1)^{r_1} K^1 (\hat{P}K^2) f.$$

Indeed $K = K^1 \circ K^2$ has type $r = r_1 + r_2$, therefore by formula (2.1.2) we obtain

$$\begin{aligned}
&(\hat{P}K^1)K^2 f + (-1)^{r_1} K^1 (\hat{P}K^2) f \\
&= \left((-1)^{r_1+1} K^1 P + P K^1\right) K^2 f + (-1)^{r_1} K^1 \left((-1)^{r_2+1} K^2 P + P K^2\right) f \\
&= (\hat{P}K) f
\end{aligned}$$

which completes the proof.

□

Corollary 2.1.3 $\hat{P}\ \mathrm{tr}\,\Delta^* = 0$.

Proof. Indeed, the kernel $\mathrm{tr}\,\Delta^* \in \mathcal{D}'((E \boxtimes E')^0)$ determines the identity operator on $\mathcal{D}(E)$ (see Subsection 1.5.2), and it remains only to invoke formula (2.1.2).

□

Definition 2.1.4 *The kernel* $K \in \mathcal{D}'((E \boxtimes E')^{-1})$ *is called a parametrix of the complex* E *if for some* $S \in \mathcal{E}((E \boxtimes E')^0)$

$$\hat{P}K = \mathrm{tr}\,\Delta^* - S. \tag{2.1.3}$$

It obviously follows from (2.1.3) that $\hat{P}S = 0$, so that by Proposition 2.1.1 the operator $S \in \mathcal{L}(\mathcal{E}'(E) \to \mathcal{E}(E))^0$ is a cochain mapping of degree 0. Sometimes we refer to S as a smooth endomorphism of the complex E. We notice also, if the kernel K is properly supported, then S is properly supported too.

Proposition 2.1.5 (homotopy formula) *The kernel* $K \in \mathcal{D}'((E \boxtimes E')^{-1})$ *satisfies equation (2.1.3) if and only if the corresponding operator* K *satisfies the equation*

$$KPf + PKf = f - Sf \quad \text{for all} \quad f \in \mathcal{D}(E^{\cdot}) \tag{2 1.4}$$

Proof. This immediately follows from Proposition 2.1.1 and Subsection 1.2.5.

□

Thus, equation (2.1.4) is simply a translation of (2.1.3) into operator language. An operator $K : \mathcal{D}(E) \to \mathcal{D}'(E)$ of degree (-1) which satisfies (2.1.4) is called a parametrix of the complex E also.

It is clear that, if the kernel K is regular, then equality (2.1.4) is pointwise. If the kernel K is continuable, then (2.1.4) is valid also for sections $f \in \mathcal{E}(E^{\cdot})$, and even for $f \in \mathcal{D}'(E^{\cdot})$ when K is properly supported.

Further, the idea of a parametrix is transferred in the obvious way to chain differential complexes.

The next statement then follows from Proposition 2.1.5.

Corollary 2.1.6 *The kernel $K \in \mathcal{D}'((E \boxtimes E')^{-1})$ is a parametrix of the complex E if and only if the kernel K' is a parametrix of the transposed complex E'*

2.1.2 *Parametrices and cohomologies of differential complexes*

Let us present the most immediate consequences of the existence of a (properly supported) parametrix concerning the cohomology spaces of the differential complex.

For a single differential operator P, the existence of a properly supported continuable left parametrix K with the pseudo-locality property implies hypoellipticity. Indeed, we have $f = KPf - (KP - 1)f$ for all $f \in \mathcal{D}'(E)$ so $\operatorname{sing\,supp} f \subset \operatorname{sing\,supp} KPf$.

On the other hand, the existence of a properly supported right parametrix K leads to existence theorems for the inhomogeneous equation $Pu = f$. Indeed, we can find $u = Kf'$ by solving the equation $f' + (PK - 1)f' = f$ which is a Fredholm type equation with a smooth kernel, at least locally.

For complexes of differential operators, the existence of a parametrix no longer implies existence theorems. However, some smoothness results follow

Proposition 2.1.7 *If a complex E has a properly supported regular parametrix then the natural mappings $H(\mathcal{E}(E)) \to H(\mathcal{D}'(E))$ and $H(\mathcal{D}(E)) \to H(\mathcal{E}'(E))$ are continuous monomorphisms.*

Proof. Let K be a properly supported regular parametrix of the complex E. Then the corresponding operator S in formula (2.1.4) is properly supported too. To prove that the mapping $\imath : H(\mathcal{E}(E)) \to H(\mathcal{D}'(E))$ is injective, it is sufficient to show that, if $f \in \mathcal{E}(E^{\cdot})$ and $f = Pu$ for some section $u \in \mathcal{D}'(E^{\cdot-1})$, then also $f = Pv$ for some section $v \in \mathcal{E}(E^{\cdot-1})$. Since the operator S is properly supported, we have $SPu = PSu$ where $Su \in \mathcal{E}(E^{\cdot-1})$ So $f = P(Kf + Su)$ by formula (2 1 4), and also $Kf \in \mathcal{E}(E^{\cdot-1})$ by Proposition 1.5.7. Therefore we can set $v = Kf + Su$. The continuity of $\imath$ follows since the initial topology of $\mathcal{E}(E^{\cdot})$ is stronger than the topology induced by $\mathcal{D}'(E^{\cdot})$, and the quotient mappings $Z^{\cdot}(\mathcal{E}(E)) \to H^{\cdot}(\mathcal{E}'(E))$ and $Z^{\cdot}(\mathcal{D}'(E)) \to H^{\cdot}(\mathcal{D}'(E))$ are continuous and open. The second part of the proposition is proved analogously.

□

Proposition 2.1.8 *If a complex E has a properly supported continuable parametrix, the natural mappings $H(\mathcal{E}(E)) \to H(\mathcal{D}'(E))$ and $H(\mathcal{D}(E)) \to H(\mathcal{E}'(E))$ are continuous epimorphisms.*

Proof. Let K be a properly supported continuable parametrix of the complex E. Then the above corresponding operator S is properly supported too. To prove that the mapping $\imath : H(\mathcal{E}(E)) \to H(\mathcal{D}'(E))$ is an epimorphism, it is sufficient to show that for each section $f \in Z^{\cdot}(\mathcal{D}'(E))$ there exists a section $f' \in Z^{\cdot}(\mathcal{E}(E))$ such that $f - f' \in B^{\cdot}(\mathcal{D}'(E))$. If $f \in Z^{\cdot}(\mathcal{D}'(E))$, then by formula (2.1.4) we have $PKf = f - Sf$, and $Kf \in \mathcal{D}'(E^{\cdot-1})$ by Proposition 1.5.7. Therefore we can set

$f' = Sf$. Continuity of $\imath$ was proved in Proposition 2.1.7. The second part of the proposition can be verified analogously.

□

Theorem 2.1.9 *Assume that a complex E has a properly supported biregular parametrix. Then*

$$\begin{array}{ccc} H(\mathcal{E}(E)) & \stackrel{top}{\cong} & H(\mathcal{D}'(E)), \\ H(\mathcal{D}(E)) & \stackrel{top}{\cong} & H(\mathcal{E}'(E)) \end{array} \tag{2.1.5}$$

Proof. Let K be a properly supported biregular parametrix of the complex E, and S be the corresponding smooth endomorphism of this complex. It follows from Propositions 2.1.7 and 2.1.8 that the natural mapping $\imath : H(\mathcal{E}(E)) \to H(\mathcal{D}'(E))$ is a continuous algebraic isomorphism. To finish the proof it remains to show that the inverse mapping $\imath^{-1} : H(\mathcal{D}'(E)) \to H(\mathcal{E}(E))$ is also continuous. It is clear from the proof of Proposition 2.1.8 that $\imath^{-1}$ decomposes into the composition

$$\imath^{-1} : H(\mathcal{D}'(E)) \stackrel{\varphi^{-1}}{\longrightarrow} Z(\mathcal{D}'(E)) \stackrel{S}{\longrightarrow} Z(\mathcal{E}(E)) \stackrel{\varphi}{\longrightarrow} H(\mathcal{E}(E)),$$

where φ is the quotient mapping. Since S is continuous, and φ is continuous and open, we may conclude that $\imath^{-1}$ is continuous as required. The second equality (2.1.5) is proved by repeating such reasonings. Thus the proof is complete.

□

The condition that the parametrix be properly supported in Theorem 2.1.9 is not too restrictive. This may be understood from the following simple statement.

Proposition 2.1.10 *If $K \in \mathcal{D}'((E \boxtimes E')^{-1})$ is a parametrix of a complex E and a function $\varphi \in \mathcal{E}(X \times X)$ is equal to 1 in a neighborhood of* sing supp K, *then φK is also a parametrix of this complex.*

Proof. Indeed

$$\begin{array}{rcl} \hat{P}(\varphi K) & = & \hat{P}K - \hat{P}((1-\varphi)K) \\ & = & \operatorname{tr}\Delta^* \pmod{\mathcal{E}((E \boxtimes E')^0)}, \end{array}$$

which completes the proof.

□

Corollary 2.1.11 *Let $K \in \mathcal{D}'((E \boxtimes E')^{-1})$ be a parametrix of a complex E with* sing supp $K = \Delta$. *Then there exists a properly supported parametrix $\widetilde{K}$ of this complex coinciding with K in a neighborhood of Δ.*

Proof. By Proposition 1.1.2, we can choose a countable locally finite open covering $U = \{U_\nu\}_{\nu \in \mathbb{Z}_+}$ of X consisting of relatively compact sets.

Let $\{\varphi_\nu\}$ be a partition of unity on X corresponding to this covering by Corollary 1.1.6. Set $\varphi(x,y) = \sum' \varphi_\mu(x)\varphi_\nu(y)$, where the stroke at the summation sign means the summation is over those indices μ and ν for which $\operatorname{supp}\varphi_\mu \cap \operatorname{supp}\varphi_\nu \neq \emptyset$.

It is obvious that $\varphi(x,y) \in \mathcal{E}(X \times X)$. Since $\sum_{\mu,\nu\in\mathbb{Z}_+} \varphi_\mu(x)\varphi_\nu(y) = 1$ on $X \times X$, we have $\varphi(x,y) = 1$ in a neighborhood of Δ. Further, it easy to see that both projections $\pi_X :\ \operatorname{supp}\varphi \to X$ and $\pi_Y :\ \operatorname{supp}\varphi \to Y\ (= X)$ are proper mappings. Hence by Proposition 2.1.10 $\widetilde{K} = \varphi K$ is the desired properly supported parametrix of the complex $E^{\cdot}$.

□

2.1.3 *Formal adjoint operator*

The concept of a (formal) adjoint operator for a differential operator appears when applying Hilbert space methods to differential problems.

First of all fix a smooth positive volume form v on X, i.e. a section $v \in \mathcal{E}(X)$ that in each coordinate neighborhood U on X with coordinates $x = x(p)$ is represented in the form $\omega(x)dx$, where $\omega \in \mathcal{E}(U)$ is a positive function on U. Such section always exist (see Wells [254]). By changing the coordinates $x = x(y)$ the representation is transformed as follows

$$\omega(x)dx \mapsto \omega(x(y)) \left|\det \frac{\partial x}{\partial y}\right| dy.$$

Let E be a differentiable $\mathbb{C}$-vector bundle over X. Fix in E a Hermitian metric h, that is, an inner product $(\ ,\ )_x$ in each fiber E_x smoothly depending on $x \in X$. By Theorem 1.2 of Wells [254, Ch. III], such metrics always exist.

This allows one to define a conjugate linear isomorphism of bundles $\star :\ E \to E'$ by $\star e = (\cdot, e)_x v(x)$.

Furthermore in the space $\mathcal{D}(E)$ we can define an inner product by the formula

$$(f,g)_X = \int_X (f(x), g(x))_x v(x).$$

Assume now that $P :\ \mathcal{D}(E) \to \mathcal{E}(F)$ is some linear operator, and the bundles E and F are endowed with Hermitian structures.

An operator $P^* :\ \mathcal{D}(F) \to \mathcal{E}(E)$ is said to be a formal adjoint for the operator P if it satisfies

$$(Pf,g)_X = (f, P^*g)_X \quad \text{for all} \quad f \in \mathcal{D}(E),\ g \in \mathcal{D}(F). \tag{2.1.6}$$

Clearly, if a formal adjoint operator exists, then it is defined uniquely. It is well-known that a formal adjoint operator exists for each operator $P \in \mathrm{do}_p(E \to F)$. Moreover, we have $P^* \in \mathrm{do}_p(F \to E)$ and $\sigma(P^*) = \sigma(P)^*$ where $\sigma(P)^*$ is the adjoint mapping for $\sigma(P)(x,z) :\ E_x \to F_x$ (see Wells [254]).

We remark that if $P(x,D) = \sum_{|\alpha|\le p} P_\alpha(x)D^\alpha$ is a local representation of the differential operator P, then the local representation of P^* is

$$P^*(x,D) = \sum_{|\alpha|\le p} D^\alpha(P_\alpha^*(x) \times \cdot).$$

More generally, the formal adjoint operator exists for each pseudo-differential operator $P \in \mathrm{pdo}_p(E \to F)$ and $P^* \in \mathrm{pdo}_p(F \to E)$. For its symbol there is a formula analogous to that in the case of differential operators (see Wells [254]).

The connection between the transposed and formal adjoint operators for a pseudo-differential operator P becomes more clear by using the bundle isomorphism $\star: E \to E'$, namely

$$P^* = \star_E{}^{-1} P' \star_F .$$

2.1.4 *Parametrices of elliptic complexes. A theorem of Atiyah and Bott*

The construction of a parametrix for elliptic differential operators has been known at least since the middle of the 50-s. Constructing a parametrix by the so-called Levi method is one of the simplest applications of (classical) pseudo-differential operator's (see, for example, Shubin [201, 5.5]).

Hörmander [110] proved the existence of a parametrix for hypoelliptic differential operators having studied for this purpose a wider class of pseudo-differential operators than the classical ones (see also Shubin [201, 5.4]).

The construction of a parametrix for (the Cauchy problem for) hyperbolic equations is based on this framework too and needs the utilization of Fourier integral operators (see Hörmander [112]).

To find a parametrix for the $\overline{\partial}$-Neumann problem, much more delicate methods are required (see Iwasaki [118] and others).

The main result on the existence of a pseudo-differential parametrix for elliptic complexes on compact manifolds is due to Atiyah and Bott [19]. The next theorem gives a slightly finer version of this result for the case of an arbitrary manifold X.

Theorem 2.1.12 *Let E be an elliptic complex on X. Then for each neighborhood $\mathcal{N}$ of the diagonal in $X \times X$ there are properly supported pseudo-differential operators $R: \mathcal{D}'(E) \to \mathcal{E}(E)$ of degree 0 and $\Pi: \mathcal{D}'(E) \to \mathcal{D}'(E)$ of degree (-1) such that:*

1) the operator R has the kernel $R(x,y) \in \mathcal{E}((E \boxtimes E')^0)$ supported in $\mathcal{N}$;

2) the operator Π has the kernel $\Pi(x,y) \in \mathcal{E}((E \boxtimes E')^{-1}|_{(X\times X)\setminus\Delta})$ supported in $\mathcal{N}$ and besides, if U is a sufficiently small coordinate neighborhood on X over which the bundles E^{i-1} and E^i are trivial, the derivatives $D_x^\alpha D_y^\beta (D_x + D_y)^\gamma \Pi_i(x,y)$ in $U \times U$ have the form $\Pi_i^{(\alpha\beta\gamma)}(x,y)/|x-y|^{n-p_{i-1}+|\alpha|+|\beta|}$ for $n - p_{i-1} + |\alpha| + |\beta| > 0$ or $\Pi_i^{(\alpha\beta\gamma)}(x,y)\log|x-y|$ for $n - p_{i-1} + |\alpha| + |\beta| = 0$ where $\Pi_i^{(\alpha\beta\gamma)}(x,y)$ are $(k_{i-1} \times k_i)$-matrices of bounded functions in $U \times U$, and for $n - p_{i-1} + |\alpha| + |\beta| < 0$ these derivatives are bounded themselves, -

and

$$\Pi P f + P \Pi f = f - Rf \quad \text{for all} \quad f \in \mathcal{D}'(E^i). \tag{2.1.7}$$

Proof. Fix a smooth positive volume form v on X and a Hermitian metric h_i in each bundle E^i.

Let p be an integer satisfying the condition $p_\imath \leq p$ for all $\imath$. For every $\imath$, we choose an elliptic differential operator $Q_\imath \in \mathrm{do}_{2(p-p_\imath)}(E^\imath \to E^\imath)$ whose symbol over each point $(x,z) \in T^*(X)$ is given by a diagonal matrix (that is by multiplication by a scalar).

For example, according to Proposition 1.11 of Wells [254, Ch. III]), for every bundle $E^\imath$ there is a connection (covariant derivative) $\nabla_\imath \in \mathrm{do}_1(E^\imath \to E^\imath \otimes T^*(X))$ compatible with the Hermitian metric $h_\imath$. Then we may set $Q_\imath = (\nabla_\imath^* \nabla_\imath)^{p-p_\imath}$.

Introduce the differential operators $\widetilde{P}_\imath$ of types $E^\imath \to E^{\imath+1}$ and orders $\widetilde{p}_\imath = 2p - p_\imath$ by means of $\widetilde{P}_\imath = P_\imath Q_\imath$, and define the differential operators $\Delta_\imath \in \mathrm{do}_{2p}(E^\imath \to E^\imath)$ by the equalities

$$\Delta_\imath = \widetilde{P}_\imath^* P_\imath + P_{\imath-1} \widetilde{P}_{\imath-1}^*. \tag{2.1.8}$$

Standard arguments from linear algebra (see Wells [254, Ch.IV, §5]) show that the $\Delta_\imath$ are elliptic differential operators of order $2p$.

By Theorem 1.5.11, there exists a properly supported parametrix

$$G_\imath \in \mathrm{pdo}_{-2p}(E^\imath \to E^\imath)$$

for $\Delta_\imath$. That is

$$G_\imath \Delta_\imath = 1 \mod (\mathrm{pdo}_{-\infty}(E^\imath \to E^\imath)), \tag{2.1.9}$$
$$\Delta_\imath G_\imath = 1 \mod (\mathrm{pdo}_{-\infty}(E^\imath \to E^\imath)). \tag{2.1.10}$$

Applying Proposition 2.1.10, one can consider that the kernel $G_\imath(x,y)$ of $G_\imath$ is supported in the a priori given neighborhood $\mathcal{N}$. Since $P_\imath \circ P_{\imath-1} = 0$, it follows from (2.1.8) that

$$P_\imath \Delta_\imath = \Delta_{\imath+1} P_\imath, \tag{2.1.11}$$

i.e. Δ defines an endomorphism of the complex E .

Multiplying (2.1.11) by $G_{\imath+1}$ from the left and by $G_\imath$ from the right and taking into consideration equalities (2.1.9) and (2.1.10), we get

$$G_{\imath+1} P_\imath = P_\imath G_\imath \pmod{\mathrm{pdo}_{-\infty}(E^\imath \to E^{\imath+1})}. \tag{2.1.12}$$

Set $\Pi_\imath = G_{\imath-1} \widetilde{P}_{\imath-1}^*$, so that $\Pi_\imath \in \mathrm{pdo}_{-p_{\imath-1}}(E^\imath \to E^{\imath-1})$ are properly supported pseudo-differential operators of degree (-1) with respect to the grading of E . Therefore they define an operator $\Pi \in \mathcal{L}(\mathcal{D}'(E) \to \mathcal{D}'(E))^{-1}$.

We claim that Π is the desired parametrix of E .

Indeed, the kernel $\Pi(x,y)$ of Π is supported in $\mathcal{N}$ and by Theorem 1.5.14 it satisfies the estimates in 2) of the theorem.

Further, equalities (2.1.12) and (2.1.9) imply

$$\begin{aligned} \Pi_{\imath+1} P_\imath + P_{\imath-1} \Pi_\imath &= G_\imath \widetilde{P}_\imath^* P_\imath + P_{\imath-1} G_{\imath-1} \widetilde{P}_{\imath-1}^* \\ &= G_\imath \Delta_\imath \pmod{\mathrm{pdo}_{-\infty}(E^\imath \to E^\imath)} \\ &= 1 \pmod{\mathrm{pdo}_{-\infty}(E^\imath \to E^\imath)}. \end{aligned} \tag{2.1.13}$$

Thus formula (2.1.7) is valid with some properly supported pseudo-differential operators $R_i \in \mathrm{pdo}_{-\infty}(E^i \to E^i)$. Statement 1) of the theorem follows then by Corollary 1.5.16.

□

If the orders of all differential operators P_i are the same, one can take $p = p_i$ and $Q_i = 1$ for all i. Then $\Delta = P^*P + PP^*$ is an obvious generalization of the Laplace operator from Hodge theory. For this reason the operators (2.1.8) will usually be called the Laplacians of the complex $E^\cdot$.

Remark 2.1.13 *By analogy with (2.1.13) we could get from equations (2.1.12) and (2.1.10) that* $\tilde{\Pi} = \tilde{P}^*G$ *is also a parametrix of the complex* $E^\cdot$ *with all required properties. But the condition for the symbols of the* Q_i *implies only that* $\Pi_i = \tilde{\Pi}_i$ *modulo pseudo-differential operators having orders* $\leq -p_{i-1} - 1$ $(i \in \mathbb{Z})$.

For a single elliptic operator between two bundles over a manifold, one constructs a parametrix by patching together the kernels of local parametrices (see Shubin [201, 5.4]). This works because the local kernels are essentially unique: any two differ by a smooth kernel.

For an elliptic complex, however, this method fails because the difference of two local kernels is in general as singular as each of them. The usual way of getting around this difficulty is to reduce the problem to a single operator - the Laplacian associated to a metric. This approach was explored in Theorem 2.1.12.

For example, for the Dolbeault complex this procedure gives a parametrix that depends not only on the complex structure but also on a metric. And even though very simple and explicit local kernels are known for the complex (see Subsection 2.5.3), the same is not true for the Laplacian of a curved metric.

As Toledo and Tong [239] observed, the Čech cohomologies give a very natural way of constructing a global parametrix from local ones without introduction of an auxiliary Laplacian (see Subsection 5.3.8).

It would be interesting to construct a family of parametrices $\{\Pi^{(\varepsilon)}\}$ of the complex $E^\cdot$, depending on a parameter $\varepsilon > 0$, such that the corresponding endomorphism $R^{(\varepsilon)}$ were a regularization, i.e. the kernel of $R^{(\varepsilon)}$ were supported in the ε-neighborhood of the diagonal Δ and $\lim_{\varepsilon\to+0} R^{(\varepsilon)}f = f$ $(f \in \mathcal{D}'(E^i))$ in the weak topology of the space $\mathcal{D}'(E^i)$. This limit is then also reached in the strong topology, by the Montel property of $\mathcal{D}'(E^i)$ and Theorems 8.4.8 and 7.1.1 of Edwards [62].

For the de Rham complex, such a parametrix $\Pi^{(\varepsilon)}$ was constructed by de Rham [50] (Theorem 12), and for the Dolbeault complex with coefficients in $\Lambda^{p,0}T^*_{\mathbb{C}}(X)$ this was done by Chirka [39]. A generalization of these results to the case of Γ-manifolds (see Remark 1.1.11) was obtained in the author's thesis [225]. See also Craioveanu and Puta [42] for a particular case.

Corollary 2.1.14 *Let* $E^\cdot$ *be a compatibility complex for a sufficiently regular differential operator* P *with injective symbol on* X. *Then* $E^\cdot$ *has a properly supported pseudo-differential parametrix.*

Proof. By Theorem 1.3.24, P is equivalent to some normalized differential operator $d_0^{(F)}$ with injective symbol on X. According to Theorem 1.3.37 together with Proposition 1.3.38, there exists an elliptic compatibility complex F for $d_0^{(F)}$. Finally, Proposition 1.2.8 implies that the complexes E and F are homotopically equivalent. So, taking into account Theorem 2.1.12, we need prove only that, if one of two homotopically equivalent complexes has a parametrix, then the second complex also has a parametrix. Let $K^{(F)}$ be a parametrix for F, and let the mappings M, M^{-1}, $h^{(F)}$ and $h^{(E)}$ be given by Definition 1.1.17. We claim that the operator $K^{(E)} = MK^{(F)}M^{-1} + h^{(E)}$ is a parametrix of the complex E. Indeed, direct calculations show that $K^{(E)}d^{(E)} + d^{(E)}K^{(E)} = 1 - S^{(E)}$ where $S^{(E)} = MS^{(F)}M^{-1}$ is obviously a smooth endomorphism of the complex E.

□

Thus, Example 1.2.19 shows that non-elliptic complexes may also have properly supported pseudo-differential parametrices.

2.1.5 *Hypoellipticity of elliptic complexes*

As was pointed out in Subsection 2.1.2, the existence of a sufficiently "good" parametrix for a differential complex always implies interior regularity of cohomologies of this complex. Let us discuss some consequences of Theorem 2.1.12 from this point of view.

Corollary 2.1.15 *Let E be an elliptic complex on X. Then the inclusion $\mathcal{E}(E) \hookrightarrow \mathcal{D}'(E)$ induces the topological isomorphisms*

$$H(\mathcal{E}(E)) \overset{top}{\cong} H(\mathcal{D}'(E)), \tag{2.1.14}$$

$$H(\mathcal{D}(E)) \overset{top}{\cong} H(\mathcal{E}'(E)). \tag{2.1.15}$$

Proof. This follows from Theorems 2.1.12 and 2.1.9.

□

The first of these isomorphisms expresses the so-called property of hypoellipticity of elliptic complexes. If the sequences of sheaves $\mathcal{E}$ and $\mathcal{D}'$ are exact at the positive degrees, then this isomorphism at least in its algebraic part follows from the Abstract de Rham Theorem (see Subsection 5.2.4) and hypoellipticity of the differential operator P_0. This is precisely how hypoellipticity was proved for elliptic Hilbert complexes (see Palamodov [173, Ch.VII, §9]). In particular, in the space of solutions of a system of differential equations with injective symbol $Pf = 0$ on X, the topologies induced from $\mathcal{E}(E)$ and $\mathcal{D}'(E)$ coincide. Hörmander [113] (4.4.2) calls this an analog of the Stieltijes-Vitali Theorem.

Corollary 2.1.14 implies that the isomorphisms (2.1.14) and (2.1.15) also take place if E is a compatibility complex for a sufficiently regular differential operator P with injective symbol on X.

Corollary 2.1.16 *Let E be an elliptic complex on X. Then for each section $f \in B^i(\mathcal{D}'(E))$ there is a section $u \in \mathcal{D}'(E^{i-1})$ satisfying $Pu = f$, whose smoothness exceeds by $(p_{i-1} - 0)$ the smoothness of f.*

Proof. Suppose that $f = Pu'$ for some section $u' \in \mathcal{D}'(E^{i-1})$. It follows from formula (2.1.7) that we have also $f = Pu$ where $u = \Pi f + Ru'$. Since the operator R is smoothing, $Ru' \in \mathcal{E}(E^i)$. On the other hand, the smoothness Πf exceeds by $(p_{i-1} - 0)$ the smoothness of f because Π_i is a properly supported pseudo-differential operator of order $(-p_{i-1})$. For example, if $f \in C^{s,\lambda}_{loc}(E^i)$ where $0 < \lambda < 1$, then $\Pi f \in C^{s+p_{i-1},\lambda}_{loc}(E^{i-1})$ (see Theorem 2.5 of Taylor [236, Ch.XI]). If $f \in W^{s,q}_{loc}(E^i)$ where $1 < q < \infty$, then $\Pi f \in W^{s+p_{i-1},q}_{loc}(E^{i-1})$ (loc. cit.). So the section $u = \Pi f + Ru'$ has the required properties which completes the proof.

□

2.1.6 *Analyticity of cohomologies of elliptic complexes*

Let us now suppose that the manifold X and all objects $\{E^i, P_i\}$ are real analytic. Then we may define the complex $\mathcal{A}(E^\cdot)$ of spaces of real analytic sections and $\mathcal{H}(E^\cdot)$ of spaces of hyperfunction section on X.

The following result completes Corollary 2.1.15 very successfully.

Proposition 2.1.17 *Suppose that $E^\cdot$ is an elliptic complex on X. Then the inclusions $\mathcal{A}(E^\cdot) \hookrightarrow \mathcal{D}'(E^\cdot) \hookrightarrow \mathcal{H}(E^\cdot)$ induce the (algebraic) isomorphisms*

$$H^\cdot(\mathcal{A}(E^\cdot)) \cong H^\cdot(\mathcal{D}'(E^\cdot)), \tag{2.1.16}$$

$$H^\cdot(\mathcal{A}(E^\cdot)) \cong H^\cdot(\mathcal{H}(E^\cdot)). \tag{2.1.17}$$

Proof. We first show that the natural mapping $i: H^i(\mathcal{A}(E^\cdot)) \to H^i(\mathcal{D}'(E^\cdot))$ is an isomorphism of vector spaces.

If the manifold X is compact, then this result easily follows from Hodge theory (see Section 2.2 below). So we are allowed to assume that X has no compact connected components.

Find a (cochain) differential complex $\tilde{E}^\cdot = \{E^i, \tilde{P}_i\}$ with (real) analytic $\tilde{P}_i$ on X such that the differential operators $\Delta_i = \tilde{P}_i^* P_i + P_{i-1}\tilde{P}_{i-1}^*$ are elliptic for all i. For example, we can take an "associated" complex for E as $\tilde{E}$ (see Subsection 3.3.1).

Let us prove that i is injective. Let $f \in Z^i(\mathcal{A}(E^\cdot))$ and $f = Pu$ for some section $u \in \mathcal{D}'(E^{i-1})$. By the well-known theorem of Malgrange [156], there is a section $v \in \mathcal{D}'(E^{i-1})$ such that $u = \Delta v = \tilde{P}^* P v + P\tilde{P}^* v$. Therefore, replacing u by $\tilde{P}^* P v$, we may consider that $\tilde{P}^* u = 0$. Then we have $\Delta u = \tilde{P}^* f \in \mathcal{A}(E^{i-1})$. Since the differential operator Δ_{i-1} is real analytic and elliptic, we conclude that $u \in \mathcal{A}(E^{i-1})$. So $f \in B^i(\mathcal{A}(E^\cdot))$ as required.

We now show that the mapping i is surjective. Suppose $f \in Z^i(\mathcal{D}'(E^\cdot))$. By a theorem of Malgrange [156], there exists a section $u \in \mathcal{D}'(E^{i-1})$ such that $\Delta u = \tilde{P}^* f$. Set $f' = f - Pu$. Since $Pf = 0$, we get $\Delta f' = P\tilde{P}^* f - P\Delta u = 0$. Hence, it follows as above that $f' \in Z^i(\mathcal{A}(E^\cdot))$. Thus, each cohomology class of $H^i(\mathcal{D}'(E^\cdot))$ contains a representative from $Z^i(\mathcal{A}(E^\cdot))$ as required.

This finishes the proof of the (algebraic) isomorphism (2.1.16).

To prove isomorphism (2.1.17), we could repeat the same reasoning except that we should refer to the theorem on solvability of non-homogeneous elliptic systems, with (real) analytic coefficients, in the spaces of hyperfunction sections (see Schapira [192, Ch.II, §4]).

□

On a compact manifold X each elliptic complex $E^{\cdot}$ has a parametrix with very special properties. Information on such a parametrix will be obtained from Hodge theory.

2.2 Hodge Theory for Elliptic Complexes on Compact Manifolds

2.2.1 *Harmonic spaces*

Assume that X is an n-dimensional differentiable compact manifold, and let $E^{\cdot}$ be an elliptic complex on X of (finite) length N.

We fix a positive volume form $v \in \mathcal{E}(\Omega)$ on X and Hermitian metrics in the bundles E^i. Then in each of the vector spaces $\mathcal{D}(E^i)$ there is the scalar product $(,)_X = \int_X (,)_x v(x)$. Define the Hilbert spaces $L^2(E^i)$ as the completions of $\mathcal{D}(E^i)$ in the norm $\|.\| = (,)_X^{1/2}$.

Set $\mathcal{H}^i(E^{\cdot}) = \{f \in \mathcal{E}(E^i) : \ Pf = P^*f = 0\}$, and endow this vector space with the topology induced from $\mathcal{E}(E^i)$. The sections $f \in \mathcal{H}^i(E^{\cdot})$ are called harmonic sections of E^i.

Proposition 2.2.1 *The natural mappings $i: \ \mathcal{H}^i(E^{\cdot}) \to H^i(\mathcal{E}(E^{\cdot}))$ are continuous monomorphisms.*

Proof. Assume that $f \in \mathcal{H}^i(E^{\cdot})$ and $f = Pu$ for some section $u \in \mathcal{E}(E^{i-1})$. Then

$$(f, f)_X = (Pu, f)_X = (u, P^*f)_X = 0,$$

hence it follows that $f = 0$. The continuity of the monomorphism i follows from the definitions of the induced and quotient topologies.

□

2.2.2 *Hodge theory*

As a matter of fact i is a topological isomorphism, and the next assertion is to prove this fact.

Theorem 2.2.2 *Let $E^{\cdot}$ be an elliptic complex of finite length N on a compact manifold X. Then there are pseudo-differential operators $H: \ \mathcal{D}'(E^{\cdot}) \to \mathcal{E}(E^{\cdot})$ of degree 0 and $\Pi: \ \mathcal{D}'(E^{\cdot}) \to \mathcal{D}'(E^{\cdot})$ of degree (-1) such that:*

1) H_i is the L^2-orthogonal projection on the finite-dimensional space $\mathcal{H}^i(E^{\cdot})$ having the kernel

$$H_i(x, y) = \sum_j h_{i,j}(x) \otimes \star h_{i,j}(y) \in \mathcal{E}(E^i \boxtimes E^{i\prime}),$$

where $\{h_{ij}\}$ *is an orthonormal basis for* $\mathcal{H}^i(E^\cdot)$, *and* $PH = HP = 0$;

2) the operator Π *has a kernel* $\Pi(x,y) \in \mathcal{E}((E^\cdot \boxtimes E^{\cdot\prime})^{-1}|_{X\times X\setminus\Delta})$ *which satisfies the estimates in part 2) of Theorem 2.1.12 near the diagonal* Δ, -

and for all $f \in \mathcal{D}'(E^i)$

$$\Pi P f + P\Pi f = f - Hf. \tag{2.2.1}$$

Theorem 2.2.2 makes Lemma 2 of Kotake [137] more exact. It was proved by Toledo [238] in the case when the orders of all differential operators P_i equal 1.

Proof. As in the proof of Theorem 2.1.12, we define the Laplacians Δ_i of the complex $E^\cdot$ according to formulas (2.1.8).

But we require in addition that the differential operators Q_i are selfadjoint and positive, i.e. $Q_i^* = Q_i$ and

$$(Q_i f, f)_X > 0 \quad \text{for all} \quad \mathcal{D}(E^i) \ni f \neq 0. \tag{2.2.2}$$

For instance we can choose a Riemannian connection ∇_i for each of the bundles E^i and define

$$Q_i = (\nabla_i^* \nabla_i)^{p-p_i} + 1.$$

We need the following Lemma 1 of Kotake [137].

Lemma 2.2.3 *1) For* $f \in \mathcal{E}(E^i)$, $\Delta f = 0$ *if and only if* $f \in \mathcal{H}^i(E^\cdot)$,

2) For $f \in \mathcal{E}(E^i)$, $\Delta^* f = 0$ *if and only if* $\tilde{P} f = P^* f = 0$.

Proof. 1) The sufficiency is obvious, and we will prove the necessity. Assume that $f \in \mathcal{E}(E^i)$ and $\Delta f = 0$. Since $P_i P_{i-1} = 0$, we have

$$(\Delta f, P^* P f)_X = (QP^* P f, P^* P f)_X = 0,$$

hence it follows that $P^* P f = 0$ by condition (2.2.2), or $Pf = 0$. Consequently $PQP^* f = 0$ and

$$(PQP^* f, f)_X = (QP^* f, P^* f)_X = 0.$$

Hence it again follows that $P^* f = 0$ by condition (2.2.2). Thus, $f \in \mathcal{H}^i(E^\cdot)$ as required.

The assertion 2) is verified in the same way.

□

Part 1) of Lemma 2.2.3 together with the well-known Finiteness Theorem for elliptic differential operators on compact manifolds (see Theorem 4.8 of Wells [254, Ch.IV] for example) imply that the spaces $\mathcal{H}^i(E^\cdot)$ are finite-dimensional.

Let us continue the proof of Theorem 2.2.2. Consider for each i the differential operator $L_i = \Delta_i^* \Delta_i$ in $\mathrm{do}_{4p}(E^i \to E^i)$ which is obviously selfadjoint and elliptic.

If follows from part 1) of Lemma 2.2.3 that $Lf = 0$ for a section $f \in \mathcal{E}(E^i)$ if and only if $f \in \mathcal{H}^i(E^\cdot)$.

Therefore, by using the fundamental Decomposition Theorem for selfadjoint elliptic differential operators on compact manifolds (see Theorem 4.12 of Wells [254,

Ch.IV]) we infer that there are selfadjoint linear mappings $G_\imath : \mathcal{E}(E^\imath) \to \mathcal{E}(E^\imath)$ which satisfy

$$G_\imath L_\imath = 1 - H_\imath \quad \text{on} \quad \mathcal{E}(E^\imath). \tag{2.2.3}$$

The operator $H_\imath$ is defined in part 1) of Theorem 2.2.2. Notice that $GH = 0$ by construction.

Let now $\mathcal{G} \in \mathrm{pdo}_{-4p}(E^\imath \to E^\imath)$ be a selfadjoint parametrix for $L_\imath$ whose existence follows for instance from Theorem 1.5.11 and the selfadjointness of $L_\imath$. We have

$$\mathcal{G}_\imath L_\imath = 1 - S_\imath \tag{2.2.4}$$

for some $S_\imath \subset \mathrm{pdo}_{-\infty}(E^\imath \to E^\imath)$.

From (2.2.3) and (2.2.4) we get

$$(G_\imath - \mathcal{G}_\imath)L_\imath = -(H_\imath - S_\imath) \quad \text{on} \quad \mathcal{E}(E^\imath).$$

Applying the operator $\mathcal{G}_\imath$ to this equality from the right, we get

$$G_\imath - \mathcal{G}_\imath = (G_\imath - \mathcal{G}_\imath)S_\imath^* - (H_\imath - S_\imath)\mathcal{G}_\imath \quad \text{on} \quad \mathcal{E}(E^\imath). \tag{2.2.5}$$

Hence, $G_\imath - \mathcal{G}_\imath \in \mathcal{L}_b(\mathcal{E}'(E^\imath) \to \mathcal{E}(E^\imath))$ has a smooth kernel by Theorem 1.5.1. Thus $G_\imath \in \mathrm{pdo}_{-4p}(E^\imath \to E^\imath)$ has the same (full) symbol as $\mathcal{G}_\imath$.

Set $\Pi_\imath = G_{\imath-1}\Delta_{\imath-1}^*\tilde{P}_{\imath-1}^*$, so that $\Pi_\imath \in \mathrm{pdo}_{-p_\imath-1}(E^\imath \to E^{\imath-1})$ $(\imath \in Z)$. By Theorem 1.5.14 the kernel $\Pi_\imath(x,y)$ of the operator $\Pi_\imath$ is differentiable in $X \times X \setminus \Delta$, and near the diagonal Δ the kernel $\Pi_\imath(x,y)$ satisfies the estimates of part 2) of Theorem 2.1.12.

By equality (2.2.3) we have for each section $f \in \mathcal{E}(E^\imath)$

$$\begin{aligned} G_\imath L_\imath f &= G_\imath\Delta_\imath^*(\tilde{P}_\imath^* P_\imath + P_{\imath-1}\tilde{P}_{\imath-1}^*)f = \Pi_{\imath+1}P_\imath f \\ &+ P_{\imath-1}\Pi_\imath f + \left((G_\imath\Delta_\imath^*)P_{\imath-1} - P_{\imath-1}(G_{\imath-1}\Delta_{\imath-1}^*)\right)\tilde{P}_{\imath-1}^* f = f - H_\imath f. \end{aligned}$$

Set $T_\imath = (G_{\imath+1}\Delta_{\imath+1}^*)P_\imath - P_\imath(G_\imath\Delta_\imath^*)$ here,so that T is a pseudo-differential operator of degree 1 with respect to the grading E. Then the proof of Theorem 2.2.2 will be complete, if we show that $T_{\imath-1}\tilde{P}_{\imath-1}^* : \mathcal{E}(E^\imath) \to \mathcal{E}(E^\imath)$ is the zero operator.

According to (2.2.3) we get for $f \in \mathcal{E}(E^{\imath-1})$

$$\begin{aligned} G_{\imath-1}\Delta_{\imath-1}^*\Delta_{\imath-1}f &= f - H_{\imath-1}f, \\ G_\imath\Delta_\imath^*\Delta_\imath P_{\imath-1}f &= P_{\imath-1}f - H_\imath P_{\imath-1}f. \end{aligned}$$

Applying the differential operator $P_{\imath-1}$ to the first of these equalities, comparing the result with the second equality, and using the commutativity relationships (2.1.11), we see that $T_{\imath-1}\Delta_{\imath-1} : \mathcal{E}(E^{\imath-1}) \to \mathcal{E}(E^\imath)$ is the zero operator.

But then $\Delta_{\imath-1}^*T_{\imath-1}^* : \mathcal{E}(E^\imath) \to \mathcal{E}(E^{\imath-1})$ is the zero operator too. By part 2) of Lemma 2.2.3 we get that $\tilde{P}_{\imath-1}T_{\imath-1}^* = 0$, or $T_{\imath-1}\tilde{P}_{\imath-1}^* = 0$ as required.

Thus we have proved formula (2.2.1) for section $f \in \mathcal{E}(E^\imath)$. But since the kernel Π is continuable, then by the continuity formula, (2.2.1) is also valid for sections $f \in \mathcal{D}(E^\imath)$. Thus the theorem is proved.

□

Corollary 2.2.4 *If $E^\cdot$ is an elliptic complex on a compact manifold X, then*

$$H^\cdot(\mathcal{E}(E^\cdot)) \overset{top}{\cong} \mathcal{H}^\cdot(E^\cdot). \qquad (2.2.6)$$

In particular the cohomology spaces $H^i(\mathcal{E}(E^\cdot))$ are finite-dimensional.

Proof. It follows immediately from formula (2.2.1) that the natural mapping $\imath: \mathcal{H}^\cdot(E^\cdot) \to H^\cdot(\mathcal{E}(E^\cdot))$ is an epimorphism. Hence by Proposition 2.2.1 it is sufficient to prove only the continuity of the inverse mapping $\imath^{-1}$. But (2.2.1) implies that $\imath^{-1}$ may be decomposed as

$$\imath^{-1}: H^\cdot(\mathcal{E}(E^\cdot)) \xrightarrow{\varphi^{-1}} Z^\cdot(\mathcal{E}(E^\cdot)) \xrightarrow{H} \mathcal{H}^\cdot(E^\cdot)$$

where φ is the quotient mapping. Hence $\imath^{-1}$ is continuous because φ is open and H is continuous.

□

Corollary 2.2.5 *If $E^\cdot$ is a compatibility complex for a sufficiently regular differential operator P_0 with injective symbol on X, then the cohomology spaces $H^i(\mathcal{E}(E^\cdot))$ ($i \in \mathbb{Z}$) are finite-dimensional.*

Proof. Indeed, it follows from Theorem 1.3.24, 1.3.37 and Propositions 1.3.38 and 1.2.8 that the complex $E^\cdot$ is homotopically equivalent to some elliptic complex of first order differential operators on X. To complete the proof use Corollaries 2.2.4 and 1.1.14 and Proposition 1.1.18.

□

2.2.3 *Orthogonal decompositions*

If the orders of all the differential operators P_i in the complex $E^\cdot$ are the same, then Theorem 2.2.2 can be made more exact. The following statement is rather a corollary of Theorem 2.2.2.

Corollary 2.2.6 (Hodge theorem) *Let $E^\cdot$ be an elliptic complex of differential operators of the same order p on a compact manifold X. Then there are selfadjoint pseudo-differential operators $H: \mathcal{E}(E^\cdot) \to \mathcal{E}(E^\cdot)$ and $G: \mathcal{E}(E^\cdot) \to \mathcal{E}(E^\cdot)$ of degree 0 such that:*

1) H is the L^2-orthogonal projection onto the finite-dimensional space $\mathcal{H}^\cdot(E^\cdot)$, and it has the kernel

$$H(x,y) = \sum_{ij} h_{ij}(x) \otimes \star h_{ij}(y) \in \mathcal{E}((E^\cdot \boxtimes E^{\cdot\prime})^0),$$

where $\{h_{ij}\}$ is an orthonormal basis for $\mathcal{H}^i(E^\cdot)$, and $PH = HP = 0$;

2) the Green operator G has the kernel $G(x,y) \in \mathcal{E}\left((E^\cdot \boxtimes E^{\cdot\prime})^0|_{X\times X\setminus\Delta}\right)$ which satisfies the estimates of part 2) of Theorem 2.1.12 with $2p$ instead of p_{i-1} near the diagonal and furthermore $PG = GP$ and $HG = GH = 0$.

This yields the L^2-orthogonal decomposition

$$f = Hf + P^*PGf + PP^*Gf \quad \text{for all} \quad f \in \mathcal{E}(E^i). \qquad (2.2.7)$$

Proof. It is sufficient to take $Q_i = 1$, $L_i = \Delta_i$, and repeat the proof of Theorem 2.2.2 with slight modifications.

□

For the de Rham complex on X this result was proved by Hodge [107] (see also Theorem 23 of de Rham [50]). For the Dolbeault complex, Corollary 2.2.6 is due to Kodaira [124]. A less complete statement is contained in the book of Wells [254] (see Theorem 5.2 of Ch.IV).

The reader could obtain useful statements from Corollary 2.2.6 like the Corollaries 1-3 of Theorem 23 from the book of de Rham [50].

In conclusion we make a remark on orthogonal decompositions (2.2.7). Cantor [35] proved the following result.

Theorem 2.2.7 *Assume that E and F are Hermitian vector bundles over a compact manifold X, and $P \in \mathrm{do}_p(E \to F)$ is an elliptic differential operator. Then, for all integer $s \geq 0$ and real q $(1 < q < \infty)$, we have the L^2-orthogonal decomposition*

$$W^{s,q}(F) = \ker\, P^* \overset{top.}{\oplus} PW^{s+p,q}(E). \tag{2.2.8}$$

We recall that the Sobolev spaces $W^{s,q}(E)$ have been introduced in Subsection 1.1.7.

It is not difficult from this theorem to get decompositions of the type (2.2.7) in classes of Sobolev sections for an elliptic complex $E^{\cdot}$. The reader can go into the details himself.

We note also the paper of Skrypnik [202] which gives a further development of Hodge theory for elliptic Koszul complexes on manifolds.

2.3 Fundamental Solutions of Differential Complexes

2.3.1 *Fundamental solutions*

Let us come back to the general situation. That is, let $E^{\cdot}$ be a (cochain) differential complex on a differentiable manifold X.

It follows from the homotopy formula (2.1.4) that, for applications, those parametrices K of the complex $E^{\cdot}$ are most important whose smooth "corrections" S are zero.

Definition 2.3.1 *A kernel $K(x,y) \in \mathcal{D}'((E^{\cdot} \boxtimes E^{\cdot\prime})^{-1})$ is said to be a fundamental solution of the complex $E^{\cdot}$ if it satisfies the fundamental equation*

$$\hat{P}K = \mathrm{tr}\,\Delta^*. \tag{2.3.1}$$

Corollary 2.1.3 implies that, if $H^0(\mathcal{D}'(E^{\cdot} \boxtimes E^{\cdot\prime})^{\cdot}) = 0$, then the complex $E^{\cdot}$ has a fundamental solution.

Moreover, for elliptic complexes on compact manifolds this condition is necessary and sufficient at the same time. Indeed, as far as the necessity is concerned, we have by Corollary 2.1.15 and Example 1.5.24 that $H^0(\mathcal{D}'((E^{\cdot} \boxtimes E^{\cdot\prime})^{\cdot})) \cong \mathcal{L}(H^{\cdot}(\mathcal{E}(E^{\cdot})) \to H^{\cdot}(\mathcal{E}(E^{\cdot})))^0$. To get the desired result from here, it suffices to invoke Proposition 2.3.11.

Proposition 2.3.2 *Let $E^{\cdot}$ be an elliptic complex on X and Π be its parametrix given by Theorem 2.1.12. Then each fundamental solution K of $E^{\cdot}$ has the form*

$$K = (\Pi + \Pi R + RKR) - \hat{P}(\Pi\Pi + RK\Pi - \Pi K).$$

Proof. Let us use Corollary 2.1.2, setting $K^1 = K$ and $K^2 = \Pi$. Since Π is a properly supported pseudo-differential operator, all required compositions are well-defined. Then (2.1.3) and (2.3.1) imply

$$\hat{P}(K\Pi) = (\hat{P}K) \circ \Pi - K \circ (\hat{P}\Pi) = (\mathrm{tr}\,\Delta^*) \circ \Pi - K \circ (tr\Delta^* - R).$$

So we get $K = \Pi + K \circ R + \hat{P}(K \circ \Pi)$ where the kernel $K \circ R$ is obviously continuable.

If K^1 and K^2 exchange roles, we obtain by analogy that $K = \Pi + R \circ K + \hat{P}(\Pi \circ K)$ where the kernel $R \circ K$ is obviously regular.

Substitute the last expression for K into the right-hand side of the first formula. Then after easy calculations we come to the required formula for K.

□

Proposition 2.3.3 *If an elliptic complex $E^{\cdot}$ has some fundamental solution, then it has also a pseudo-differential fundamental solution Φ with the regularity properties, as in Theorem 2.1.12(2).*

Proof. Let K be a some fundamental solution of $E^{\cdot}$. Since $\hat{P}^2 = 0$, it follows from Proposition 2.3.2 that the kernel $\Phi = \Pi + \Pi \circ R + R \circ K \circ R$ is also a fundamental solution of this complex. But Theorem 1.5.1 implies that the difference $(\Phi - \Pi)$ is a smoothing operator. So Φ is a pseudo-differential operator with the same full symbol as Π.

□

We are also in need of an operator translation of the fundamental equation (2.3.1).

Proposition 2.3.4 (homotopy formula) *A kernel $K(x,y) \in \mathcal{D}'((E^{\cdot} \boxtimes E^{\cdot\prime})^{-1})$ is a fundamental solution of the complex $E^{\cdot}$ if and only if*

$$KPf + PKf = f \quad \textit{for all} \quad f \in \mathcal{D}(E^{\cdot}). \tag{2.3.2}$$

Proof. This follows from Proposition 2.1.5.

□

For homogeneous components, equations (2.3.1) and (2.3.2) have the forms

$$P_{i-1}(x)K_i(x,y) + P_i'(y)K_{i+1}(x,y) = \mathrm{tr}\,\Delta^* \quad (i \in \mathbb{Z}); \tag{2.3.3}$$

$$P_{i-1}K_i + K_{i+1}P_i = 1_{\mathcal{D}(E^i)} \quad (i \in \mathbb{Z}). \tag{2.3.4}$$

2.3.2 *Existence of local fundamental solutions*

Following Toledo [238], we shall say that a (cochain) differential complex E is sheaf exact at degree q if the sequence of sheaves $\tilde{E}$ is exact at this degree.

Notice this means that given any point $x \in X$ and a section $f \in Z^q(\mathcal{E}(E|_U))$ where U is a neighborhood of x, there exist a possibly smaller neighborhood U' of x and a section $u \in \mathcal{E}(E^{q-1}|_{U'})$ such that $Pu = f$ in U'.

Proposition 2.3.5 *Let E be an elliptic complex on a compact manifold X. If E is sheaf exact at positive degrees, then for each point $x \in X$ there is a neighborhood U such that the complex $E|_U$ has a fundamental solution.*

A particular case of this proposition was proved by Toledo [238]

To prove Proposition 2.3.5, we require the following simple statement.

Lemma 2.3.6 *Assume $P \in \mathrm{do}_p(E \to F)$ is a differential operator with surjective symbol on a (not necessary compact) manifold X. Then the sequence of sheaves $\tilde{E} \xrightarrow{P} \tilde{F} \to 0$ is exact.*

Proof. Exactness of the sequence $\tilde{E} \xrightarrow{P} \tilde{F} \to 0$ means that given any point $x \in X$ and a section $f \in \mathcal{E}(F|_U)$ where U is a neighborhood of x, there is a possibly smaller neighborhood U' of x and a section $u \in \mathcal{E}(E|_{U'})$ such that $Pu = f$ in U'.

As by the proof of Theorem 2.1.12, we define the Laplacian $\Delta = PP^*$. It is clear that Δ is an elliptic differential operator of type $F \to F$ and order $(2p)$ on X.

Use now Theorem 13.3.3 of Hörmander [114], according to which each elliptic differential operator on X has locally a (bilateral) fundamental solution. Precisely, for each point $x \in X$ there exists a neighborhood U' and a biregular operator $G: \mathcal{E}'(F) \to \mathcal{E}'(F)$ such that

$$\begin{aligned} G\Delta f &= f \quad \text{in} \quad U' \quad \text{for all} \quad f \in \mathcal{E}'(F|_{U'}), \\ \Delta G f &= f \quad \text{in} \quad U' \quad \text{for all} \quad f \in \mathcal{E}'(F). \end{aligned} \tag{2.3.5}$$

We can assume $U' \subset\subset U$, and let us choose a function $\varphi \in \mathcal{D}(U)$ such that $\varphi \equiv 1$ in U'. Then for a section $f \in \mathcal{E}(F|_U)$ we have $(\varphi f) \in \mathcal{D}(F|_U)$. Set $u = P^* G(\varphi f)$, so that $u \in \mathcal{D}(F)$. It follows from the second equation (2.3.5) that $Pu = f$ in U'. □

By Lemma 2.3.6, if E is an elliptic complex of length N on a differentiable manifold X, then E is sheaf exact at degree N. Sheaf exactness of E at degrees $< N$ is a very hard problem. For more information on these questions see Section 1.3.

Proof of Proposition 2.3.5. Let Π be the parametrix of E given by Theorem 2.2.2.

It follows from the condition of this proposition and Lemma 2.3.6 that for each point $x \in X$ there exists a neighborhood U, section $m_{i-1}^{(\nu)} \in \mathcal{E}(E^{i-1}|_U)$, and $t_1^{(\nu)} \in \mathcal{E}\left(E^{1'}|_U\right)$ such that $Pm_{i-1}^{(\nu)} = h_i^{(\nu)}$ $(i \geq 1)$ and $P't_1^{(\nu)} = \star h_0^{(\nu)}$ in U.

Let us define the kernel $M \in \mathcal{E}\left((E^{\cdot} \boxtimes E^{\cdot\prime})^{-1}|_{U\times U}\right)$ by the formulas

$$\begin{aligned} M_1(x,y) &= \sum_\nu m_0^{(\nu)}(x) \otimes \star h_1^{(\nu)}(y) + \sum_\nu h_0^{(\nu)}(x) \otimes t_1^{(\nu)}(y), \\ M_i(x,y) &= \sum_\nu m_i^{(\nu)}(x) \otimes \star h_i^{(\nu)}(y) \quad (i \neq 1). \end{aligned}$$

It is easy to see that $\hat{P}M = H$ on $U \times U$, or equivalently

$$P_{i-1}(x)M_i(x,y) + P_i'(y)M_{i+1}(x,y) = H_i(x,y) \qquad (i \in \mathbb{Z}).$$

Therefore, the kernel $\Phi^{loc} = \Pi + M$ is a fundamental solution of the complex $E^{\cdot}|_U$, which was to be proved.

□

More complicated reasonings, like those which will be used in the proof of Theorem 2.3.12, show that Proposition 2.3.5 is also valid without the assumption that X be compact.

Remark 2.3.7 *The elliptic complex $E^{\cdot}$ in $\mathbb{R}^n$ consisting of the bundles $E^i = \Lambda^i T^*_{\mathbb{C}}(\mathbb{R}^n)$ and differential operators $P_i = \Delta^i d_i \in \mathrm{do}_{2i+1}(E^i \to E^{i+1})$, where Δ is the Laplace operator in $\mathbb{R}^n$, has a fundamental solution; however it is not sheaf exact at any degree $0 \leq q < n$.*

2.3.3 *Fundamental solution at degree q. A generalization of a theorem of Malgrange*

If E is a differential complex of length $N = 1$, then a fundamental solution of $E^{\cdot}$ is simply a bilateral fundamental solution of the differential operator $P = P_0$. It is well-known that locally for each elliptic differential operator P such a fundamental solution exists (see Theorem 13.3.3 of Hörmander [114]). An obstruction to existence of a global fundamental solution for P is non-uniqueness of solutions of the local Cauchy problem for the differential operator P itself or its transposed operator P' (see Theorem 13.6.15 therein). By this simple example we have already seen that it is convenient to distinguish the notions of a left fundamental solution and a right one. More generally, formulas (2.3.3) and (2.3.4) allow one to define fundamental solutions for a complex $E^{\cdot}$ at degree q.

Definition 2.3.8 *A parametrix $K \in \mathcal{D}'((E^{\cdot} \boxtimes E^{\cdot\prime})^{-1})$ of the complex $E^{\cdot}$ is said to be a fundamental solution of $E^{\cdot}$ at degree q if K satisfies equation (2.3.3) for $i = q$.*

In other words, K is a fundamental solution of $E^{\cdot}$ at degree q if $\hat{P}K = \mathrm{tr}\,\Delta^* - S$ where the kernel $S \in \mathcal{E}((E^{\cdot} \boxtimes E^{\cdot\prime})^0)$ has zero homogeneous component at degree q, i.e. $S_q = 0$.

Propositions 2.3.3 and 2.3.5 can be carried over with obvious modifications for fundamental solutions at degree q.

Proposition 2.3.9 *If an elliptic complex $E^\cdot$ has some fundamental solution at degree q, then it has also a pseudo-differential fundamental solution Φ at degree q with regularity properties, as in Theorem 2.1.12 (n.2).*

Proof. Indeed, a delicate but quite elementary analysis of the proof of Proposition 2.3.3 shows that the parametrix $\Phi = \Pi + \Pi \circ R + R \circ K \circ R$ constructed there is a fundamental solution of the complex $E^\cdot$ at degree q together with K.

□

Proposition 2.3.10 *Let $E^\cdot$ be an elliptic complex on a compact manifold X. If $E^\cdot$ is sheaf exact at degree q, then for each point $x \in X$ there is a neighborhood U such that the complex $E^\cdot|_U$ has a fundamental solution at degree q.*

Proof. The proof is an obvious modification of the proof of Proposition 2.3.5.

□

Proposition 2.3.11 *An elliptic complex $E^\cdot$ on a compact manifold X has a fundamental solution at degree q if and only if $H^q(\mathcal{E}(E^\cdot)) = 0$.*

Proof. Necessity follows immediately from Proposition 2.3.9 and the homotopy formula (2.3.4) for $i = q$. Sufficiency is an easy consequence of formula (2.2.1) and Corollary 2.2.4.

□

The condition of triviality of the cohomology group $H^q(\mathcal{E}(E^\cdot))$ is not necessary in general (see Remark 2.3.7). As regards sufficiency, there is the following result.

Taking into account known geometrical conditions on the manifold X which are sufficient for the cohomology groups $H^i(\mathcal{E}(E^\cdot))$ to be trivial for certain complexes $E^\cdot$ (see Palamodov [175, Ch.VII], and others), we say that X is q-convex with respect to $E^\cdot$ if $H^i(\mathcal{E}(E^\cdot)) = 0$ for all $i \geq q$.

Theorem 2.3.12 *Let $E^\cdot$ be a differential complex on X having a parametrix K. If the manifold X is q-convex with respect to $E^\cdot$ for some $q \leq N$, then $E^\cdot$ has a fundamental solution at degrees ($\geq q$) of the form $K^{(q)} = K + M^{(q)}$, where $M^{(q)} \in \mathcal{E}((E^\cdot \boxtimes E^{\cdot\prime})^{-1})$.*

Proof. Let us write equation (2.1.3) for components of the parametrix K:

$$P_{i-1}(x)\, K_i(x,y) + P_i'(y)\, K_{i+1}(x,y) = \operatorname{tr} \Delta^* - S_i(x,y) \quad (i = 0, \dots, N). \qquad (2.3.6)$$

Hence it follows that the kernels $S_i \in \mathcal{E}(E^i \boxtimes E^{i\prime})$ satisfy the system

$$P_i(x)\, S_i(x,y) - P_i'(y)\, S_{i+1}(x,y) = 0 \quad (i = 0, \dots, N-1). \qquad (2.3.7)$$

In our case $P_i = 0$ for $i \geq N$, and $K_i = S_i = 0$ for $i > N$.

The idea of the proof consists of successively solving system (2.3.6) outside the diagonal Δ in the class of smooth kernels on $X \times X$, in order to get rid of the errors S_i by correcting K_i. We start with $i = N$.

Consider the complex $\mathcal{E}(E \boxtimes E^{N'})$ endowed with the differential $P(x)$. Since $H^N(\mathcal{E}(E)) = 0$, we get by Theorem 1.4.9 that $H^N(\mathcal{E}(E \boxtimes E^{N'})) = 0$. So for the kernel $S_N \in \mathcal{E}(E^N \boxtimes E^{N'})$ we can find a kernel $M_N \in \mathcal{E}(E^{N-1} \boxtimes E^{N'})$ such that $P_{N-1}(x)M_N = S_N$.

Define now the kernel $K^{(N)} \in \mathcal{D}'((E \boxtimes E')^{-1})$ by the formulas

$$\begin{aligned} K_i^{(N)}(x,y) &= K_i(x,y) \quad (i \neq N), \\ K_N^{(N)}(x,y) &= K_N(x,y) + M_N(r,y). \end{aligned} \tag{2.3.8}$$

It is clear that $K^{(N)}$ is a parametrix of the complex E together with K, that is $\hat{P}K^{(N)} = \operatorname{tr}\Delta^* - S^{(N)}$ where $S^{(N)} \in \mathcal{E}((E \boxtimes E')^0)$. Moreover, it follows from (2.3.8) that $S_N^{(N)} = 0$, so $K^{(N)}$ is a fundamental solution of the complex E at degree N. Further, from equation (2.3.7) for $i = N-1$ we get that $P_{N-1}(x)S_{N-1}^{(N)} = 0$.

Consider the complex $\mathcal{E}(E \boxtimes E^{N-1'})$ endowed with the differential $P(x)$. Since $H^{N-1}(\mathcal{E}(E)) = 0$, we get by Theorem 1.4.9 that $H^{N-1}(\mathcal{E}(E \boxtimes E^{N-1'})) = 0$. Therefore, for the kernel $S_{N-1}^{(N)} \in Z^{N-1}(\mathcal{E}(E \boxtimes E^{N-1'}))$ there exists a kernel $M_{N-1} \in \mathcal{E}(E^{N-2} \boxtimes E^{N-1'})$ such that $P_{N-2}(x)M_{N-1} = S_{N-1}^{(N)}$ everywhere on $X \times X$.

Define the kernel $K^{(N-1)} \in \mathcal{D}'((E \boxtimes E')^{-1})$ by means of the equalities

$$\begin{aligned} K_i^{(N-1)}(x,y) &= K_i^{(N)}(x,y) \quad (i \neq N-1), \\ K_{N-1}^{(N-1)}(x,y) &= K_{N-1}^{(N)}(x,y) + M_{N-1}(x,y). \end{aligned} \tag{2.3.9}$$

We again notice that $K^{(N-1)}$ is a parametrix of the complex E together with K. That is, $\hat{P}K^{(N-1)} = \operatorname{tr}\Delta^* - S^{(N-1)}$ where $S^{(N-1)} \in \mathcal{E}((E \boxtimes E')^0)$. Besides, (2.3.9) implies that $S_i^{(N-1)} = 0$ for $i \geq N-1$. In other words, $K^{(N-1)}$ is a fundamental solution of $E^\cdot$ at degrees $\geq (N-1)$

Continuing this process by induction, in $(N-q+1)$ steps we shall obtain the desired fundamental solution $K^{(q)}$ at degrees $\geq q$, which completes the proof.

□

Corollary 2.3.13 *Let E be a differential complex on X having a parametrix K If $H^i(\mathcal{E}(E')) = 0$ for all $i \leq q$ with some integer $q \geq 0$, then E has a fundamental solution at degrees $(\leq q)$ of the form $K^{(q)} = K + M^{(q)}$, where $M^{(q)} \in \mathcal{E}((E \boxtimes E')^{-1})$.*

Proof. It is sufficient to apply Theorem 2.3.12 for the transposed complex E' because by Corollary 2.1.6 the kernel K is a fundamental solution of a complex E at degree q if and only if the transposed kernel K' is a fundamental solution of the complex E' at dimension q.

□

For $q = 0$, the conditions of Theorem 2.3.12 are usually not fulfilled, at least if the manifold X is not compact. So it gives no conditions for existence of a fundamental solution on the whole manifold. The same is true concerning Corollary 2.3.13. But we can combine these two results.

Corollary 2.3.14 *Let E be a differential complex on X having parametrix K. If the manifold X is 1-convex with respect to E and $H_0(\mathcal{E}(E')) = 0$, then E has a fundamental solution of the form $K^{(0)} = K + M^{(0)}$, where $M^{(0)} \in \mathcal{E}((E \boxtimes E')^0)$.*

Proof. Assume that $K^{(1)}$ is the fundamental solution of the complex E at the positive degrees which is given by Theorem 2.3.12.

We have $\hat{P}K^{(1)} = \operatorname{tr}\Delta^* - S^{(1)}$ where $S^{(1)} \in \mathcal{E}((E \boxtimes E')^0)$ and $S_i^{(1)} = 0$ for all $i \geq 1$. Hence it follows from equation (2.3.7) for $i = 0$ that $P_0(x)S_0^{(1)}(x,y) = 0$.

Consider the (chain) complex L on $X \times X$ endowed with the differential $P'(y)$ where L_i is the subspace of $\mathcal{E}(E^0 \boxtimes E^{i\prime})$ formed by all sections f which satisfy $P_0(x)f = 0$. Since $H_0(\mathcal{E}(E')) = 0$, we get by Theorem 1.4.9 that $H_0(L) = 0$. Therefore, for the kernel $S_0^{(1)} \in L_0$ there exists a kernel $\widetilde{M}_1 \in L_1$ such that $P_0'(y)\widetilde{M}_1 = S_0^{(1)}$. We have $\widetilde{M}_1 \in \mathcal{E}(E^0 \boxtimes E^{1\prime})$ and $P_0(x)\widetilde{M}_1 = 0$.

Define the kernel $K^{(0)} \in \mathcal{D}'((E \boxtimes E')^{-1})$ by the formulas

$$\begin{aligned} K_i^{(0)}(x,y) &= K_i^{(1)}(x,y) \quad (i \neq 1), \\ K_1^{(0)}(x,y) &= K_1^{(1)}(x,y) + \widetilde{M}_1(x,y). \end{aligned}$$

It is easy to see that $K^{(0)}$ is the desired fundamental solution of the complex E, which completes the proof.

□

Example 2.3.15 Let X have no compact connected components and P be a sufficiently regular differential operator of type $E \to F$ with injective symbol on X. We assume that objects X and P are (real) analytic. The formal theory of Subsection 1.3.9 implies that the differential operator P has a compatibility complex E in the (real) analytic category. By Corollary 2.1.14, there exists a pseudo-differential parametrix Π for E. Further, it follows from a well-known theorem of Malgrange [156, p.341] that the mapping $\Delta' : \mathcal{E}(E') \to \mathcal{E}(E')$ is surjective, where $\Delta = P^*P$. Therefore, the mapping $P' : \mathcal{E}(F') \to \mathcal{E}(E')$ is also surjective. Thus, $H_0(\mathcal{E}(E')) = 0$. We have seen that the conditions of Corollary 2.3.13 are fulfilled for $q = 0$. According to this corollary, the complex E has a fundamental solution at degree 0 of the form $\Phi = \Pi + M$, where $M \in \mathcal{E}((E \boxtimes E')^{-1})$. In particular, the component Φ_1 gives a special left fundamental solution of the differential operator P. This is an extension of the well-known result of Malgrange [156, p.349]. Similarly, the conditions of Corollary 2.3.14 are fulfilled for $N = 1$, i.e. if P is elliptic. So this corollary guarantees existence of a bilateral fundamental solution for elliptic differential operators P with (real) analytic coefficients on X. This result is due to Malgrange [156, p.349].

□

We present also an extension of Proposition 2.3.11 to the case of non-elliptic complexes E.

Proposition 2.3.16 *For a differential complex E on a compact manifold X to have a regular (respectively, continuable) fundamental solution, it is necessary and sufficient that the following conditions should be fulfilled: 1) the complex $\mathcal{E}(E)$ (respectively $\mathcal{E}'(E)$) is exact and 2) for every i, the subspace $Z^i(\mathcal{E}(E))$ (respectively $Z^i(\mathcal{E}'(E^i))$) has a topological complement in $\mathcal{E}(E^i)$ (respectively $\mathcal{E}'(E^i)$).*

Proof. This easily follows from Propositions 1.1.15 and 1.5.2.

□

Corollary 2.3.14 together with Theorems 2.1.12 and 1.2.21 imply that each elliptic Hilbert complex in $\mathbb{R}^n$ has a pseudo-differential fundamental solution. Our next purpose is to generalize this result.

2.3.4 *Kernels of convolution type*

Let $X = \mathbb{R}^n$. We fix the volume form $v = dx$ on X where $dx = dx_1 \cdot \ldots \cdot dx_n$.

Given (trivial) vector bundles $E = \mathbb{R}^n \times \mathbb{C}^k$ and $F = \mathbb{R}^n \times \mathbb{C}^l$ over $\mathbb{R}^n$, we endow each of the fibers $\mathbb{C}^k$ and $\mathbb{C}^l$ with the standard Hermitian inner product. This defines the Hermitian metrics on E and F.

Sections of E are interpreted as k-columns, and sections of $E^* = \mathbb{R}^n \times (\mathbb{C}^k)^*$ are interpreted as k-rows of complex-valued functions.

Since there is the canonical isomorphism $E^* \to E'$ given by $g \in E^* \to (g\,dx) \in E'$, it will be more convenient to use the algebraic adjoint bundle E^* instead of the dual bundle E'.

The natural pairing $\mathrm{tr} = \langle , \rangle_x$ of $E^*_x \otimes E_x \to \mathbb{C}$ determines the pairing $\mathrm{tr} = \langle , \rangle_X$ of $\mathfrak{C}(E^*) \otimes \mathfrak{C}'(E) \to \mathbb{C}$ by

$$\langle , \rangle_X = \int_X \langle , \rangle_x dx.$$

Sections $K \in \mathfrak{C}(E \otimes F^*)$ may be considered as $(k \times l)$-matrices of functions of class $\mathfrak{C}$ on X, and sections $K \in \mathfrak{C}(E \boxtimes F^*)$ may be interpreted as $(k \times l)$-matrices of functions of the same class on $X \times X$.

In this situation the isomorphism of Theorem 1.5.1 becomes

$$\mathfrak{C}(E \boxtimes F^*) \cong \mathcal{L}_b(\mathfrak{C}'(F) \to \mathfrak{C}(E))$$

where the case $\mathfrak{C} = \mathcal{D}$ requires the usual remark.

Definition 2.3.17 *Given a section $K(x) \in \mathcal{D}'(E \otimes F^*)$, the kernel $K(x-y) \in \mathcal{D}'(E \boxtimes F^*)$ is said to be a kernel of convolution type associated to $K(x)$.*

The corresponding operator $K : \mathcal{D}(F) \to \mathcal{D}'(E)$ is seen to be a convolution operator, namely $Kf = K * f$ for $f \in \mathcal{D}(F)$. Hence it follows that kernels of convolution type are biregular (see Hörmander [113, 1.3]).

One can easily conclude from the relationship (1.5.1) between kernels and operators that an operator $K : \mathcal{D}(F) \to \mathcal{D}'(E)$ is an operator of convolution with some section $K \in \mathcal{D}'(E \otimes F^*)$ if and only if it is *invariant under translations*, that is, $Kt^*_x = t^*_x K$ for all $x \in \mathbb{R}^n$, where t^*_x is the operator of converse image under the mapping $t_x : \mathbb{R}^n \to \mathbb{R}^n$ given by $t_x(y) = y + x$.

Notice that, if $K \in \mathcal{D}'(E \boxtimes F^*)$ is a kernel of convolution type associated to $K(x) \in \mathcal{D}'(E \otimes F^*)$, then K' is a kernel of convolution type associated to $K'(x) \in \mathcal{D}'(E \otimes F^*)$ where $K'(x) = (K(-x))^t$.

2.3.5 Fundamental solutions of convolution type of complexes of differential operators with constant coefficients in $\mathbb{R}^n$

Let a (cochain) differential complex E be given in $\mathbb{R}^n$ consisting of (trivial) bundles $E^i = \mathbb{R}^n \times \mathbb{C}^{k_i}$ and differential operators $P_i \in \mathrm{do}_{p_i}(E^i \to E^{i+1})$ with constant coefficients.

In this case it would be natural to seek a fundamental solution for E of convolution type, i.e. in the form $Kf = K * f$ where $K(x) \in \oplus \mathcal{D}'(E^{i-1} \otimes E^{i*})$. For this purpose we need a suitable translation of fundamental equation (2.3.1) which will be given by Proposition 2.3.18.

Let us denote by $\delta(= \delta_0)$ the value functional (of Dirac) in $\mathbb{R}^n$, and by I_k the unit $(k \times k)$-matrix.

Proposition 2.3.18 *A kernel $K \in \mathcal{D}'((E \boxtimes E^*)^{-1})$ of convolution type associated to $K(x) \in \oplus\mathcal{D}'(E^{i-1} \otimes E^{i*})$ is a fundamental solution of the complex E if and only if*

$$P(D)K(x) + K(x)P(D) = \delta(x)(\oplus I_{k_i}). \tag{2.3.10}$$

In the above expression, by $K(x)P(D)$ we understand $(P(D)^t K(x)^t)^t$.

Proof. Let us rewrite equation (2.3.10) for the components of the section $K(x)$:

$$P_{i-1}(D)K_i(x) + K_{i+1}(x)P_i(D) = \delta(x)I_{k_i} \quad (i = 0, \ldots, N). \tag{2.3.11}$$

Since

$$\begin{aligned} (P_{i-1}(D)K_i)(x-y) &= P_{i-1}(x)K_i(x-y), \\ (K_{i+1}P_i(D))(x-y) &= K_{i+1}(x-y)(P_i'(D_y))^t = P_i'(y)K_{i+1}(x-y) \end{aligned}$$

and

$$\operatorname{tr} \Delta^*|_{\mathcal{D}(E^{i*} \boxtimes E^i)} = \delta(x-y)I_{k_i},$$

equalities (2.3.3) and (2.3.11) coincide.

□

Consider the complex of (full) symbols corresponding to E:

$$\mathbb{C}_z^k : 0 \longrightarrow \mathbb{C}^{k_0} \xrightarrow{P_0(z)} \mathbb{C}^{k_1} \xrightarrow{P_1(z)} \ldots \xrightarrow{P_{N-1}(z)} \mathbb{C}^{k_N} \longrightarrow 0, \tag{2.3.12}$$

where $z \in \mathbb{C}^n$, and $P_i(z)$ is obtained from $P_i(D)$ by the substitution $D \to z$ (see Subsection 1.2.2).

Definition 2.3.19 *A complex E is said to be non-degenerate if there exists a point $z^0 \in \mathbb{R}^n$ such that the complex $\mathbb{C}_{z^0}^k$ is exact.*

To formulate the next result, we will make a few remarks. Given a $\mathcal{P}$-module M, we choose for it some representation $M \cong \mathcal{P}^k/(P(z))^t\mathcal{P}^l$, where $P(z)$ is an $(l \times k)$-matrix of polynomials. Consider the algebraic variety $\mathfrak{M}(M)$ in $\mathbb{C}^n$ consisting of all points $z \in \mathbb{C}^n$ such that $\mathrm{rank}_{\mathbb{C}} P(z) < k$. This variety is known not to depend on the choice of the representation of M (see Palamodov [175, Ch.IV, §1]). It is called the algebraic variety associated to M. Further information on $\mathfrak{M}(M)$ may be found in the book of Palamodov [175].

Proposition 2.3.20 *A Hilbert complex $E^\cdot$ associated with a $\mathcal{P}$-module M is non-degenerate if and only if*

$$\mathfrak{M}(M) \neq \mathbb{C}^n. \tag{2.3.13}$$

In the terminology of the book of Palamodov [175, Ch.VIII, §14], condition (2.3.13) means that the $\mathcal{P}$-module M (and any differential operator $P(D)$ with $M \cong \mathcal{P}^k/(P(z))^t\mathcal{P}^l$) is determined. Equivalent conditions are $\dim \mathfrak{M}(M) < n$, or $\mathrm{Ext}^0(M, \mathcal{P}) = 0$.

Proof. *Necessity.* This is clear because, if complex (2.3.12) is exact at $\mathbb{C}^{k_0}$ for $z = z^0$, then $\mathrm{rank}_{\mathbb{C}} P_0(z) = k_0$ and so $z^0 \notin \mathfrak{M}(M)$.

Sufficiency. Suppose that $\mathfrak{M}(M) \neq \mathbb{C}^n$. Then there exists a point $z^0 \in \mathbb{R}^n$ such that $\mathrm{rank}_{\mathbb{C}} P_0(z^0) = k_0$. Hence it follows that

$$k_0 = \mathrm{rank}_{\mathbb{C}} P_0(z^0) \leq \mathrm{rank}_{\mathcal{R}} P_0(z) \leq k_0,$$

so in fact the equalities hold here.

By Proposition 1.2.17, sequence (2.3.12) is exact for $z = z^0$ which completes the proof. □

As one can see from the proof, for Hilbert complexes E the set of those $z \in \mathbb{C}^n$ such that the sequence (2.3.12) is not exact coincides with $\mathfrak{M}(M)$.

Exercise 2.3.21 *Prove or refute, that in general the set of those $z \in \mathbb{C}^n$, for which sequence (2.3.12) is not exact, is an algebraic subvariety of $\mathbb{C}^n$ too.*

The following result was announced by the author [234].

Theorem 2.3.22 *Each non-degenerate complex $E^\cdot$ of differential operators with constant coefficients in $\mathbb{R}^n$ has a fundamental solution of convolution type.*

Proof. We shall present a construction of a fundamental solution which for elliptic complexes gives a fundamental solution with regularity properties, as in Theorem 2.1.12 (2). Though to prove the theorem, a more simple reasoning would be sufficient. Namely, one could take all Q_i to equal 1 below.

Let us agree to consider any vector space of differential operators of type $E \to F$ with constant coefficients in $\mathbb{R}^n$ as a module over the ring of scalar differential operators with constant coefficients in $\mathbb{R}^n$ in the natural way.

Choose an integer p such that $p \geq p_i$ for all i. For every i, let Q_i be a non-zero scalar differential operator of order $2(p - p_i)$ with constant coefficients.

Introduce the differential operators $\widetilde{P}_i$ of type $E^i \to E^{i+1}$ and order $\widetilde{p}_i = 2p - p_i$ by means of $\widetilde{P}_i = P_i Q_i$. Obviously, $\widetilde{P}_{i+1} \circ \widetilde{P}_i = 0$ for all i, so we obtain the new differential complex $\widetilde{E}^\cdot = \{E^i, \widetilde{P}^i\}$ in $\mathbb{R}^n$.

For every i, define a differential operator $\Delta_i \in \mathrm{do}_{2p}(E^i \to E^i)$ by the equality

$$\Delta_i = \widetilde{P}_i^* P_i + P_{i-1} \widetilde{P}_{i-1}^*.$$

We notice at once that

$$\begin{aligned} P_i \Delta_i &= \Delta_{i+1} P_i, \\ \widetilde{P}_{i-1}^* \Delta_i &= \Delta_{i-1} \widetilde{P}_{i-1}^*. \end{aligned} \tag{2.3.14}$$

Let us show that every differential operator $\Delta_\imath$ is determined, i.e. $\det \Delta_\imath(z) \not\equiv 0$.

First of all we observe that simple arguments from linear algebra (which will partly be recalled below) show that the exactness of (2.3.12) at $\mathbb{C}^{k_\imath}$ is equivalent to

$$\operatorname{rank}_{\mathbb{C}} \left\| \begin{array}{c} P_\imath(z) \\ (P_{\imath-1}(z))^* \end{array} \right\| = k_\imath .$$

Of course, this does not answer the question of Exercise 2.3.21 since $(P_{\imath-1}(z))^* = P_{\imath-1}{}^*(\overline{z})$. But we have seen that the set of those points $z \in \mathbb{R}^n$, for which the sequence (2.3.12) is not exact is the intersection of an algebraic subvariety of $\mathbb{C}^n$ with the real space $\mathbb{R}^n$. Hence, if the complex E is non-degenerate, then this set has zero measure in $\mathbb{R}^n$.

Now let a point $z \in \mathbb{R}^n$ be such that $Q_\imath(z)Q_{\imath-1}(z) \neq 0$, and

$$\operatorname{rank}_{\mathbb{C}} \left\| \begin{array}{c} P_\imath(z) \\ (P_{\imath-1}(z))^* \end{array} \right\| = k_\imath .$$

As we have seen, this is valid for almost all points $z \in \mathbb{R}^n$.

Suppose that $e \in \mathbb{C}^{k_\imath}$ and $\Delta_\imath(z)e = 0$. Since $P_\imath(z)P_{\imath-1}(z) = 0$, we have

$$(\Delta_\imath(z)e, (P_\imath(z))^* P_\imath(z)e) = \overline{Q_\imath(z)}\, | (P_\imath(z))^* P_\imath(z)e |^2 = 0$$

so that $(P_\imath(z))^* P_\imath(z)e = 0$. Then we get

$$((P_\imath(z))^* P_\imath(z)e, e) = | P_\imath(z)e |^2 = 0$$

so that $P_\imath(z)e = 0$. Hence $P_{\imath-1}(z)\widetilde{P}^*_{\imath-1}(z)e = 0$, and we obtain

$$\left(P_{\imath-1}(z)\widetilde{P}^*_{\imath-1}(z)e, e\right) = \overline{Q_{\imath-1}(z)} |(P_{\imath-1}(z))^* e|^2 = 0,$$

so that $(P_{\imath-1}(z))^* e = 0$. Therefore, $e = 0$. Thus,

$$\operatorname{rank}_{\mathbb{C}} \Delta_\imath(z) = k_\imath \quad \text{for almost all} \quad z \in \mathbb{C}^n . \tag{2.3.15}$$

For a real number r, we denote by $E^{(r)}$ the vector space of all entire functions f in $\mathbb{C}^n$ for which all the next norms are finite:

$$p_\nu^{(r)} = \sup_{z \in \mathbb{C}^n} \left((1 + |z|)^\nu e^{-r|\operatorname{Im} z|} |f(z)| \right) \quad (\nu = 0, 1, \ldots).$$

With the topology defined by this family of seminorms, each $E^{(r)}$ is a Fréchet space.

Further, the vector space $E = \cup_r E^{(r)}$ is endowed with the inductive limit topology induced by the Fréchet spaces $E^{(r)}$.

It follows from the Paley-Wiener-Schwartz Theorem that the Fourier transform $F : f \to \hat{f}$ determines a topological isomorphism of the spaces E and $\mathcal{D}(\mathbb{R}^n)$. Therefore their (strong) dual spaces $\mathcal{D}'(\mathbb{R}^n)$ and E' are also isomorphic as topological vector spaces.

One of the ways to determine this isomorphism was suggested by Ehrenpreis [65]. It consists in defining the Fourier transform of distributions as the transpose of the mapping $F: E \to \mathcal{D}(\mathbb{R}^n)$. Namely, the Fourier transformation of a distribution $g \in \mathcal{D}'(\mathbb{R}^n)$ is the element $\widehat{g} \in E'$ defined by the equality

$$\langle \widehat{g}, f \rangle = \langle g, Ff \rangle \quad \text{for} \quad f \in E.$$

The mapping $F': g \to \widehat{g}$ determines a topological isomorphism $\mathcal{D}'(\mathbb{R}^n) \overset{\cong}{\to} E'$, and $F'g$ coincides with the usual Fourier transform of g for $g \in L^1(\mathbb{R}^n)$.

Let us return to our problem. By Proposition 2.3.18 one must find a sequence of sections $\Phi_i(x) \in \mathcal{D}'(E^{i-1} \otimes E^{i*})$ such that $\Phi_i \neq 0$ only for $i = 1, \dots, N$ and

$$P_{i-1}(D)\Phi_i(x) + \Phi_{i+1}(x)P_i(D) = \delta(x)I_{k_i} \quad (i = 0, \dots, N).$$

First we wish to construct fundamental matrices $G_i(x) \in \mathcal{D}'(E^i \otimes E^{i*})$ of convolution type for the differential operators Δ_i $(i = 0, \dots, N)$, i.e.

$$\begin{aligned} G_i(x)\Delta_i(D) &= \delta(x)I_{k_i}, \\ \Delta_i(D)G_i(x) &= \delta(x)I_{k_i}, \end{aligned} \tag{2.3.16}$$

satisfying in addition the commutativity relations

$$\begin{aligned} G_{i+1}(x)P_i(D) &= P_i(D)G_i(x), \\ G_{i-1}(x)\widetilde{P}^*_{i-1}(D) &= \widetilde{P}^*_{i-1}(D)G_i(x). \end{aligned} \tag{2.3.17}$$

By passing to the Fourier transform we get the following system of equations for the matrices $\widehat{G}_i \in [E']^{(k_i \times k_i)}$:

$$\begin{aligned} \widehat{G}_i(z)\Delta_i(z) &= I_{k_i}, \\ \Delta_i(z)\widehat{G}_i(z) &= I_{k_i}, \end{aligned} \tag{2.3.18}$$

and

$$\begin{aligned} \widehat{G_{i+1}}(z)P_i(z) &= P_i(z)\widehat{G}_i(z), \\ \widehat{G_{i-1}}(z)\widetilde{P}^*_{i-1}(z) &= \widetilde{P}^*_{i-1}(z)\widehat{G}_i(z). \end{aligned} \tag{2.3.19}$$

One can show that this system is solvable. Then by making the inverse Fourier transform action we shall get matrices $G_i(x)$ with the desired properties (2.3.16) and (2.3.17).

Let us associate with each matrix $M(z)$ of $[E']^{(k \times l)} = \mathbb{C}^k \widehat{\otimes} [E']^{(1 \times l)}$ the continuous linear mapping $M: E^l \to \mathbb{C}^k$ given by $Mf = M(z) \times f$, where after multiplying the column $f \in E^l$ by the matrix $M(z)$ from the left, elements of $M(z)$ act on elements of f as linear functionals. This correspondence between matrices and operators is a topological isomorphism because by Theorem 1.4.2

$$\mathbb{C}^k \widehat{\otimes} [E']^{(1 \times l)} \overset{top.}{\cong} \mathcal{L}_b([E]^l \to \mathbb{C}^k).$$

The matrix $M(z)$ will also be called a *kernel* of the operator M.

Notice that, since E is a $\mathcal{P}$-module, the mappings of $\mathcal{L}_b([E]^l \to \mathbb{C}^k)$ may be multiplied by $\mathcal{P}$-matrices of suitable sizes both from the left and from the right.

Assume now that the polynomials $\det \Delta_i(z)$ ($i = 0, \dots, N$) have no roots for real z (i.e. $z \in \mathbb{R}^n$) and moreover, theirs modules are bounded from below on $\mathbb{R}^n$ by a positive constant. Then we define the operators $\widehat{G_i} : [E]^{k_i} \to \mathbb{C}^{k_i}$ by the formulas

$$\widehat{G_i} f = \int_{\mathbb{R}^n} (\Delta_i(z))^{-1} f(z) dz \quad \text{for} \quad f \in [E]^{k_i}. \tag{2.3.20}$$

It is clear that $\widehat{G_i} \in \mathcal{L}_b([E]^{k_i} \to \mathbb{C}^{k_i})$ and we get for all $f \in [E]^{k_i}$:

$$\begin{aligned} \left(\widehat{G_i}\Delta_i(z)\right) f &= \textstyle\int_{\mathbb{R}^n} (\Delta_i(z))^{-1} \Delta_i(z) f(z) dz &= \textstyle\int_{\mathbb{R}^n} f(z) dz, \\ \left(\Delta_i(z)\widehat{G_i}\right) f &= \textstyle\int_{\mathbb{R}^n} \Delta_i(z) (\Delta_i(z))^{-1} f(z) dz &= \textstyle\int_{\mathbb{R}^n} f(z) dz. \end{aligned}$$

Hence it follows that the kernels $\widehat{G_i}(z) \in [E']^{(k_i \times k_i)}$ of the operators $\widehat{G_i}$ satisfy equations (2.3.18).

Further, in view of (2.3.14)

$$\begin{aligned} (\Delta_{i+1}(z))^{-1} P_i(z) &= P_i(z) (\Delta_i(z))^{-1}, \\ (\Delta_{i-1}(z))^{-1} \widetilde{P}^*_{i-1}(z) &= \widetilde{P}^*_{i-1}(z) (\Delta_i(z))^{-1}, \end{aligned}$$

so we obtain for all $f \in [E]^{k_i}$:

$$\begin{aligned} \left(\widehat{G_{i+1} P_i(z)}\right) f &= \int_{\mathbb{R}^n} (\Delta_{i+1}(z))^{-1} P_i(z) f(z) dz \\ &= \int_{\mathbb{R}^n} P_i(z) (\Delta_i(z))^{-1} f(z) dz = \left(P_i(z)\widehat{G_i}\right) f, \\ \left(\widehat{G_{i-1} \widetilde{P}^*_{i-1}(z)}\right) f &= \int_{\mathbb{R}^n} (\Delta_{i-1}(z))^{-1} \widetilde{P}^*_{i-1}(z) f(z) dz \\ &= \int_{\mathbb{R}^n} \widetilde{P}^*_{i-1}(z) (\Delta_i(z))^{-1} f(z) dz = \left(\widetilde{P}^*_{i-1}(z)\widehat{G_i}\right) f. \end{aligned}$$

Therefore, the kernels $\widehat{G_i}(z)$ satisfy also equations (2.3.19).

Finally, in the general case we should replace the cycle $\mathbb{R}^n$ in definition (2.3.20) of the operators $\widehat{G_i}$ by a suitable "Hörmander staircase" Σ (see Shilov [200, §17,3]) in a region $|\text{Im } z_1| \le C$ of the space $\mathbb{C}^1 \times \mathbb{R}^{n-1}$ where $|\det \Delta_i(z)| \ge c > 0$ for all $i = 0, \dots, N$. To prove the existence of such a "Hörmander staircase" Σ, common for all polynomials ($\det \Delta_i(z)$), we can repeat the reasoning from the book of Shilov [200, §17,2-4]) word for word. Since the number of polynomials ($\det \Delta_i(z)$) is finite and condition (2.3.15) is fulfilled, no additional difficulties arise. So the existence of the matrices $G_i(x) \in \mathcal{D}'(E^i \otimes E^{i*})$, satisfying (2.3.16) and (2.3.17) has been proved.

For every i, we set

$$\begin{aligned} \Phi_i(x) &= G_{i-1}(x)\widetilde{P}^*_{i-1}(D) \\ &= \widetilde{P}^*_{i-1}(D) G_i(x), \end{aligned} \tag{2.3.21}$$

so that $\Phi_i(x) \in \mathcal{D}'(E^{i-1} \otimes E^{i*})$.

We claim that $\Phi(x) = \oplus\Phi_i(x)$ satisfies equation (2.3.10). Indeed, by (2.3.17) and (2.3.16) we have for $i = 0, \ldots, N$

$$\begin{aligned} &P_{i-1}(D)\Phi_i(x) + \Phi_{i+1}(x)P_i(D) \\ &\quad = P_{i-1}(D)G_{i-1}(x)\widetilde{P}^*_{i-1}(D) + G_i(x)\widetilde{P}^*_i(D)P_i(D) \\ &\quad = G_i(x)\Delta_i(D) \\ &\quad = \delta(x)I_{k_i}, \end{aligned}$$

as required.

To complete the proof simply invoke Proposition 2.3.18.

□

The construction of a fundamental solution, presented in Theorem 2.3.22, was first suggested in the author's paper [223] for elliptic complexes $E^{\cdot}$. In this case one can take

$$\widetilde{P}_i = P_i\ \Delta^{p-p_i},$$

where Δ is the Laplace operator in $\mathbb{R}^n$. Then the Laplacians Δ_i are elliptic differential operators of order $2p$. Hence the fundamental solution Φ given by (2.3.21) has regularity properties as in Theorem 2.1.12 (2).

Our construction admits a considerable flexibility for the choice of a "Hörmander staircase" Σ which may be used to determine more precisely the properties of fundamental solutions (see Shilov [200, §17,5]).

Exercise 2.3.23 *Prove that every non-degenerate complex $E^{\cdot}$ of differential operators with constant coefficients has a temperately growing fundamental solution of convolution type.*

Hint. Use the construction of Hörmander [114, §10.2].

□

Exercise 2.3.24 *Find a rough formula of Green type for solutions $f \in \mathcal{D}'(E^q|_Y)$ of $Pf = 0$ where Y is an open subset of $\mathbb{R}^n$.*

Hint. Follow §16 of Shilov's book [200]. Use Proposition 2.3.4 and Theorem 2.3.22.

□

In conclusion we note a simple consequence of Theorem 2.3.22

Corollary 2.3.25 *A Hilbert complex $E^{\cdot}$ associated with a $\mathcal{P}$-module M has a fundamental solution of convolution type if and only if the module M is determined.*

Proof. *Necessity.* Let K be a fundamental solution of $E^{\cdot}$. By Corollary 2.1.6, K' is then a fundamental solution of the transposed complex $E^{\cdot\prime}$. In particular,

$$P_0{}'K_1{}'g = g \qquad \text{for all } g \in \mathcal{D}(E^{0\prime}).$$

Hence it follows that the compatibility conditions for the non-homogeneous system $P_0{}'v = g$ are zero. In other words, the equality $P_0(z)p(z) = 0$ implies $p(z) = 0$ for a column $p(z) \in \mathcal{P}^{k_0}$. So $\operatorname{Ext}^0(M, \mathcal{P}) = 0$ which was required.

Another proof is to note that the identity

$$K_1 P_0 = 1 \qquad \text{on } \mathcal{E}'(E^0)$$

implies that the system $P_0 f = 0$ has no non-trivial generalized solutions with compact supports in $\mathbb{R}^n$, and then to use Proposition 1 of Palamodov [173, Ch.VIII, §14].

Sufficiency. This follows from Theorem 2.3.22 and Proposition 2.3.20.

□

Corollary 2.3.25 extends the famous result of Ehrenpreis [64] and Malgrange [156, p.288] that every non-zero scalar differential operator with constant coefficients in $\mathbb{R}^n$ has a fundamental solution in the class of distributions.

2.4 Green Operators for Differential Operators and the Homotopy Formula on Manifolds with Boundary

2.4.1 *Polydifferential operators with values in the space of differential forms*

Let X be an n-dimensional orientable differentiable manifold countable at infinity.

Already the simplest applications of the homotopy formula (2.1.4) lead to the necessity of computing the commutators $[P, \chi_{\mathcal{O}}]$, where $\chi_{\mathcal{O}}$ is the characteristic function of an open set $\mathcal{O} \subset X$. Here the so-called polydifferential operators with values in the space of differential forms appear.

Let $B^1, \dots, B^e$ and B be differentiable $\mathbb{C}$-vector bundles over X. Denote by

$$\mathrm{do}_g((B^1, \dots, B^e) \to B)$$

the space of all e-differential operators of type $(B^1, \dots, B^e) \to B$ and order g on X (see Bourbaki [29, 14.1.11]).

The inclusion $G \in \mathrm{do}_g((B^1, \dots, B^e) \to B)$ means that for fixed sections $f_\iota \in \mathcal{E}(B^j)$ $(j \neq q)$ the correspondence $f_q \to G(f_1, \dots, f_e)$ is a differential operator of type $B^q \to B$ and the total order of "derivatives" of $f_1, \dots, f_e$ is not more then g.

The cases $B = \Lambda^q$ $(q = 0, \dots, n)$ are of particular interest where $\Lambda^q = \Lambda^q T^*_{\mathbb{C}}(x)$ is the bundle of complex-valued exterior forms of degree q over X.

Assume now we are given an e-differential operator $G \in \mathrm{do}_g((B^1, \dots, B^e) \to \Lambda^q)$. Then by taking the composition of G with the exterior derivative d, we obtain the operator

$$dG \in \mathrm{do}_{g+1}((B^1, \dots, B^e) \to \Lambda^{q+1}).$$

Since $d^2 = 0$, we get $d(dG) = 0$. Thus the following complex on X appears:

$$\begin{array}{l} \mathrm{do}_{g-n+\cdot}((B^1,\ldots,B^e) \to \Lambda^{\cdot}): \\ 0 \to \mathrm{do}_{g-n}((B^1,\ldots,B^e) \to \Lambda^0) \xrightarrow{d} \mathrm{do}_{g-n+1}((B^1,\ldots,B^e) \to \Lambda^1) \xrightarrow{d} \ldots \\ \xrightarrow{d} \mathrm{do}_g((B^1,\ldots,B^e) \to \Lambda^n) \to 0. \end{array} \tag{2.4.1}$$

Here we define $\mathrm{do}_g((B^1,\ldots,B^e) \to B)$ to be zero if $g < 0$.

The following theorem is the basis for application of integral formulas to the investigation of differential complexes.

Theorem 2.4.1 *1) Complex (2.4.1) is exact at degrees $q < n$, and*
2) $G \in \mathrm{do}_g((B^1,\ldots,B^e) \to \Lambda^n)$ is a differential if and only if

$$\int_X G(f_1,\ldots,f_e) = 0 \quad \textit{for all} \quad f_j \in \mathcal{D}(B^j) \quad (j = 1,\ldots,e). \tag{2.4.2}$$

For the scalar case this result was announced by Grothendieck [89].

Proof. Let $U = \{U_\nu\}_{\nu\in\mathcal{N}}$ be a locally finite open covering of X formed by coordinate neighborhoods over which all bundles $B^1,\ldots,B^e$ are trivial and $\{\varphi_\nu\}$ be a C^∞_{loc} partition of unity on X subordinated to the covering U (see Corollary 1.1.6).

Suppose that for an operator $G \in \mathrm{do}_g((B^1,\ldots,B^e) \to \Lambda^q)$ there exist operators $T^{(\nu)} \in \mathrm{do}_{g-1}((B^1,\ldots,B^e)|_{U_\nu} \to \Lambda^{q-1}|_{U_\nu})$ such that

$$dT^{(\nu)}(f_1,\ldots,f_e) = G(f_1,\ldots,f_e) \quad \text{for all} \quad f_j \in \mathcal{E}(B^j|_{U_\nu}) \quad (j = 1,\ldots,e) \tag{2.4.3}$$

Define now the operator $T \in \mathrm{do}_{g-1}((B^1,\ldots,B^e) \to \Lambda^{q-1})$ by the following:

$$T(f_1,\ldots,f_e) = \sum_{\nu_1,\ldots,\nu_e} T^{(\nu_1)}(\varphi_{\nu_1} f_1,\ldots,\varphi_{\nu_e} f_e) \quad \text{for} \quad f_j \in \mathcal{E}(B^j) \quad (j = 1,\ldots,e). \tag{2.4.4}$$

Since $\operatorname{supp} T^{(\nu)}(f_1\ldots,f_e) \subset \cap_{j=1}^e(\operatorname{supp} f_j)$, the mapping T is well-defined, and according to (2.4.3) and (2.4.4) we have:

$$\begin{aligned} dT(f_1,\ldots,f_e) &= \sum_{\nu_1,\ldots,\nu_e} dT^{(\nu_1)}(\varphi_{\nu_1} f_1,\ldots,\varphi_{\nu_e} f_e) \\ &= \sum_{\nu_1,\ldots,\nu_e} G(\varphi_{\nu_1} f_1,\ldots,\varphi_{\nu_e} f_e) \\ &= G\left(\left(\sum_{\nu_1}\varphi_{\nu_1}\right) f_1,\ldots,\left(\sum_{\nu_e}\varphi_{\nu_e}\right) f_e\right) \\ &= G(f_1,\ldots,f_e). \end{aligned}$$

That is, $G = dT$.

Thus the exactness question for the global complex (2.4.1) has been reduced to the exactness question for the local complexes $\mathrm{do}_{g-n+\cdot}((B^1,\ldots,B^e)|_{U_\nu} \to \Lambda^{\cdot}|_{U_\nu})$.

Therefore, as the exterior derivative d is invariant under change of variables, we can assume X to be an open set in $\mathbb{R}^n$. Then all the bundles B^j $(j = 1,\ldots,e)$ are trivial and the vector spaces $\mathrm{do}_g((B^1,\ldots,B^e)\to\Lambda^q)$ are identified with the spaces of e-differential operators with partial derivatives of order g from $\mathcal{E}(B^1)\otimes\ldots\otimes\mathcal{E}(B^e)$ to $\mathcal{E}(\Lambda^q)$ (Peetre's theorem).

Let us verify that the complex (2.4.1) is exact at the first non-zero member. Consider the case where $g<n$, then the initial fragment of complex (2.4.1) is

$$0\longrightarrow \mathrm{do}_0((B^1,\ldots,B^e)\to\Lambda^{n-g})\xrightarrow{d}\ldots. \tag{2.4.5}$$

If $G\in\mathrm{do}_0((B^1,\ldots,B^e)\to\Lambda^{n-g})$ and $dG=0$, then $dG(f_1,\ldots,f_e)=0$ for all $f_j\in\mathcal{E}(B^j)$ $(j=1,\ldots,e)$. Since G is e-linear over the ring $\mathcal{E}(X)$, we get $G=0$ That is, sequence (2.4.5) is exact.

Let now $g\geq n$, $G\in\mathrm{do}_{g-n}((B^1,\ldots,B^e)\to\Lambda^0)$, and $dG=0$. This means that $dG(f_1,\ldots,f_e)=0$ for all $f_j\in\mathcal{E}(B^j)$ $(j=1,\ldots,e)$. Then $G(f_1,\ldots,f_e)=const$, and again we get $G\equiv0$. So complex (2.4.1) is exact at degree 0.

Further, let us prove the exactness of complex (2.4.1) at degrees $0<q<n$,

$$\ldots\longrightarrow \mathrm{do}_{g-1}((B^1,\ldots,B^e)\to\Lambda^{q-1})\xrightarrow{d}\mathrm{do}_g((B^1,\ldots,B^e)\to\Lambda^q)\xrightarrow{d}\ldots \tag{2.4.6}$$

where $g\geq1$.

We shall write operators $G\in\mathrm{do}_g((B^1,\ldots,B^e)\to\Lambda^q)$ in the form

$$G(f_1,\ldots,f_e)={\sum}'_{\substack{\#I=q\\|\alpha|\leq g}}D^\alpha(G_{I,\alpha}(f_1,\ldots,f_{e-1})f_e)dx_I, \tag{2.4.7}$$

where $G_{I,\alpha}\in\mathrm{do}_{g-|\alpha|}((B^1,\ldots,B^{e-1})\to(B^e)^*)$, and some $(e-1)$-differential operators.

Notice, that such a representation is always possible. In fact, it is sufficient to write G in the form

$$G(f_1,\ldots,f_e)={\sum}'_{\substack{\#I=q\\|\beta|\leq g}}\widetilde{G}_{I,\beta}(f_1,\ldots,f_{e-1})D^\beta f_e)dx_I,$$

where $\widetilde{G}_{I,\beta}\in\mathrm{do}_{g-|\beta|}((B^1,\ldots,B^{e-1})\to(B^e)^*)$, and then to use the transposed Leibniz formula

$$g(D^\beta f)=\sum_\alpha(-1)^{|\beta-\alpha|}\frac{\beta!}{\alpha!(\beta-\alpha)!}D^\alpha(D^{\beta-\alpha}g)f.$$

Since the representation of $G\in\mathrm{do}_g((B^1,\ldots,B^e)\to\Lambda^q)$ in the form (2.4.7) is unique, one can identify G with the system of operators $\{G_{I,\alpha}\}$, which we shall do. By regulating the system $\{G_{I,\alpha}\}$ over the multi-indices $\#I=q$ and $|\alpha|\leq g$, for fixed sections $f_j\in\mathcal{E}(B^j)$ $(j=1,\ldots,e-1)$, one can consider $\{G_{I,\alpha}(f_1,\ldots,f_{e-1})\}$ as a differentiable function on X with values in some space $\mathbb{C}^{l_q}$. Of course, l_q depends on g too, but for simplicity we do not show it in the notation.

Further,

$$dG(f_1,\ldots,f_e)$$
$$={\sum}'_{\substack{\#J=q+1\\|\beta|\leq g+1}}D^\beta\left(\left(\sqrt{-1}{\sum}'_{\substack{I,j,\alpha\\I\cup j=J\\\alpha+1_j=\beta}}\sigma(I,j)\,G_{I,\alpha}(f_1,\ldots,f_{e-1})\right)f_e\right)dx_J, \tag{2.4.8}$$

where the constant $\sigma(I,j) = 0, \pm 1$ is determined from the equality $dx_j \wedge dx_I = \sigma(I,j)\, dx_J$.

It follows from (2.4.7) and (2.4.8) that, by the identification above, the operator $G \to d \circ G$ from $\mathrm{do}_g((B^1,\dots,B^e) \to \Lambda^q)$ to $\mathrm{do}_{g+1}((B^1,\dots,B^e) \to \Lambda^{q+1})$ is interpreted as $G \to M_q \cdot G$, where $M_q : \mathbb{C}^{d_q} \to \mathbb{C}^{d_{q+1}}$ is a fixed linear mapping whose matrix has entries 0 and ± 1 and does not depend on x and $f_1,\dots,f_{e-1}$.

Simple algebraic arguments, using the representation (2.4.8) for M, show that the sequence

$$\dots \longrightarrow \mathbb{C}^{d_{q-1}} \xrightarrow{M} \mathbb{C}^{d_q} \xrightarrow{M} \dots$$

is exact for $q < n$. Hence, the linear mappings $L_q = M_{q-1}M_{q-1}^* + M_q^* M_q$ of $\mathbb{C}^{d_q}$ are invertible for $q < n$, and the following homotopy formula is valid:

$$\left(L_q^{-1}M_q^*\right) M_q + M_{q-1}\left(L_{q-1}^{-1}M_{q-1}^*\right) = 1 \quad \text{on} \quad \mathbb{C}^{d_q} \quad (q = 0,\dots,n-1). \tag{2.4.9}$$

Let now $G \in \mathrm{do}_g((B^1,\dots,B^e) \to \Lambda^q)$ $(q < n)$ and $dG = 0$. Then by the interpretation above, we get $MG = 0$, and formula (2.4.9) gives

$$G = M_{q-1}(L_{q-1}^{-1}M_{q-1}^*)G.$$

By setting $T = (L_{q-1}^{-1}M_{q-1}^*)G$ we have $T \in \mathrm{do}_{g-1}((B^1,\dots,B^e) \to \Lambda^{q-1})$ and $dT = G$. This proves the exactness of complex (2.4.6), as required.

Finally, we investigate the last fragment of the complex (2.4.1):

$$\dots \longrightarrow \mathrm{do}_{g-1}((B^1,\dots,B^e) \to \Lambda^{n-1}) \xrightarrow{q} \mathrm{do}_g((B^1,\dots,B^e) \to \Lambda^n) \longrightarrow 0. \tag{2.4.10}$$

Obviously, condition (2.4.2) for the operator $G \in \mathrm{do}_g((B^1,\dots,B^e) \to \Lambda^n)$ is necessary for G to be in $\mathrm{im}\, d$. We show that it is also sufficient.

We have

$$G(f_1,\dots,f_e) = \sum_{|\alpha|\le g} D^\alpha(G_\alpha(f_1,\dots,f_{e-1})f_e)dx,$$

where $G_\alpha \in \mathrm{do}_{g-|\alpha|}((B^1,\dots,B^{e-1}) \to (B^e)^*)$ are some $(e-1)$-differential operators. Then for all $f_j \in \mathcal{D}(B^j)$ $(j = 1,\dots,e)$, we get by condition (2.4.2)

$$\int_X G(f_1\dots,f_e) = \int_X G_0(f_1\dots,f_{e-1})f_e\, dx = 0,$$

so that $G_0 \equiv 0$.

Hence it follows at once that there exists an operator $T \in \mathrm{do}_{g-1}((B^1,\dots,B^e) \to \Lambda^{n-1})$ such that $dT = G$ which completes the proof.

□

Remark 2.4.2 *As we have seen from the proof of this theorem, if G is in $\mathrm{do}_g((B^1,\dots,B^e) \to \Lambda^q)$ and $G = dT$, then the operator $T \in \mathrm{do}_{g-1}((B^1,\dots,B^e) \to \Lambda^{q-1})$ can be chosen so that (locally) its coefficients are linear combinations of derivatives up to order $(g-1)$ of the coefficients of G.*

Often, e-differential operators $G \in \mathrm{do}_g((B^{1''}, \ldots, B^{e''}) \to B)$ on X are used in the following situation.

Let $B^{j'}$ $(j = 1, \ldots, e)$ be differentiable $\mathbb{C}$-vector bundles over manifolds X_j. Then the operator G determines the unique continuous e-linear mapping

$$\tilde{G} : \mathcal{E}(B^{1'} \boxtimes B^{1''}) \times \ldots \times \mathcal{E}(B^{e'} \boxtimes B^{e''}) \to \mathcal{E}((B^{1'} \boxtimes \ldots \boxtimes B^{e'}) \boxtimes B),$$

given on elements of the form $f_j{}' \otimes f_j{}'' \in \mathcal{E}(B^{j'}) \otimes \mathcal{E}(B^{j''})$ $(j = 1, \ldots, e)$ by the equality

$$\tilde{G}(f_1{}' \otimes f_1{}'', \ldots, f_e{}' \otimes f_e{}'') = (f_1{}' \otimes \ldots \otimes f_e{}') \otimes G(f_1{}'', \ldots, f_e{}'').$$

Obviously, $\tilde{G} \in \mathrm{do}_g\left((B^{1'} \boxtimes B^{1''}, \ldots, B^{e'} \boxtimes B^{e''}) \to (B^{1'} \boxtimes \ldots \boxtimes B^{e'}) \boxtimes B\right)$ and we keep the same notation G for $\tilde{G}$ for simplicity. There will be no confusion because the sense is clear from context.

2.4.2 *Green operator of a differential operator*

In this subsection E, F and G will be some differentiable $\mathbb{C}$-vector bundles over X.

Definition 2.4.3 *A Green operator for a differential operator $P \in \mathrm{do}_p(E \to F)$ is a bidifferential operator $G(g, f) \in \mathrm{do}_{p-1}((F', E) \to \Lambda^{n-1})$ such that, for each open set $U \subset X$ and all sections $g \in \mathcal{E}(F'|_U)$ and $f \in \mathcal{E}(E|_U)$, we have*

$$\langle g, Pf \rangle_x - \langle P'g, f \rangle_x = dG(g, f) \quad (x \in U). \tag{2.4.11}$$

Proposition 2.4.4 *For each differential operator $P \in \mathrm{do}_p(E \to F)$, a Green operator always exists, and if G^1 and G^2 are two Green operators for P, then there exists a bidifferential operator $T \in \mathrm{do}_{p-2}((F', E) \to \Lambda^{n-2})$ such that $G^2 - G^1 = dT$.*

For the purposes of this book one can consider two such Green operators to be equivalent. Denote this equivalence class by $G_P(g, f)$.

The first part of Proposition 2.4.4, that is, the existence of a Green operator for every differential operator $P \in \mathrm{do}_p(E \to F)$, is well-known (see, for instance, Bourbaki [29, 14.3.5]. An original method of constructing a Green operator was suggested by Vladimirov and Zharinov [248] in connection with the investigation of conservation laws.

Proof. Since the manifold X is orientable by assumption, there is the canonical isomorphism $\Omega \cong \Lambda^n$. Thus, we have $D \in \mathrm{do}_p((F', E) \to \Lambda^n)$ where $D(g, f)$ is the operator in the left-hand side of (2.4.11). From Theorem 2.4.1 and the definition of a transposed operator (1.5.2), it follows that there exists $G \in \mathrm{do}_{p-1}((F', E) \to \Lambda^{n-1})$ such that $D = dG$. Obviously, G is the desired Green operator for P. Moreover, if G^1 and G^2 are two such operators, then $d(G^2 - G^1) = 0$. By Theorem 2.4.1, there exists an operator $T \in \mathrm{do}_{p-2}((F', E) \to \Lambda^{n-2})$ such that $G^2 - G^1 = dT$. The proof is complete.

□

Locally, for a differential operator $P \in \text{do}_p(E \to F)$ with the local representation $P(x,D) = \sum_{|\alpha|\le p} P_\alpha(x)\, D^\alpha$, a Green operator may be written in the form

$$G_P(g,f) = \frac{1}{\sqrt{-1}} {\sum}'_{|\beta+\gamma+1_j|\le p} (-1)^{|\beta|} D^\beta(g \cdot P_{\beta+\gamma+1_j})(D^\gamma f) \star dx_j, \tag{2.4.12}$$

where $\sum'$ means that the summation is taken over the multi-indices β, γ and 1_j in some particular (non-unique) way, and $\star$ is the star operator of Hodge.

In particular, this summation can be chosen so that the derivatives of order $(p-1)$ of f enter into $G_P(g,f)$ only through the expression

$$\frac{1}{\sqrt{-1}}\, g \left(\frac{1}{p} \sum_{j=1}^{n} (P^{(j)}(x,D)f) \star dx_j \right) \tag{2.4.13}$$

where $P^{(j)}(x,D)$ is obtained by substituting $D \to z$ into $P^{(j)}(x,z) = \frac{\partial}{\partial z_j} P(x,z)$ $(j = 1, \ldots, n)$.

Thus, for every first order differential operator P there exists only one Green operator (see also Bourbaki [29, 14.3.7]).

The following technical statement can be verified immediately (see Bourbaki [29, 14.3.6]).

Proposition 2.4.5 *If $P \in \text{do}_p(E \to F)$ and $Q \in \text{do}_q(F \to G)$ then*

$$G_{P'}(f,g) = \overline{G_{P^*}(\star f, \star^{-1} g)} = -G_P(g,f), \tag{2.4.14}$$
$$G_{Q\circ P}(g,f) = G_Q(g,Pf) + G_P(Q'g,f). \tag{2.4.15}$$

Later on we shall often use the following version of the formula of integration by parts given in the author's paper [223].

Proposition 2.4.6 *Let $P \in \text{do}_p(E \to F)$, $Q \in \text{do}_q(F \to G)$, and $Q \circ P = 0$. Then there exists a bidifferential operator $T(g,f) \in \text{do}_{p+q-2}((G',E) \to \Lambda^{n-2})$ such that, for each open set $U \subset X$ and all sections $g \in \mathcal{E}(G'|_U)$ and $f \in \mathcal{E}(E|_U)$, we have*

$$G_Q(g,Pf) + G_P(Q'g,f) = dT(g,f) \quad \text{in} \quad U. \tag{2.4.16}$$

Proof. Denote the left side of (2.4.16) by $D(g,f)$ so that D is a bidifferential operator of type $(G',E) \to \Lambda^{n-1}$ and order $p+q-1$ on X. Since $Q \circ P = 0$, we get $dD = 0$ by formula (2.4.15). Hence, by Theorem 2.4.1 there exists an operator $T \in \text{do}_{p+q-2}((G',E) \to \Lambda^{n-2})$ such that $D = dT$. Obviously, T has the required properties.

□

2.4.3 The homotopy formula on manifolds with boundary

The homotopy formula on manifolds with boundary is a useful link in the applications of Proposition 2.1.5 to boundary value problems for elliptic complexes. We need an auxiliary result having also an independent interest.

Denote by $\pi_1 : X \times X \to X$ the projection onto the first multiplier. That is, $\pi_1(x,y) = x$.

Theorem 2.4.7 *For every bidifferential operator $G(\cdot,\cdot) \in \mathrm{do}_g((B^1,B^2) \to B)$ on X there exists a unique differential operator $G(\cdot) \in \mathrm{do}_g(B^1 \boxtimes B^2 \to \pi_1^* B)$ on $X \times X$ such that $\Delta^* G(f_1 \otimes f_2) = G(f_1, f_2)$ for all $f_j \in \mathcal{E}(B^j)$ $(j = 1,2)$.*

Proof. Since the operator $G(f_1,f_2)$ determines a continuous linear mapping $\mathcal{E}(B^1) \times \mathcal{E}(B^2) \to \mathcal{E}(B)$, it can be extended to a continuous linear mapping from the completed tensor product $\mathcal{E}(B^1)\hat{\otimes}\mathcal{E}(B^2)$ to $\mathcal{E}(B)$ (see Grothendieck [92]). However, $\mathcal{E}(B^1)\hat{\otimes}\mathcal{E}(B^2)$ can be easily interpreted as $\mathcal{E}(B^1 \boxtimes B^2)$ (see Example 1.4.1). Therefore, $G(\cdot,\cdot)$ determines a unique continuous linear mapping $\tilde{G}(\cdot) : \mathcal{E}(B^1 \boxtimes B^2) \to \mathcal{E}(B)$ with the property that $\tilde{G}(f_1 \otimes f_2) = G(f_1,f_2)$ for all sections $f_j \in \mathcal{E}(B^j)$ $(j = 1,2)$.

We claim that $\tilde{G}$ can be expressed in the form $\tilde{G}(\cdot) = \Delta^* G(\cdot)$, where $G(\cdot) \in \mathrm{do}_g(B^1 \boxtimes B^2 \to \pi_1^* B)$.

Let $U = \{U_\nu\}_{\nu \in \mathcal{N}}$ be a locally finite open covering of X consisting of coordinate neighborhoods, over which the bundles B^1, B^2 and B are trivial, and $\{\varphi_\nu\}$ be a partition of unity on X subordinated to U.

We may represent sections of $B^1|_{U_\mu}$ by (l_1-) rows, sections of $B^2|_{U_\nu}$ by (l_2-) columns, and sections of $B^1 \boxtimes B^2|_{U_\mu \times U_\nu}$ by $((l_2 \times l_1)-)$ matrices of functions on a corresponding set, where l_j is the rank of B^j $(j = 1,2)$.

Fix a repère $\{b_i^{(\mu)}\}_{i=1,\dots,l}$ for $B|_{U_\mu}$ where l is the rank of B. Then by Peetre's theorem the operator $G(\cdot,\cdot)$ in U_μ can be represented in the form

$$G^{(\mu)}(f_1,f_2) = \sum_{i=1}^{l} \left(\sum_{|\beta+\gamma| \le g} D^\beta f_1(x) G_{\beta,\gamma}^{(\mu,i)}(x) D^\gamma f_2(x) \right) b_i^{(\mu)}(x),$$

where $G_{\beta,\gamma}^{(\mu,i)}$ are $(l_1 \times l_2)$-matrices of differentiable functions in U_μ and $f_j \in \mathcal{E}(B^j|_{U_\mu})$ $(j = 1,2)$.

Set now for a section $f \in \mathcal{E}(B^1 \boxtimes B^2|_{U_\mu \times U_\nu})$

$$G^{(\mu)}(f) = \sum_{i=1}^{l} \mathrm{tr} \left(\sum_{|\beta+\gamma| \le g} G_{\beta,\gamma}^{(\mu,i)}(x) D_x^\beta D_y^\gamma f(x,y) \right) b_i^{(\mu)}(x), \tag{2.4.17}$$

where tr means the trace of a matrix, that is, the sum of its diagonal elements.

It is easy to see, that this equality defines operators $G^{(\mu)}(\cdot) \in \mathrm{do}_g((B^1 \boxtimes B^2)|_{U_\mu \times U_\nu} \to \pi_1^* B|_{U_\mu \times U_\nu})$. Moreover, if $f_1 \in \mathcal{E}(B^1|_{U_\mu})$ and $f_2 \in \mathcal{E}(B^2|_{U_\nu})$, then on $U_\mu \times U_\nu$ we have

$$\Delta^* G^{(\mu)}(f_1(x) \otimes f_2(y))$$

$$= \sum_{i=1}^{l} \operatorname{tr} \Delta^* \left(\sum_{|\beta+\gamma|\le g} G^{(\mu,i)}_{\beta,\gamma}(x) D^\beta f_1(x) \otimes D^\gamma f_2(y) \right) b^{(\mu)}_i(x) = G^{(\mu)}(f_1, f_2). \tag{2.4.18}$$

Finally for a section $f \in \mathcal{E}(B^1 \boxtimes B^2)$ we define

$$G(f) = \sum_{\mu,\nu} G^{(\mu)}\left(\varphi_\mu(x)\varphi_\nu(y) f(x,y)\right). \tag{2.4.19}$$

Since $\operatorname{supp} \varphi_\mu \subset U_\mu$ for all μ, equality (2.4.19) determines a well-defined operator $G(\cdot) \in \operatorname{do}_g(B^1 \boxtimes B^2 \to \pi_1^* B)$.

Moreover, if $f_j \in \mathcal{E}(B^j)$ ($j = 1, 2$), then in view of (2.4.18), we get

$$\begin{aligned} \Delta^* G(f_1(x) \otimes f_2(y)) &= \sum_{\mu,\nu} \Delta^* G^{(\mu)}((\varphi_\mu f_1)(x), (\varphi_\nu f_2)(y)) \\ &= \sum_{\mu,\nu} G(\varphi_\mu f_1, \varphi_\nu f_2) \\ &= G(f_1, f_2). \end{aligned}$$

That is, the operator G has the required properties.

□

We emphasize that formulas (2.4.17) and (2.4.19) provide a direct construction of the operator $G(\cdot)$.

Let S be an oriented closed piecewise smooth hypersurface in X. This means that S is the union of a finite number of pieces of submanifolds of X of codimension 1 (hypersurfaces) and of class C^1_{loc} with transversal intersections.

Suppose further we are given a (cochain) differential complex E on X.

Then with help of Theorem 2.4.7 we can associate to S the kernel $[S]^P \in \mathcal{D}'((E \boxtimes E')^1)$, defined by

$$\langle [S]^P, K \rangle_{X \times X} = \sum_{i=0}^{N-1} \int_S \Delta^* G_{P_i}(K) \quad \text{for} \quad K \in \oplus\, \mathcal{D}'(E^{i+1'} \boxtimes E^i). \tag{2.4.20}$$

Since $\operatorname{supp}[S]^P = \{(x,x) : \ x \in S\}$ is a subset of the diagonal Δ, the kernel $[S]^P$ is properly supported.

The corresponding operator from $\mathcal{L}(\mathcal{D}(E) \to \mathcal{D}'(E))^1$ is denoted by $[S]^P$ also.

If $f \in C^{p_i-1}_{loc}(E^i|_S)$, then $[S]^P f$ is the element of $\mathcal{D}'(E^{i+1})$ determined by the equality

$$\langle g, [S]^P f \rangle_X = \int_S G_{P_i}(g, f) \quad \text{for} \quad g \in \mathcal{D}(E^{i+1'}). \tag{2.4.21}$$

The following elementary result plays an important roll both in partial differential equations (see Hadamard [97], John [119], Shilov [200], Bers, John and Schechter [22], Bitsadze [23], Friedman [68], Wildenhain [255], Schulze and Wildenhain [193] and others) and in complex analysis (see Aizenberg and Yuzhakov [6], Henkin [104] and others). It forms the basis for the parametrix method.

Let us denote by χ_σ the characteristic function of a set $\sigma \subset X$.

Theorem 2.4.8 *Let K be a continuable parametrix of the complex $E^{\cdot}$, and $Y \subset X$ be an open set with a piecewise smooth boundary. Then for all sections $f \in C^{p_i}_{comp}(E^i|_{\overline{Y}})$ we have*

$$- K([\partial Y]^P f) + K(\chi_Y(Pf)) + PK(\chi_Y f) = \chi_Y f - S(\chi_Y f). \tag{2.4.22}$$

Notice that, if in addition, the parametrix K is properly supported, then formula (2.4.22) also holds for all sections $f \in C^{p_i}_{loc}(E^i|_{\overline{Y}})$.

Proof. Notice first of all that both $[\partial Y]^P f$ and $\chi_Y(Pf)$ are in $\mathcal{E}'(E^{i+1})$ because $f \in C^{p_i}_{comp}(E^i|_{\overline{Y}})$. Therefore the left-hand side of (2.4.22) is well-defined.

Further, since K is a continuable parametrix, homotopy formula (2.1.4) is valid for sections from $\mathcal{E}'(E^i)$ also. In particular, by applying it to $(\chi_Y f) \in \mathcal{E}'(E^i)$ we obtain

$$KP(\chi_Y f) + PK(\chi_Y f) = \chi_Y f - S(\chi_Y f).$$

But under the hypotheses of the theorem the following equality holds:

$$P(\chi_Y f) - \chi_Y(Pf) = -[\partial Y]^P f \quad \text{in} \quad \mathcal{E}'(E^{i+1}). \tag{2.4.23}$$

Indeed, applied to $g \in \mathcal{E}(E^{i+1'})$, (2.4.23) has the form

$$\int_Y \langle P'g, f\rangle_X - \int_Y \langle g, Pf\rangle_X = - \int_{\partial Y} G_P(g, f),$$

where this last equality follows from the definition of the Green operator by (2.4.11) and Stokes' formula.

By substitution of (2.4.23) into the homotopy formula, we get (2.4.22), which completes the proof.

□

Corollary 2.4.9 *Let K be a continuable fundamental solution of the complex E at degree i, and $Y \subset X$ be an open set with a piecewise smooth boundary. Then for all sections $f \in C^{p_i}_{comp}(E^i|_{\overline{Y}})$ the following formula holds:*

$$- K([\partial Y]^P f) + K(\chi_Y(Pf)) + PK(\chi_Y f) = \chi_Y f. \tag{2.4.24}$$

Formula (2.4.24) was first presented in the author's paper [223], where the following interesting corollary of (2.4.24) was formulated.

Corollary 2.4.10 *Under the conditions of Corollary 2.4.9, the homology class in $H^i(\mathcal{D}'(E^{\cdot}|_Y))$ of a section $f \in C^{p_i-1}_{comp}(E^i|_{\overline{Y}})$, satisfying $Pf = 0$ in Y, is completely determined by its component $K([\partial Y]^P f)$, and it depends only on the $(p_i - 1)$-jet of f on the boundary ∂Y.*

In view of Theorem 2.3.22, these corollaries can be applied to every non-degenerate complex of differential operators with constant coefficients in $\mathbb{R}^n$.

2.4.4 *Conservation laws for solutions of overdetermined systems*

Theorem 2.4.1 can be used to obtain a family of conservation laws for solutions of the overdetermined system $P_\iota f = 0$.

By a conservation law of degree q, we mean any differential operator G_ι of type $E^\iota \to \Lambda^q$ with the property that the differential form $G_\iota(f)$ is closed for solutions of the system $P_\iota f = 0$, that is for $f \in \mathcal{E}_{P_\iota}(X)$ (see Vinogradov [243] and others).

This definition generalizes the notion of the first integral of a system of differential equations because, if G_ι is a conservation law of degree q for solutions of the system $P_\iota f = 0$, then for $f \in \mathcal{E}_{P_\iota}(X)$ the integral $\int_C G_\iota(f)$ depends on the class $[c] \in H_q^{sign}(X)$ of q-dimensional singular homologies of a cycle $c \subset X$ only.

A conservation law G_ι is said to be trivial if the form $G_\iota(f)$ is exact for $f \in \mathcal{E}_{P_\iota}(X)$.

Conservation laws of degrees n and $(n-1)$ always exist. These are the form $(\langle g, P_\iota f\rangle_X - \langle P_\iota' g, f\rangle_X)$ of $\mathrm{do}_{p_\iota}((E^{\iota+1'}, E^\iota) \to \Lambda^n)$ and its primitive $G_{P_\iota}(g, f)$ of $\mathrm{do}_{p_\iota - 1}((E^{\iota+1'}, E^\iota) \to \Lambda^{n-1})$, where as g any element of $\mathcal{E}_{P_\iota'}(X)$ can been taken.

It appears that this family of conservation laws can be extended. In other words, for solutions of the system $P_\iota f = 0$ there are conservation laws of lesser degrees, and their range is so much the richer the more overdetermined the differential operator P_ι is, i.e. the more non-zero objects are in the complex $\mathcal{E}(E^\iota) \xrightarrow{P} \mathcal{E}(E^{\iota+1}) \xrightarrow{P} \dots$. One special result of such a type is contained in the paper of Vladimirov and Zharinov [248].

Of course, the last reasoning is of a heuristic character, however for many cases it can be given an exact sense. In particular, if $P \in \mathrm{do}_p(X \times \mathbb{C}^k \to X \times \mathbb{C}^l)$ is a differential operator with constant coefficients on an open set $X \subset \mathbb{R}^n$, then the number $o(P) = n - d - 1$ is said to be the overdeterminence degree of P where d is the dimension of the algebraic variety $\mathfrak{M}(M)$ associated to the $\mathcal{P}$-module $M = \mathcal{P}^k/(P(z))^t\mathcal{P}^l$ (see Subsection 2.3.5 and Subsection 5.3.6).

Theorem 2.4.11 *Assume that $G_\iota \in \mathrm{do}_g(E^\iota \to \Lambda^q)$ is an arbitrary conservation law for solutions of the system $P_\iota f = 0$. Then there exists a conservation law $G_{\iota-1} \in \mathrm{do}_{g+(p_{\iota-1}-1)}(E^{\iota-1} \to \Lambda^{q-1})$ for solutions of the system $P_{\iota-1}u = 0$ such that $dG_{\iota-1} = G_\iota \circ P_{\iota-1}$, and it is unique up to trivial conservation laws.*

Proof. Consider the differential operator $G = G_\iota \circ P_{\iota-1}$ of $\mathrm{do}_{g+p_{\iota-1}}(E^{\iota-1} \to \Lambda^q)$. Since sections of the type $P_{\iota-1}u$ are in $\mathcal{E}_{P_\iota}(X)$ for $u \in \mathcal{E}(E^{\iota-1})$, we get $dG = 0$. By Theorem 2.4.1, there exists an operator $G_{\iota-1} \in \mathrm{do}_{g+p_{\iota-1}-1}(E^{\iota-1} \to \Lambda^{q-1})$ such that $dG_{\iota-1} = G$. It immediately follows from this equality that $G_{\iota-1}$ is a conservation law for solutions of the system $P_{\iota-1}u = 0$.

It remains to show that this law is unique with accuracy up to trivial conservation laws for solutions of the system $P_{\iota-1}u = 0$. In fact, let $G^j \in \mathrm{do}_{g+p_{\iota-1}-1}(E^{\iota-1} \to \Lambda^{q-1})$ $(i = 1, 2)$ be differential operators such that $dG^j = G_\iota \circ P_{\iota-1}$ for $j = 1, 2$. Then $d(G^2 - G^1) = 0$, and we conclude by Theorem 2.4.1 that $G^2 - G^1 = dT$ for some differential operator $T \in \mathrm{do}_{g+p_{\iota-1}-2}(E^{\iota-1} \to \Lambda^{q-2})$. This completes the proof. □

Corollary 2.4.12 *Every conservation law $G_\imath \in \mathrm{do}_g(E^\imath \to \Lambda^q)$ for solutions of the system $P_\imath f = 0$ determines the pairing $\langle [c], [f] \rangle = \int_c G_\imath(f)$ between the groups $H_q^{sing}(X)$ and $H^\imath(\mathcal{E}(E^\cdot))$, and $G_\imath$ is trivial if and only if this pairing is zero.*

Proof. It follows from Theorem 2.4.11 that the bilinear form $\langle c, f \rangle = \int_c G_\imath(f)$, given a priori for q-dimensional cycles $c \subset X$ and sections $f \in \mathcal{E}_{P_\imath}(X)$, determines a pairing between the groups $H_q^{sing}(X)$ and $H^\imath(\mathcal{E}(E^\cdot))$. Moreover, by Theorem 17' of de Rham [50] this pairing $\langle [c], [f] \rangle$ is zero if and only if the differential form $G_\imath(f)$ is exact for $f \in \mathcal{E}_{P_\imath}(X)$, which completes the proof.

□

Investigation of this relation seems to be a promising problem having a great deal of physical sense (see Vinogradov [243]).

In particular, if $H^\imath(\mathcal{E}(E^\cdot)) = 0$, then all conservation laws for solutions of the system $P_\imath f = 0$ are trivial.

A more careful analysis of the construction of a primitive $G_{\imath-1}$ in Theorem 2.4.11 using Remark 2.4.2 allows one to obtain a further development of Proposition 2.4.6. It is convenient to understand this via the examples of the classical de Rham and Dolbeault complexes, where we have $G_\imath(g, f) = g \wedge f$ with accuracy up to a sign.

Corollary 2.4.13 *There exist sequences of bidifferential operators*

$$G_\imath \in \mathrm{do}_{p_0+\ldots+p_\imath-(\imath+1)}((E^{\imath+1'}, E^0) \to \Lambda^{n-\imath-1})$$

and

$$\tilde{G}_\imath \in \mathrm{do}_{p_1+\ldots+p_{\imath+1}-(\imath+1)}((E^{\imath+2'}, E^1) \to \Lambda^{n-\imath-1})$$

($i \in \mathbb{Z}_+$) such that:

1) $G_0 = G_{P_0}$ and

2) $dG_\imath(g, f) = G_{\imath-1}(P'g, f) + \tilde{G}_{\imath-1}(g, P_0 f)$ for all $g \in \mathcal{E}(E^{\imath+1'})$ and $f \in \mathcal{E}(E^0)$.

Exercise 2.4.14 *Prove Corollary 2.4.13.*

Hint. Use Theorem 2.4.1 and Remark 2.4.2.

□

Exercise 2.4.15 *Apply the approach above to explore non-linear systems of differential equations.*

Hint. Given some non-linear system of differential equations

$$P_\imath(x, \{D^\alpha f_\jmath\}) = 0 \quad (i = 1, \ldots, l),$$

let V denote the variety in the space of variables $(x, \{f_\imath^\alpha\})$ corresponding to the $(\infty-)$ prolongation of the system, i.e.,

$$V = \left\{ \left(x, \{f_\jmath^\alpha\}\right) : \ D^\beta P_\imath \left(x, \{f_\jmath^\alpha\}\right) = 0 \quad (\beta \in \mathbb{Z}_+^n) \right\}$$

where

$$D_\nu = \frac{1}{\sqrt{-1}}\frac{\partial}{\partial x_\nu} + \sum_{j,\alpha} f_j^{\alpha+1_\nu}\frac{\partial}{\partial f_j^\alpha} \quad (\nu = 1,\ldots,n)$$

are the full derivative operators.

In the general position situation, the variables $(x, \{f_j^\alpha\})$ on V may be expressed through finitely many of them which we shall consider as coordinates on V.

The differentiations D_ν admit restrictions to $\mathcal{E}(V)$ because $F = 0$ on V implies $D_\nu F = 0$ on V for every $F = F(x, \{f_j^\alpha\})$.

Consider the linear differential operator $P: [\mathcal{E}(V)]^k \to [\mathcal{E}(V)]^l$ given by

$$PF = \sum_\alpha \left\| \frac{\partial P_i}{\partial f_j^\alpha} \right\| D^\alpha F.$$

The system $PF = 0$ is known to describe infinitesimal symmetries of the initial non-linear system. Apply to the differential operator P the results above (cf. Zharinov [261]).

□

2.5 The Most Immediate Corollaries and Examples

2.5.1 *The homotopy formula for elliptic complexes on a manifold with boundary*

Let $E^\cdot$ be an elliptic complex on a differentiable manifold X countable at infinity.

By Theorem 2.1.12, this complex has a pseudo-differential parametrix $K \in \mathcal{D}'((E^\cdot \boxtimes E^{\cdot\prime})^{-1}$ with homogeneous components $K_i \in \mathrm{pdo}_{-p_{i-1}}(E^i \to E^{i-1})$ $(i = 1,\ldots,N)$. Fix such a parametrix.

Corollary 2.5.1 *Let Y be an open subset of X with a piecewise smooth boundary. Then for all sections $f \in C^{p_i}_{comp}(E^i|_{\overline{Y}})$ the following formula holds:*

$$-\int_{\partial Y} G_{P_i}(K(x,y), f(y)) + \int_Y K(x,y)Pf(y) + P\int_Y K(x,y)f(y) + \int_Y S(x,y)f(y) = \begin{cases} f(x), & x \in Y, \\ 0, & x \notin Y. \end{cases} \tag{2.5.1}$$

Notice that the juxtaposition Kf here means the evaluation pairing $\langle K, f\rangle_y$. We will be using this convention unabashedly below.

Proof. In view of formula (2.4.22) it is sufficient to show, that for $x \notin \partial Y$ we have

$$K([\partial Y]^P f) = \int_{\partial Y} G_{P_i}(K(x,y), f(y)). \tag{2.5.2}$$

In fact, let the section $g \in \mathcal{D}(E^{\iota\prime})$ be such that $\sigma = \operatorname{supp} g$ does not intersect with ∂Y. Since $K \in \mathcal{E}((E^{\cdot} \boxtimes E^{\prime})^{-1}|_{\sigma \times \partial Y})$, we get

$$\begin{aligned}\langle g, K([\partial Y]^P f)\rangle_X &= \langle K'g, [\partial Y]^P f\rangle_X \\ &= \int_{\partial Y} G_{P_\iota}\left(\left(\int_X \langle g(x), K(x,y)\rangle_X\right), f(y)\right) = \\ &= \left\langle g(x), \int_{\partial Y} G_{P_\iota}(K(x,y), f(y))\right\rangle_X .\end{aligned}$$

This proves (10.2) as required.

□

Observe that, since $K(x,\cdot) \in \oplus L^1_{loc}(E_x^{\iota-1} \otimes E^{\iota\prime})$ for each $x \in X$, the integrals over Y on the left-hand side of (10.1) converge absolutely.

2.5.2 *Fundamental solution of convolution type for the de Rham complex in* $\mathbb{R}^n$

Let X be an orientable differential manifold of dimension n.

It follows from Theorem 2.3.22 that locally the de Rham complex on X has a fundamental solution of convolution type. To describe this fundamental solution it is sufficient to consider the complex $\Lambda^{\cdot} = \Lambda^{\cdot} T^*_{\mathbb{C}}(\mathbb{R}^n)$.

There is a natural pairing $\langle,\rangle_x : \Lambda^{n-q} \otimes \Lambda^q \to \Lambda^n$, given by the wedge product in each fiber, $\langle g, f\rangle_x = f \wedge g$. Since $\Omega \cong \Lambda^n$, this allows one to identify $(\Lambda^q)'$ with Λ^{n-q}.

Here the transpose to the exterior derivative operator is $(d_q)' = (-1)^{q+1} d_{n-q-1}$, and the corresponding Green operator $G_{d_q} \in \mathrm{do}_0((\Lambda^{n-q-1}, \Lambda^q) \to \Lambda^{n-1})$ is given by $G_{d_q}(g, f) = f \wedge g$.

A Hermitian metric in Λ^q is usually introduced by means of the inner product of Hodge in $\mathcal{D}(\Lambda^q)$,

$$(f, g) = \int f \wedge \star g \quad \text{for} \quad f, g \in \mathcal{D}(\Lambda^q).$$

Here $\star : \Lambda^q \to \Lambda^{n-q}$ is the so-called star operator of Hodge (see de Rham [50, §24]). This is the conjugate linear bundle isomorphism given on basic forms by

$$\star dx_I = \sigma(I)\, dx[I] \quad (\# I = q),$$

where $dx[I]$ is the wedge product of the differentials $dx_1, \ldots, dx_n$ one after another excepting dx_ι for $i \in I$, and the constant $\sigma(I) = \pm 1$ is determined from the condition $dx_I \wedge \star dx_I = dx$.

Then the (formal) adjoint d_q^* to the exterior derivative operator is the operator $\star^{-1} d_q' \star = (-1)^{nq+1} \star d_{n-q-1} \star$ usually denoted by δ (see de Rham [50, §25]).

Proposition 2.5.2 *The Laplacian $\Delta = dd^* + d^*d$ of the de Rham complex in $\mathbb{R}^n$ is the same as the usual Laplace operator with the opposite sign acting on each coefficient of a differential form.*

Proof. Indeed, an easy calculation shows that

$$\Delta_q = d_{q-1}d^*_{q-1} + d^*_q d_q = \left(\sum_{j=1}^n D_j^2\right) I_{k_q},$$

where $k_q = \binom{n}{q}$. This is our assertion.

□

After these explanations, let us begin constructing a fundamental solution of convolution type for the complex $\Lambda^{\cdot}$.

We define more precisely the notion of convolution of two currents. Let g_j $(j = 1, 2)$ be currents of degrees q_j in $\mathbb{R}^n$, and let one of them have compact support. Then the convolution $g_1 * g_2$ of g_1 and g_2 is defined to be the current of degree $(q_1 + q_2 - n)$ in $\mathbb{R}^n$ given by

$$\langle g_1 * g_2, f\rangle_{\mathbb{R}^n} = \langle g_1(x) \wedge g_2(y), f(x+y)\rangle_{\mathbb{R}^n\times\mathbb{R}^n} \quad \text{for} \quad f \in \mathcal{D}(\Lambda^{2n-q_1-q_2}),$$

where $f(x+y)$ is the pull-back of a form f under the mapping $(x, y) \to x + y$ of $\mathbb{R}^n \times \mathbb{R}^n \to \mathbb{R}^n$.

Notice that $\deg(g_1 * g_2) = \deg g_1 - \dim g_2$.

Consider the Newtonian potential on $\mathbb{R}^n$ as the n-form $(-g(x)dx)$ where $g(x)$ is the standard fundamental solution of the Laplace equation in $\mathbb{R}^n$. It is easy to see that

$$\Delta(-g(x)dx) = dd^*(-g(x)dx) = \delta(x)dx.$$

Set $\Phi(x) = d^*(-g(x)dx)$. Then $d\Phi(x) = \delta(x)dx$. Let $\Phi(x, y)$ denote the associated kernel of convolution type, i.e. $\Phi(x, y) = (-1)^n\Phi(x - y)$ so that $\Phi f = \Phi * f$.

Let us denote by $[\Delta]$ the current of integration over the diagonal submanifold of $\mathbb{R}^n \times \mathbb{R}^n$. Obviously, $[\Delta]$ is of $\mathcal{D}'(\Lambda^n T^*_{\mathbb{C}}(\mathbb{R}^n \times \mathbb{R}^n))$, and its representation by a differential form is $[\Delta] = \delta(x - y)\, d(x - y)$.

Proposition 2.5.3 *The kernel $\Phi(x, y)$ satisfies the fundamental equation on $\mathbb{R}^n \times \mathbb{R}^n$:*

$$d\Phi = (-1)^n[\Delta]. \tag{2.5.3}$$

Proof. Indeed, we get by Proposition 1.2.38

$$d\Phi(x, y) = (-1)^n(d\Phi)(x - y) = (-1)^n\delta(x - y)\, d(x - y)$$

which completes the proof.

□

Every solution $K \in \mathcal{D}'(\Lambda^{n-1}T^*_{\mathbb{C}}(X \times X))$ of the fundamental equation $dK = (-1)^n[\Delta]$ on $X \times X$ is called a fundamental solution of the exterior derivative operator on the manifold X.

It is clear that $\Phi(x, y) = (-1)^n\Phi(y, x)$. In general, if $K(x, y)$ is a fundamental solution of the de Rham complex on X then $K'(x, y) = (-1)^n K(y, x)$ is also a fundamental solution of this complex. This follows from the above identification of the transposed complex for the de Rham complex with itself.

Denote by $\Phi_q(x,y)$ the component of $\Phi(x,y)$ of degree $(n-q)$ in y, multiplied by $(-1)^{n(q-1)+q}$ and considered already as a double differential form (see de Rham [50, §7]), where one must first rearrange all differentials in x after the differentials in y.

Equality (2.5.3) means

$$d_{q-1}(x)\Phi_q(x,y) + (-1)^{q+1} d_{n-q-1}(y)\Phi_{q+1}(x,y) = \delta(x-y)\left(\sum\nolimits'_{\#I=q}(\star\, dy_I)\, dx_I\right). \tag{2.5.4}$$

The following simple properties of the double forms Φ_q will be of use later on. They were noted in the author's paper [219].

Let σ_n denote the "area" of the unit sphere in $\mathbb{R}^n$, namely $\sigma_n = 2\pi^{n/2}/\Gamma(n/2)$.

Proposition 2.5.4 *For* $1 \le q \le n$,

$$\begin{aligned}\Phi_q(x,y) &= -\sum\nolimits'_{\#I=q-1} \star_y \left(d_y g(x-y) \wedge dy_I\right) dx_I \\ &= \frac{1}{\sigma_n} \sum\nolimits'_{\#I=q-1} \sum_{\substack{\imath=1 \\ \imath \notin I}}^{n} \frac{x_\imath - y_\imath}{|x-y|^n} \star_y \left(dy_\imath \wedge dy_I\right) dx_I .\end{aligned}$$

Proposition 2.5.5 *If* $1 \le q \le n$, *then*

$$\Phi_q(x,y) = (-1)^{n(q-1)+1}\ \Phi_{n-q+1}(y,x).$$

Formula (2.5.1) implies the next result due to Norguet [172].

Corollary 2.5.6 *If Y is an open subset of $\mathbb{R}^n$ with a piecewise smooth boundary, and $f \in C^1_{comp}(\Lambda^q|_{\overline{Y}})$, then*

$$\begin{aligned} &-\int_{\partial Y} f(y) \wedge \Phi_{q+1}(x,y) \\ &\quad + \int_Y df(y) \wedge \Phi_{q+1}(x,y) + d\int_Y f(y) \wedge \Phi_q(x,y) = \begin{cases} f(x), & x \in Y, \\ 0, & x \in \mathbb{R}^n \setminus \overline{Y}. \end{cases} \end{aligned} \tag{2.5.5}$$

Some applications of formula (2.5.5) for studying the equation $du = f$ and close questions have been given by the author [219].

2.5.3 *Fundamental solution of convolution type for the Dolbeault complex in $\mathbb{C}^n$*

Assume that X is a complex manifold of dimension n.

By Theorem 2.3.22, the Dolbeault complex on X with coefficients from $\Lambda^{p,0}$ has locally a fundamental solution of convolution type. To construct it, let us consider the complex $\Lambda^{p,\cdot} = \Lambda^{p,\cdot} T^*_{\mathbb{C}}(\mathbb{C}^n)$ in $\mathbb{C}^n$.

First of all there is a natural pairing $\langle,\rangle_x$ (or tr) : $\Lambda^{n-p,n-q}\otimes\Lambda^{p,q}\to\Lambda^{n,n}$ defined by the wedge product in each fiber, $\langle g,f\rangle_x = f\wedge g$. Since $\Omega\cong\Lambda^{n,n}$, this allows one to identify the dual bundle for $\Lambda^{p,q}$ with $\Lambda^{n-p,n-q}$.

Here the transpose to the $\overline{\partial}^{(p)}_q$ operator is $(-1)^{p+q+1}\overline{\partial}^{(n-p)}_{n-p-1}$, and the corresponding Green operator of do$_0((\Lambda^{n-p,n-q-1},\Lambda^{p,q})\to\Lambda^{n,n-1})$ is given by $G_{\overline{\partial}^{(p)}_q}(g,f) = f\wedge g$.

In particular, the transposed complex to the complex $\Lambda^{p,\cdot}$ is isomorphic to $\Lambda^{n-p,\cdot}$.

The Hermitian metric on Λ^{p+q} induces a Hermitian metric on $\Lambda^{p,q}$, and the star operator of Hodge gives a conjugate linear bundle isomorphism $\star:\ \Lambda^{p,q}\to\Lambda^{n-p,n-q}$. Then the (formal) adjoint $(\overline{\partial}^{(p)}_q)^*$ to the operator $(\overline{\partial}^{(p)}_q)$ is determined by $\star^{-1}(\overline{\partial}^{(p)}_q)'\star = -\star\,\overline{\partial}^{(n-p)}_{n-q-1}\star$.

Proposition 2.5.7 *The Laplacian $\Delta = \overline{\partial}^{(p)}\overline{\partial}^{(p)^*} + \overline{\partial}^{(p)^*}\overline{\partial}^{(p)}$ of the complex $\Lambda^{p,\cdot}$ is the same as the usual Laplace operator in $\mathbb{R}^{2n}$ acting on each coefficient of a differential form, with accuracy up to the factor $(-\frac{1}{2})$.*

Proof. Indeed, an easy calculation shows that

$$\Delta_q = \overline{\partial}^{(p)}_{q-1}\,\overline{\partial}^{(p)\,*}_{q-1} + \overline{\partial}^{(p)*}_q\,\overline{\partial}^{(p)}_q = \left(-2\sum_{j=1}^{n}\frac{\partial^2}{\partial z_j\,\partial\overline{z}_j}\right) I_{k_q},$$

where $k_q = \binom{n}{p}\binom{n}{q}$. This is our statement.

□

Newton's potential $\left(-2g(z)\,(\frac{\sqrt{-1}}{2})^n\,dz\wedge d\overline{z}\right)$ on $\mathbb{C}^n$ satisfies

$$\overline{\partial}\overline{\partial}^*(\cdot) = \delta(z)\left(\frac{\sqrt{-1}}{2}\right)^n dz\wedge d\overline{z}.$$

Set $\Phi(z) = \overline{\partial}^*\left(-2g(z)\left(\frac{\sqrt{-1}}{2}\right)^n dz\wedge d\overline{z}\right)$, then $\overline{\partial}\Phi(z) = \delta(z)\left(\frac{\sqrt{-1}}{2}\right)^n dz\wedge d\overline{z}$. Denote by $\Phi(z,\zeta)$ the associated kernel of convolution type, i.e. $\Phi(z,\zeta) = \Phi(z-\zeta)$, so that $\Phi(z,\zeta)$ is in $\mathcal{D}'(\Lambda^{n,n-1}T^*_{\mathbb{C}}(\mathbb{C}^n\times\mathbb{C}^n))$ and $\Phi f = \Phi * f$.

Proposition 2.5.8 *The kernel $\Phi(z,\zeta)$ satisfies on $\mathbb{C}^n\times\mathbb{C}^n$ the fundamental equation:*

$$\overline{\partial}\Phi = [\Delta]. \tag{2.5.6}$$

Proof. Since the mapping $(z,\zeta)\to z-\zeta$ of $\mathbb{C}^n\times\mathbb{C}^n\to\mathbb{C}^n$ is holomorphic, we get by Proposition 1.2.46

$$\overline{\partial}\Phi(z,\zeta) = (\overline{\partial}\Phi)(z-\zeta) = \delta(z-\zeta)\left(\frac{\sqrt{-1}}{2}\right)^n d(z-\zeta)\wedge d(\overline{z-\zeta})$$

as required.

□

Every solution $K\in\mathcal{D}'(\Lambda^{n,n-1}T^*_{\mathbb{C}}(X\times X))$ of the fundamental equation $\overline{\partial}K = [\Delta]$ on $X\times X$ is called a fundamental solution of the $\overline{\partial}$ operator on a (complex) manifold X.

This terminology is due to Harvey and Polking [100].

Let $\Phi_q^{(p)}(z,\zeta)$ be the component of $\Phi(z,\zeta)$ of bidegree $(n-p, n-q)$ in ζ multiplied by $(-1)^{p+q}$ and considered as a double differential form, where one must first rearrange all differentials in z and $\bar{z}$ after the differentials in ζ and $\bar{\zeta}$.

Equality (2.5.6) means that

$$\begin{aligned}\bar{\partial}_{q-1}^{(p)}(z)\ \Phi_q^{(p)}(z,\zeta) + (-1)^{p+q+1}\bar{\partial}_{n-q-1}^{(n-p)}(\zeta)\ \Phi_{q+1}^{(p)}(z,\zeta) \\ = \delta(z-\zeta)\left(\frac{1}{2^{p+q}}{\sum}'_{\substack{\#J=p\\ \#I=q}} \star_\zeta \left(d\bar{\zeta}_J \wedge d\bar{\zeta}_I\right) dz_J \wedge d\bar{z}_I\right)\end{aligned} \qquad (2.5.7)$$

The following results are completely analogous to Propositions 2.5.4 and 2.5.5.

Proposition 2.5.9 *For* $1 \le q \le n$,

$$\begin{aligned}&\Phi_q^{(p)}(z,\zeta) \\ &= \frac{(-1)}{2^{p+q-2}}{\sum}'_{\substack{\#J=p\\ \#I=q-1}} \star_\zeta \left(\bar{\partial}_\zeta\, g(z-\zeta) \wedge d\zeta_J \wedge d\bar{\zeta}_I\right)\ dz_J \wedge d\bar{z}_I \\ &= \frac{1}{\sigma_{2n}\, 2^{p+q-1}}{\sum}'_{\substack{\#J=p\\ \#I=q-1}} \sum_{\substack{\imath=1\\ \imath\notin I}}^{n} \frac{\bar{z}_\imath - \bar{\zeta}_\imath}{|z-\zeta|^{2n}} \star_\zeta \left(d\bar{\zeta}_\imath \wedge d\zeta_J \wedge d\bar{\zeta}_I\right) dz_J \wedge d\bar{z}_I\end{aligned}$$

Proposition 2.5.10 *Given* $1 \le q \le n$, *then*

$$\Phi_q^{(p)}(z,\zeta) = -\Phi_{n-q+1}^{(n-p)}(\zeta, z).$$

The following result was proved by Koppelman [136] for differential forms of bidegree $(0,q)$. The generalization to forms of bidegree (p,q) with arbitrary p is immediate (see Aizenberg and Dautov [5, §1]).

Corollary 2.5.11 *Let* Y *be an open subset of* $\mathbb{C}^n$ *with a piecewise smooth boundary and* $f \in C^1_{comp}(\Lambda^{p,q}|_{\overline{Y}})$. *Then*

$$\begin{aligned}&-\int_{\partial Y} f(\zeta)\wedge\Phi_{q+1}^{(p)}(z,\zeta) \\ &\quad + \int_Y \bar{\partial} f(\zeta)\wedge\Phi_{q+1}^{(p)}(z,\zeta) + \bar{\partial}\int_Y f(\zeta)\wedge\Phi_q^{(p)}(z,\zeta) = \begin{cases} f(z), & z \in Y, \\ 0, & z \in \mathbb{C}^n \setminus \overline{Y}. \end{cases}\end{aligned} \qquad (2.5.8)$$

Fruitful applications of formula (2.5.8) in multidimensional complex analysis have been elucidated in the monograph of Aizenberg and Dautov [5].

2.5.4 Green formulas

Let $P \in \mathrm{do}_p(E \to F)$ be a differential operator with injective symbol on a differentiable manifold X countable at infinity.

It follows from Theorem 1.5.11 that P has a left parametrix $K \in \mathrm{pdo}_{-p}(F \to E)$, i.e. $KP = 1 - S$ for some smoothing operator $S \in \mathrm{pdo}_{-\infty}(E \to E)$.

For example, one could take $\Pi = GP^*$ where $G \in \mathrm{pdo}_{-2p}(E \to E)$ is a parametrix of the Laplacian P^*P, an elliptic differential operator of type $E \to E$ and order $2p$ on X.

Fix some left parametrix K for P with the properties above.

Proposition 2.5.12 (Green formula) *If Y is an open subset of X with a piecewise smooth boundary, then for all sections $f \in C^p_{comp}(E|_{\overline{Y}})$ the following formula holds:*

$$-\int_{\partial Y} G_P(K(x,y), f(y)) + \int_Y K(x,y)Pf(y) + \int_Y S(x,y)f(y) = \begin{cases} f(x), & x \in Y, \\ 0, & x \in X \setminus \overline{Y}. \end{cases} \quad (2.5.9)$$

Proof. Apply the equality $KP = 1 - S$ to the section $\chi_Y f \in \mathcal{E}'(E)$. Since also $\chi_Y(Pf) \in \mathcal{E}'(F)$ we obtain

$$-K(\chi_Y(Pf) - P(\chi_Y f)) + K(\chi_Y(Pf)) + S(\chi_Y f) = \chi_Y f.$$

To end the proof, use formulas (2.4.23) and (2.5.2) and take into account that $K(x,\cdot) \in L^1_{loc}(E_x \otimes F')$ for each $x \in X$.

□

Remark 2.5.13 *Of course, we could construct a compatibility complex $E^{\cdot}$ for P as in Subsection 1.3.9, then take a pseudo-differential parametrix K for $E^{\cdot}$ guaranteed by Corollary 2.1.14, and obtain (2.5.9) as special case of (2.5.1) for $i = 0$.*

In particular, let us write formula (2.5.9) for the Laplacian $\Delta_i = P_{i-1}\tilde{P}^*_{i-1} + \tilde{P}^*_i P_i$ (see (2.1.8)) of an elliptic complex $E^{\cdot}$ on X.

Let G_i be a parametrix for Δ_i, so that $G_i\Delta_i = 1 - S_i$ with some smoothing operator $S_i \in \mathrm{pdo}_{-\infty}(E^i \to E^i)$.

Further, according to Proposition 2.4.5 we have

$$\begin{aligned} G_{\Delta_i}(g,f) &= G_{P_i}(\tilde{P}^{*\prime}_i g, f) + G_{\tilde{P}^*_{i-1}}(P'_{i-1} g, f) \\ &+ G_{P_{i-1}}(g, \tilde{P}^*_{i-1} f) + G_{\tilde{P}^*_i}(g, P_i f). \end{aligned} \quad (2.5.10)$$

Corollary 2.5.14 *If Y is an open subset of X with piecewise smooth boundary, then for every section $f \in C^{2p}_{comp}(E^i|_{\overline{Y}})$, the following formula holds:*

$$\begin{aligned} -\int_{\partial Y} \Big(& G_{P_i}\left(\tilde{P}^{*\prime}_i(y)G_i(x,y), f(y)\right) + G_{\tilde{P}^*_{i-1}}\left(P'_{i-1}(y)G_i(x,y), f(y)\right) \\ & + G_{P_{i-1}}\left(G_i(x,y), \tilde{P}^*_{i-1} f(y)\right) + G_{\tilde{P}^*_i}\left(G_i(x,y), P_i f(y)\right)\Big) \\ & + \int_Y G_i(x,y)\Delta_i f(y) + \int_Y S_i(x,y) f(y) = \begin{cases} f(x), & x \in Y, \\ 0, & x \in X \setminus \overline{Y}. \end{cases} \end{aligned} \quad (2.5.11)$$

Apparently, such a formula was first used for the de Rham complex by Duff and Spencer [59].

Corollary 2.5.15 *Let Y be an open subset of X with piecewise smooth boundary. Then, for every section $f \in C^{\max(\widetilde{p}_{i-1}, p_i)}_{comp}(E^i|_{\overline{Y}})$, the following formula holds:*

$$\begin{aligned}
&-\int_{\partial Y} \left(G_{P_i}\left(\widetilde{P}_i^{*\prime}(y)G_i(x,y), f(y)\right) + G_{\widetilde{P}^*_{i-1}}\left(P'_{i-1}(y)G_i(x,y), f(y)\right)\right)\\
&+\int_Y \left(\left(\widetilde{P}_i^{*\prime}(y)G_i(x,y)\right) P_i f(y) + \left(P'_{i-1}(y)G_i(x,y)\right)\widetilde{P}^*_{i-1} f(y)\right)\\
&+\int_Y S_i(x,y)f(y) = \begin{cases} f(x), & x \in Y, \\ 0, & x \in X \setminus \overline{Y}. \end{cases}
\end{aligned} \qquad (2.5.12)$$

Proof. Apply Stokes' formula to the last two integrals over the boundary in (2.5.11).

□

2.5.5 *Fundamental solution of convolution type for the Koszul complex*

Let $E^\cdot$ be a Koszul complex in $\mathbb{R}^n$ associated with an $(N-)$ column P of scalar differential operators with constant coefficients (see Subsection 1.2.7).

It follows from Theorem 2.3.22 that E has a fundamental solution of convolution type. Let us explain more precisely its construction. To do this, we need the following technical statement.

Lemma 2.5.16 *If d is the exterior derivative operator corresponding to the space $\mathbb{R}^N$, then for all $z^j \in \mathbb{C}^N$ $(j = 1, 2)$*

$$d_{q-1}(z^1)\ d^*_{q-1}(z^2) + d^*_q(z^2)\ d_q(z^1) = \langle z^1,\ z^2\rangle\ I_{k_q},$$

where $k_q = \binom{N}{q}$.

Proof. This is obtained by direct calculations.

□

Fix now any $(N-)$ column $\widetilde{P}$ of scalar differential operators with constant coefficients in $\mathbb{R}^n$ such that $\widetilde{P}^*P \not\equiv 0$. For example, one can choose $\widetilde{P} = P$.

Let us denote by $\widetilde{E} = \{E^q, \widetilde{P}_q\}$ the Koszul complex in $\mathbb{R}^n$ associated with the column $\widetilde{P}$.

Proposition 2.5.17 *The Laplacian $\Delta = P\widetilde{P}^* + \widetilde{P}^*P$ of the complex $E^\cdot$ is the same as the scalar differential operator $\widetilde{P}^*P$ acting on each component of a test section.*

Proof. Indeed, by Lemma 2.5.16 we get

$$\begin{aligned}\Delta_q &= P_{q-1}\tilde{P}^*_{q-1} + \tilde{P}^*_q P_q \\ &= d_{q-1}\left(\frac{1}{\sqrt{-1}}P(D)\right)\left(d_{q-1}\left(\frac{1}{\sqrt{-1}}\tilde{P}(D)\right)\right)^* + \left(d_q\left(\frac{1}{\sqrt{-1}}\tilde{P}(D)\right)\right)^* d_q\left(\frac{1}{\sqrt{-1}}P(D)\right) \\ &= (\tilde{P}^*P)I_{k_q},\end{aligned}$$

where $k_q = \binom{N}{q}$. This completes the proof.

□

Let $g(x)$ be a fundamental solution of convolution type for the scalar differential operator $\tilde{P}^*P$. For every $q \in \mathbb{Z}$, we set $\Phi_q(x) = \tilde{P}^*_{q-1}(D)g(x)$ so that $\Phi_q \in \mathcal{D}'(E^{q-1} \otimes E^{q*})$.

Proposition 2.5.18 *The kernel $\Phi \in \mathcal{D}'((E^\cdot \boxtimes E^{\cdot *})^{-1})$ of convolution type, associated to $(\otimes\Phi_q(x))$, is a fundamental solution of the complex $E^\cdot$.*

Proof. Indeed, equality (2.3.10) is easily verified by using Proposition 2.5.17.

□

Exercise 2.5.19 *Give formula (2.5.1) more precisely for elliptic Koszul complexes.*

Hint. Use Proposition 1.2.51.

□

2.5.6 *Homotopy formula on manifolds with boundary for Asada complexes*

Consider some Asada complex $E^\cdot$ in $\mathbb{R}^n$, associated with the column $P(D) = MD$, where the $(N \times n)$-matrix M satisfies condition (1.2.23) (see Subsection 1.2.8).

Set $\tilde{P}(D) = \tilde{M}D$. Then (1.2.23) implies that $\tilde{P}^*P = -\Delta$ where Δ is the Laplace operator in $\mathbb{R}^n$.

Let $\tilde{E}^\cdot = \{E^q, \tilde{P}_q\}$ be the Asada complex in $\mathbb{R}^n$ associated with the column $\tilde{P}$.

It follows from Proposition 2.5.17 that the Laplacian $\Delta = P\tilde{P}^* + \tilde{P}^*P$ of the complex $E^\cdot$ is the same as the Laplace operator with the opposite sign acting on each component of a test section.

For very $q \in \mathbb{Z}$, we set $\Phi_q(x) = \tilde{P}^*_{q-1}(D)(-g(x))$ where $g(x)$ is the standard fundamental solution of the Laplace operator in $\mathbb{R}^n$.

By Proposition 2.5.18, the kernel $\Phi \in \mathcal{D}'((E^\cdot \boxtimes E^{\cdot *})^{-1})$ of convolution type, associated to $(\oplus\Phi_q(x))$, is a fundamental solution of $E^\cdot$.

Now it is clear how to sharpen Corollary 2.5.1 for Asada complexes.

Corollary 2.5.20 *If Y is an open subset of $\mathbb{R}^n$ with a piecewise smooth boundary and $f \in C^1_{comp}(E^q|_{\overline{Y}})$, then the following formula holds:*

$$- \int_{\partial Y} \Phi_{q+1}(x,y) \left(\frac{1}{\sqrt{-1}} \sigma(P_q)(y, \nu(y))\, f(y) \right) ds$$
$$+ \int_Y \Phi_{q+1}(x,y) P_q f(y) dy + P_{q-1} \int_Y \Phi_q(x,y) f(y) dy = \begin{cases} f(x), & x \in Y, \\ 0, & x \in \mathbb{R}^n \setminus \overline{Y}. \end{cases} \tag{2.5.13}$$

Proof. Use formula (2.5.1). Then, to complete the proof, it remains to observe from (2.4.13) that, if a differential operator P has order 1, for the restriction of the Green operator $G_P(gf)$ to a smooth piece S of the hypersurface ∂Y, we have the formula $G_P(g,f)|_S = g\left(\frac{1}{\sqrt{-1}}\, \sigma(P)(y, \nu(y)) f\right) ds$ where $\nu(y)$ is the unit outward normal vector to S at the point y, and ds is the induced volume form on S.

□

The particular case of formula (2.5.13) for $q = 0$ is due to Asada [1].

It is clear that this formula contains Norguet's formula (2.5.5) and Koppelman's formula (2.5.8) as very special cases.

Chapter 3

Sokhotskii-Plemelj Formulas for Elliptic Complexes

3.1 Formally Non-characteristic Hypersurfaces for Differential Complexes. The Tangential Complex

3.1.1 *Preliminaries*

Suppose that X is an oriented differentiable manifold of dimension n countable at infinity and $E^{\cdot}$ is a (cochain) differential complex on X.

Let S be a closed differentiable submanifold of X of codimension 1 (i.e. a hypersurface). We assume that S has two sides in X.

This latter condition means the following: 1) S is oriented; 2) $X \setminus S = X^- \cup X^+$, where X^+ and X^- are disjoint non-void open subsets of X, and 3) the orientation of S is induced from the orientation of X^-, i.e. at each point of S the positive normal to S points into X^+.

Then $\overline{X^{\pm}}$ are oriented differentiable manifolds with boundaries, and $\partial \overline{X^{\pm}} = \mp S$.

Under the assumptions listed above, S has a global definition.

Proposition 3.1.1 *These exists a real-valued function $\rho \in C^{\infty}_{loc}(X)$ such that $X^{\pm} = \{x \in X : \pm\rho(x) > 0\}$ and $S = \{x \in X : \rho(x) = 0\}$, and $d\rho|_S \neq 0$.*

Proof. This is quite simple. See Andreotti and Hill [9, p.328].

□

In particular, ρ can be used as one of the coordinates in a local coordinate system near any point of S.

Moreover, any such a function ρ can be used to define a volume form ds on S in accordance with any fixed volume form dv (in $\Lambda^n T^*(X)$) on X.

Remark 3.1.2 *Let a vector field ν be given in a neighborhood of S by the condition that $\nu \rfloor \frac{d\rho}{|d\rho|} = 1$ where $\rfloor$ means the operation of interior product of a differential form by a vector field. Then $ds = \nu \rfloor dv$ is a volume form on S.*

The sense of this definition becomes more clear from the following exercise.

Exercise 3.1.3 *Prove that the restriction of ds to S does not depend on the choice of a function ρ with the properties listed in Proposition 3.1.1 and $\frac{d\rho}{|d\rho|} \wedge dS = dv$.*

Hint. Use the formula

$$v \rfloor (f_1 \wedge f_2) = (v \rfloor f_1) \wedge f_2 + (-1)^{\deg f_1} f_1 \wedge (v \rfloor f_2) \qquad (3.1.1)$$

which holds for each vector field v and differential forms f_j $(j = 1, 2)$ with a common domain on X.

□

Notice, we can always replace ρ by $\frac{\rho}{|d\rho|}$ in a neighborhood of S to obtain $|d\rho| = 1$ on S.

3.1.2 *Quotient complexes*

Here we recall the well-known construction of a quotient complex from homological algebra (see Dold [56, Ch.II,2.1]).

Suppose we are given a (cochain) complex of topological vector spaces $L^\cdot$ whose differential is d_L.

Suppose that for every $i \in \mathbb{Z}$ some subspace Σ^i of L^i is distinguished. If the differential d preserves this sequence of subspaces, i.e. $d(\Sigma^i) \subset \Sigma^{i+1}$ for all i, then the sequence Σ^i $(i \in \mathbb{Z})$ is a complex of topological vector spaces itself, endowed with the same differential d_L.

Remark 3.1.4 *This complex $\Sigma^\cdot$ is said to be a subcomplex of $L^\cdot$.*

If $\Sigma^\cdot$ is a subcomplex of $L^\cdot$, we can construct a new complex of topological vector spaces $K^\cdot$ in the following way.

Set $K^i = L^i/\Sigma^i$ $(i \in \mathbb{Z})$. The differential d_K is defined so as to make the next diagram, whose rows are exact, commutative:

$$\begin{array}{ccccccccc}
 & & \cdots & & \cdots & & \cdots & & \\
 & & \downarrow & & \downarrow & & \downarrow & & \\
0 & \longrightarrow & \Sigma^0 & \xrightarrow{\iota} & L^0 & \xrightarrow{\varphi} & K^0 & \longrightarrow & 0 \\
 & & \downarrow d_L & & \downarrow d_L & & \downarrow d_K & & \\
0 & \longrightarrow & \Sigma^1 & \xrightarrow{\iota} & L^1 & \xrightarrow{\varphi} & K^1 & \longrightarrow & 0 \\
 & & \downarrow d_L & & \downarrow d_L & & \downarrow d_K & & \\
0 & \longrightarrow & \Sigma^2 & \xrightarrow{\iota} & L^2 & \xrightarrow{\varphi} & K^2 & \longrightarrow & 0 \\
 & & \downarrow d_L & & \downarrow d_L & & \downarrow d_K & & \\
 & & \cdots & & \cdots & & \cdots & &
\end{array} \qquad (3.1.2)$$

Here the mappings labeled i represent inclusion and those labeled φ represent projection onto the quotient.

It is easy to verify that d_K is continuous and ${d_K}^2 = 0$.

Remark 3.1.5 *The complex $K^\cdot$ obtained this way is called the quotient complex of $L^\cdot$ over $\Sigma^\cdot$ and it is denoted by $\frac{L}{\Sigma}$.*

Consider an important example.

Example 3.1.6 We take $L^\cdot = \mathfrak{C}(E^\cdot)$ where $\mathfrak{C}$ is a subclass of $\mathcal{E}$ stable under differentiations. Let σ be some closed subset of X. Denote by $\Sigma^\cdot$ the (closed) subspace of $L^\cdot$ consisting of all sections $f \in \mathfrak{C}(E^\cdot)$ which vanish to infinite order on σ. This means that $D^\alpha f_\nu|_{\sigma\cap U_\nu} = 0$ for all $\alpha \in \mathbb{Z}_+^n$, where $\{f_\nu\}$ represents the section f with respect to a locally trivializing coordinate covering $\{U_\nu\}$ of $E^\cdot$. Since $f_\mu = t_{\mu\nu} f_\nu$ on $U_\mu \cap U_\nu$ with some non-singular transition matrices $\{t_{\mu\nu}\}$, it is a well-defined concept for sections $f \in \mathcal{E}(E^\cdot)$. The differential P of $\mathfrak{C}(E^\cdot)$ is easily verified to preserve the sequence of subspaces $\Sigma^\cdot$, i.e. $\Sigma^\cdot$ is a subcomplex of $L^\cdot$. So we can construct the quotient complex $\frac{K}{\Sigma}$, as above. In keeping with general notations of this book, it is denoted by $\mathfrak{C}(E^\cdot|_\sigma)$, or $\mathfrak{C}^\cdot(\sigma)$.

□

3.1.3 *Complexes of jet spaces on S*

Let E be a differentiable $\mathbb{C}$-vector bundle over X.

Definition 3.1.7 *A section $f \in \mathcal{E}(E)$ is said to be flat on S (write $j_E(f) = 0$) iff f vanishes to infinite order on S.*

Given any subclass $\mathfrak{C}$ of $\mathcal{E}$ and a closed subset σ of X containing S, every class $[f] \in \mathfrak{C}(E|_\sigma)$ is represented by a section $f \in \mathfrak{C}(E)$. We say that $[f]$ is flat on S iff some representative f is flat. Since any two representatives of $[f]$ differ by a section vanishing to infinite order on σ, it is a well-defined concept.

Let us denote by $\mathfrak{C}_{j_E}(\sigma)$ the subspace of $\mathfrak{C}(E|_\sigma)$ consisting of sections flat on S.

Proposition 3.1.8 *For each subclass $\mathfrak{C}$ of $\mathcal{E}$ and closed subset σ of X containing S, we have*

$$\frac{\mathfrak{C}(E|_\sigma)}{\mathfrak{C}_{j_E}(\sigma)} \cong \mathfrak{C}(E|_S).$$

Proof. Use the definition of $\mathfrak{C}(E|_\sigma)$ of Subsection 1.1.5.

□

Thus the quotient space on the left-hand side is concentrated on S. Its elements can be interpreted as ($\infty-$) jets on S of sections from $\mathfrak{C}(E|_\sigma)$. The particular choice of σ does not matter, and we may consider j_E to be the quotient mapping $\mathfrak{C}(E|_\sigma) \to \mathfrak{C}(E|_S)$.

Proposition 3.1.9 *Let U be an open set in $\mathbb{R}^n$, and $S = \{x \in U : x_n = 0\}$. Given any sequence $f_j \in \mathcal{E}(S)$ ($j \in \mathbb{Z}_+$) there exists a function $f \in \mathcal{E}(U)$ such that $D_n^j f|_S = f$ for all $j \in \mathbb{Z}_+$.*

Proof. With the notations $x = (x', x_n)$ and $\alpha = (\alpha', \alpha_n)$ we set $f_\alpha = D_{x'}^{\alpha'} f_{\alpha_n}$ ($\alpha \in \mathbb{Z}_+^n$). From the Taylor expansion of $D_{x'}^{\alpha'} f_{\alpha_n}$, one can see that $\{f_\alpha\}$ forms a Whitney function on S of class C_{loc}^∞. By the *Extension Theorem of Whitney*, there exists a function $f \in \mathcal{E}(U)$ such that $D^\alpha f|_S = f_\alpha$ for all $\alpha \in \mathbb{Z}_+^n$. In particular, $D_n^j f|_S = f_j$ for all j which completes the proof.

□

Let us denote by $E[\rho]$ the bundle of formal power series at ρ with coefficients in E over S. Given a local coordinate system $x = (x', \rho)$ in a neighborhood U of a point on S, sections f of $E[\rho]$ are represented by $\left(\sum_{j=0}^\infty f_j(x')\ (\rho^j/j!)\right)$ in $S \cap U$ where f_j ($j \in \mathbb{Z}_+$) are sections of $E|_{S\cap U}$. The (infinite) transition matrices which define the bundle $E[\rho]$ are obtained in the obvious way by differentiating the transition relations $f_\mu = t_{\mu\nu} f_\nu$ for E with respect to ρ.

Corollary 3.1.10 *Suppose $\mathfrak{C}$ is $\mathcal{E}$ or $\mathcal{D}$. Then we have*

$$\mathfrak{C}(E|_S) \cong \mathfrak{C}(E[\rho]).$$

Proof. Given any class $[f] \in \mathfrak{C}(E|_S)$, we choose a representation $f \in \mathfrak{C}(E)$ of $[f]$. Consider the following mapping from $\mathfrak{C}(E|_S)$ to $\mathfrak{C}(E[\rho])$. The image of $[f]$ is defined to be

$$\sum_{j=0}^{\infty} D_n^j f(x', 0)\ (\rho^j/j!)$$

in a trivializing coordinate neighborhood U of a point on S, with local coordinates $x = (x', \rho)$. It is easy to see that this mapping is well-defined and injective. From Proposition 3.1.9 we conclude that it is surjective.

□

Returning to the complex $E^\cdot$, we shall write j_i instead of j_{E^i}.

Moreover, if $f \in \mathcal{E}_{j_i}(X)$, then $Pf \in \mathcal{E}_{j_{i+1}}(X)$, i.e. the differential P of the complex $E^\cdot$ preserves the spaces of flat sections. Thus, for $\mathfrak{C} = \mathcal{E}, \mathcal{D}$ and others, we get a subcomplex $\mathfrak{C}_j^\cdot(X)$ of the complex $\mathfrak{C}^\cdot(X)$.

So we can use the construction of Subsection 3.1.2 to set up the quotient complex $\frac{\mathfrak{C}^\cdot(X)}{\mathfrak{C}_j^\cdot(X)}$ denoted by $\mathfrak{C}^\cdot(S)$, just as in Example 3.1.6.

Its differential (again denoted by P) is defined so as to make the following diagram having exact rows commutative:

$$\begin{array}{ccccccccc}
 & & 0 & & 0 & & 0 & & \\
 & & \downarrow & & \downarrow & & \downarrow & & \\
0 & \longrightarrow & \mathfrak{C}_{j0}(X) & \xrightarrow{i} & \mathfrak{C}^0(X) & \xrightarrow{j} & \mathfrak{C}^0(S) & \longrightarrow & 0 \\
 & & \downarrow P & & \downarrow P & & \downarrow P & & \\
0 & \longrightarrow & \mathfrak{C}_{j1}(X) & \xrightarrow{i} & \mathfrak{C}^1(X) & \xrightarrow{j} & \mathfrak{C}^1(S) & \longrightarrow & 0 \\
 & & \downarrow P & & \downarrow P & & \downarrow P & & \\
0 & \longrightarrow & \mathfrak{C}_{j2}(X) & \xrightarrow{i} & \mathfrak{C}^2(X) & \xrightarrow{j} & \mathfrak{C}^2(S) & \longrightarrow & 0 \\
 & & \downarrow P & & \downarrow P & & \downarrow P & & \\
 & & \cdots & & \cdots & & \cdots & &
\end{array} \tag{3.1.3}$$

Here the mappings labeled i represent inclusion and those labeled j represent projection onto the quotient.

Proposition 3.1.11 *For $\mathfrak{C} = \mathcal{E}, \mathcal{D}$ and others, the following are exact sequences of cochain complexes:*

$$\begin{array}{ccccccccc}
0 & \longrightarrow & \mathfrak{C}_{\overline{X^{\mp}}}(E^\cdot) & \xrightarrow{i} & \mathfrak{C}^\cdot(X) & \xrightarrow{\sigma} & \mathfrak{C}^\cdot(\overline{X^{\pm}}) & \longrightarrow & 0, \\
0 & \longrightarrow & \mathfrak{C}_j^\cdot(X) & \xrightarrow{i} & \mathfrak{C}^\cdot(X) & \xrightarrow{j} & \mathfrak{C}^\cdot(S) & \longrightarrow & 0, \\
0 & \longrightarrow & \mathfrak{C}_j^\cdot(\overline{X^{\pm}}) & \xrightarrow{i} & \mathfrak{C}^\cdot(\overline{X^{\pm}}) & \xrightarrow{j} & \mathfrak{C}^\cdot(S) & \longrightarrow & 0.
\end{array} \tag{3.1.4}$$

Proof. Exactness of the first two sequences in (3.1.4 is clear from the definition of a complex $\mathfrak{C}^\cdot(\sigma)$ where $\sigma \subset X$ is closed (see Example 3.1.6). Exactness of the third sequence follows from Proposition 3.1.8.

□

3.1.4 *Cauchy data complexes on S*

Let $P \in \mathrm{do}_p(E \to F)$ be some differential operator on a manifold X.

Definition 3.1.12 *We say that a section $f \in C^{p-1}_{loc}(E)$ has zero Cauchy data on S with respect to P and write $c_P(f) = 0$ if*

$$\int_S G_P(g, f) = 0 \quad \textit{for all} \quad g \in \mathcal{D}(F'). \tag{3.1.5}$$

In other words, $c_P(f) = 0$ iff $[S]^P f = 0$.

In follows from Proposition 2.4.4 that this definition does not depend on the choice of a Green operator G for P.

Moreover, by (2.4.23) $f \in C^p_{loc}(E)$ has zero Cauchy data on S with respect to P if and only if $P(\chi^\pm f) = \chi^\pm Pf$ in $\mathcal{D}'(F)$, where $\chi^\pm$ is the characteristic function of $X^\pm$.

Proposition 3.1.13 *Let the symbol mapping $\sigma(P)(x, d\rho(x))$ of $E_x \to F_x$ be injective for each $x \in S$ and $f \in C^{p-1}_{loc}(E)$. Then $c_P(f) = 0$ if and only if f is $(p-1)$-flat on S, i.e. it vanishes up to order $(p-1)$ on S.*

Proof. We prove only the *necessity* because the sufficiency is obvious. Fix any point $x \in S$ and a local coordinate system $x = (x', \rho)$ in a neighborhood U of x over which the bundles E and F are trivial.

Let us write the differential operator P in U in the form

$$P(x, D) = \sum_{i=0}^{p} D_n^i \circ P_i(x, D'),$$

where the differential operators $P_i(x, D')$ contain differentiations up to order $(p-i)$ only in the variables $x' = (x_1, \ldots, x_{n-1})$.

Then

$$G_P(g, f) = \frac{1}{\sqrt{-1}} \sum_{j=0}^{p-1} (-1)^j D_n^j g \left(\sum_{i=j}^{p-1} D_n^{i-j} P_{i+1}(x, D') f \right) \star dx_n + dx_n \wedge G_n(g, f) \tag{3.1.6}$$

is a Green operator for P in U, where G_n is some bidifferential operator of type $(F', E) \to \Lambda^{n-2}$ and order $\leq p-1$ in U.

It follows from (3.1.6) and Proposition 3.1.9 that if a section $f \in C^{p-1}_{loc}(E)$ has zero Cauchy data on S with respect to P, then

$$\left(\sum_{i=j}^{p-1} D_n^{i-j} P_{i+1}(x, D') f \right) \Bigg|_{S \cap U} = 0 \quad \text{for} \quad j = 0, \ldots, p-1. \tag{3.1.7}$$

Since $P_p(x, D') = \sigma(P)(x, d\rho(x))$ is injective for all $x \in S \cap U$, we get from (3.1.7) that f vanishes to order $(p-1)$ on $S \cap U$.

□

Given any subclass $\mathfrak{C}$ of C^{p-1}_{loc} and a closed subset σ of X containing S, every class $[f] \in \mathfrak{C}(E|_\sigma)$ is represented by a section $f \in \mathfrak{C}(E)$. We say that $[f]$ has zero Cauchy data on S with respect to P iff some representative f has zero Cauchy data. Since any two representatives of $[f]$ differ by a section at least $(p-1)$-flat on S, this is a well-defined concept.

Let us denote by $\mathfrak{C}_{c_P}(\sigma)$ the subspace of $\mathfrak{C}(E|_\sigma)$ consisting of sections with zero Cauchy data on S with respect to P.

Proposition 3.1.14 *For each subclass $\mathfrak{C}$ of C^{p-1}_{loc} and any closed subset σ of X containing S, we have*

$$\frac{\mathfrak{C}(E|_\sigma)}{\mathfrak{C}_{c_P}(\sigma)} \cong \frac{\mathfrak{C}(E|_S)}{\mathfrak{C}_{c_P}(S)}. \tag{3.1.8}$$

Proof. Use the definition of the spaces $\mathfrak{C}(E|_\sigma)$ given in Subsection 1.1.5.

□

Thus the quotient space on the left-hand side of (3.1.8) is concentrated on S. Its elements can be interpreted as Cauchy data on S with respect to P of sections from $\mathfrak{C}(E|_\sigma)$. The particular choice of σ does not matter. Denoting this quotient by $\mathfrak{C}_{[c_P]}(S)$, we can arrange c_P to be the quotient mapping $\mathfrak{C}(E|_\sigma) \to \mathfrak{C}_{[c_P]}(S)$.

Remark 3.1.15 *For elements of $\mathfrak{C}_{[c_P]}(S)$, the notion of support can be introduced in the following way. Let $[f] \in \mathfrak{C}_{[c_P]}(S)$ and $f \in \mathfrak{C}(E)$ be some representative of $[f]$. We set $\operatorname{supp}[f] = S \setminus U$ where U is the largest open subset of S such that f has zero Cauchy data on U with respect to P. This definition is easily verified to be well-defined.*

Let us return to the complex $E^\cdot$. The Cauchy data on S with respect to the differential operator P_ι are traditionally denoted by $t_\iota(f)$, i.e. $t_\iota = c_{P_\iota}$. And the Cauchy data on S with respect to the differential operator $\tilde{P}^*_{\iota-1}$ are traditionally denoted by $n_\iota(f)$, i.e. $n_\iota = c_{\tilde{P}^*_{\iota-1}}$ where $\tilde{P}$ is the differential of some associated complex for $E^\cdot$ which has been stipulated a priori.

Lemma 3.1.16 *If $f \in \mathcal{E}_{t_\iota}(X)$, then $Pf \in \mathcal{E}_{t_{\iota+1}}(X)$.*

Proof. Indeed, by Proposition 2.4.6 and Stokes' formula we get for all $g \in \mathcal{D}(E^{\iota+2'})$

$$\int_S G_{P_{\iota+1}}(g, Pf) = -\int_S G_{P_\iota}(P'g, f) = 0$$

which completes the proof.

□

We have seen that the differential P of the complex $E^\cdot$ preserves the spaces of sections having zero Cauchy data on S. Thus, for $\mathfrak{C} = \mathcal{E}, \mathcal{D}$ and others, we get a subcomplex $\mathfrak{C}_t^\cdot(X)$ of the complex $\mathfrak{C}^\cdot(X)$.

Hence we can use the construction of Subsection 3.1.2 to construct the quotient complex $\frac{\mathfrak{C}^\cdot(X)}{\mathfrak{C}_t^\cdot(X)}$ denoted by $\mathfrak{C}_{[t]}^\cdot(S)$.

Its differential P_b is defined so as to make the following diagram with exact rows commutative:

$$\begin{array}{ccccccccc}
 & & 0 & & 0 & & 0 & & \\
 & & \downarrow & & \downarrow & & \downarrow & & \\
0 & \longrightarrow & \mathfrak{C}_{t_0}(X) & \xrightarrow{\iota} & \mathfrak{C}^0(X) & \xrightarrow{t} & \mathfrak{C}_{[t_0]}(S) & \longrightarrow & 0 \\
 & & \downarrow P & & \downarrow P & & \downarrow P_b & & \\
0 & \longrightarrow & \mathfrak{C}_{t_1}(X) & \xrightarrow{\iota} & \mathfrak{C}^1(X) & \xrightarrow{t} & \mathfrak{C}_{[t_1]}(S) & \longrightarrow & 0 \\
 & & \downarrow P & & \downarrow P & & \downarrow P_b & & \\
0 & \longrightarrow & \mathfrak{C}_{t_2}(X) & \xrightarrow{\iota} & \mathfrak{C}^2(X) & \xrightarrow{t} & \mathfrak{C}_{[t_2]}(S) & \longrightarrow & 0 \\
 & & \downarrow P & & \downarrow P & & \downarrow P_b & & \\
 & & \cdots & & \cdots & & \cdots & &
\end{array} \tag{3.1.9}$$

The notation $\mathfrak{C}_{[t]}^\cdot(S)$ is unambiguous.

Proposition 3.1.17 *For $\mathfrak{C} = \mathcal{E}, \mathcal{D}$ and others, the following are exact sequences of cochain complexes:*

$$\begin{array}{ccccccccc} 0 & \longrightarrow & \mathfrak{C}_t(S) & \xrightarrow{i} & \mathfrak{C}^{\cdot}(S) & \xrightarrow{t} & \mathfrak{C}_{[t\,]}(S) & \longrightarrow & 0, \\ 0 & \longrightarrow & \mathfrak{C}_t(X) & \xrightarrow{i} & \mathfrak{C}^{\cdot}(X) & \xrightarrow{t} & \mathfrak{C}_{[t\,]}(S) & \longrightarrow & 0, \\ 0 & \longrightarrow & \mathfrak{C}_t(\overline{X^{\pm}}) & \xrightarrow{i} & \mathfrak{C}^{\cdot}(\overline{X^{\pm}}) & \xrightarrow{t} & \mathfrak{C}_{[t\,]}(S) & \longrightarrow & 0. \end{array} \qquad (3.1.10)$$

Proof. The exactness of the first sequence in (3.1.10) follows from the definition of the complex $\mathfrak{C}^{\cdot}(S)$ given by (3.1.3) and Proposition 3.1.14. The exactness of the second sequence was established by the commutative diagram (3.1.9) as a definition. The exactness of the third sequence in (3.1.10) follows directly from Proposition 3.1.14.

□

3.1.5 *Tangential complexes*

Thus we have obtained the following complex of Cauchy data spaces on S:

$$\mathfrak{C}_{[t\,]}(S): \quad 0 \to \mathfrak{C}_{[t_0]}(S) \xrightarrow{P_b} \mathfrak{C}_{[t_1]}(S) \xrightarrow{P_b} \ldots \xrightarrow{P_b} \mathfrak{C}_{[t_{N-1}]}(S) \longrightarrow 0. \qquad (3.1.11)$$

The next definition is due to Andreotti [8] (see also Andreotti et al. [11]).

Definition 3.1.18 *The complex (3.1.11) is said to be a tangential or boundary complex on S induced by the complex $\mathfrak{C}^{\cdot}(X)$.*

The tangential operator $P_b : \mathfrak{C}_{[t_\cdot]}(S) \to \mathfrak{C}_{[t_{\cdot+1}]}(S)$ acts as a local operator.

Proposition 3.1.19 *For every $f \in \mathfrak{C}_{[t_\cdot]}(S)$,*

$$\text{supp } P_b f \subset \text{supp } f.$$

Proof. This follows from the definition of the support of $f \in \mathfrak{C}_{[t_\cdot]}(S)$ and Lemma 3.1.16.

□

So P_b is certainly a differential operator by Peetre's theorem provided the spaces $\mathcal{E}_{[t_\cdot]}(S)$ are locally free $\mathcal{E}(S)$-modules. In general, $\mathcal{E}_{[t_\cdot]}(S)$ is not the space of smooth sections of any vector bundle over S. It is seen from (3.1.7) that $\mathcal{E}_{t_\cdot}(X)$ consists of sections in $\mathcal{E}^{i}(X)$ which usually satisfy some system of differential equations on S. Thus $\mathcal{E}^{i}(X)/\mathcal{E}_{t_\cdot}(X)$ can be very complicated. Nevertheless $\mathcal{E}_{[t_\cdot]}(S)$ are locally free $\mathcal{E}(S)$-modules provided the hypersurface S is non-characteristic for the complex $E^{\cdot}$.

Definition 3.1.20 *The hypersurface S is said to be non-characteristic for the complex $E^{\cdot}$ at a point $x \in S$ if the vector $d\rho(x) \in T^*_x(X)$ is non-characteristic for $E^{\cdot}$ at x.*

Otherwise S is said to be characteristic for $E^{\cdot}$ at x.

Clearly the particular choice of a determining function ρ for S does not matter.

We say that S is non-characteristic for $E^{\cdot}$ iff it is non-characteristic for $E^{\cdot}$ at every point.

In particular, if the complex $E^{\cdot}$ is elliptic, each hypersurface S is non-characteristic for $E^{\cdot}$.

Example 3.1.21 Let $P \in \mathrm{do}_p(E \to F)$ be some differential operator on X. Consider a differential complex $E^{\cdot}$ of length $N = 1$ on X obtained by setting $E^0 = E$, $E^1 = F$ and $P_0 = P$. Then S is non-characteristic for $E^{\cdot}$ at a point $x \in S$ iff $\sigma(P)(x, d\rho(x)) : E_x \to F_x$ is an isomorphism. This is just the classical definition of non-characteristic.

□

Andreotti and Nacinovich [14] proved the following remarkable result.

Theorem 3.1.22 *If the hypersurface S is non-characteristic for $E^{\cdot}$, then there exists a differential complex $E_t^{\cdot}$ on S such that $\mathfrak{C}_{[t\,]}(S) = \mathfrak{C}(E_t^{\cdot})$.*

Proof. The proof consists in finding a local canonical form of $E^{\cdot}$ to derive conditions (3.1.7) as simply as possible. See Andreotti and Nacinovich [14] for details.

□

Remark 3.1.23 *This theorem is quite easy when $E^{\cdot}$ is a complex of first order differential operators. The proof in this case will be given in Subsection 3.2.9.*

The tangential Cauchy-Riemann complex appeared first in the paper of Kohn and Rossi [133]. Andreotti and Hill [9] got its systematic study off the ground.

Sweeney [211] extended the results of Kohn and Rossi [133] to the second Spencer sequence for a differential operator with injective symbol on X.

MacKichan [155] considered tangential complexes on surfaces S of codimension > 1 induced by complexes $E^{\cdot}$ of first order differential operators.

Kuranishi [141] studied a version of a tangential complex with $E^{\cdot}$ an elliptic complex of first order differential operators. Andreotti [8] was the first to introduce the construction of a tangential complex for general complexes $E^{\cdot}$. Nacinovich [168] investigated boundary Hilbert differential complexes. The author (see [226, 230, 231, 233]) had found a satisfactory description of tangential complexes for elliptic complexes.

Let us consider two well-known examples.

Example 3.1.24 Let $E^{\cdot} = \Lambda^{\cdot} T^*_{\mathbb{C}}(X)$ be the de Rham complex on X. The inclusion map $i : S \to X$ induces the pull-back operator on forms $i^\sharp : \mathfrak{C}(\Lambda^q T^*_{\mathbb{C}}(X)) \to \mathfrak{C}(\Lambda^q T^*_{\mathbb{C}}(S))$ which is surjective. By Stokes' Theorem, a form f in $\mathfrak{C}(\Lambda^q T^*_{\mathbb{C}}(X))$ has zero Cauchy data on S with respect to d_q if and only if $i^\sharp(f) = 0$. Since also $d(i^\sharp f) = i^\sharp df$ by Proposition 1.2.38, the sequence of complexes

$$0 \longrightarrow \mathfrak{C}_t^{\cdot}(X) \stackrel{\iota}{\longrightarrow} \mathfrak{C}^{\cdot}(X) \stackrel{i^\sharp}{\longrightarrow} \mathfrak{C}(\Lambda^{\cdot} T^*_{\mathbb{C}}(S)) \longrightarrow 0$$

is exact. This means that $\mathfrak{C}_{[t\,]}(S) \cong \mathfrak{C}(\Lambda^{\cdot} T^*_{\mathbb{C}}(S))$, where the isomorphism is constructed from the pull-back operator $i^\sharp$.

□

Example 3.1.25 Suppose that $E^{\cdot} = \Lambda^{p,\cdot} T^*_{\mathbb{C}}(X)$ is the Dolbeault complex with coefficients in $\Lambda^{p,0}$ on a complex manifold X. It is easy to see that a form $f \in \mathfrak{C}(\Lambda^{p,q} T^*_{\mathbb{C}}(X))$ has zero Cauchy data on S with respect to $\overline{\partial}^{(p)}_q$ if and only if f is in the ideal of $\oplus_{p,q}\mathfrak{C}(\Lambda^{p,q} T^*_{\mathbb{C}}(X))$ generated by ρ and $\overline{\partial}\rho$. Observe that this ideal is actually independent of the function ρ used to define S globally. The complex $\mathfrak{C}_{[t\,]}(S)$ is called the tangential Cauchy-Riemann complex on S induced by the complex $\mathfrak{C}(\Lambda^{p,\cdot} T^*_{\mathbb{C}}(X))$ (see Andreotti and Hill [9]). This differential complex on S is known to be a non-elliptic complex (see Kohn [127]).

□

The last example shows that the tangential complex $E^{\cdot}_t$ may be non-eliptic while the initial complex $E^{\cdot}$ is elliptic.

There is a deep relation between the cohomologies of the complex $\mathcal{E}_{[t\,]}(S)$ and $\mathcal{D}'^{\cdot}(X)$. The roots of this relation are hidden in the next statement.

Theorem 3.1.26 *The correspondence $f \to [S]^P f$ determines an injective cochain mapping of the complexes $\mathcal{E}_{[t\,]}(S) \to \mathcal{D}'^{\cdot}(X)$ of degree* 1.

Proof. Given any class $[f] \in \mathcal{E}_{[t\,]}(S)$, we choose a representative $f \in \mathcal{E}^{\iota}(X)$ of $[f]$. Since $[S]^P \in \mathcal{D}'((E^{\cdot} \boxtimes E^{\cdot\prime})^1)$ we have $[S]^P f \in D'^{\iota+1}(X)$.

If f_j $(j = 1, 2)$ are two different representatives of $[f]$, then $t(f_1 - f_2) = 0$. By Definition 3.1.12, this means that $[S]^P f_1 = [S]^P f_2$.

Hence, it follows that we obtain a well-defined mapping $\mathcal{E}_{[t,\iota]}(S) \to \mathcal{D}'^{\iota+1}(X)$ by setting $[S]^P([f]) = [S]^P f$.

To complete the proof, we need only show that

$$[S]^P(P_b[f]) = -P[S]^P([f]) \quad \text{for all} \quad [f] \in \mathcal{E}_{[t,\iota]}(S). \tag{3.1.12}$$

Let $f \in \mathcal{E}^{\iota}(X)$ then be a representative of the class $[f]$, i.e. $t(f) = [f]$. By the commutative diagram (3.1.9), $P_b[f] = t(Pf)$. Therefore, (3.1.12) is reduced to

$$[S]^P(Pf) = -P[S]^P(f) \quad \text{in} \quad \mathcal{D}'^{\iota+2}(X). \tag{3.1.13}$$

Using (2.4.21), Proposition 2.4.6 and Stokes' Theorem, we get for all $g \in \mathcal{D}_{\iota+2}(X)$

$$\begin{aligned} \langle g, [S]^P(Pf)\rangle_X &= \int_S G_{P_{\iota+1}}(g, Pf) \\ &= -\int_S G_{P_\iota}(P'g, f) \\ &= -\langle P'g, [S]^P(f)\rangle \\ &= -\langle g, P[S]^P(f)\rangle, \end{aligned}$$

i.e. (3.1.13).

□

Notice that $[S]^P$ maps $\mathcal{D}_{[t\,]}(S)$ into $\mathcal{E}'^{\cdot}(X)$ because this kernel is properly supported.

3.1.6 *Formally non-characteristic hypersurfaces*

For each subclass $\mathfrak{C}$ of $\mathcal{E}$ stable under differentiations, the complex $\mathfrak{C}_J(X)$ is easily seen to be a subcomplex of $\mathfrak{C}_t(X)$. So we may use the construction of 3.1.2 to construct the quotient complex $\frac{\mathfrak{C}_t(X)}{\mathfrak{C}_J(X)}$.

The cohomology of the complex $\frac{\mathfrak{C}_t(X)}{\mathfrak{C}_J(X)}$ is called the Cauchy-Kovalevskaya cohomology of $\mathfrak{C}^\cdot(X)$ on S because it is the obstruction to the formal Cauchy-Kovalevskaya procedure on S.

Proposition 3.1.27 *For $\mathfrak{C} = \mathcal{E}, \mathcal{D}$ and others, the following is an exact sequence of cochain complexes:*

$$0 \longrightarrow \mathfrak{C}_J(X) \stackrel{\iota}{\longrightarrow} \mathfrak{C}_t(X) \stackrel{J}{\longrightarrow} \mathfrak{C}_t(S) \longrightarrow 0. \tag{3.1.14}$$

Proof. This follows from Proposition 3.1.11 because (3.1.14) is merely the restriction of the second sequence in (3.1.4) to sections with zero Cauchy data on S.

□

So $\frac{\mathfrak{C}_t(X)}{\mathfrak{C}_J(X)} \cong \mathfrak{C}_t(S)$, and the next definition is justified.

Definition 3.1.28 *The hypersurface S is said to be formally non-characteristic for the complex $\mathfrak{C}^\cdot(X)$ if the Cauchy-Kovalevskaya cohomology is trivial at all degrees, i.e. $H^\cdot(\mathfrak{C}_t(S)) = 0$.*

The notion of S being formally non-characteristic is more general than the classical notion of S being non-characteristic.

Example 3.1.29 Let $X = \mathbb{R}^2$, $E^0 = E^1 = X \times \mathbb{R}^2$ and $E^i = 0$ for $i \neq 0, 1$, and $P_0 = \frac{\partial}{\partial t} - \frac{\partial^2}{\partial x^2}$ be the heat operator. Then the hypersurface $S = \{(t,x) \in \mathbb{R}^2 : t = 0\}$ is characteristic for the complex $E^\cdot$ at each of its points. But S is formally non-characteristic, as is easily checked.

□

The following basic fact is due to Andreotti, Hill Łojasiewicz and MacKichan [11], who used Guillemin's [94] program of constructing formal solutions to a local Cauchy problem for overdetermined systems.

Theorem 3.1.30 *Let $\mathfrak{C}$ be $\mathcal{E}$ or $\mathcal{D}$. If a hypersurface S is non-characteristic for the complex $E^\cdot$, then it is formally non-characteristic for the complex $\mathfrak{C}^\cdot(X)$.*

As a matter of fact Andreotti et al. [8] proved a stronger result.

Exercise 3.1.31 *Define the transversal symbol of a differential operator near S so that each transversally non-characteristic hypersurface for the complex $E^\cdot$ is formally non-characteristic for the complex $\mathfrak{C}^\cdot(X)$.*

Hint. See Andreotti et al. [8].

□

Note in Example 3.1.29 that S is transversally non-characteristic for $E^\cdot$.

Proof of Theorem 3.1.30. We will deduce this theorem from a more precise one whose proof will be divided into several steps.

3.1.7 Decomposition of a complex

A decomposition of a complex modeled on that of Guillemin [94] reduces the problem of constructing formal solutions to a Cauchy problem for overdetermined systems of differential equations to one of constructing a formal solution to a determined Cauchy problem.

Let us denote by $\sigma_\imath(x)$ the symbols of the differential operators $P_\imath$ computed at the point $\left(x, \frac{d\rho(x)}{|d\rho(x)|}\right)$ of $T^*(X)$ $(\imath \in \mathbb{Z})$.

Assume that $x \in S$ and $\imath \in \mathbb{Z}$ are fixed, and the sequence $E_x^{\imath-1} \xrightarrow{\sigma} E_x^\imath \xrightarrow{\sigma} E_x^{\imath+1}$ is exact. Then there is a neighborhood U of x such that $d\rho \neq 0$ in U and the following sequence of smooth bundle homomorphisms is exact:

$$E^{\imath-1}|_U \xrightarrow{\sigma_{\imath-1}} E^\imath|_U \xrightarrow{\sigma_\imath} E^{\imath+1}|_U. \tag{3.1.15}$$

Exactness of (3.1.15) is easily seen to be equivalent to the homomorphism $L_\imath = \sigma_{\imath-1}\sigma_{\imath-1}^* + \sigma_\imath^*\sigma_\imath$ being a bundle isomorphism of $E^\imath|_U$

We show that the homomorphism $\pi_\imath = (\sigma_{\imath-1}\sigma_{\imath-1}^*)L_\imath^{-1}$ is a projector in $E^\imath|_U$. First of all, since

$$\begin{aligned} L_\imath(\sigma_{\imath-1}\sigma_{\imath-1}^*) &= (\sigma_{\imath-1}\sigma_{\imath-1}^*)L_\imath, \\ L_\imath(\sigma_\imath^*\sigma_\imath) &= (\sigma_\imath^*\sigma_\imath)L_\imath, \end{aligned} \tag{3.1.16}$$

we get

$$\begin{aligned} (\sigma_{\imath-1}\sigma_{\imath-1}^*)L_\imath^{-1} &= L_\imath^{-1}(\sigma_{\imath-1}\sigma_{\imath-1}^*), \\ (\sigma_\imath^*\sigma_\imath)L_\imath^{-1} &= L_\imath^{-1}(\sigma_\imath^*\sigma_\imath). \end{aligned} \tag{3.1.17}$$

Therefore

$$\pi_\imath^2 = (\sigma_{\imath-1}\sigma_{\imath-1}^*)L_\imath^{-1} \circ \pi_\imath = (L_\imath - \sigma_\imath^*\sigma_\imath)L_\imath^{-1} \circ \pi_\imath = \pi_\imath,$$

i.e $\pi_\imath$ is a projector.

Hence it follows that $E_1^\imath = (1 - \pi_\imath)E^\imath|_U$ and $E_2^\imath = \pi_\imath E^\imath|_U$ are vector subbundles of $E^\imath|_U$, and $E^\imath|_U = E_1^\imath \oplus E_2^\imath$ (see Atiyah [18, §1.3]).

Using (3.1.17), one can easily see that

$$\begin{aligned} E_1^\imath &= \ker \sigma_{\imath-1}^* : \; E^\imath|_U \to E^{\imath-1}|_U, \\ E_2^\imath &= \ker \sigma_\imath : \; E^\imath|_U \to E^{\imath+1}|_U, \end{aligned}$$

and $L_\imath^{-1}$ acts on $E_1^\imath$ and $E_2^\imath$ in an invariant way.

Therefore the bundle homomorphisms $\sigma_\imath : \; E_1^\imath \to E^{\imath+1}|_U$ and $\sigma_{\imath-1}^* L_\imath^{-1} : \; E_2^\imath \to E^{\imath-1}|_U$ are monomorphisms. By Lemma 1.3.1 of Atiyah [18], $E_2^{\imath+1} = \sigma_\imath E_1^\imath$ and $E_1^{\imath-1} = \sigma_{\imath-1}^* L_\imath^{-1} E_2^\imath$ are vector subbundles of $E^{\imath+1}|_U$ and $E^{\imath-1}|_U$ respectively.

Denote by $E_1^{\imath+1}$ the orthogonal complement of $E_2^{\imath+1}$ in $E^{\imath+1}|_U$, and by $E_2^{\imath-1}$ the orthogonal complement of $E_1^{\imath-1}$ in $E^{\imath-1}|_U$. Then $E_1^{\imath+1}$ and $E_2^{\imath-1}$ are smooth subbundles of $E^{\imath+1}|_U$ and $E^{\imath-1}|_U$ respectively, and $E^\jmath|_U = E_1^\jmath \oplus E_2^\jmath$ $(\jmath = \imath - 1, \imath + 1)$.

It is easy to see that

$$\begin{aligned} E_1^{\imath+1} &= \ker \sigma_\imath^* : \; E^{\imath+1}|_U \to E^\imath|_U, \\ E_2^{\imath-1} &= \ker \sigma_{\imath-1} : \; E^{\imath-1}|_U \to E^\imath|_U. \end{aligned}$$

So we get the following diagram of bundle homomorphisms over U:

$$\begin{array}{ccccc} E_1^{i-1} & \overset{\sigma_{i-1}}{\searrow} & E_1^i & \overset{\sigma_i}{\searrow} & E_1^{i+1} \\ \oplus & & \oplus & & \oplus \\ E_2^{i-1} & \underset{\sigma_{i-1}^* L_i^{-1}}{\nwarrow} & E_2^i & \underset{L_i^{-1}\sigma_i^*}{\nwarrow} & E_2^{i+1} \end{array} . \tag{3.1.18}$$

In accordance with (3.1.18) we have the decompositions of spaces of sections $\mathfrak{C}^j(U) = \mathfrak{C}(E_1^j) \oplus \mathfrak{C}(E_2^j)$ $(j = i-1, i, i+1)$.

Let us agree to represent sections $f \in \mathfrak{C}(E^j)$ by the columns $f = \begin{pmatrix} f_1 \\ f_2 \end{pmatrix}$ where $f_\nu \in \mathfrak{C}(E_\nu^j)$ $(\nu = 1, 2)$. Then the differential operators P_j $(j = i-1, i)$ are represented by the matrices $\begin{pmatrix} P_j^{11} & P_j^{12} \\ P_j^{21} & P_j^{22} \end{pmatrix}$, where $P_j^{\mu\nu}$ is a differential operator of type $E_\nu^j \to E_\mu^{j+1}$ and order $\le p_j$, so that

$$P_j f = \begin{pmatrix} P_j^{11} f_1 & + & P_j^{12} f_2 \\ P_j^{21} f_1 & + & P_j^{22} f_2 \end{pmatrix}.$$

Thus, we get the following diagram of spaces of sections:

$$\begin{array}{ccccc} \mathfrak{C}(E_1^{i-1}) & & \mathfrak{C}(E_1^i) & & \mathfrak{C}(E_1^{i+1}) \\ \oplus & \xrightarrow{P_{i-1}} & \oplus & \xrightarrow{P_i} & \oplus \\ \mathfrak{C}(E_2^{i-1}) & & \mathfrak{C}(E_2^i) & & \mathfrak{C}(E_2^{i+1}) \end{array} . \tag{3.1.19}$$

The relation $P_i \circ P_{i-1}$ yields

$$P_i^{21} P_{i-1}^{22} + P_i^{22} P_{i-1}^{21} = 0 \tag{3.1.20}$$

as well as three similar relations which we will not need.

It is very important that the hypersurface $S \cap U$ turns out to be non-characteristic for the differential operators P_j^{21} $(j = i-1, i)$.

Lemma 3.1.32 *For $j = i-1, i$, the following sequence of bundle homomorphisms over U is exact:*

$$o \longrightarrow E_1^j \xrightarrow{\sigma(P_j^{21}(x, d\rho(x)))} E_2^{j+1} \longrightarrow 0.$$

We recall that $\sigma(P_j^{21})$ denotes the symbol of order p_j of the differential operator P_j^{21}.

Proof. Let x be a fixed point of U, $e \in E_x^j$ and $f = \begin{pmatrix} f_1 \\ f_2 \end{pmatrix} \in \mathcal{E}(E^j|_U)$ be such that $f(x) = e$.

Denoting $\varphi = \rho - \rho(x)$ we get $\varphi(x) = 0$ and $d\varphi(x) = d\rho(x)$ so that

$$\begin{aligned} \sigma(P_j)(x, d\rho(x))\, e &= \frac{(\sqrt{-1})^{p_j}}{p_j!} P_j(\varphi^{p_j} f)(x) \\ &= \frac{(\sqrt{-1})^{p_j}}{p_j!} \begin{pmatrix} P_j^{11}(\varphi^{p_j} f_1)(x) + P_j^{12}(\varphi^{p_j} f_2)(x) \\ P_j^{21}(\varphi^{p_j} f_1)(x) + P_j^{22}(\varphi^{p_j} f_2)(x) \end{pmatrix}. \end{aligned}$$

Since σ_j maps $E^j|_U$ into E_2^j, and E_2^j into 0,we conclude that

$$\sigma(P_j^{21})(x, d\rho(x))e = \sigma(P_j)(x, d\rho(x))e \quad \text{for} \quad e \in (E_1^j)_x.$$

To complete the proof simply note that $\sigma_j : E_1^j \to E_2^{j+1}$ is an isomorphism.

□

3.1.8 *Existence and uniqueness of formal solutions to a non-characteristic local Cauchy problem*

Let $P \in \mathrm{do}_p(E \to F)$ be any differential operator on the manifold X.

For a point $x \in S$, choose some coordinate neighborhood U of x in X with local coordinates $x = (x', \rho)$ such that E and F are trivial over U.

Lemma 3.1.33 *Suppose that $\sigma(P)(x, d\rho(x)) : E_x \to F_x$ is an isomorphism for all $x \in S \cap U$. Then for every $f \in \frac{\mathcal{E}(F|_U)}{\mathcal{E}_{J_F}(U)}$ there exists a unique solution $u \in \frac{\mathcal{E}_{c_P}(U)}{\mathcal{E}_{J_E}(U)}$ of the equation $Pu = f$ (mod $\mathcal{E}_{J_F}(U)$).*

Proof. By the identification $\frac{\mathcal{E}(E|_U)}{\mathcal{E}_{J_E}(U)}$ with $\mathcal{E}(E|_U[\rho])$ given in Corollary 3.1.10, we may replace f by $\sum_{j=0}^{\infty} f_j(\rho^j/j!)$ and u by $\sum_{j=0}^{\infty} u_j(\rho^j/j!)$ where f_j and u_j are sections of F and E over $S \cap U$, and $f_j = D_n^j f|_{S \cap U}$

The equation $Pu = f$ (mod $\mathcal{E}_{J_F}(U)$) becomes a formal power series equation

$$\left(\sigma(P)(x, d\rho(x)) \left(\sqrt{-1}\frac{\partial}{\partial \rho}\right)^p + \dots\right) \left(\sum_{j=0}^{\infty} u_j(\rho^j/j!)\right) = \sum_{j=0}^{\infty} f_j(\rho j/j!) \qquad (3.1.21)$$

where omitted summands have orders less than p in $\left(\frac{\partial}{\partial \rho}\right)$.

By Proposition 3.1.13, $\sum_{j=0}^{\infty} u_j(\rho^j/j!)$ represents a section in $\frac{\mathcal{E}_{c_P}(U)}{\mathcal{E}_{J_E}(U)}$ if and only if $u_j = 0$ for $j = 0, \dots, p-1$.

Since $\sigma(P)(x, d\rho(x))$ is an isomorphism of $E|_{S\cap U}$ with $F|_{S\cap U}$, (3.1.21) gives a unique solution for u_j in terms of $\{u_0, \dots, u_{p-1}\}$ if $j \geq p$. Therefore there is a unique solution of (3.1.21) with $u_j = 0$ for $j = 0, \dots, p-1$.

Thus there exists a unique solution u in $\frac{\mathcal{E}_{c_P}(U)}{\mathcal{E}_{J_E}(U)}$ to the equations $Pu = f$ (mod $\mathcal{E}_{J_F}(U)$), which completes the proof.

□

3.1.9 *A condition for a hypersurface to be formally non-characteristic locally*

The result of this subsection is of interest itself.

Lemma 3.1.34 *Let $x \in S$ and the sequence $E_x^{i-1} \xrightarrow{\sigma} E_x^i \xrightarrow{\sigma} E_x^{i+1}$ be exact. Then there is a neighborhood U of x in X such that $H^i(\mathcal{E}_t(S \cap U')) = 0$ for each open set $U' \subset U$.*

Proof. Take as U the neighborhood of x which has been chosen in Subsection 3.1.7. We may always decrease it so as to assume that U is provided with a local coordinate system $x = (x', \rho)$, and E^j ($j = i-1, i, i+1$) are trivial over U.

Given any open set $U' \subset U$, we need to show that the following sequence is exact

$$\frac{\mathcal{E}_{t_{i-1}}(U')}{\mathcal{E}_{j_{i-1}}(U')} \xrightarrow{P} \frac{\mathcal{E}_{t_i}(U')}{\mathcal{E}_{j_i}(U')} \xrightarrow{P} \frac{\mathcal{E}_{t_{i+1}}(U')}{\mathcal{E}_{j_{i+1}}(U')},$$

i.e., if f is in $\mathcal{E}_{t_i}(U')$ and $Pf = 0 \pmod{\mathcal{E}_{j_{i+1}}(U')}$, then there is a u in $\mathcal{E}_{t_{i-1}}(U')$ with $Pu = f \pmod{\mathcal{E}_{j_i}(U')}$.

We shall construct u by solving the formal non-characteristic Cauchy problem for P^{21}_{i-1}.

Consider the decomposition (3.1.19). Since f is in $\mathcal{E}^i(U')$ we may write $f = \begin{pmatrix} f_1 \\ f_2 \end{pmatrix}$ where f_j is in $\mathcal{E}(E^i_j|_{U'})$, $j = 1, 2$. Then $Pf = 0 \pmod{\mathcal{E}_{j_{i+1}}(U')}$ implies that

$$P^{11}_i f_1 + P^{12}_i f_2 = 0 \qquad \pmod{\mathcal{E}_{j_{E_1^{i+1}}}(U')}, \tag{3.1.22}$$

$$P^{21}_i f_1 + P^{22}_i f_2 = 0 \qquad \pmod{\mathcal{E}_{j_{E_2^{i+1}}}(U')}. \tag{3.1.23}$$

Since f is in $\mathcal{E}_{t_i}(U')$, for each $g \in \mathcal{D}_{i+1}(U')$

$$\int_{U'^{\pm}} \langle g, Pf \rangle_x = \int_{U'^{\pm}} \langle P'g, f \rangle_x.$$

where $U'^{\pm} = U' \cap X^{\pm}$.

Writing $g = (g_1, g_2)$ where g_j is in $\mathcal{D}\left(E^{i+1'}_j|_{U'}\right)$, $j = 1, 2$, we obtain

$$\int_{U'^{\pm}} \langle g_1, P^{11}_i f_1 + P^{12}_i f_2 \rangle_x$$
$$= \int_{U'^{\pm}} \langle P^{11'}_i g_1, f_1 \rangle_x + \int_{U'^{\pm}} \langle P^{12'}_i g_1, f_2 \rangle_x \quad \text{for each} \quad g_1 \in \mathcal{D}(E^{i+1'}_1|_{U'}), \tag{3.1.24}$$

and

$$\int_{U'^{\pm}} \langle g_2, P^{21}_i f_1 + P^{22}_i f_2 \rangle_x$$
$$= \int_{U'^{\pm}} \langle P^{21'}_i g_2, f_1 \rangle_x + \int_{U'^{\pm}} \langle P^{22'}_i g_2, f_2 \rangle_x \quad \text{for each} \quad g_2 \in \mathcal{D}(E^{i+1'}_2|_{U'}). \tag{3.1.25}$$

Let $u_1 \in \mathcal{E}\left(E^{i-1}_1|_{U'}\right)$ be the unique solution modulo $\mathcal{E}_{j_{E_1^{i-1}}}(U')$ of the equation

$$P^{21}_{i-1} u_1 = f_2 \pmod{\mathcal{E}_{j_{E_2^i}}(U')}, \tag{3.1.26}$$

with zero Cauchy data on $S \cap U'$ for P^{21}_{i-1}.

The existence and uniqueness of u_1 follow from Lemmas 3.1.32 and 3.1.33.

We shall prove

$$P_{i-1}^{11}u_1 = f_1 \pmod{\mathcal{E}_{j_{E_1^i}}(U')}. \tag{3.1.27}$$

Apply P_i^{22} to both sides of (3.1.26) and apply (3.1.23) and (3.1.20) to obtain

$$P_i^{21}\left(P_{i-1}^{11}u_1 - f_1\right) = 0 \pmod{\mathcal{E}_{j_{E_2^{i+1}}}(U')}.$$

If $(P_{i-1}^{11}u_1 - f_1)$ has zero Cauchy data on $S \cap U'$ with respect to P_i^{21}, the uniqueness given in Lemma 3.1.33 implies (3.1.27). Thus we need only show that for every $g \in \mathcal{D}(E_i^{i+1'}|_{U'})$.

$$\int_{U'\pm}\langle g, P_i^{21}(P_{i-1}^{11}u_1 - f_1)\rangle_x = \int_{U'\pm}\langle P_i^{21'}g, P_{i-1}^{11}u_1 - f_1\rangle_x. \tag{3.1.28}$$

By (3.1.20),

$$\begin{aligned}\int_{U'\pm}&\langle g, P_i^{21}(P_{i-1}^{11}u_1 - f_1)\rangle_x \\ &= -\int_{U'\pm}\langle g, P_i^{22}(P_{i-1}^{21}u_1 - f_2)\rangle_x - \int_{U'\pm}\langle g, P_i^{21}f_1 + P_i^{22}f_2\rangle_x.\end{aligned}$$

It follows from (3.1.26) that the section $(P_{i-1}^{21}u_1 - f_2)$ has zero Cauchy data on $S \cap U'$ with respect to any operator. So by (3.1.25) we get

$$\begin{aligned}\int_{U'\pm}&\langle g, P_i^{21}(P_{i-1}^{11}u_1 - f_1)\rangle_x \\ &= -\int_{U'\pm}\langle P_i^{22'}g, P_{i-1}^{21}u_1 - f_2\rangle_x - \int_{U'\pm}\langle P_i^{21'}g, f_1\rangle_x - \int_{U'\pm}\langle P_i^{22'}g, f_2\rangle_x.\end{aligned}$$

Cancellation yields

$$\int_{U'\pm}\langle g, P_i^{21}(P_{i-1}^{11}u_1 - f_1)\rangle_x = -\int_{U'\pm}\langle P_i^{22'}g, P_{i-1}^{21}u_1\rangle_x - \int_{U'\pm}\langle P_i^{21'}g, f_1\rangle_x.$$

Since u_1 has zero Cauchy data on $S \cap U'$ with respect to P_{i-1}^{21}, we may bring P_{i-1}^{21} to the other side and apply the transpose version (3.1.20) to obtain

$$\int_{U'\pm}\langle g, P_i^{21}(P_{i-1}^{11}u_1 - f_1)\rangle_x = \int_{U'\pm}\langle P_{i-1}^{11}{}'P_i^{21'}g, u_1\rangle_x - \int_{U'\pm}\langle P_i^{21'}g, f_1\rangle_x.$$

By Proposition 3.1.13, u_1 has zero Cauchy data on $S \cap U'$ with respect to the differential operator P_{i-1}^{11} also. So the last equation implies (3.1.28) as required.

To complete the proof, we consider $u = \begin{pmatrix} u_1 \\ 0 \end{pmatrix}$ so that u is in $\mathcal{E}^{i-1}(U')$. By Proposition 3.1.13, since u_1 has zero Cauchy data on $S \cap U'$ with respect to P_{i-1}^{21}, this section vanishes to order $(p_{i-1} - 1)$ on $S \cap U'$. Therefore u has zero Cauchy data on $S \cap U'$ with respect to the differential operator P_{i-1}, i.e. u is in $\mathcal{E}_{t_{i-1}}(U')$. Finally we obtain

$$Pu = \begin{pmatrix} P_{i-1}^{11}u_1 \\ P_{i-1}^{21}u_1 \end{pmatrix} = \begin{pmatrix} f_1 \\ f_2 \end{pmatrix} = f \pmod{\mathcal{E}_{j_i}(U')},$$

and the proof is complete.

□

3.1.10 *Completion of the proof of Theorem 3.1.30*

The following is the goal of this subsection.

Lemma 3.1.35 *Suppose that* $\mathfrak{C}$ *is* $\mathcal{E}$ *or* $\mathcal{D}$. *If the sequence* $E_x^{\imath-1} \xrightarrow{\sigma} E_x^{\imath} \xrightarrow{\sigma} E_x^{\imath+1}$ *is exact for all* $x \in S$, *then* $H^{\imath}((\mathfrak{C}_t\,(S)) = 0$.

Proof. We will show that the sequence

$$\mathfrak{C}_{t_{\imath-1}}(S) \xrightarrow{P} \mathfrak{C}_{t_\imath}(S) \xrightarrow{P} \mathfrak{C}_{t_{\imath+1}}(S)$$

is exact.

Let $f \in \mathfrak{C}_{t_\imath}(S)$ with $Pf = 0$ be given. Then f can be represented by a section $f \in \mathfrak{C}_{t_\imath}(X)$ with $Pf = 0 \pmod{\mathfrak{C}_{t_\imath}(X)}$.

According to Lemma 3.1.34, S has an open covering $\{U_\nu\}_{\nu\in\mathbb{N}}$ such that for every $\nu \in \mathbb{N}$ there exists $u^{(\nu)} \in \mathcal{E}_{t_{\imath-1}}(U_\nu)$ with $Pu^{(\nu)} = f \pmod{\mathcal{E}_{\jmath_\imath}(U_\nu)}$ in U_ν.

Moreover, $u^{(\nu)} = \begin{pmatrix} u_1^{(\nu)} \\ 0 \end{pmatrix}$ satisfies

$$P^{11}_{\imath-1}u_1^{(\nu)} = f_1 \pmod{\mathcal{E}_{\jmath_{E_1^\imath}}(U_\nu)} \quad \text{in} \quad U_\nu,$$

$$P^{21}_{\imath-1}u_1^{(\nu)} = f_2 \pmod{\mathcal{E}_{\jmath_{E_2^\imath}}(U_\nu)} \quad \text{in} \quad U_\nu,$$

where $f = \begin{pmatrix} f_1 \\ f_2 \end{pmatrix}$ is the decomposition of f in U_ν corresponding to (3.1.19) and $u_1^{(\nu)}$ has zero Cauchy data on $S \cap U_\nu$ with respect to $P^{21}_{\imath-1}$.

Consider any two of the sets $\{U_\nu\}$ which overlap, say U_μ and U_ν with their corresponding $u^{(\mu)}$ and $u^{(\nu)}$. Since the decompositions $\mathcal{E}^{\imath}(U_\nu) = \mathcal{E}(E_1^\imath|_{U_\nu}) \oplus \mathcal{E}(E_2^\imath|_{U_\nu})$ are unique, we have

$$P^{21}_{\imath-1}(u_1^{(\mu)} - u_1^{(\nu)}) = 0 \pmod{\mathcal{E}_{\jmath_{E_2^\imath}}(U_\mu \cap U_\nu)} \quad \text{on} \quad U_\mu \cap U_\nu.$$

Since $(u_1^{(\mu)} - u_1^{(\nu)})$ has zero Cauchy data on $S \cap (U_\mu \cap U_\nu)$ with respect to $P^{21}_{\imath-1}$, it follows from Lemma 3.1.33 that $u_1^{(\mu)} - u_1^{(\nu)} = 0 \pmod{\mathcal{E}_{\jmath_{E_1^{\imath-1}}}(U_\mu \cap U_\nu)}$ on $U_\mu \cap U_\nu$. Thus we get $u^{(\mu)} - u^{(\nu)} = 0 \pmod{\mathcal{E}_{\jmath_{\imath-1}}(U_\mu \cap U_\nu)}$ on $U_\mu \cap U_\nu$.

So we may define an element $u \in \mathcal{E}_{t_{\imath-1}}(S)$ consistently by defining it on each $S \cap U_\nu$ to be the image of $u^{(\nu)}$ by the jet mapping. Clearly $Pu = f$ on S which completes the proof in the case $\mathfrak{C} = \mathcal{E}$.

Moreover the above construction of the $u \in \mathcal{E}_{t_{\imath-1}}(S)$ from the $f \in \mathcal{E}_{t_\imath}(S)$ shows that $\operatorname{supp} u \subset \operatorname{supp} f$. Therefore u is in $\mathcal{D}_{t_{\imath-1}}(S)$ provided f is in $\mathcal{D}_{t_\imath}(S)$. This proves Lemma 3.1.35 in the case $\mathfrak{C} = D$ also.

□

Lemma 3.1.35 is easily seen to be slightly stronger than Theorem 3.1.30. Hence Theorem 3.1.30 follows.

□

3.2 Sokhotskii-Plemelj Formulas for Elliptic Complexes of First Order Differential Operators

3.2.1 *Historical reference*

Now our aim is to study jumps when crossing S of the integral $\Pi([S]^P f)$ associated to an elliptic complex $E^{\cdot}$. In view of Theorem 2.4.8, this is a far reaching generalization of the Sokhotskii-Plemelj formulas for the Cauchy type integral (see, for example, Muskhelishvili [165, §16]). Jumps of Martinelli-Bochner type integrals have been studied by Look and Zhony [152] (for a Hölder density), and by Dautov and Kytmanov [45] (for a continuous density). The continuity of boundary values of Martinelli-Bochner type integrals was considered by Kakichev [120]. Chirka [38] and Gasiev [69] investigated the behaviour of such integrals near by the surface of integration. The classical Sokhotskii-Plemelj formulas have been extended to various analogues of the Cauchy-Riemann system in the works of Bitsadse (see his book [23, Ch.VII, §4]), V.S.Vinogradov [245], Shevchenko [199] and others. Generalizations of these formulas to any systems of differential equations of one of the classical types are also known within potential theory (see Friedman [68], Ch.V, §2, 4, 6, and the bibliographical remarks to Ch.V). Investigation of jump behavior of the integral $\Pi([S]^P f)$ is difficult in the general case because the usual Sokhotskii-Plemelj formulas fail for this integral. For the Dolbeault complex with coefficients from $\Lambda^{n,0}$ in $\mathbb{C}^n$, the jump of this integral was considered by Aizenberg and Dautov [5, Theorem 2.5] Henkin [103] completed this result by considering the Dolbeault complex with coefficients in $\Lambda^{p,0}T^*_{\mathbb{C}}(\mathbb{C}^n)$ where $p < n$ was arbitrary. But the most complete results for the Dolbeault complex have been obtained by Harvey and Polking [101]. The author [226, 230, 231, 233] found explicit formulas of Sokhotskii-Plemelj type for arbitrary elliptic complexes on manifolds.

3.2.2 *Tangential and normal components of sections*

In this section we consider the instructive situation when X is an open subset of $\mathbb{R}^n$, and the complex $E^{\cdot}$ consists of differential operators P_i of the first order.

As mentioned in Subsection 2.1.4, for this case the Laplacians $\Delta_i = P_{i-1}P^*_{i-1} + P^*_i P_i$ are selfadjoint elliptic differential operators in $\mathrm{do}_2(E^i \to E^i)$.

Fix some properly supported selfadjoint parametrix $G_i \in \mathrm{pdo}_{-2}(E^i \to E^i)$ for Δ_i, i.e. $\Delta_i G_i = 1 - S_i$ for a suitable $S_i \in \mathrm{pdo}_{-\infty}(E^i \to E^i)$. For instance, one can take $G_i = (K_i + K^*_i)/2$ where K_i is some properly supported parametrix for Δ_i.

Then by Theorem 2.1.12 $\Pi_i = G_{i-1}P^*_{i-1}$ $(i \in \mathbb{Z})$ is a properly supported paramertix of the complex $E^{\cdot}$, i.e. equality (2.1.7) is fulfilled.

As in Subsection 3.1.5, let $\sigma_i(x) = \sigma(P_i)(x, \frac{d\rho(x)}{|d\rho(x)|})$ and $L_i = \sigma_{i-1}\sigma^*_{i-1} + \sigma^*_i\sigma_i$ $(i \in \mathbb{Z})$. Since the differential operators Δ_i are elliptic, the matrices $L_i(x) = \sigma(\Delta_i)(x, \frac{d\rho(x)}{|d\rho(x)|})$ are non-singular in the neighborhood of the hypersurface S where $d\rho(x) \neq 0$.

Hence we have in accordance with decomposition (3.1.18)

$$E^{\imath}|_S = E^{\imath}_t \oplus \sigma_{\imath-1}(E^{\imath-1}_t), \tag{3.2.1}$$

where $E^{\imath}_t = \sigma^*_{\imath}\sigma_{\imath}L^{-1}_{\imath}(E^{\imath}|_S)$ are differentiable vector subbundles of $E^{\imath}|_S$.

Further every section $f \in \mathcal{D}'(E^{\imath}|_S)$ has the unique decomposition

$$f = (\sigma^*_{\imath}\sigma_{\imath}L^{-1}_{\imath}f) + \sigma_{\imath-1}(\sigma^*_{\imath-1}L^{-1}_{\imath}f) = t(f) + \sigma_{\imath-1}n(f). \tag{3.2.2}$$

Definition 3.2.1 *$t(f) \in \mathcal{D}'(E^{\imath}_t)$ and $n(f) \in \mathcal{D}'(E^{\imath-1}_t)$ are said to be the tangential and the normal components of a section $f \in \mathcal{D}'(E^{\imath}|_S)$ respectively.*

These components $t(f)$ and $n(f)$ are easily seen to be independent of the particular choice of a function ρ defining S.

It is easy to verify that the bundle homomorphisms $t : E^{\imath}|_S \to E^{\imath}_t$ and $n : E^{\imath}|_S \to E^{\imath-1}_t$ have the following properties:

$$\begin{array}{rcl} t \circ t &=& t, \\ t \circ n &=& n, \\ n \circ t &=& 0, \\ n \circ n &=& 0. \end{array} \tag{3.2.3}$$

Example 3.2.2 Consider the de Rham complex $E = \Lambda^{\cdot} T^*_{\mathbb{C}}(X)$ on X. Let a vector field ν be given in a neighborhood of S by the condition that $\nu \rfloor \frac{d\rho}{|d\rho|} = 1$. Then every current f of degree q in the neighborhood of S can be decomposed into the sum $f = t(f) + \frac{d\rho}{|d\rho|} \wedge n(f)$ where $t(f) = \nu \rfloor (\frac{d\rho}{|d\rho|} \wedge f)$ is the tangential component of f, and $n(f) = \nu \rfloor f$ is the normal component of f. To prove this, it suffices to use formula (3.1.1). Notice that this decomposition is just the one given by (3.2.2). If f is continuous (i.e. f is a differential form), then the restriction $t(f)$ to S equals $i^{\sharp} f$ where $i^{\sharp} f$ is the pull-back of f under the inclusion $S \to X$.

□

Example 3.2.3 Consider the Dolbeault complex $E^{\cdot} = \Lambda^{p,\cdot} T^*_{\mathbb{C}}(X)$ on X, supposing X be a complex manifold. Let a $(0,1)$-vector field ν be given in a neighborhood of S by the condition that $\nu \rfloor \frac{\bar\partial\rho}{|d\rho|} = 1$. Then every current f of bidegree (p,q) in the neighborhood of S can be decomposed into the sum $f = t(f) + \frac{\bar\partial\rho}{|d\rho|} \wedge n(f)$ where $t(f) = \nu \rfloor (\frac{\bar\partial\rho}{|d\rho|} \wedge f)$ is the complex tangential component of f, and $n(f) = \nu \rfloor f$ is the complex normal component of f. To prove this, use formula (3.1.1) again. Obviously, this decomposition is just the one given by (3.2.2).

□

Proposition 3.2.4 *If $f \in C_{loc}(E^{\imath}|_{\overline{X^{\pm}}})$, then $t(f) = 0$ (respectively, $n(f) = 0$) if and only if f has zero Cauchy data on S with respect to $P_{\imath}$ (respectively $P^*_{\imath-1}$).*

Proof. This immediately follows from Definition 3.1.12, since by formula (2.4.13) we have

$$\begin{array}{rcccl} G_{P_i}(g,f)|_S &=& \frac{1}{\sqrt{-1}}g(\sigma_i f)ds &=& \frac{1}{\sqrt{-1}}(n(g^*))^* L_i t(f)ds, \\ G_{P^*_{i-1}}(g,f)|_S &=& \frac{1}{\sqrt{-1}}g(\sigma^*_{i-1}f)ds &=& \frac{1}{\sqrt{-1}}(n(g^*))^* L_{i-1} n(f)ds, \end{array}$$

where ds denotes the Euclidean element of hypersurface area on S.

□

One can observe from the proof that $[S]^P f = \frac{1}{\sqrt{-1}}[S]\sigma_i t(f)$ where $[S]$ is the distribution of integration over S.

3.2.3 *The Sokhotskii-Plemelj formulas*

By Proposition 3.2.4 the mapping $f \to [S]^P f$ determines an embedding $\mathcal{D}'(E^i_t) \to \mathcal{D}'_s(E^{i+1})$.

Since Π is a properly supported pseudo-differential parametrix of $E^{\cdot}$, we can define the section $\Pi([S]^P f)$ in $\mathcal{D}'(E^i)$ for every section $f \in \mathcal{D}'(E^i_t)$.

Definition 3.2.5 *$\Pi([S]^P f)$ is called the double layer potential with density f associated to the complex $E^{\cdot}$.*

If $E^{\cdot}$ is the de Rham complex on X and $f \in \mathcal{D}'(S)$ (i.e. i=0), then $\Phi([S]f)$ is just the classical double layer potential with density f on S.

It follows from the regularity properties of the parametrix Π (see of Theorem 2.1.12) that $\Pi([S]^P f)^{\pm} \in \mathcal{E}(E^i|_{X^{\pm}})$, and

$$\Pi([S]^P f)(x) = \int_S G_{P_i}(\Pi(x,y), f(y)) \quad (x \notin S).$$

Here, for a section $f \in \mathcal{D}'(E^i)$, we denote by $f^{\pm}$ its restrictions to $X^{\pm}$.

For $x \in S$, we define $\Pi([S]^P f)^{(s)}(x)$ to be the Cauchy principal value of the singular integral $\Pi([S]^P f)(x)$ provided it exists:

$$\Pi([S]^P f)^{(s)}(x) = \lim_{\varepsilon \to +0} \int_{\substack{y \in S \\ |y-x|>\varepsilon}} G_{P_i}(\Pi(x,y), f(y))$$

The following result for the Dolbeault complex with coefficients in $\Lambda^{p,0}T^*_{\mathbb{C}}(\mathbb{C}^n)$ is due to Harvey and Polking [101]. For arbitrary elliptic complexes it was proved in the author's paper [226].

Theorem 3.2.6 *Assume $f \in \mathcal{E}(E^i_t)$, then:*

1) $\Pi([S]^P f)^{\pm}$ have differentiable continuations to $\overline{X^{\pm}}$;

2) for every $x \in S$, the Cauchy principal value of the integral $\Pi([S]^P f)(x)$ exists, and $\Pi([S]^P f)^{(s)} \in \mathcal{E}(E^i|_S)$;

3) on S the values of $\Pi([S]^P f)^{\pm}$ and $\Pi([S]^P f)^{(s)}$ are connected by the Sokhotskii-Plemelj jump formulas

$$\Pi([S]^P f)^{\pm} = \Pi([S]^P f)^{(s)} \pm \frac{1}{2} f. \qquad (3.2.4)$$

To prove this theorem, we need three lemmas which are also of independent interest.

3.2.4 Structure of the Cauchy data for the differential operator Δ_i

For $f \in \mathcal{E}(E^i)$, we let $c_\Delta(f) = \{t(f), n(f), t(P^*f), n(Pf)\}$. This defines a mapping

$$c_\Delta : \mathcal{E}(E^i) \to \mathcal{E}(E^i_t) \oplus \mathcal{E}(E^{i-1}_t) \oplus \mathcal{E}(E^{i-1}_t) \oplus \mathcal{E}(E^i_t).$$

Lemma 3.2.7 *The mapping c_Δ is surjective and $c_\Delta f = 0$ if and only if $D^\alpha f|_S = 0$ for all $|\alpha| \le 1$.*

Proof. Given $f \in \mathcal{E}(E^i)$, we obviously have $D^\alpha f|_S = 0$ for $|\alpha| \le s$ if and only if $f = \rho^{s+1} f'$ for some section $f' \in \mathcal{E}(E^i)$.

Let $D^\alpha f|_S = 0$ for $|\alpha| \le 1$, then $f = \rho^2 f'$. We have

$$\begin{aligned} P^*f &= \rho^2(P^*f') + \tfrac{2}{\sqrt{-1}}\rho|d\rho|\,(L_{i-1}n(f')), \\ Pf &= \rho^2(Pf') + \tfrac{2}{\sqrt{-1}}\rho|d\rho|\,(\sigma_i t(f')). \end{aligned}$$

Hence $P^*f|_S = Pf|_S = 0$ and $c_\Delta(f) = 0$.

Inversely if $c_\Delta(f) = 0$ then $f|_S = 0$ by (3.2.2). Hence it follows that $f = \rho f'$ for some section $f' \in \mathcal{E}(E^i)$, and

$$\begin{aligned} P^*f &= \rho(P^*f') + \tfrac{1}{\sqrt{-1}}|d\rho|\,(L_{i-1}n(f')), \\ Pf &= \rho(Pf') + \tfrac{1}{\sqrt{-1}}|d\rho|\,(\sigma_i t(f')). \end{aligned}$$

Therefore, we get on S

$$\begin{aligned} t(P^*f) &= \tfrac{1}{\sqrt{-1}}|d\rho|\,(L_{i-1}n(f')), \\ n(Pf) &= \tfrac{1}{\sqrt{-1}}|d\rho|\,t(f'). \end{aligned} \tag{3.2.5}$$

Equalities (3.2.5) imply $t(f') = n(f') = 0$, so that $f' = \rho f''$ for same $f'' \in \mathcal{E}(E^i)$.

Finally, to prove that the mapping c_Δ is surjective, let

$$\{f_1, f_2, f_3, f_4\} \in \mathcal{E}(E^i_t) \oplus \mathcal{E}(E^{i-1}_t) \oplus \mathcal{E}(E^{i-1}_t) \oplus \mathcal{E}(E^i_t)$$

be given, and let $\tilde{f}_j$ $(j = 1, \ldots, 4)$ denote differentiable extensions of f_j to X.

If $f \in \mathcal{E}(E^i)$ coincides with the section $\tilde{f}_1 + \sigma_{i-1}\tilde{f}_2$ in some neighborhood of S, then $t(f) = f_1$ and $n(f) = f_2$. Hence one can assume that $f_1 = 0$ and $f_2 = 0$.

In this case let $f \in \mathcal{E}(E^i)$ coincide with the section

$$\sqrt{-1}\frac{\rho}{|d\rho|}\,(f_4 + \sigma_{i-1}L^{-1}_{i-1}\tilde{f}_3)$$

in some neighborhood of S. Then $t(f) = 0$, $n(f) = 0$, and according to (3.2.5) $t(P^*f) = f_3$ and $n(f) = f_4$, as required.

□

Remark 3.2.8 *The four dimensional vector $\{t(f), n(f), t(P^*f), n(Pf)\}$ thus gives a representation of the Cauchy data of f on S with respect to the Laplacian Δ_i. The tangential part of the Cauchy data, $\{t(f), t(P^*f)\}$ is called the $(P-)$ Dirichlet data, and the normal part of the Cauchy data, $\{n(f), n(Pf)\}$ is called the $(P-)$ Neumann data.*

3.2.5 *Formal version of the Cauchy-Kovalevskaya Theorem for the differential operator* Δ

The following lemma is a particular version of Theorem 3.1.30 for the complex $0 \to E^{\cdot} \xrightarrow{\Delta_\cdot} E^{\cdot} \to 0$.

Lemma 3.2.9 *For each* $f_0 \in \mathcal{E}(E^{\cdot})$ *and*

$$\{f_1, f_2, f_3, f_4\} \in \mathcal{E}(E_t^{\cdot}) \oplus \mathcal{E}(E_t^{\cdot-1}) \oplus \mathcal{E}(E_t^{\cdot-1}) \oplus \mathcal{E}(E_t^{\cdot})$$

there exists a section $f \in \mathcal{E}(E^{\cdot})$ *such that*

$$\begin{cases} D^\alpha(\Delta_\cdot f - f_0)|_S = 0 \quad \text{for all } \alpha, \\ c_\Delta f = \{f_1, f_2, f_3, f_4\}. \end{cases} \tag{3.2.6}$$

Moreover, if f^1 *and* f^2 *satisfy (3.2.6), then* $D^\alpha(f^1 - f^2)|_S = 0$ *for all* α.

Proof. Firstly, conditions (3.2.6) determine all the derivatives of f on S uniquely (see Subsection 3.1.7). Thus the second part of the lemma has been proved.

On the other hand, since all the derivatives of f on S have been determined, the Whitney Extension Theorem guarantees the existence of $f \in \mathcal{E}(E^{\cdot})$ with these derivatives. Hence the first part of the lemma follows.

□

3.2.6 *Green's identity for the differential operator* Δ

Denote by χ^- the characteristic function of X^-.

Lemma 3.2.10 *If* $f \in \mathcal{E}(E^{\cdot})$, *then in* $\mathcal{D}'(E^{\cdot})$

$$\begin{aligned} &\Delta_\cdot(\chi^- f) - \chi^-(\Delta_\cdot f) \\ &\quad = -P_\cdot^*([S]^P t(f)) - P_{\cdot-1}([S]^{P^*}\sigma_{\cdot-1} n(f)) \\ &\quad\quad -[S]^P t(P^* f) - [S]^{P^*}\sigma_\cdot n(Pf). \end{aligned}$$

Proof. This follows immediately from equalities (2.4.21) and (2.5.10) together with the definitions of subsection 3.2.2.

□

3.2.7 *Proof of Theorem 3.2.6*

Assume $f \in \mathcal{E}(E_t^{\cdot})$. By Lemma 3.2.9 there exists a section $\tilde{f} \in \mathcal{E}(E^{\cdot})$ such that

$$\begin{aligned} t(\tilde{f}) = f, &\quad n(\tilde{f}) = 0, \\ t(P^*\tilde{f}) = 0, &\quad n(P\tilde{f}) = 0 \end{aligned} \tag{3.2.7}$$

and

$$D^\alpha(\Delta_\cdot \tilde{f})|_S = 0 \quad \text{for all} \quad \alpha. \tag{3.2.8}$$

By using Green's identity for the differential operator $\Delta_\imath$ and (3.2.7), we have

$$\Delta_\imath(\chi^- \tilde{f}) = \chi^-(\Delta_\imath \tilde{f}) - P_\imath^*([S]^P f). \tag{3.2.9}$$

We conclude from (3.2.8)

$$\chi^-(\Delta_\imath \tilde{f}) \in \mathcal{E}(E^\imath). \tag{3.2.10}$$

Let us calculate $\Delta_\imath(\Pi([S]^P f))$, and then we will compare the answer with $\Delta_\imath(\chi^- \tilde{f})$.

Since $G_\imath$ is a properly supported pseudo-differential parametrix of $\Delta_\imath$ we get

$$\Delta_\imath G_\imath P_\imath^*([S]^P f) = P_\imath^*([S]^P f) - S_\imath P_\imath^*([S]^P f) \tag{3.2.11}$$

where

$$S_\imath P_\imath^*([S]^P f) \in \mathcal{E}(E^\imath). \tag{3.2.12}$$

It follows from (3.2.9) and (3.2.11) that

$$\Delta_\imath(\Pi([S]^P f) + \chi^- \tilde{f}) = \chi^-(\Delta_\imath \tilde{f}) - S_\imath P_\imath^*([S]^P f). \tag{3.2.13}$$

By the interior regularity property of solutions of $\Delta_\imath(\cdot) = 0$, we can conclude from (3.2.10), (3.2.12) and (3.2.13) that

$$(\Pi([S]^P f) + \chi^- \tilde{f}) \in \mathcal{E}(E^\imath). \tag{3.2.14}$$

Hence, $\Pi([S]^P f)^\pm$ have infinitely differentiable extensions to $\overline{X^\pm}$ together with $\chi^- \tilde{f}$.

Moreover, (3.2.14) implies

$$\Pi([S]^P f)^+ - (\Pi([S]^P f)^- + \tilde{f}) = 0 \quad \text{on} \quad S,$$

so by (3.2.7)

$$(\Pi([S]^P f)^+ - \Pi([S]^P f)^-)|_S = f. \tag{3.2.15}$$

To complete the proof of the theorem, it remains to show the existence of the Cauchy principal value of the integral $\Pi([S]^P f)$ on S, and one of the sides of formula (3.2.4).

Denote by $B(x,\varepsilon)$ the open ball in X with center x and radius ε, and let $x \in S$ be a fixed point.

Using Corollary 2.5.14 and equalities (3.2.7), we get

$$\begin{aligned}
\Pi([S]^P f)^{(s)}(x) &= \lim_{\varepsilon \to +0} \left(\int_{S \setminus B(x,\varepsilon)} G_{\Delta_\imath}(G_\imath(x,y), \tilde{f}(y)) \right) \\
&= \lim_{\varepsilon \to +0} \left(\int_{\partial(X^- \cup B(x,\varepsilon))} G_{\Delta_\imath}(G_\imath(x,y), \tilde{f}(y)) - \int_{\partial B(x,\varepsilon) \cap X^+} G_{\Delta_\imath}(G_\imath(x,y), \tilde{f}(y)) \right) \\
&= \lim_{\varepsilon \to +0} \left(-\tilde{f}(x) + \int_{X^- \cup B(x,\varepsilon)} S_\imath(x,y) \tilde{f}(y) + \int_{X^- \cup B(x,\varepsilon)} G_\imath(x,y) \Delta_\imath \tilde{f}(y) \right. \\
&\quad \left. - \int_{\partial B(x,\varepsilon) \cap X^+} G_{\Delta_\imath}(G_\imath(x,y), \tilde{f}(y)) \right).
\end{aligned} \tag{3.2.16}$$

Use now the description of kernels of pseudo-differential operators given by Theorem 1.5.14. Since $\tilde{f}$ and the hypersurface S are smooth, it follows that

$$\begin{aligned}
&\lim_{\varepsilon\to+0}\int_{\partial B(x,\varepsilon)\cap X^+} G_{\Delta_*}(G_*(x,y),\tilde{f}(y)) \\
&= \lim_{\varepsilon\to+0}\int_{\partial B(x,\varepsilon)\cap X^+} G_{\Delta_*}(G_*(x,y),\tilde{f}(x)) \\
&= \frac{1}{2}\lim_{\varepsilon\to+0}\int_{\partial B(x,\varepsilon)} G_{\Delta_*}(G_*(x,y),\tilde{f}(x)) \\
&= -\frac{1}{2}\tilde{f}(x).
\end{aligned}$$

Therefore, by taking the limit in the right-hand side of (3.2.16), we obtain

$$\Pi([S]^P f)^{(s)}(x) = -\frac{1}{2}f(x) + S_*(\chi^-\tilde{f})(x) + G_*(\chi^-\Delta_*\tilde{f})(x). \tag{3.2.17}$$

It immediately follows from (3.2.17) and (3.2.10) that $\Pi([S]^P f)^{(s)}$ exists on S and it is in $\mathcal{E}(E^*|_S)$.

Use again Corollary 2.5.14 and properties (3.2.7) of the section f, to get for $x^- \in X^-$:

$$\Pi([S]^P f)(x^-) = -\tilde{f}(x^-) + S_*(\chi^-\tilde{f})(x^-) + G_*(\chi^-\Delta_*\tilde{f})(x^-).$$

Since the sections $S_*(\chi^-\tilde{f})$ and $G_*(\chi^-\Delta_*\tilde{f})$ are continuous on X, we can let x^- tend to x to see that

$$\Pi([S]^P f)^-(x) = -f(x) + S_*(\chi^-\tilde{f})(x) + G_*(\chi^-\Delta_*\tilde{f})(x). \tag{3.2.18}$$

If we subtract (3.2.17) from (3.2.18), we have that

$$\Pi([S]^P f)^-(x) = \Pi([S]^P f)^{(s)}(x) - \frac{1}{2}f(x).$$

Comparing this equality with (3.2.15), we obtain the second correlation (3.2.4) which completes the proof.

□

3.2.8 *Further results*

Prenov and Tarkhanov [184] extended Theorem 3.2.6 to the case when the hypersurface S is assumed to be piecewise smooth, and the section $f \in C_{loc}(E_t^*)$ is only assumed to satisfy a Dini condition for its modulus of continuity.

In this case the coefficient $(\pm\frac{1}{2})$ in formula (3.2.4) is replaced by $\pm a(C_x(X^\mp))$ where $a(C_x(X^\pm))$ is the angular measure of the tangential cone to $X^\pm$ at the point x, i.e., the ratio

$$\frac{\operatorname{vol}\,(C_x(X^\pm)\cap\partial B(x,\varepsilon))}{\operatorname{vol}\,(\partial B(x,\varepsilon))}$$

which is independent of $\varepsilon > 0$.

Exercise 3.2.11 *Prove that, if $f \in L^1_{loc}(E^{\cdot}_i)$ and $x^0 \in S$ is a Lebesgue point of f, then for every path $w \subset X^{\pm}$ with end x^0, tending non-tangentially to S at x^0 we have*

$$\begin{aligned}\Pi([S]^P f)(x) &= \int_{S\setminus B(x^0,|x-x^0|)} G_{P_i}(\Pi_i(x^0,y), f(y)) \pm a(C_{x^0}(X^{\mp}))\cdot f(x^0)\\ &\quad +o(1) \quad (x\in w).\end{aligned} \tag{3.2.19}$$

Hint. Use Corollary 2.5.15.

□

This is a straightforward extension of a well-known lemma of Privalov for Cauchy type integrals (see Kytmanov [145, §2]).

It follows from formula (3.2.19) that the Cauchy principal value of the singular integral $\Pi([S]^P f)$ at $x^0 \in S$ exists if and only if the limit of $\Pi([S]^P f)(x)$ exists as $x \to x^0$ $(x \in w)$. In particular, $\Pi([S]^P f)^{(s)}$ exists at $x^0 \in S$ provided the modulus of continuity of f at this point satisfies a Dini condition.

Given any smooth mapping m from a neighborhood of a point $x \in S$ to $\mathbb{R}^n$ such that x is a simple zero point for m, we let

$$\Pi([S]^P f)^{(m)}(x) = \lim_{\varepsilon\to+0}\int_{S\setminus\{m(y)<\varepsilon\}} G_{P_i}(\Pi(x,y), f(y)). \tag{3.2.20}$$

Exercise 3.2.12 *Prove that, if the modulus of continuity of $f \in L^1_{loc}(E^{\cdot}_i)$ at the point x satisfied the Dini condition, then the limit (3.2.20) exists, and write the suitable Sokhotskii-Plemelj type formulas.*

Hint. Calculate

$$\lim_{\varepsilon\to+0}\int_{\substack{y\in X^{\pm}\\ |m(y)|=\varepsilon}} G_{\Delta_i}(G_i(x,y), I_{k_0}).$$

□

For a classical double layer potential, the limit (3.2.20) is independent of the particular choice of a mapping m with the above properties. However, for the singular Martinelli-Bochner integral, this is so if and only if m is holomorphic. This result has been proved by Prenov and Tarkhanov [185].

For arbitrary complexes $E^{\cdot}$, only the general result of Mikhlin [161, §5] is known in this direction.

One of the direct corollaries of Theorem 3.2.6 is the Poincaré-Bertrand type rearrangement formula (see Muskhelishvili [165, §28]) for the double layer potential $\Pi([S]^P f)$ associated to an elliptic complex of first order differential operators.

Exercise 3.2.13 *Prove that for every section $f \in [C^{0,\lambda}_{loc}(S\times S)]^{k_0}$ $(\lambda > 0)$ the following Poincaré-Bertrand type formula holds:*

$$\begin{aligned}&\int_{S_z} G_{P_0(z)}\left(\Pi_0(x,z), \int_{S_y} G_{P_0(y)}(\Pi_0(z,y), f(z,y))\right)\\ &\quad -\int_{S_y}\int_{S_z} G_{P_0(z)}\left(\Pi_0(x,z), G_{P_0(y)}(\Pi_0(z,y), f(z,y))\right) = \frac{1}{4}f(x,x) \quad (x\in S).\end{aligned} \tag{3.2.21}$$

Hint. Use the rearrangement formula for a *SIO* and an integral operator with weak singularity (see Mikhlin [161, §9]).

Observe that

$$\int_{S_z} G_{P_0(z)}\left(\Pi_0(x,z), \int_{S_y} G_{P_0(y)}(\Pi_0(z,y), I_{k_0})\right)$$
$$- \int_{S_y} G_{P_0(y)}\left(\int_{S_z} G_{P_0(z)}(\Pi_0(x,z), \Pi_0(z,y)), I_{k_0}\right) = \frac{1}{4} I_{k_0} \quad (x \in S).$$

Imitate the proof of Theorem 22.6 in the book of Kytmanov [145].

□

For the Moisil-Theodoresco system such a formula has been found by Bitsadze (see his book [23, Ch.VII, §4]).

A theory of singular integral equations with matrix-valued kernels, based on a similar formula, was outlined in the thesis of Krendelev [140].

3.2.9 *A description of the tangential complex*

Given any non-elliptic complex $E^{\cdot}$ of first order differential operators on X, all the results of Subsections 3.2.2, 3.2.4 and 3.2.5 continue to be valid provided the hypersurface S is non-characteristic for $E^{\cdot}$. It follows from Proposition 3.2.4 and Lemma 3.2.7 that the rows are exact in the following diagram:

$$\begin{array}{ccccccccc}
 & & 0 & & 0 & & 0 & & \\
 & & \downarrow & & \downarrow & & \downarrow & & \\
0 & \longrightarrow & \mathfrak{C}_{t_0}(X) & \xrightarrow{\iota} & \mathfrak{C}^0(X) & \xrightarrow{t} & \mathfrak{C}(E_t^0) & \longrightarrow & 0 \\
 & & \downarrow P & & \downarrow P & & \downarrow P_b & & \\
0 & \longrightarrow & \mathfrak{C}_{t_1}(X) & \xrightarrow{\iota} & \mathfrak{C}^1(X) & \xrightarrow{t} & \mathfrak{C}(E_t^1) & \longrightarrow & 0 \\
 & & \downarrow P & & \downarrow P & & \downarrow P_b & & \\
0 & \longrightarrow & \mathfrak{C}_{t_2}(X) & \xrightarrow{\iota} & \mathfrak{C}^2(X) & \xrightarrow{t} & \mathfrak{C}(E_t^2) & \longrightarrow & 0 \\
 & & \downarrow P & & \downarrow P & & \downarrow P_b & & \\
 & & \cdots & & \cdots & & \cdots & &
\end{array} \qquad (3.2.22)$$

Let us define the differential P_b so as to make this diagram commutative. Then $(P_b)_i \in \mathrm{do}_1(E_t^i \to E_t^{i+1})$ $(i \in \mathbb{Z})$, and $P_b^2 = 0$.

Thus we have arrived at a differential complex $E_t^{\cdot}$ on S.

Corollary 3.2.14 *If the hypersurface S is non-characteristic for the complex $E^{\cdot}$, then*

$$\mathfrak{C}_{[t^{\cdot}]}(S) \cong \mathfrak{C}(E_t^{\cdot})$$

Proof. This immediately follows from the commutative diagram (3.2.22).

□

As mentioned in Section 3.1, the complex $E_t^{\cdot}$ is not elliptic in the general case, even if the initial complex $E^{\cdot}$ is elliptic.

3.3 Generalization of the Sokhotskii-Plemelj Formulas to the Case of Arbitrary Elliptic Complexes

3.3.1 *Special parametrix of an elliptic complex*

Suppose we are given an arbitrary elliptic complex $E^{\cdot}$ on the manifold X.

A differential complex $\tilde{E}^{\cdot} = \{E^i, \tilde{P}_i\}_{i\in\mathbb{Z}}$, with $\tilde{P}_i \in \mathrm{do}_{\tilde{p}_i}(E^i \to E^{i+1})$, is said to be an associated complex for E if the differential operators $\Delta_i = \tilde{P}_i^* P_i + P_{i-1}\tilde{P}_{i-1}^*$ are elliptic of order $2p$, where $2p = \tilde{p}_i + p_i$ is the same for all $i = 0, \dots, N$.

If the orders of all the differential operators P_i are the same (and equal p), one can take the complex $E^{\cdot}$ itself as an associated complex.

In the general case the existence of an associated complex for $E^{\cdot}$ is equivalent to the ellipticity of $E^{\cdot}$.

Proposition 3.3.1 *For an associated complex for $E^{\cdot}$ to exist, it is necessary and sufficient that $E^{\cdot}$ should be elliptic.*

Proof. The necessity is obvious. Let us prove the sufficiency. Fix some integer p such that $p_i \le p/2$ for all i. For every i, we choose an elliptic differential operator $Q_i \in \mathrm{do}_{p-2p_i}(E^i \to E^i)$ whose symbol over each point $(x, z) \in T^*(X)$ is given by a diagonal matrix. For example, we may take $Q_i = (\nabla_i^*\nabla_i)^{p/2-p_i}$ where $\nabla_i \in \mathrm{do}_1(E^i \to E^i \otimes T^*(X))$ is a connection for the bundle E^i compatible with a Hermitian metric. Let $\tilde{P}_i = P_i Q_i^2 P_i^* P_i$ so that $\tilde{P}_i$ are differential operators of types $E^i \to E^{i+1}$ and orders $\tilde{p}_i = 2p - p_i$ satisfying $\tilde{P}_{i+1} \circ \tilde{P}_i = 0$ for all i. Standard reasonings from linear algebra (see Wells [254, Ch.IV, §5]) show that $\Delta_i = \tilde{P}_i^* P_i + P_{i-1}\tilde{P}_{i-1}^*$ are elliptic differential operators of order $2p$ for all i. In other words, $\{E^i, \tilde{P}_i\}_{i\in\mathbb{Z}}$ is an associated complex for $E^{\cdot}$.

□

Remark 3.3.2 *If, in defining an associated complex, all the conditions on the orders of the differential operators Δ_i are omitted, we obtain a generalization of the classical notion of an elliptic complex.*

Fix some associated complex $\tilde{E}^{\cdot} = \{E^i, \tilde{P}^i\}$ for $E^{\cdot}$. Let $G_i \in \mathrm{pdo}_{-2p}(E^i \to E^i)$ be properly supported parametrices for the Laplacians Δ_i.

As it was shown in the proof of Theorem 2.1.12, both $\Pi = G\tilde{P}^*$ and $\tilde{\Pi} = \tilde{P}^* G$ are parametrices of the complex $E^{\cdot}$. Indeed they coincide modulo smoothing operators because the equality $\tilde{P}^*\Delta = \Delta\tilde{P}^*$ implies now

$$G_{i-1}\tilde{P}_{i-1}^* = \tilde{P}_{i-1}^* G_i \pmod{\mathrm{pdo}_{-\infty}(E^i \to E^{i-1})} \tag{3.3.1}$$

Exercise 3.3.3 *Check that $\Pi^2 = 0$ modulo smoothing operators.*

Hint. Use (3.3.1).

□

In other words, the curvature of the sequence $\{E^{\cdot}, \Pi_{\cdot}\}$ is a smoothing operator (cf. Wells [254, Ch.III]).

Exercise 3.3.4 *Show that $(GP)^*$ is a parametrix of the associated complex $\widetilde{E}^{\cdot}$.*

Hint. Follow the proof of Theorem 2.1.12. Use (3.3.1) instead of (2.1.12).

□

3.3.2 *Behaviour of potentials near the surface of integration*

Notice first that every differential operator $P \in \mathrm{do}_p(E \to F)$ on X may be lowered to a mapping $C^s_{loc}(E|_S) \to C^{s-p}_{loc}(F|_S)$ provided $s \geq p$. This mapping will be denoted by P also.

For a jet $f \in C^{2p-1}_{loc}(E^{\cdot}|_S)$, we shall be interested in the jump behaviour while crossing S of the sections $f_{I-IV} \in \mathcal{D}'(E^{\cdot})$ defined by the boundary integral of formula (2.5.11), namely

$$\begin{array}{rclrcl} f_I &=& G\widetilde{P}^*([S]^P f), & f_{II} &=& GP([S]^{\widetilde{P}^*} f), \\ f_{III} &=& G([S]^P \widetilde{P}^* f), & f_{IV} &=& G([S]^{\widetilde{P}^*} Pf). \end{array} \tag{3.3.2}$$

Let us denote by $t(f)$ the Cauchy data of f on S with respect to the differential operator $P_{\cdot}$, and by $n(f)$ the Cauchy data of f on S with respect to the differential operator $\widetilde{P}^*_{\cdot-1}$ (see Subsection 3.1.4). Then we have

$$\begin{array}{rclrcl} f_I &=& G\widetilde{P}^*([S]^P t(f)), & f_{II} &=& GP([S]^{\widetilde{P}^*} n(f)), \\ f_{III} &=& G([S]^P t(\widetilde{P}^* f)), & f_{IV} &=& G([S]^{\widetilde{P}^*} n(Pf)). \end{array} \tag{3.3.3}$$

Remark 3.3.5 *It follows from (3.3.1) and (2.1.12) that $f_I = \widetilde{P}^* G([S]^P f)$ and $f_{II} = PG([S]^{\widetilde{P}^*} f)$ modulo smooth sections of $E^{\cdot}$ on X. Thus the jump behaviours of both the left and right-hand sides are the same.*

This observation is useful in applying Theorem 3.3.9 below to boundary value problems for elliptic complexes (see the author's [226] paper).

Let us develop an analogy with the classical potential theory.

Definition 3.3.6 • *$G([S]^P\cdot)$ and $G([S]^{\widetilde{P}^*}\cdot)$ are called the single layer potentials associated to the complex $E^{\cdot}$.*

• *$G\widetilde{P}^*([S]^P\cdot)$ and $GP([S]^{\widetilde{P}^*}\cdot)$ are called the double layer potentials associated to the complex $E^{\cdot}$.*

Since pseudo-differential operators are pseudo-local operators by Proposition 1.5.12, all the potentials f_{I-IV} are (infinitely) differentiable sections of the bundle $E^{\cdot}$ everywhere on X outside the hypersurface S.

To describe their behaviour near S, it is convenient to digress from the situation under consideration. Namely, we suppose that $P \in \mathrm{do}_p(E \to F)$ is an arbitrary differential operator on X, and $\Pi \in \mathrm{pdo}_\pi(F \to G)$ is a properly supported parametrix of some elliptic differential operator of order $(-\pi)$ on X. Let us agree, for a section $f \in \mathcal{D}'(E)$, to denote by $f^\pm$ the restrictions of f to the sets $X^\pm$. Then the following holds.

Proposition 3.3.7 *1) If $S \in C^{s+1,\lambda}_{loc}$ and $f \in C^{s,\lambda}_{loc}(E|_S)$, where $s \geq p-1$ and $0 < \lambda < 1$, then each of the sections $\Pi([S]^P f)^\pm$ is of class $C^{s-p-\pi,\lambda}_{loc}(G|_{\overline{X^\pm}})$ and depends on f continuously in the corresponding topologies.*

2) If $S \in C^{s+1,\lambda}_{loc}$ and $f \in C^{s,\lambda}_{loc}(F|_{\overline{X^-}})$, where $0 < \lambda < 1$, then each of the sections $\Pi(\chi^- f)^\pm$ is of class $C^{s-\pi,\lambda}_{loc}(G|_{\overline{X^\pm}})$ and depends on f continuously in the corresponding topologies.

Proof. 1) for $s = \infty$ was proved by Hörmander [111, Theorem 2.1.4]. The fact, that under the conditions of 2) $\Pi(\chi^- f)^\pm$ is of class $C^{s-\pi,\lambda}_{loc}(G|_{X^\pm})$ and depends on f continuously in the corresponding topologies, is a well-known property of pseudo-differential operators (see Taylor [236, Ch. XI, Theorem 2.5]). In full generality the proposition follows from the Privalov type theorem of Agmon, Douglis and Nirenberg [2, Theorem 3.1] and the description of kernels of pseudo-differential operators given by Theorem 1.5.14. We refer also to the book of Rempel and Schulze [188] where the boundedness properties of the so-called potential (or coboundary) operators in Hölder spaces have been proved (ibid 2.3.2.6 and 2.3.3.3). Proposition 3.3.7 is embraced by this theory.

□

Corollary 3.3.8 *Let $S \in C^{s+1,\lambda}_{loc}$ and $f \in C^{s,\lambda}_{loc}(E^{\cdot}|_S)$, where $s \geq 2p-1$ and $0 < \lambda < 1$. Then each of the potentials $f^\pm_{I-IV}$ is of class $C^{s,\lambda}_{loc}(E^{\cdot}|_{\overline{X^\pm}})$ and depends on f continuously in the corresponding topologies.*

Proof. This follows from Proposition 3.3.7 and Remark 3.3.5 immediately.

□

3.3.3 *Jump formulas*

The following theorem was first proved by the author [226].

Theorem 3.3.9 *If $S \in C^{s+1,\lambda}_{loc}$ and $f \in C^{s,\lambda}_{loc}(E^{\cdot}|_S)$ with $s \geq 2p-1$ and $0 < \lambda < 1$, then $f^\pm_{I-IV} \in C^{s,\lambda}_{loc}(E^{\cdot}|_{\overline{X^\pm}})$ and*

$$\begin{cases} t(f_I^+ - f_I^-) &= t(f), \\ n(f_I^+ - f_I^-) &= 0, \\ t(\tilde{P}^* f_I^+ - \tilde{P}^* f_I^-) &= 0, \\ n(P f_I^+ - P f_I^-) &= 0; \end{cases} \tag{3.3.4}$$

$$\begin{cases} t(f_{II}^+ - f_{II}^-) &= 0, \\ n(f_{II}^+ - f_{II}^-) &= n(f), \\ t(\tilde{P}^* f_{II}^+ - \tilde{P}^* f_{II}^-) &= 0, \\ n(P f_{II}^+ - P f_{II}^-) &= 0; \end{cases} \tag{3.3.5}$$

$$\begin{cases} t(f_{III}^+ - f_{III}^-) &= 0, \\ n(f_{III}^+ - f_{III}^-) &= 0, \\ t(\tilde{P}^* f_{III}^+ - \tilde{P}^* f_{III}^-) &= t(\tilde{P}^* f), \\ n(P f_{III}^+ - P f_{III}^-) &= 0; \end{cases} \tag{3.3.6}$$

$$\begin{cases} t(f_{IV}^+ - f_{IV}^-) &= 0, \\ n(f_{IV}^+ - f_{IV}^-) &= 0, \\ t(\tilde{P}^* f_{IV}^+ - \tilde{P}^* f_{IV}^-) &= 0, \\ n(P f_{IV}^+ - P f_{IV}^-) &= n(Pf). \end{cases} \tag{3.3.7}$$

For complexes E of first order differential operators the formulas (3.3.4) follow from (3.2.14) and (3.2.7) immediately.

Proof. Firstly we have obtained in Corollary 3.3.8 that each of the section $f_{I-IV}^{\pm}$ is of class $C_{loc}^{s,\lambda}(E^{\cdot}|_{\overline{X^{\pm}}})$.

Choose some representative $f \in C_{loc}^{s,\lambda}(E^{\cdot})$ of the jet f. If $0 < \lambda' < \lambda$, then f can be approximated in the topology of $C_{loc}^{s,\lambda'}(E^{\cdot})$ by elements of $\mathcal{E}(E^{\cdot})$. Let $\{f_\nu\} \subset \mathcal{E}(E^{\cdot})$ be some sequence which approximates f. According to Corollary 3.3.8, all the potentials $(f_\nu)_{I-IV}^{\pm}$ are of class $C_{loc}^{s,\lambda}(E^{\cdot}|_{\overline{X^{\pm}}})$, and they converge in the topology of $C_{loc}^{s,\lambda'}(E^{\cdot}|_{\overline{X^{\pm}}})$ to the potentials $f_{I-IV}^{\pm}$ respectively. Since the relationships (3.3.4) - (3.3.7) contain derivatives up to order $(2p-1)$ of the sections f and $f_{I-IV}^{\pm}$ on S, we may conclude that it is sufficient to prove Theorem 3.3.9 only for the case when $f \in \mathcal{E}(E^{\cdot})$.

Moreover, because of the local character of the relationships (3.3.4) - (3.3.7) one can assume that X is an open set in $\mathbb{R}^n$ and the hypersurface $S \subset X$ is compact.

We have $X^{\pm} = \{x \in X : \pm\rho(x) > 0\}$ where $\rho \in C_{loc}^{s+1,\lambda}(X)$ is a real-valued function on X with $d\rho|_S \neq 0$. Let $U_\varepsilon = \{x \in X : |\rho(x)| < \varepsilon\}$. If $\varepsilon > 0$ is sufficiently small, we get $U_\varepsilon \subset\subset X$ and $\partial U_\varepsilon \in C_{loc}^{s+1,\lambda}$. Besides, for each differential form δ of degree $(n-1)$ continuous in a neighborhood of S, we can write

$$\lim_{\varepsilon \to +0} \int_{\partial U_\varepsilon \cap X^{\pm}} \delta = \pm \int_S \delta.$$

Thus, if $g \in \mathcal{D}(E^{\cdot+1'})$, we get, by using equality (2.4.14),

$$\int_S G_{P_\cdot}(g, f_I^+ - f_I^-)$$
$$= \lim_{\varepsilon \to +0} \left\{ \int_{\partial U_\varepsilon \cap X^+} G_{P_\cdot}(g, \Pi([S]^P f)) + \int_{\partial U_\varepsilon \cap X^-} G_{P_\cdot}(g, \Pi([S]^P f)) \right\}$$
$$= \lim_{\varepsilon \to +0} \int_S G_{P_\cdot}(-\Pi'([\partial U_\varepsilon]^{P'} g), f).$$

To transform the integral in the right-hand side, we apply the transposed variant of homotopy formula (2.4.22), and then Proposition 2.4.6. We obtain

$$\int_S G_{P_\cdot}(g, f_I^+ - f_I^-) = \int_S G_{P_\cdot}(g, f)$$

$$- \lim_{\varepsilon\to+0} \int_S \left\{ G_{P_\imath}(\Pi'(\chi_{U_\varepsilon}(P'g)), f) - G_{P_{\imath+1}}(\Pi'(\chi_{U_\varepsilon}g), Pf) \right\}.$$

By Theorem 2.1.12, we obtain

$$\int_S G_{P_\imath}(g, f_I^+ - f_I^-)$$
$$= \int_S G_{P_\imath}(g, f) - \lim_{\varepsilon\to+0} \int_S ds \left\{ \int_{U_\varepsilon} \frac{\varphi(x,y)}{|x-y|^{n-1}} dx \right\} \qquad (3.3.8)$$

where $\varphi(x, y)$ is a bounded function in some neighborhood of $S \times S$ in the product $X \times X$.

Simple arguments show that the integral $\int_{S\times U_\varepsilon} \frac{1}{|x-y|^{n-1}} ds dx$ converges. Hence it follows by the Fubini Theorem that the limit in (3.3.8) equals zero. This gives the first equality (3.3.4).

To prove the second of these equalities, we apply Remark 3.3.5 and formula (2.4.23). We obtain for $g \in \mathcal{D}(E^{\imath+1'})$

$$\int_S G_{\tilde{P}^*_{\imath-1}}(g, f_I^+ - f_I^-)$$
$$= \lim_{\varepsilon\to+0} \left\{ \int_{\partial U_\varepsilon \cap X^+} G_{\tilde{P}^*_{\imath-1}}(g, \tilde{P}^* G([S]^P f)) + \int_{\partial U_\varepsilon \cap X^-} G_{\tilde{P}^*_{\imath-1}}(g, \tilde{P}^* G([S]^P f)) \right\}$$
$$= \lim_{\varepsilon\to+0} \int_S G_{P_\imath}(-G' \tilde{P}^{*\prime}([\partial U_\varepsilon]^{\tilde{P}^{*\prime}} g), f)$$
$$= \lim_{\varepsilon\to+0} \int_S G_{P_\imath}(G' \tilde{P}^{*\prime}(\chi_{U_\varepsilon}(\tilde{P}^{*\prime} g)), f). \qquad (3.3.9)$$

Using Theorem 2.1.12 we see that the expression under the last limit sign of (3.3.9) has the form $\int_S ds \left\{ \int_{U_\varepsilon} \frac{\varphi(x,y)}{|x-y|^{n-1}} dx \right\}$ where $\varphi(x, y)$ is a bounded function in a suitable neighborhood of $S \times S$. Hence it follows that the limit in (3.3.9) equals zero. So the second equality of (3.3.4) has been proved.

The third equality of (3.3.4) follows immediately from Remark 3.3.5 because $\tilde{P}^*(\tilde{P}^* G) = 0$, so $\tilde{P}^* f_I = 0$ modulo smooth sections of $E^{\imath-1}$ over X.

The last of the equalities (3.3.4) can be reduced to the second one because the definition of a parametrix and formula (2.4.23) together imply that

$$P f_I = P\Pi([S]^P f)$$
$$= [S]^P f + (Pf)_I \pmod{\mathcal{E}(E^{\imath+1})}. \qquad (3.3.10)$$

Let us now prove relationships (3.3.6).

Let $g \in \mathcal{D}(E^{\imath+1'})$. Using formula (2.4.23) and the fact that $P'G' = G'P'$ modulo smoothing operators, we get as above

$$\int_S G_{P_\imath}(g, f_{III}^+ - f_{III}^-)$$
$$= \lim_{\varepsilon\to+0} \int_S G_{P_{\imath-1}}(G'(-[\partial U_\varepsilon]^{P'} g), \tilde{P}^* f)$$
$$= \lim_{\varepsilon\to+0} \int_S G_{P_{\imath-1}} \left(\{ G'P'(\chi_{U_\varepsilon} g) - G'(\chi_{U_\varepsilon}(P'g)) \}, \tilde{P}^* f \right)$$
$$= \lim_{\varepsilon\to+0} \int_S \left\{ -G_{P_\imath}(G'(\chi_{U_\varepsilon} g), P\tilde{P}^* f) - G_{P_{\imath-1}}(G'(\chi_{U_\varepsilon}(P'g)), \tilde{P}^* f) \right\}$$
$$= 0.$$

Thus is precisely the first equality of (3.3.6).

Further, by Theorem 2.1.12 the derivatives up to order $(2p-2)$ of the kernel $G(x,y)$ are locally summable over y on S. Therefore the potential f_{III} is of class $C_{loc}^{\tilde{p}_{i-1}-1}(E^{i})$ on X. In particular, the restrictions f_{III}^{+} and f_{III}^{-} coincide up to order $(\tilde{p}_{i-1}-1)$ on S. This is more than the second equality of (3.3.6) states.

Since $\tilde{P}^{*}f_{III}=(\tilde{P}^{*}f)_{I}$ modulo smooth sections of E^{i-1} over X (use Remark 3.3.5 !), the third equality (3.3.6) follows from the first one of (3.3.4).

Finally, to prove the last equality of (3.3.6), we apply relationships (2.1.12) and (2.4.23). We obtain

$$\begin{aligned} Pf_{III} &= GP([S]^{P}\tilde{P}^{*}f) \pmod{\mathcal{E}(E^{i+1})} \\ &= -G([S]^{P}P\tilde{P}^{*}f). \end{aligned}$$

Since the derivatives up to order $(2p-2)$ of $G(x,y)$ are locally summable over y on S, it follows that the section Pf_{III} is of class $C_{loc}^{\tilde{p}_{i}-1}(E^{i+1})$ on X. In particular, the last equality of (3.3.6) holds.

To complete the proof of the theorem, note that relationships (3.3.5) and (3.3.7) are just (3.3.4) and (3.3.6) written for the (chain) complex $\{E^{i},\tilde{P}_{i-1}^{*}\}$.

□

The arguments given at the beginning of the proof of Theorem 3.3.9 show that the first formula of (3.3.4) is valid under lesser restrictions on s, namely for $s\geq p_i-1$. Analogous sharpenings can be made easily, for other relationships. For $i=0$, i.e. for a differential operator P_0 with injective symbol, the first formula of (3.3.4) gives a jump relationship of the usual Sokhotskii-Plemelj type because the equality $t_0(f)=0$ means simply that f vanishes up to order (p_0-1) on S. The last equality of (3.3.4) generalizes the classical Lyapunov theorem on the jump of the normal derivative of the double layer potential. The second formula of (3.3.4) and the third one are non-trivial for $i>0$ only.

We shall see later that the following result completes Theorem 3.3.9 successfully.

Corollary 3.3.10 *If $S\in C_{loc}^{s+1,\lambda}$ and $f\in C_{loc}^{s,\lambda}(E^{i}|_{S})$ with $s\geq 2p-1$ and $0<\lambda<1$, then $G([S]^{\Delta}f)^{\pm}\in C_{loc}^{s,\lambda}(E^{i}|_{\overline{X^{\pm}}})$ and*

$$D^{\alpha}\left\{(G([S]^{\Delta}f)^{+}-G([S]^{\Delta}f)^{-})-f\right\}|_{S}=0 \quad (|\alpha|\leq 2p-1). \tag{3.3.11}$$

Proof. The fact that the restrictions $G([S]^{\Delta}f)^{\pm}$ are of class $C_{loc}^{s,\lambda}(E^{i}|_{\overline{X^{\pm}}})$ follows from Theorem 3.3.9 immediately because

$$G([S]^{\Delta}f)=f_{I}+f_{II}+f_{III}+f_{IV}. \tag{3.3.12}$$

As in Theorem 3.3.9, the proof of (3.3.11) can be reduced to the case $f\in\mathcal{E}(E^{i})$ where by Corollary 2.5.14 we get

$$G([S]^{\Delta}f)=-\chi^{-}f+G(\chi^{-}(\Delta f)) \pmod{\mathcal{E}(E^{i})}.$$

Hence (3.3.11) follows because $G(\chi^{-}(\Delta f))$ is of class $C_{loc}^{2p-1}(E^{i})$ on X.

□

3.3.4 *Formal version of the Cauchy-Kovalevskaya Theorem for the differential operator Δ (general case)*

Theorem 3.3.9 and Corollary 3.3.10 together are quite sufficient to generalize Lemma 3.2.7.

For a section $f \in C^s_{loc}(E^\cdot)$, we denote by $\jmath^s_S(f)$ the s-jet of f on S. In other words, $\jmath^s_S$ is the quotient mapping from $C^s_{loc}(E^\cdot)$ to $C^s_{loc}(E^\cdot|_S)$.

Corollary 3.3.11 *Let $S \in C^{s+1,\lambda}_{loc}$ where $s \geq 2p-1$ and $0 < \lambda < 1$ Then for every $f_0 \in C^{s-2p,\lambda}_{loc}(E^\cdot|_{\overline{X^\pm}})$ and $f_\jmath \in C^{s,\lambda}_{loc}(E^\cdot|_S)$ ($\jmath = 1, \ldots, 4$) there exists a section $f \in C^{s,\lambda}_{loc}(E^\cdot)$, unique modulo sections s-flat on S, such that*

$$\begin{array}{rclrcl} & & \jmath^{s-2p}_S(\Delta f - f_0) = 0, & & & \\ t(f) & = & t(f_1), & n(f) & = & n(f_2), \\ t(\tilde{P}^* f) & = & t(\tilde{P}^* f_3), & n(Pf) & = & n(Pf_4). \end{array} \tag{3.3.13}$$

Proof. We let

$$\begin{aligned} F \;=\; & \left\{ G\tilde{P}^*([S]^P f_1) + GP([S]^{\tilde{P}^*} f_2) + G([S]^P \tilde{P}^* f_3) + G([S]^{\tilde{P}^*} P f_4) \right\} \\ & \pm G(\chi^\pm f_0). \end{aligned}$$

It follows from Proposition 3.3.7 that $F^\pm$ is of class $C^{s,\lambda}_{loc}(E^\cdot|_{\overline{X^\pm}})$.

Moreover, since the derivatives of the kernel $G(x,y)$ up to order $(2p-1)$ are locally summable over y on X, the potential $G(\chi^\pm f_0)$ is of class $C^{2p-1}_{loc}(E^\cdot)$ on X. Therefore Theorem 3.3.9 implies

$$\begin{array}{rclrcl} t(F^+ - F^-) & = & t(f_1), & n(F^+ - F^-) & = & n(f_2), \\ t(\tilde{P}^* F^+ - \tilde{P}^* F^-) & = & t(\tilde{P}^* f_3), & n(PF^+ - PF^-) & = & n(Pf_4). \end{array}$$

Now we use the fact that G is a properly supported parametrix of the differential operator Δ, to get

$$\begin{aligned} \Delta F \;=\; & \left\{ \tilde{P}^*([S]^P f_1) + P([S]^{\tilde{P}^*} f_2) + ([S]^P \tilde{P}^* f_3) + ([S]^{\tilde{P}^*} P f_4) \right\} \\ & \pm(\chi^\pm f_0) \pmod{\mathcal{E}(E^\cdot)}. \end{aligned}$$

Hence it follows that

$$\jmath^{s-2p}_S(\Delta F^+ - \Delta F^- - f_0) = 0.$$

Thus, if we continue the sections $F^\pm$ to X while keeping the same smoothness, then we obtain a solution of problem (3.3.13) by letting $f = F^+ - F^-$.

On the other hand, if f' and f'' are two solution of (3.3.13) from $C^{s,\lambda}_{loc}(E^\cdot)$, then for $f = f' - f''$ we have

$$\begin{array}{rclrcl} & & \jmath^{s-2p}_S(\Delta f) = 0, & & & \\ t(f) & = & 0, & n(f) & = & 0, \\ t(\tilde{P}^* f) & = & 0, & n(Pf) & = & 0. \end{array}$$

It follows from formulas (3.3.3) and (3.3.12) that $G([S]^{\Delta} f) = 0$. So we get $j_S^{2p-1}(f) = 0$ by Corollary 3.3.10.

Since $\Delta_\cdot$ is an elliptic differential operator, the hypersurface S is non-characteristic for it. Thus $j_S^s(f) = 0$ which completes the proof.

□

Thus, the Cauchy data on S for the Laplacian Δ consist of the $(P-)$ Dirichlet data $\{t(f), t(\tilde{P}^* f)\}$ and the $(P-)$ Neumann data $\{n(f), n(Pf)\}$.

Corollary 3.3.11, when $E^\cdot$ is the Dolbeault complex and $s = \infty$, is due to Harvey and Polking [101].

3.3.5 *Decomposition in jet spaces*

To formulate the following result we require additional notations.

For a subclass $\mathfrak{C}$ of C^s_{loc} $(s \geq 2p-1)$, we let

$$\begin{array}{lll}
\mathfrak{C}_{[t,]}(S) & = \{f \in \mathfrak{C}(E^\cdot|_S): & n(f) = t(\tilde{P}^* f) = n(Pf) = j_S^{s-2p}(\Delta f) = 0\}, \\
\mathfrak{C}_{[n,]}(S) & = \{f \in \mathfrak{C}(E^\cdot|_S): & t(f) = t(\tilde{P}^* f) = n(Pf) = j_S^{s-2p}(\Delta f) = 0\}, \\
\mathfrak{C}_{[t\tilde{P}^*_{\cdot-1}]}(S) & = \{f \in \mathfrak{C}(E^\cdot|_S): & t(f) = n(f) = n(Pf) = j_S^{s-2p}(\Delta f) = 0\}, \\
\mathfrak{C}_{[nP_\cdot]}(S) & = \{f \in \mathfrak{C}(E^\cdot|_S): & t(f) = n(f) = t(\tilde{P}^* f) = j_S^{s-2p}(\Delta f) = 0\}, \\
\mathfrak{C}_{[j_S^{s-2p}\Delta_\cdot]}(S) & = \{f \in \mathfrak{C}(E^\cdot|_S): & t(f) = n(f) = t(\tilde{P}^* f) = n(Pf) = 0\}.
\end{array}$$

Though the spaces $\mathfrak{C}_{[t,]}(S)$ have already been defined in Subsection 3.1.4 in a different way, there need be no misunderstanding in view of the result below.

Corollary 3.3.12 *Let $S \in C^{s+1,\lambda}_{loc}$ where $s \geq 2p-1$ and $0 < \lambda < 1$ and let $\mathfrak{C}$ be one of the classes $C^{s,\lambda}_{loc}$ or $C^{s,\lambda}_{comp}$. Then*

$$\begin{aligned}
&\mathfrak{C}(E^\cdot|_S) \\
&= \mathfrak{C}_{[t,]}(S) \overset{top.}{\oplus} \mathfrak{C}_{[n,]}(S) \overset{top.}{\oplus} \mathfrak{C}_{[t\tilde{P}^*_{\cdot-1}]}(S) \overset{top.}{\oplus} \mathfrak{C}_{[nP_\cdot]}(S) \overset{top.}{\oplus} \mathfrak{C}_{[j_S^{s-2p}\Delta_\cdot]}(S).
\end{aligned} \tag{3.3.14}$$

Proof. Let us denote by L_j $(j = 1, \ldots, 5)$ the summands in the right-hand side of (3.3.14) respectively.

Corollary 3.3.11 implies that $L_j \cap L_k = \{0\}$ for $j \neq k$, i.e., the algebraic sum in the right-hand side of (3.3.14) is really a direct one.

Using Theorem 3.3.9 we get that the operators

$$\begin{array}{rcl}
t(f) & = & j_S^s(f_I^+ - f_I^-), \\
n(f) & = & j_S^s(f_{II}^+ - f_{II}^-), \\
t\tilde{P}^*(f) & = & j_S^s(f_{III}^+ - f_{III}^-), \\
nP(f) & = & j_S^s(f_{IV}^+ - f_{IV}^-)
\end{array} \tag{3.3.15}$$

and

$$\begin{aligned}
&j_S^{s-2p}\Delta(f) \\
&= f - j_S^s(f_I^+ - f_I^-) - j_S^s(f_{II}^+ - f_{II}^-) - j_S^s(f_{III}^+ - f_{III}^-) - j_S^s(f_{IV}^+ - f_{IV}^-)
\end{aligned} \tag{3.3.16}$$

map $\mathfrak{C}(E^{\cdot}|_S)$ continuously onto L_j ($j = 1, \ldots, 5$) respectively.

To complete the proof, it is sufficient to notice that these operators decompose the identity operator on $\mathfrak{C}(E^{\cdot}|_S)$.

□

Remark 3.3.13 *We showed that $t, n, t \circ \widetilde{P}^*, n \circ P$ and $j_S^{s-2p} \circ \Delta$ are projectors both in $C^{s,\lambda}_{loc}(E^{\cdot}|_S)$ and $C^{s,\lambda}_{comp}(E^{\cdot}|_S)$ which are orthogonal in pairs and decompose the identity operator.*

Using Corollary 2.5.15 one can show that, if $t(f) = n(f) = 0$ for a jet $f \in C_{loc}^{\max(\widetilde{p}_{\cdot-1}, p_{\cdot})-1,\lambda}(E^{\cdot}|_S)$, then $j_S^{\min(\widetilde{p}_{\cdot-1}, p_{\cdot})-1}(f) = 0$. Hence it follows that non-zero $(\min(\widetilde{p}_{\cdot-1}, p_{\cdot}) - 1)$-jets on S are concentrated only in the first two summands on the right-hand side of (3.3.14). When $E^{\cdot}$ is a complex of first order differential operators (and $\widetilde{E}^{\cdot} = E^{\cdot}!$), the components of zero degree of the jets $t(f)$ and $n(f)$ give precisely the operators of (3.2.2). It is interesting that, though the parametrix G of Δ participates in constructing the operators (3.3.15) and (3.3.16), the latter do not depend on G because they are projectors for the direct sum (3.3.14).

Exercise 3.3.14 *Prove or refute that Corollary 3.3.12 is valid for $\lambda = 0$ also.*

Commentary. For this exercise the analytical approach is no longer of use because pseudo-differential operators of order 0 are not bounded in the spaces $C^s_{loc}(E)$ where $s \in \mathbb{Z}_+$.

□

3.3.6 *Dual decompositions*

If we start with the complex $\widetilde{E}^{\cdot}$ and take E as an associated complex, we shall analogously get projectors $\tilde{t}$, $\tilde{n}$, $\tilde{t}P^*$, $\tilde{t}\widetilde{P}$, and $j_S^{s-2p}\Delta^*$ on $C^{s,\lambda}_{loc}(E^{\cdot}|_S)$ or $C^{s,\lambda}_{comp}(E^{\cdot}|_S)$ which are pairwise orthogonal and decompose the identity operator.

Corollary 3.3.15 *Let $S \in C^{s+1,\lambda}_{loc}$ where $s \geq 2p - 1$ and $0 < \lambda < 1$, and let $\mathfrak{C}$ be one of the classes $C^{s,\lambda}_{loc}$ or $C^{s,\lambda}_{comp}$. Then*

$$\mathfrak{C}(E^{\cdot}|_S)$$
$$= \mathfrak{C}_{[\tilde{t}_{\cdot}]}(S) \overset{top.}{\oplus} \mathfrak{C}_{[\tilde{n}_{\cdot}]}(S) \overset{top.}{\oplus} \mathfrak{C}_{[\tilde{t}P^*_{\cdot-1}]}(S) \overset{top.}{\oplus} \mathfrak{C}_{[\tilde{n}\widetilde{P}_{\cdot}]}(S) \overset{top.}{\oplus} \mathfrak{C}_{[j_S^{s-2p}\Delta_{\cdot}^{*}]}(S). \tag{3.3.17}$$

Proof. One repeats the proof of Corollary 3.3.12 word for word.

□

The following result proved by the author [231] is of interest in the theory of overdetermined boundary value problems.

Theorem 3.3.16 *If $S \in C^{2p,\lambda}_{loc}$ where $0 < \lambda < 1$, then the following semilinear form is non-degenerate on the product $C^{2p-1,\lambda}_{comp}(E^\cdot|_S) \otimes C^{2p-1,\lambda}_{loc}(E^\cdot|_S)$:*

$$\begin{aligned}[S]^\Delta(\star g \otimes f) &= [S]^P(\star \tilde{P}(\tilde{n}\tilde{P}(g)) \otimes t(f)) + [S]^{\tilde{P}^*}(\star P^*(\tilde{t}P^*(g)) \otimes n(f)) \\ &\quad + [S]^P(\star \tilde{n}(g) \otimes \tilde{P}^*(t\tilde{P}^*(f))) + [S]^{\tilde{P}^*}(\star \tilde{t}(g) \otimes P(nP(f))).\end{aligned}$$

Proof. This follows from formulas (2.5.10) and (2.4.14), and Corollaries 3.3.12, 3.3.15 and 3.3.10.

□

This theorem allows us to answer a technical question of Kohn and Nirenberg [132] concerning the (P-) Neumann problem (see Section 4.1 below).

Corollary 3.3.17 *Let $S \in C^{2p,\lambda}_{loc}$ $(0 < \lambda < 1)$ and $f \in C^{2p-1,\lambda}_{loc}(E^\cdot|_S)$. Then $t(f) = 0$ iff $[S]^P(\star g \otimes f) = 0$ for all $g \in \mathcal{D}(E^{\cdot+1})$ such that $\tilde{t}(g) = 0$.*

Proof. The necessity is obvious, let us prove the sufficiency. Suppose that $[S]^P(\star g \otimes f) = 0$ for all $g \in \mathcal{D}(E^{\cdot+1})$ such that $\tilde{t}(g) = 0$. Fix some arbitrary section $g \in \mathcal{D}(E^{\cdot+1})$. The projection $\tilde{n}\tilde{P}(g)$ of $j^{2p-1}_S(g)$ in accordance with the decomposition (3.3.17) satisfies $\tilde{t}(\tilde{n}\tilde{P}(g)) = 0$. By Lemma 3.1.16, $\tilde{t}(\tilde{P}(\tilde{n}\tilde{P}(g))) = 0$. Therefore we get $[S]^P(\star \tilde{P}(\tilde{n}\tilde{P}(g)) \otimes t(f)) = 0$ from the condition. Use Theorem 3.3.16 now implies that $[S]^\Delta(\star g \otimes t(f)) = 0$ for all $g \in \mathcal{D}(E^{\cdot+1})$, and hence $t(f) = 0$. We have omitted inessential details concerning the smoothness of $\tilde{n}\tilde{P}(g)$. No difficulties appear if $S \in C^\infty_{loc}$, and in the general case they are easily overcome with the help of approximations. The proof is complete.

□

3.3.7 *Formal version of the Cartan-Kähler Theorem for elliptic complexes*

Theorem 3.3.9 gives also the possibility of sharpening Theorem 3.1.30 for elliptic complexes.

Corollary 3.3.18 *Let $S \in C^{s+p_\cdot-1+1,\lambda}_{loc}$ where $s \geq p_\cdot - 1$ and $0 < \lambda < 1$. Then for each sections $f \in C^{s,\lambda}_{loc}(E^\cdot|_{\overline{X^\pm}})$ and $u_0 \in C^{s+p_\cdot-1,\lambda}_{loc}(E^{\cdot-1}|_{\overline{X^\pm}})$ satisfying*

$$\left\{\begin{array}{rcl} j^{s-p_\cdot}_S(Pf) &=& 0, \\ t(Pu_0 - f) &=& 0, \end{array}\right. \tag{3.3.18}$$

there exists a section $u \in C^{s+p_\cdot-1,\lambda}_{loc}(E^{\cdot-1})$ such that

$$\left\{\begin{array}{rcl} j^s_S(Pu - f) &=& 0, \\ t(u) &=& t(u_0). \end{array}\right. \tag{3.3.19}$$

Proof. First of all we observe that (3.3.18) are necessary conditions for the solvability of problem (3.3.19).

We let

$$F = G([S]^P u_0) \pm G(\chi^{\pm} f).$$

It is easy to see as in the proof of Corollary 3.3.11, that $F^{\pm} \in C^{s+2p_\cdot,\lambda}_{loc}(E^\cdot|_{\overline{X^\pm}})$ and

$$\begin{array}{rclcrcl} t(F^+ - F^-) & = & 0, & & n(F^+ - F^-) & = & 0, \\ t(\widetilde{P}^* F^+ - \widetilde{P}^* F^-) & = & t(u_0), & & n(PF^+ - PF^-) & = & 0, \end{array} \tag{3.3.20}$$

and

$$j^s_S(\Delta F^+ - \Delta F^- - f) = 0. \tag{3.3.21}$$

Let us extend $\widetilde{P}^* F^{\pm} \in C^{s+p_\cdot-1,\lambda}_{loc}(E^{\cdot-1}|_{\overline{X^\pm}})$ to sections of $E^{\cdot-1}$ of the same smoothness. Since we shall be interested in the behaviour of these continuations near the hypersurface S only, we retain for them the notations $\widetilde{P}^* F^{\pm}$.

Consider $u = \widetilde{P}^* F^+ - \widetilde{P}^* F^-$. Then u is of class $C^{s+p_\cdot-1,\lambda}_{loc}(E^{\cdot-1})$, and $t(u) = t(u_0)$ by the third equality (3.2.19). Since

$$j^s_S(Pu - f) = j^s_S(\Delta F^+ - \Delta F^- - f) - j^s_S(\widetilde{P}^* PF^+ - \widetilde{P}^* PF^-)$$

and because of (3.3.21), it remains only to verify that $j^s_S(\widetilde{P}^* PF^+ - \widetilde{P}^* PF^-) = 0$.

In fact, this follows easily from the equality

$$j^{s+\widetilde{p}_\cdot}_S(PF^+ - PF^-) = 0, \tag{3.3.22}$$

and so our goal is to prove (3.3.22).

The first and the last equations in (3.3.20) imply that

$$\begin{array}{rcl} t(PF^+ - PF^-) & = & 0, \\ n(PF^+ - PF^-) & = & 0. \end{array}$$

Moreover the third equation of (3.3.20) implies $t(P\widetilde{P}^* F^+ - P\widetilde{P}^* F^-) = t(Pu_0)$, so in view of (3.3.21) and the second condition of (3.3.18), we arrive at

$$\begin{aligned} & t(\widetilde{P}^* PF^+ - \widetilde{P}^* PF^-) \\ & = t\left\{(\Delta F^+ - \Delta F^-) - (P\widetilde{P}^* F^+ - P\widetilde{P}^* F^-)\right\} \\ & = t(f - Pu_0) \\ & = 0. \end{aligned}$$

Obviously,

$$n(PPF^+ - PPF^-) = 0.$$

Now, by (3.3.21) and the first condition of (3.3.18)

$$j^{s-p_\cdot}_S\{\Delta(PF^+ - PF^-)\} = j^{s-p_\cdot}_S\{P(\Delta F^+ - \Delta F^- - f) + Pf\} = 0,$$

and we may use Corollary 3.3.11 to derive (3.3.22).

□

Corollary 3.3.18, when E is the Dolbeault complex and $s = \infty$, is due to Andreotti and Hill [9, Lemma 2.1]. For arbitrary elliptic complexes $E^\cdot$ it was published by the author [226].

Remark 3.3.19 *One sees from the proof of Corollary 3.3.18 that the solution u of (3.3.19) can be chosen to be compactly supported provided the data f and u_0 are compactly supported.*

3.3.8 *A description of the tangential complex for an arbitrary elliptic complex*

Another application of Theorem 3.3.9 is a highly useful description of the tangential complex $\mathfrak{C}_{[t\cdot]}(S)$ (see Subsection 3.1.5). We limit ourselves to two of the more important cases $\mathfrak{C} = \mathcal{E}$ or $\mathcal{D}$.

For $s \in \mathbb{Z}_+$, we denote by $\mathfrak{C}^{(s)}(E^\cdot|_S)$ the topological vector space of s-jets on S of sections from $\mathfrak{C}(E^\cdot)$, i.e, $\mathfrak{C}^{(s)}(E^\cdot|_S)$ is the quotient space of $\mathfrak{C}(E^\cdot)$ over the subspace of all sections s-flat on S, endowed with the quotient topology. We continue to use the symbol j_S^s for the quotient mapping $\mathfrak{C}(E^\cdot) \to \mathfrak{C}^{(s)}(E^\cdot|_S)$.

Define the mapping $t: \ \mathfrak{C}^{(p_\cdot-1)}(E^\cdot|_S) \to \mathfrak{C}^{(p_\cdot-1)}(E^\cdot|_S)$ by setting

$$t(f) = j_S^{p_\cdot-1}(f_I^+ - f_I^-)$$

for a jet $f \in \mathfrak{C}^{(p_\cdot-1)}(E^\cdot|_S)$.

This does not contradict Definition 3.1.12 as the first of the relationships (3.3.4) implies that $t(f) = 0$ if and only if $j_S^{p_\cdot-1}(f_I^+ - f_I^-) = 0$, at least for $f \in C_{loc}^{p_\cdot-1,\lambda}(E^\cdot|_S)$.

It follows from Corollary 3.3.8 that the mapping t is continuous. Moreover, $t^2 = t$ by the first equality (3.3.4), i.e., t is a projector in $\mathfrak{C}^{(p_\cdot-1)}(E^\cdot|_S)$. Therefore, this space is decomposed into the topological direct sum of the range of t denoted by $\mathfrak{C}_{[t_\cdot]}^{(p_\cdot-1)}(S)$, and the kernel of t denoted by $\mathfrak{C}_{t_\cdot}^{(p_\cdot-1)}(S)$.

We see that the space $\mathfrak{C}_{[t_\cdot]}^{(p_\cdot-1)}(S)$ on the whole does not depend on a particular choice of the parametrix Π, though Π participates in the construction of t. Indeed, $\mathfrak{C}_{[t_\cdot]}^{(p_\cdot-1)}(S)$ is isomorphic (as a topological vector space) to the quotient space $\mathfrak{C}^{(p_\cdot-1)}(E^\cdot|_S)/\mathfrak{C}_{t_\cdot}^{(p_\cdot-1)}(S)$. Moreover, if $p_\cdot = \tilde{p}_{\cdot-1}$, then we have $\mathfrak{C}_{[t_\cdot]}^{(p_\cdot-1)}(S) = \left\{f \in \mathfrak{C}^{(p_\cdot-1)}(E^\cdot|_S) : \ n(f) = 0\right\}$ (use Corollary 2.5.15).

We collect the spaces $\mathfrak{C}_{[t_\cdot]}^{(p_\cdot-1)}(S)$ into a new differential complex $\mathfrak{C}_{[t\cdot]}^{(p_\cdot-1)}(S)$ on S whose differential P_b is defined so as to make the following diagram with exact rows commutative:

$$\begin{array}{ccccccccc} & & 0 & & 0 & & 0 & & \\ & & \downarrow & & \downarrow & & \downarrow & & \\ 0 & \longrightarrow & \mathfrak{C}_{t_0}(X) & \stackrel{\iota}{\longrightarrow} & \mathfrak{C}^0(X) & \stackrel{t}{\longrightarrow} & \mathfrak{C}^{(p_0-1)}_{[t_0]}(S) & \longrightarrow & 0 \\ & & \downarrow P & & \downarrow P & & \downarrow P_b & & \\ 0 & \longrightarrow & \mathfrak{C}_{t_1}(X) & \stackrel{\iota}{\longrightarrow} & \mathfrak{C}^1(X) & \stackrel{t}{\longrightarrow} & \mathfrak{C}^{(p_1-1)}_{[t_1]}(S) & \longrightarrow & 0 \\ & & \downarrow P & & \downarrow P & & \downarrow P_b & & \\ 0 & \longrightarrow & \mathfrak{C}_{t_2}(X) & \stackrel{\iota}{\longrightarrow} & \mathfrak{C}^2(X) & \stackrel{t}{\longrightarrow} & \mathfrak{C}^{(p_2-1)}_{[t_2]}(S) & \longrightarrow & 0 \\ & & \downarrow P & & \downarrow P & & \downarrow P_b & & \\ & & \cdots & & \cdots & & \cdots & & \end{array} \qquad (3.3.23)$$

The complex $\mathfrak{C}^{(p-1)}_{[t]}(S)$ is precisely the desired realization of the tangential complex $\mathfrak{C}_{[t]}(S)$.

Corollary 3.3.20 *If $S \in C^\infty_{loc}$ and $\mathfrak{C} = \mathcal{E}$ or $\mathcal{D}$, then*

$$\mathfrak{C}_{[t]}(S) \stackrel{top}{\cong} \mathfrak{C}^{(p-1)}_{[t]}(S).$$

Proof. This follows immediately from the commutative diagram (3.3.23). □

3.3.9 *Jump formulas for the volume potential*

We describe some more jump behaviour of the integral $\Pi(\chi^- f)$ (see the last summand in the left-hand side of (2.4.22)), although this will not be used later.

Theorem 3.3.21 *If $S \in C^{s+1,\lambda}_{loc}$ and $f \in C^{s,\lambda}_{loc}(E^\iota|_{\overline{X^-}})$, where $s \geq \tilde{p}_{\iota-1} - 1$ and $0 < \lambda < 1$, then $\Pi(\chi^- f)^\pm \in C^{s+p_{\iota-1},\lambda}_{loc}(E^{\iota-1}|_{\overline{X^\pm}})$ and*

$$\left\{ \begin{array}{rcl} t(\Pi(\chi^- f)^+ - \Pi(\chi^- f)^-) & = & 0, \\ n(\Pi(\chi^- f)^+ - \Pi(\chi^- f)^-) & = & 0, \\ t(\tilde{P}^*\Pi(\chi^- f)^+ - \tilde{P}^*\Pi(\chi^- f)^-) & = & 0, \\ n(P\Pi(\chi^- f)^+ - P\Pi(\chi^- f)^-) & = & -n(f). \end{array} \right. \qquad (3.3.24)$$

Proof. The fact that the restrictions $\Pi(\chi^- f)^\pm$ are of class $C^{s+p_{\iota-1},\lambda}_{loc}(E^{\iota-1}|_{\overline{X^\pm}})$ follows from Proposition 3.3.7.

As in Theorem 3.3.9, proving (3.3.24) can be reduced to the case where $f \in \mathcal{E}(E^\iota)$ and X is an open subset of $\mathbb{R}^n$.

By Theorem 2.1.12 the derivatives up to order $(p_{\iota-1}-1)$ of the kernel $\Pi_\iota(x,y)$ are locally summable in y on X. Therefore the potential $\Pi(\chi^- f)$ is of class $C^{p_{\iota-1}-1}_{loc}(E^{\iota-1})$ on X. In particular, the restrictions $\Pi(\chi^- f)^+$ and $\Pi(\chi^- f)^-$ coincide up to order $(p_{\iota-1} - 1)$ on S. This implies the first equality in (3.3.24).

Let us prove the second of these equalities. Fix some test section $g \in \mathcal{D}(E^{i-2'})$. Applying equality (3.3.1), Proposition 2.4.6 and Stokes' Theorem we get

$$\begin{aligned}
&\int_S G_{\widetilde{P}^*_{i-2}}(g, \Pi(\chi^- f)^+ - \Pi(\chi^- f)^-) \\
&= \lim_{\varepsilon\to+0}\left\{\int_{\partial U_\varepsilon \cap X^+} G_{\widetilde{P}^*_{i-2}}(g, \widetilde{P}^* G(\chi^- f)) + \int_{\partial U_\varepsilon \cap X^-} G_{\widetilde{P}^*_{i-2}}(g, \widetilde{P}^* G(\chi^- f))\right\} \\
&= \lim_{\varepsilon\to+0}\left\{-\int_{\partial U_\varepsilon} G_{\widetilde{P}^*_{i-1}}(\widetilde{P}^{*\prime} g, G(\chi^- f))\right\} \\
&= -\lim_{\varepsilon\to+0}\int_{X^-}\left\{\int_{\partial U_\varepsilon} G_{\widetilde{P}^*_{i-1}}(\widetilde{P}^{*\prime} g, G)\right\} f(y) \\
&= -\lim_{\varepsilon\to+0}\int_{X^-} G' \widetilde{P}^{*\prime}(\chi_{U_\varepsilon}(\widetilde{P}^{*\prime} g)) f \\
&= 0
\end{aligned}$$

(cf (3.3.9)). This is precisely what we need.

The third equality (3.3.24) follows from (3.3.1) because $\widetilde{P}^*\Pi = (\widetilde{P}^*)^2 G = 0$ modulo smoothing operators.

Finally, the last equality (3.3.24) is a consequence of the second relationship (3.3.5) and Remark 3.3.5. Indeed, by formula (2.4.23) we have

$$\Pi(\chi^- f) = G\widetilde{P}^*(\chi^- f) = -G([S]^{\widetilde{P}^*} f) + G(\chi^-(\widetilde{P}^* f)),$$

and $G(\chi^-(\widetilde{P}^* f)) \in C^{2p-1}_{loc}(E^{i-1})$.

□

For the Dolbeault complex a similar result was noted by Harvey and Polking [101].

3.3.10 *Jump formulas for potentials with smooth densities*

Theorem 3.3.9 asserts in particular that, if $S \in C^{p_i,\lambda}_{loc}$ and $f \in C^{p_i-1,\lambda}_{loc}(E^i|_S)$ with $0 < \lambda < 1$, then the potential $f_I = \Pi([S]^P f)$ can be continued up to S from the sides of S in the same Hölder class $C^{p_i-1,\lambda}_{loc}(E^i|_{\overline{X^\pm}})$, and the jump formula $t(f_I^+ - f_I^-) = t(f)$ holds.

If the jet f is only of class $C^{p_i-1}_{loc}(E^i|_S)$, then the simplest examples show that the potential f_I may be unbounded near S (see Example 2.8 of Aizenberg and Dautov [5]). Therefore Theorem 3.3.9 is not valid for such densities f.

Nevertheless the first formula of (3.3.4) can have some sense in this case also, as Sokhotskii and Plemelj observed already for the Cauchy type integral (see Muskhelishvili [165, §17]).

Let $x^0 \in S$ and let U be a coordinate neighborhood of x^0 in X over which all bundles E^i are trivial. Let K_{x^0} be a right circular cone with vertex x^0 whose axis coincides with the normal to S at x^0 and whose generator forms an angle α of magnitude less than $\pi/2$ with the axis. We choose two points $x^+ \in U \cap X^+$ and $x^- \in U \cap X^-$ inside K_{x^0} such that $c_1|x^+ - x^0| \le |x^- - x^0| \le c_2|x^+ - x^0|$, where $c_1 < c_2$ are fixed positive constants. Thus the points x^+ and x^- may tend to x^0 in a co-ordinated manner only.

Theorem 3.3.22 *(see Tarkhanov [226]). Suppose that* $S \in C^{p_i}_{loc}$, *and* $f \in C^{\max\,(\widetilde{p}_{i-1},p_i)}_{loc}(E^i|_S)$ *satisfies* $n(f) = 0$. *Then the limit*

$$\lim_{x^\pm \to x^0} \left(D^\alpha f_I(x^+) - D^\alpha f_I(x^-) \right) = D^\alpha f(x^0)$$

exists for every multi-index α *with* $|\alpha| \le p_i - 1$, *and it is attained uniformly in* x *in a relatively compact neighborhood of* x^0 *in* S *if the angle* α *and the constants* c_1 *and* c_2 *are fixed.*

Proof. Since the proof is rather cumbersome and standard, we limit ourselves to new considerations only. First of all, the theorem is easily reduced to the case where X is an open subset of $\mathbb{R}^n$. Further, we have $f_{II} = 0$ because $n(f) = 0$. Hence it follows that the jump behaviours of the potentials f_I and $(f_I + f_{II})$ in crossing S are the same. However, in the only difficult case $\widetilde{p}_{i-1} = p_i$ it turns out, that the sum $(f_I + f_{II})$ is again a potential of the form f_I constructed for some new elliptic complex E at degree 0. More precisely, the component of the differential of E at degree 0 is $P_i \oplus \widetilde{P}^*_{i-1}$. Thus it is sufficient to consider only the case $i = 0$ in Theorem 3.3.22. Standard considerations like those used by Dautov and Kytmanov [45] show that, if $S \in C^{p_0}_{loc}$ and $f \in C^{p_0-1}_{loc}(E^0|_S)$, then for every multi-index α with $|\alpha| \le p_0 - 1$ there exists the limit

$$\lim_{x^\pm \to x^0} \left((D^\alpha f)_I(x^+) - (D^\alpha f)_I(x^-) \right) = D^\alpha f(x^0),$$

and it is attained uniformly in x in a relatively compact neighborhood of x^0 in S provided the angle α and the constants c_1 and c_2 are fixed. Then it only remains to prove that the potentials $D^\alpha f_I$ and $(D^\alpha f)_I$ differ by an expression which has zero jump while crossing S in the sense above. If the coefficients of the differential operator P_0 are constant and the jet f is sufficiently smooth, then the required result follows from Theorem 2.4.1 and the identity

$$d\left(G_P(D^{\alpha\prime} g, f) - G_P(g, D^\alpha f) - G_{D^\alpha}(P' g, f) + G_{D^\alpha}(g, Pf) \right) = 0.$$

In the general case we need some delicate but quite simple considerations. They make use of the decomposition of the tangent space to X at x^0 into the direct sum of the subspaces respectively tangent and normal to S at x^0, and the induced decomposition in the space of differential operators on X. The details can be found in the author's thesis [225].

□

For Martinelli-Bochner type integrals Theorem 3.3.22 is due to Dautov and Kytmanov [45]. However in the case $i > 0$ this result is new for the Dolbeault complex also.

Corollary 3.3.23 *Under the conditions of Theorem 3.3.22, if some derivative* $D^\alpha f_I$, *where* $|\alpha| \le p_i - 1$, *can be continued continuously to* $S \cap U$ *from one side of* S, *then it can be continued continuously to* $S \cap U$ *from the other side* S *also.*

Proof. See, for example, Muskhelishvili [165, §9].

□

The Sokhotskii-Plemelj type formulas of this section are of use in reducing boundary value problems for elliptic systems (like the ($P-$) Neumann problem) to systems of pseudo-differential equations on the boundary (see Hörmander [111, Ch.II]).

3.4 Integral Formulas for Elliptic Complexes. Morera's Theorem

3.4.1 *Historical remarks*

Let X be a differentiable compact manifold of dimension n, and E be an elliptic complex on X.

For an open set $Y \subset X$ with a piecewise smooth boundary, we set

$$\mathcal{H}^{\cdot}(E^{\cdot}|_{\overline{Y}}) = \left\{ f \in \mathcal{E}(E^{\cdot}|_{\overline{Y}}) : \; Pf = P^{*}f = n(f) = 0 \right\}.$$

Here $n(f) = 0$ means that f has zero Cauchy data on ∂Y with respect to the differential operator P^*.

The spaces $\mathcal{H}^{\cdot}(E\,|_Y)$ are natural generalizations of the harmonic spaces of Subsection 2.2.1 to the case of compact manifolds with boundaries (see Section 4.1 below).

In this section we shall obtain integral formulas which characterize the sections of $Z^{\cdot}(\mathcal{E}(E^{\cdot}|_{\overline{Y}}))$ and $\mathcal{H}^{\cdot}(E^{\cdot}|_{\overline{Y}})$. The history of this question is the following.

The Martinelli-Bochner kernel in complex analysis (see $\Phi_1^{(0)}(z,\zeta)$ in Subsection 2.5.3) is not holomorphic in the external variable z. So it is not obvious whether a function representable by the Martinelli-Bochner integral (formula (2.5.8) with $p = q = 0$) is holomorphic. Aronov and Kytmanov [16] answered this question in the affirmative. Then the second of the authors obtained an analogous statement for the general formula (2.5.8). In the author's note [218] a similar question was considered for Norguet's formula (2.5.5).

Such a problem, applied to general complexes of differential operators, appears to be artificial. But it turns out sometimes to be a useful tool (see, for example, Dairbekov and Kopylov [43]). The jump formulas (3.3.4) provide a simple solution to this problem.

3.4.2 *Preliminary results*

Fix some integer p such that $p_\imath \le p/2$ for all $\imath$. For every i, we choose a differential operator $D_\imath \in \mathrm{do}_{p-2p_{\imath-1}}(E^\imath \to F^\imath)$ with injective symbol where $F^\imath$ is some vector bundle over X. Let $\widetilde{P}_\imath = P_\imath((D_{\imath+1}P_\imath)^*(D_{\imath+1}P_\imath) + 1)$ so that $\widetilde{P}_\imath$ is a differential operator of type $E^\imath \to E^{\imath+1}$ and order $\tilde{p}_\imath = 2p - p_\imath$.

Lemma 3.4.1 *The complex $\widetilde{E}^{\cdot} = \{E^\imath, \widetilde{P}_\imath\}$ is an associated complex for E.*

Proof. We need to show that the Laplacians $\Delta_i = P_{i-1}\tilde{P}^*_{i-1} + \tilde{P}^*_i P_i$ are elliptic differential operators of order $2p$.

For $i \in \mathbb{Z}$, we let $Q_i = (D_{i+1}P_i)^*(D_{i+1}P_i)+1$. Since D_{i+1} has an injective symbol it follows that $(\sigma(Q_i)^*\sigma(P_i)^*e, \sigma(P_i)^*e)_x = 0$ for $e \in E^{i+1}_x$ if and only if $\sigma(P_i)^*e = 0$.

We now suppose that $e \in E^i_x$ and $\sigma(\Delta_i)e = 0$. Since $P_i P_{i-1} = 0$, we get

$$(\sigma(Q_i)^*\sigma(P_i)^*\sigma(P_i)e,\ \sigma(P_i)^*\sigma(P_i)e)_x = 0.$$

Therefore $\sigma(P_i)^*\sigma(P_i)e) = 0$ and $\sigma(P_i)e = 0$.

Hence it follows that

$$\sigma(P_{i-1})\sigma(Q_{i-1})^*\sigma(P_{i-1})^*e = 0.$$

Then

$$(\sigma(Q_{i-1})^*\sigma(P_{i-1})^*e,\ \sigma(P_{i-1})^*e)_x = 0,$$

and we may conclude again that $\sigma(P_{i-1})^*e = 0$.

Consequently $\sigma(\Delta_i)(x,z)$ is an isomorphism for all $z \in T^*_x(X) \setminus \{0\}$ because the complex E is elliptic.

□

Thus $\Delta = P\tilde{P}^* + \tilde{P}^*P$ is an elliptic differential operator of type $(\oplus E^i) \to (\oplus E^i)$ and order $2p$ on X.

Lemma 3.4.2 *1) For $f \in \mathcal{E}(E^i)$, we have $\Delta f = 0$ if and only if $f \in \mathcal{H}^i(E)$ i.e., $Pf = P^*f = 0$.*

2) For $f \in \mathcal{E}(E^i)$, we have $\Delta^* f = 0$ if and only if $\tilde{P}f = P^*f = 0$.

Proof. It follows from Lemma 2.2.3 because the differential operators $Q_i = (D_{i+1}P_i)^*(D_{i+1}P_i) + 1$ are selfadjoint and positive.

□

As in part 1) of Theorem 2.2.2, we denote by H_i the L^2-orthogonal projection onto the finite-dimensional space $\mathcal{H}^i(E)$.

By the fundamental Decomposition Theorem for selfadjoint elliptic differential operators on compact manifolds (see Section 2 2), there are selfadjoint operators $G_i \in \mathrm{pdo}_{-4p}(E^i \to E^i)$ with satisfy $G_i(\Delta^*_i\Delta_i) = 1 - H_i \quad (i \in \mathbb{Z})$ on X.

We now let $\Pi_i = G_{i-1}\Delta^*_{i-1}\tilde{P}^*_{i-1}$ so that Π_i is a pseudo-differential operator of type $E^i \to E^{i-1}$ and order $(-p_{i-1})$ on X.

Lemma 3.4.3 *The operator $\Pi = \{\Pi_i\}$ satisfies the homotopy equation*

$$\Pi P + P\Pi = 1 - H \quad \text{on} \quad \mathcal{E}'(E^i).$$

Proof. We have simply rewritten part 2) of Theorem 2.2.2.

□

Unlike Theorem 2.2.2, the operator Π has the following special property.

Lemma 3.4.4 *The operator $P^*P\Pi$ is local, namely $P^*P\Pi = P^*$.*

Proof. Using Lemma 3.4.3 and the definition of H, we get

$$P^*P\Pi = P^*(1 - H - \Pi P) = P^* - P^*G\Delta^*\tilde{P}^*P.$$

To complete the proof, we need to show that $P^*G\Delta^*\tilde{P}^*P = 0$, or $P^*\tilde{P}\Delta GP = 0$. In view of Lemma 3.4.2, the last equality follows from the equality $\Delta^*(\tilde{P}\Delta GP) = 0$. We prove this equality.

Since $\tilde{P}^*\Delta = \Delta\tilde{P}^*$ and $\tilde{P} = \tilde{Q}P$ for a suitable differential operator $\tilde{Q}$, we have

$$\Delta^*(\tilde{P}\Delta GP) = \tilde{P}(\Delta^*\Delta G)P = \tilde{P}(1 - H)P = 0$$

as required.

□

3.4.3 *A complete description of cocycles of the complex* $\mathcal{E}(E|_{\overline{Y}})$

It follows from Corollary 2.5.1 with the help of standard regularization considerations that, if $f \in C^{p_\iota - 1}_{loc}(E^\iota|_{\overline{Y}})$ satisfies $Pf = 0$ weakly on Y, then

$$f(x) = -\Pi([\partial Y]^P f) + P\Pi(\chi_Y f) + H(\chi_Y f) \quad (x \in Y). \tag{3.4.1}$$

Consider the following problem. To what extent does formula (3.4.1) characterize sections $f \in C^{p_\iota-1}_{loc}(E^\iota|_{\overline{Y}})$ satisfying $Pf = 0$? I.e., does f necessarily satisfy $Pf = 0$ in Y provided it can be represented by formula (3.4.1)?

Theorem 3.4.5 *Suppose that $Y \subset X$ is an open set whose boundary is of class $C^{2p_\iota,\lambda}_{loc}$ ($\lambda > 0$), and $f \in C^{2p_\iota-1,\lambda}_{loc}(E^\iota|_{\overline{Y}})$. Then $Pf = 0$ if and only if f can be represented in Y by formula (3.4.1).*

Proof. We need to prove the sufficiency only. Let $f \in C^{2p_\iota-1,\lambda}_{loc}(E^\iota|_{\overline{Y}})$ be represented by formula (3.4.1).

We shall apply Theorem 3.3.9 for the case where $S = \partial Y$, $X^- = Y$ and $X^+ = X \setminus \overline{Y}$. As before, we let $f_I = \Pi([\partial Y]^P f)$.

More exactly we shall use only the fact that $f_I^\pm \in C^{2p_\iota-1,\lambda}_{loc}(E^\iota|_{\overline{X^\pm}})$, and $n(Pf_I^+ - Pf_I^-) = 0$ on S. Theorem 3.3.9 asserts that the last relationship holds provided $n = c_{\tilde{P}_\iota^*}$. But the same arguments show that by n we can understand the Cauchy data on S with respect to the differential operator P_ι^* if the differential operator $\tilde{P}_\iota$ has been chosen as in Subsection 3.4.2.

By (3.4.1), $Pf = -Pf_I^-$. Therefore it suffices to show that $Pf_I^- = 0$.

First of all by Lemma 3.4.4 we get $P^*Pf_I = P^*([\partial Y]^P f)$, so that

$$\begin{aligned} P^*Pf_I^+ &= 0, \\ P^*Pf_I^- &= 0. \end{aligned} \tag{3.4.2}$$

Since $PH = 0$, formula (2.5.1) for $x \in X \setminus \overline{Y}$ implies that

$$Pf_I^+ = P\Pi(\chi_Y(Pf))^+. \tag{3.4.3}$$

Moreover, comparing (3.4.1) with (2.5.1) for $x \in Y$, we derive

$$\Pi(\chi_Y(Pf))^- = 0. \tag{3.4.4}$$

The potential $\Pi(\chi_Y(Pf))$ is of class $C^{p_\imath-1}_{loc}(E^\imath)$ on X because $\Pi_{\imath+1}$ is a pseudo-differential operator of type $E^{\imath+1} \to E^\imath$ and order $(-p_\imath)$. Hence (3.4.4) implies

$$j^{p_\imath-1}_S(\Pi(\chi_Y(Pf)))^+ = 0. \tag{3.4.5}$$

It follows from the first equality in (3.4.2) and (3.4.3) that $P^*P\Pi(\chi_Y(Pf))^+ = 0$. Therefore in view of (3.4.5) we obtain

$$\begin{aligned}
&\|P\Pi(\chi_Y(Pf))^+\|^2_{L^2(E^{\imath+1}|_{X\setminus Y})} \\
&= \int_{X\setminus Y} (P\Pi(\chi_Y(Pf))^+, P\Pi(\chi_Y(Pf))^+)_x v(x) \\
&= \int_{X\setminus Y} \langle \star P\Pi(\chi_Y(Pf))^+, P\Pi(\chi_Y(Pf))^+\rangle_x \\
&= \int_{X\setminus Y} \langle \star P^*P\Pi(\chi_Y(Pf))^+, \Pi(\chi_Y(Pf))^+\rangle_x \\
&\quad - \int_{\partial Y} G_P(\star P\Pi(\chi_Y(Pf))^+, \Pi(\chi_Y(Pf))^+) \\
&= 0,
\end{aligned}$$

so $P\Pi(\chi_Y(Pf))^+ = 0$.

According to (3.4.3) we have $Pf_I^+ = 0$. The jump relationship for $n(Pf_I)$ on S now implies that $n(Pf_I^-) = 0$.

Hence we derive in view of the second equality (3.4.2)

$$\begin{aligned}
&\|Pf_I^-\|^2_{L^2(E^{\imath+1}|_Y)} \\
&= \int_Y \langle \star Pf_I^-, Pf_I^-\rangle_x \\
&= \int_Y \langle \star f_I^-, P^*Pf_I^-\rangle_x - \int_{\partial Y} G_{P^*}(\star f_I^-, Pf_I^-) \\
&= 0.
\end{aligned}$$

Thus $Pf_I^- = 0$. This completes the proof.

□

Theorem 3.4.5 characterizes the space $Z^\imath(\mathcal{E}(E^\cdot|_{\overline{Y}}))$ as the eigen subspace of the operator $-\Pi([\partial Y]^P\cdot) + P\Pi(\chi_Y\cdot) + H(\chi_Y\cdot)$ in $\mathcal{E}(E^\imath|_{\overline{Y}})$ which corresponds to the eigenvalue 1.

Remark 3.4.6 *For elliptic complexes of differential operators with constant coefficients in $\mathbb{R}^n$, Theorem 3.4.5 was obtained in the author's thesis [225]. As one can see from the proof of Theorem 2.3.22, such complexes have parametrices like those given by the Hodge theory. Hence the same proof carries over.*

3.4.4 *The spaces* $\mathcal{H}^i(E|_{\overline{Y}})$

We start with two technical lemmas.

Lemma 3.4.7 *The operator* $\Delta\Pi$ *is local, namely* $\Delta\Pi = \widetilde{P}^*$.

Proof. First, $\Delta\Pi = \widetilde{P}^*$ if and only if $(\Delta G\Delta^* - 1)\widetilde{P}^* = 0$, or equivalently $\widetilde{P}(\Delta G\Delta^* - 1) = 0$.

By Lemma 3.4.2, the last equality follows from the equality $\Delta^*(\Delta G\Delta^* - 1) = 0$. Also

$$\Delta^*(\Delta G\Delta^* - 1) = (\Delta^*\Delta G - 1)\Delta^* = -(\Delta H)^* = 0,$$

which finishes the proof.

□

We now let $\mathfrak{P}_i = G_{i+1}\Delta^*_{i+1}P_i$ $(i \in \mathbb{Z})$ so that $\mathfrak{P}_i$ is a pseudo-differential operator of type $E^i \to E^{i+1}$ and order $(-\widetilde{p}_i)$ on X.

Lemma 3.4.8 *The following commutativity relations hold:*

$$\begin{aligned} P\Pi &= \mathfrak{P}\widetilde{P}^*, \\ \Pi P &= \widetilde{P}^*\mathfrak{P}. \end{aligned}$$

Proof. We have already seen in the proof of Theorem 2.2.2 that $PG\Delta^*\widetilde{P}^* = G\Delta^*P\widetilde{P}^*$. This is just the first of the required relations.

To prove the second relation we need to show that $(G\Delta^*\widetilde{P}^* - \widetilde{P}^*G\Delta^*)P = 0$, or equivalently $P^*(\widetilde{P}\Delta G - \Delta G\widetilde{P}) = 0$. By Lemma 3.4.2, the last equality follows from the fact that $\Delta^*(\widetilde{P}\Delta G - \Delta G\widetilde{P}) = 0$.

Since $\Delta^*\widetilde{P} = \widetilde{P}\Delta^*$ and $\widetilde{P}H = H\widetilde{P} = 0$, we get

$$\Delta^*(\widetilde{P}\Delta G - \Delta G\widetilde{P}) = \widetilde{P}(\Delta^*\Delta G) - (\Delta^*\Delta G)\widetilde{P} = \widetilde{P}(1 - H) - (1 - H)\widetilde{P} = 0$$

with completes the proof.

□

Corollary 3.4.9 *The operator* $\mathfrak{P} = \{\mathfrak{P}_i\}$ *is a parametrix for the (chain) complex* $\{E^i, \widetilde{P}^*\}$, *i.e.,*

$$\mathfrak{P}\widetilde{P}^* + \widetilde{P}^*\mathfrak{P} = 1 - H \quad \text{on} \quad \mathcal{E}'(E^i).$$

Proof. This follows immediately from Lemmas 3.4.3 and 3.4.8.

□

Before formulating the next result, we emphasize that $n(f) = c_{P^*}(f)$ means the Cauchy data of f on ∂Y with respect to the differential operator P^* (not $\widetilde{P}^*$!).

Lemma 3.4.10 *For* $f \in C^{2p}_{loc}(E^i|_{\overline{Y}})$, *we have* $\Delta f = c_{\widetilde{P}^*}(f) = c_{\widetilde{P}^*}(Pf) = 0$ *if and only if* $Pf = P^*f = n(f) = 0$.

Proof. *Necessity.* Suppose that $f \in C^{2p}_{loc}(E^{\cdot}|_{\overline{Y}})$ and $\Delta f = c_{\tilde{P}^*}(f) = c_{\tilde{P}^*}(Pf) = 0$.

Choose a sequence $\{f_\nu\}$ in $\mathcal{E}(E^{\cdot}|_{\overline{Y}})$ such that it converges to f in the topology of $C^{2p}_{loc}(E^{\cdot}|_{\overline{Y}})$. Since $\Delta f = 0$, we have $\tilde{P}^* P f = -P\tilde{P}^* f$. Using now the condition $c_{\tilde{P}^*}(Pf) = 0$ and the equality $\tilde{P} \circ P = 0$, we have

$$(\tilde{P}^* Pf, \tilde{P}^* Pf)_Y = -(P\tilde{P}^* f, \tilde{P}^* Pf)_Y = -\lim_{\nu \to \infty} (P\tilde{P}^* f_\nu, \tilde{P}^* Pf)_Y = 0$$

so that $\tilde{P}^* Pf = 0$. Hence it follows that also $P\tilde{P}^* f = 0$.

Since $\tilde{P} = \tilde{Q}P$ for a suitable differential operator $\tilde{Q}$, we conclude that $\tilde{P}\tilde{P}^* f = 0$. Therefore $(\tilde{P}^* f, \tilde{P}^* f)_Y = (\tilde{P}\tilde{P}^* f, f)_Y = 0$ because $c_{\tilde{P}^*}(f) = 0$. We have shown that $\tilde{P}^* f = 0$.

Since $\tilde{P}^* f = c_{\tilde{P}^*}(f) = 0$, we have $\tilde{P}^*(\chi_Y f) = 0$ (see (2.4.23)). Applying Corollary 3.4.9, we get $\chi_Y f = \tilde{P}^* \mathfrak{P}(\chi_Y f) + H(\chi_Y f)$. Therefore $P^*(\chi_Y f) = 0$ because $P^* \tilde{P}^* = 0$ and $P^* H = 0$. Notice now that the equality $P^*(\chi_Y f) = 0$ is equivalent to $P^* f = n(f) = 0$.

It remain to check that $Pf = 0$. We reason as above. Since $\tilde{P}^*(Pf) = c_{\tilde{P}^*}(Pf) = 0$, we have $\tilde{P}^*(\chi_Y(Pf)) = 0$. Apply Corollary 3.4.9 to obtain $\chi_Y(Pf) = \tilde{P}^* \mathfrak{P}(\chi_Y(Pf)) + H(\chi_Y(Pf))$. Hence it follows that $P^*(\chi_Y(Pf)) = 0$, or equivalently $P^* Pf = n(Pf) = 0$. And finally $(Pf, Pf)_Y = (P^* Pf, f)_Y = 0$, so that $Pf = 0$ as required.

Sufficiency. Suppose that $f \in C^{\tilde{p}_{\cdot}-1}_{loc}(E^{\cdot}|_{\overline{Y}})$ and $Pf = P^* f = n(f) = 0$. Since $\tilde{P} = PQ$ for a suitable differential operator Q, we obtain $\Delta f = 0$. To finish the proof, it remains to show that $c_{\tilde{P}^*}(f) = 0$. Since $\tilde{P}^* f = 0$, the last equality means that $(f, \tilde{P}g)_Y = 0$ for all sections $g \in \mathcal{E}(E^{\cdot-1})$. But we have even $(f, Pg)_Y = 0$ for all sections $g \in \mathcal{E}(E^{\cdot-1})$ because $\tilde{P}^* f = n(f) = 0$. This completes the proof.

□

Thus the space $\mathcal{H}^{\cdot}(E^{\cdot}|_{\overline{Y}})$ may be equivalently described as the space of all $f \in \mathcal{E}(E^{\cdot}|_Y)$ satisfying $\Delta f = c_{\tilde{P}^*}(f) = c_{\tilde{P}^*}(Pf) = 0$.

Remark 3.4.11 *It follows from Lemma 3.4.10 and Theorem 3.3.16 that the* ($\tilde{P}-$) *Neumann problem*

$$\begin{cases} \Delta^* u &= f \\ n(u) &= n(\tilde{P}u) = 0 \end{cases}$$

is of importance in studying the non-homogeneous system $Pu = f$ (see Section 4.1).

3.4.5 *A complete description of harmonic sections of* $E|_{\overline{Y}}$

It follows from Corollary 2.5.15 with the help of standard reqularization considerations that, if $f \in C^{\max(\tilde{p}_{\cdot}-1, p_{\cdot})-1}_{loc}(E^{\cdot}|_{\overline{Y}})$ satisfies $Pf = \tilde{P}^* f = 0$ weakly in Y and $c_{\tilde{P}^*}(f) = 0$, then

$$f(x) = -\Pi([\partial Y]^P f) + H(\chi_Y f) \quad (x \in Y). \tag{3.4.6}$$

Consider the following problem. Must a section $f \in C_{loc}^{\max(\widetilde{p}_i-1,p_i)-1}(E^i|_{\overline{Y}})$ necessarily satisfy $Pf = \widetilde{P}^*f = c_{\widetilde{P}^*}(f) = 0$ provided it can be represented by formula (3.4.6)?

Theorem 3.4.12 *Suppose that $Y \subset X$ is an open set with boundary of class $C_{loc}^{2p_i,\lambda}$ $(\lambda > 0)$, and $f \in C_{loc}^{2p_i-1,\lambda}(E^i|_{\overline{Y}})$. Then $Pf = P^*f = n(f) = 0$ if and only if f can be represented in Y by formula (3.4.6).*

Proof. We need only to prove the sufficiency. Let $f \in C_{loc}^{2p_i-1,\lambda}(E^i|_{\overline{Y}})$ be represented by formula (3.4.6).

We shall apply Theorem 3.3.9 to the case where $S = \partial Y$, $X^- = Y$ and $X^+ = X \setminus \overline{Y}$. As before, we let $f_I = \Pi([\partial Y]^P f)$.

According to Theorem 3.3.9 both restrictions $f_I^\pm$ are of class $C_{loc}^{2p_i-1,\lambda}(E^i|_{\overline{X^\pm}})$, and the following jump formulas hold:

$$\begin{cases} t(f_I^+ - f_I^-) &= t(f), \\ n(f_I^+ - f_I^-) &= 0, \\ n(Pf_I^+ - Pf_I^-) &= 0. \end{cases} \tag{3.4.7}$$

Formula (3.4.6) means that $f = -f_I^- + H(\chi_Y f)$. Therefore we have $t(-f_I^- + H(\chi_Y f)) = t(f)$. Since the section $H(\chi_Y f)$ is of class $\mathcal{E}(E^i)$ on X, we derive from the first relationship (3.4.7) that

$$t(-f_I^+ + H(\chi_Y f)) = 0. \tag{3.4.8}$$

Hence it follows in view of the first equality in (3.4.2) that

$$\begin{aligned} &\|Pf_I^+\|^2_{L^2(E^{i+1}|_{X\setminus Y})} \\ &= \left(P(-f_I^+ + H(\chi_Y f)), P(-f_I^+ + H(\chi_Y f))\right)_{X\setminus Y} \\ &= \left(P^*P(-f_I^+ + H(\chi_Y f)), (-f_I^+ + H(\chi_Y f))\right)_{X\setminus Y} \\ &= 0, \end{aligned}$$

i.e., $Pf_I^+ = 0$.

Using the third relationship of (3.4.7) we conclude that $n(Pf_I^-) = 0$. Since in view of (3.4.2), $P^*Pf_I^- = 0$, we obtain $(Pf_I^-, Pf_I^-)_Y = 0$. Therefore $Pf_I^- = 0$, and $Pf = 0$ in Y because $Pf = -Pf_I^-$.

Now applying Lemma 3.4.8 yields

$$f_I = \Pi([\partial Y]^P f) = -\Pi P(\chi_Y f) = -\widetilde{P}^*\mathfrak{P}(\chi_Y f) \quad \text{on} \quad X. \tag{3.4.9}$$

Hence it follows that $\widetilde{P}^*f = 0$ in Y, since $P^*\widetilde{P}^* = 0$ and $P^*H = 0$.

To complete the proof it remains to check that $n(f) = 0$.

Since $Pf = 0$ in Y, we derive from formula (3.4.1) that

$$-P\Pi(\chi_Y f) = \begin{cases} 0 & \text{in} \quad Y, \\ (-f_I^+ + H(\chi_Y f)) & \text{in} \quad X \setminus \overline{Y}. \end{cases} \tag{3.4.10}$$

It follows from equalities (3.4.9) and (3.4.10) that

$$\| - f_I^+ + H(\chi_Y f)\|^2_{L^2(E^\cdot|_{X\setminus Y})} = \left(\tilde{P}^*\mathfrak{P}(\chi_Y f) + H(\chi_Y f), -P\Pi(\chi_Y f)\right)_X .$$

By the boundedness theorem for pseudo-differential operators in Sobolev spaces (see theorem 2.5 of Taylor [236, Ch.XI]) we have $\Pi(\chi_Y f) \in W^{p_\cdot-1,q}_{loc}(E^{\cdot-1})$ for any $q < \infty$ because $\chi_Y f \in L^\infty_{comp}(E^\cdot)$. Fix now some q' with $1 < q' < \infty$, and choose a sequence $\{f_\nu\}$ in $\mathcal{D}(E^\cdot)$ which converges to $(\chi_Y f)$ in the topology of $L^{q'}_{comp}(E^\cdot)$. Then by the same boundedness theorem the sequence of potentials $\{\mathfrak{P}(f_\nu)\}$ converges to $\mathfrak{P}(\chi_Y f)$ in the topology of $W^{\tilde{p}_\cdot,q'}_{loc}(E^{\cdot+1})$. Therefore we get

$$\begin{aligned}
&\| - f_I^+ + H(\chi_Y f)\|^2_{L^2(E^\cdot|_{X\setminus Y})} \\
&= \lim_{\nu\to\infty} \left(\tilde{P}^*\mathfrak{P}(f_\nu) + H(\chi_Y f), -P\Pi(\chi_Y f)\right)_X \\
&= - \lim_{\nu\to\infty} \left(P^*\tilde{P}^*\mathfrak{P}(f_\nu) + P^*H(\chi_Y f), \Pi(\chi_Y f)\right)_X \\
&= 0.
\end{aligned}$$

Thus $(-f_I^+ + H(\chi_Y f)) \equiv 0$ in $X \setminus Y$, and we can use the second relationship of (3.4.7) to conclude that $n(-f_I^- + H(\chi_Y f)) = 0$. Hence it follows that $n(f) = 0$ which completes the proof.

□

Theorem 3.4.12 characterizes the space $\mathcal{H}^\cdot(E^\cdot|_{\overline{Y}})$ as the eigen subspace of the operator $-\Pi([\partial Y]^P\cdot) + H(\chi_Y\cdot)$ in $\mathcal{E}(E^\cdot|_{\overline{Y}})$ corresponding to the eigenvalue 1.

Exercise 3.4.13 *Prove that, if* $\partial Y \in C^{2p_\cdot,\lambda}_{loc}$ $(\lambda > 0)$ *and* $f \in C^{2p_\cdot-1,\lambda}_{loc}(E^\cdot|_{\overline{Y}})$, *then* $t(f) = H(\chi_Y f) = 0$ *if and only if* f *is representable by the formula* $f = \Pi(\chi_Y(Pf)) + P\Pi(\chi_Y f)$ *in* Y.

Hint. To prove the sufficiency, notice first that $H(\chi_Y f) \equiv 0$ for $f \in L^q_{loc}(E^\cdot|_{\overline{Y}})$ $(q > 1)$ if and only if $H(\chi_Y f) = 0$ in Y.

Indeed, we suppose that $H(\chi_Y f) = 0$ in Y. Since by Lemmas 3.4.3 and 3.4.8 $H(\chi_Y f) = -\left(\tilde{P}^*\mathfrak{P}(\chi_Y f) + P\Pi(\chi_Y f)\right)$ in $X \setminus Y$, we get

$$(H(\chi_Y f), H(\chi_Y f))_{X\setminus Y} = (H(\chi_Y f), -\{\tilde{P}^*\mathfrak{P}(\chi_Y f) + P\Pi(\chi_Y f)\})_X = 0.$$

Thus, also $H(\chi_Y f) = 0$ in $X \setminus Y$.

Now show that $-f_I^+ + H(\chi_Y f) = 0$ in $X \setminus Y$ provided f is representable by the formula $f = \Pi(\chi_Y(Pf)) + P\Pi(\chi_Y f)$ in Y.

□

Suppose $\partial Y \in C^{s+1}_{loc}$ $(s \geq p_\cdot)$. We consider two operators T_j $(j = 1, 2)$ in $W^{s,q}_{loc}(E^\cdot|_{\overline{Y}})$ given by

$$\begin{aligned}
T_1(f) &= -\Pi([\partial Y]^P f) + H(\chi_Y f), \\
T_2(f) &= \Pi(\chi_Y(Pf)) + P\Pi(\chi_Y f).
\end{aligned}$$

It follows from Theorem 1.1.9 that for $W^{s,q}_{loc}(E^\cdot|_{\overline{Y}})$ $(q > 1)$ the section $[\partial Y]^P f \in W^{-p_\cdot,q}_{loc}(E^{\cdot+1})$ is well-defined. One can see, from the boundedness theorems for pseudo-differential operators in Sobolev spaces on manifolds with boundaries (see Rempel and Schulze [188], 2.3.2.5 and 2.3.3.3), that T_j $(j = 1,2)$ are continuous linear operators in $W^{s,q}_{loc}(E^i|_{\overline{Y}})$ provided $1 < q < \infty$.

The following conjecture could sharpen the statements of Theorem 3.4.12 and Exercise 3.4.13, unless of course it is wrong.

Exercise 3.4.14 *Let ∂Y be of class C^{s+1}_{loc} where $s \geq \max(p_{i-1}, p_i)$. Prove that, for every $f \in W^{s,q}_{loc}(E^i|_{\overline{Y}})$ $(1 < q < \infty)$, the iterations $\{T_1^\nu f\}_{\nu\in\mathbb{Z}_+}$ converge in $W^{s,q}_{loc}(E^i|_{\overline{Y}})$ to a section f_1 satisfying $Pf_1 = P^* f_1 = n(f_1) = 0$, and the iterations $\{T_2^\nu f\}_{\nu\in\mathbb{Z}_+}$ converge to a section f_2 satisfying $t(f_2) = H(\chi_Y f_2) = 0$.*

Hint. Analyze the proof of A.V.Romanov's theorem on iterations of the Martinelli-Bochner integral in the book of Kytmanov [145, §16].

□

3.4.6 *Morera's theorem*

We supplement Theorems 3.4.5 and 3.4.12 with an interesting description of sections $f \in C^{p_i-1}_{loc}(E^i)$ satisfying $Pf = 0$ weakly on X.

Let P be a differential operator of type $E \to F$ and order p on X. For every integer $l' > 0$, the transposed operator P' gives rise to a differential operator of type $\oplus_1^{l'} F' \to \oplus_1^{l'} E'$ by $\oplus_1^{l'} g_j \to \oplus_1^{l'} P' g_j$, and we denote this operator by P' again.

Suppose that g is a differentiable section of $\left(\oplus_1^{l'} F'\right)$ satisfying $P'g = 0$, and moreover that $\operatorname{rank}_{\mathbb{C}} g = l$ on X where l is the rank of F.

Let us explain the latter. We say that $\operatorname{rank}_{\mathbb{C}} g(x) = l$ if, for every local representation of g in a neighborhood U of the point x by a $(l \times l')$-matrix g_U of functions of $\mathcal{E}(U)$, we have $\operatorname{rank}_{\mathbb{C}} g_U(x) = l$. Since different representations of g may be obtained one from another by multiplying by non-degenerate $(l \times l)$-matrices, this notion is well-defined.

We shall not discuss when such a section g exists, but consider only an example.

Example 3.4.15 Let P be a partial differential operator on an open set $X \subset \mathbb{R}^n$ given by $\left(\sum_{|\alpha|\leq p} P_\alpha(x) D^\alpha\right)$, where P_α $(|\alpha| \leq p)$ are $(l \times k)$-matrices of differentiable functions on X. If $\sum_{|\alpha|\leq p} (-1)^{|\alpha|} D^\alpha P_\alpha(x) = 0$ on X, then the section $g = I_l$ satisfies $P'g = 0$, and obviously $\operatorname{rank}_{\mathbb{C}} g = l$ on X.

□

Fix some base $\mathcal{B}$ for the topology on X, all the sets of which have piecewise smooth boundaries and compact closures in X.

As far as other conditions on $\mathcal{B}$ are concerned, we require that X can be packed by sets of $\mathcal{B}$ as small as we please. More precisely, for every compact set $K \subset X$ and every $\delta > 0$, there are pairwise disjoint sets σ_ν $(\nu = 1, \dots, N)$ of $\mathcal{B}$ such that: 1) diam $\sigma_\nu < \delta$; 2) meas $\partial(\cup\overline{\sigma_\nu}) < c(K)$ where $c(K)$ does not depend on δ, and 3) $K \subset (U\overline{\sigma_\nu})$.

Example 3.4.16 Suppose that X is an open set in $\mathbb{R}^n$. Then one can take as $\mathcal{B}$ the set of all cubes in X with edges parallel to the coordinate axes.

□

Example 3.4.17 Suppose that X is an open set on the plane $\mathbb{R}^2$. Then one can take as $\mathcal{B}$ the set of all triangles in X similar to a fixed triangle.

□

We are ready to formulate Morera's theorem for solutions of the system $Pf = 0$.

Theorem 3.4.18 *Let $f \in C^{p-1}_{loc}(E)$. Then $Pf = 0$ weakly on X if and only if*

$$\int_{\partial\sigma} G_P(g, f) = 0 \quad \text{for every set} \quad \sigma \in \mathcal{B}. \tag{3.4.11}$$

When $P = \overline{\partial}$ is the Cauchy-Riemann operator in the plane, Theorem 3.4.18 coincides with the classical Morera theorem because this operator is elliptic. For holomorphic functions of several variables this theorem was noticed by Aizenberg [4], and Dautov and Znamenskii [46]. For the Moisil-Theodoresko system this result appeared in the paper of Moisil and Theodoresko [162], and for arbitrary elliptic systems of first order differential equations it appeared in the author's paper [220] (see Dairbekov and Kopylov [43] also). For $\overline{\partial}$-closed differential forms of bidegree $(p, n-1)$ Theorem 3.4.18 is due to Dautov [44], and finally for arbitrary differential operators P with constant coefficients it was proved in the author's thesis [225].

Notice that in general a section f satisfying condition (3.4.11) may not be a classical solution of the equation $Pf = 0$. For example, if $P = P_i$ $(i \geq 1)$, then according to Proposition 2.4.6, condition (3.4.11) is fulfilled for any section of the form $f = P_{i-1}u$ where $u \in C^{p_{i-1}+p_i-1}_{loc}(E^{i-1})$.

Proof. *Necessity.* One can assume without loss of generality that X is an open subset of $\mathbb{R}^n$. Let σ be a set in X with a piecewise smooth boundary and a compact closure.

Fix some non-negative function $\omega \in \mathcal{D}(\mathbb{R}^n)$, with support in the unit ball $B(0,1)$, which depends on $|x|$ only and is normalized by the condition $\int \omega(x)dx = 1$.

Set $\omega^\varepsilon(x) = \frac{1}{\varepsilon^n}\omega\left(\frac{x}{\varepsilon}\right)$. The operator given by convolution with ω^ε, i.e.,

$$R^{(\varepsilon)}f = \omega^\varepsilon \star f = \int \omega^\varepsilon(x-y)f(y)dy,$$

is called the standard regularization in $\mathbb{R}^n$. Notice that $R^{(\varepsilon)}f$ is considered on the set $X_\varepsilon = \{x \in X : \ B(x,\varepsilon) \subset X\}$ only.

The properties of the standard regularization are well-known (see, for instance, Desin [52, Ch.II, §5]). In particular, if $f \in C^{p-1}_{loc}(E)$, then $R^{(\varepsilon)}f \in \mathcal{E}(E|_{X_\varepsilon})$ and $R^{(\varepsilon)}f$ converges to f in the topology of $C^{p-1}_{loc}(E)$ provided $\varepsilon \to +0$.

For sufficiently small ε we have $\sigma \subset\subset X_\varepsilon$, so by Stokes' Theorem

$$\int_{\partial\sigma} G_P(g, f) = \lim_{\varepsilon\to+0} \int_{\partial\sigma} G_P(g, R^{(\varepsilon)}f) = \lim_{\varepsilon\to+0} \int_\sigma g^t P(R^{(\varepsilon)}f). \tag{3.4.12}$$

Let us represent the differential operator P by a $(l \times k)$-matrix $P(x, D)$ of scalar differential operators with partial derivatives on X, and denote

$$P^{(j)}(x, D) = \frac{\partial}{\partial z_j} P(x, z)|_{z \to D}$$

for $j = 1, \ldots, n$. It follows from the Euler formula for homogeneous functions that the difference

$$R(x, D) = P(x, D) - \frac{1}{p} \sum_{j=1}^{n} D_j \circ P^{(j)}(x, D)$$

is a differential operator of type $E \to F$ and order $\leq p-1$ on X. Then we can write for $x \in \sigma$:

$$\begin{aligned}
&P(R^{(\varepsilon)} f)(x) \\
&= \int_{\mathbb{R}^n} (P(x, D)\omega^{\varepsilon}(x-y))\, f(y)\, dy \\
&= \frac{1}{p} \sum_{j=1}^{n} D_j \int_X \omega^{\varepsilon}(x-y) \left(P^{(j)}(x, D_y) - P^{(j)}(y, D_y)\right) f(y)\, dy \\
&\quad + \int_X \omega^{\varepsilon}(x-y)\, (R(x, D_y) - R(y, D_y))\, f(y)\, dy \\
&\quad + \int_X (P'(y, D_y)\omega^{\varepsilon}(x-y))^t\, f(y)\, dy. \qquad (3.4.13)
\end{aligned}$$

Since we have $\sigma \subset X_\varepsilon$, the function $\omega^{\varepsilon}(x - \cdot)$ is supported in X for all $x \in \sigma$. Hence it follows that the last summand in the right-hand side of (3.4.13) vanishes since $Pf = 0$ weakly in X.

As far as the first and second summands on the right-hand side of (3.4.13) are concerned, they converge to zero in the topology of the space $L^2(F|_{\bar{\sigma}})$. To prove this it suffices to use Friedrichs' Lemma and Property $J4$ of the standard regularization from the book of Desin [52, Ch.II, §5].

Therefore the limit in the right-hand side of (3.4.12) equals zero, and condition (3.4.11) is fulfilled.

Sufficiency. We assume that X is an open subset of $\mathbb{R}^n$, and $f \in C^{p-1}_{loc}(E)$ satisfies condition (3.4.11).

Let $R^{(\varepsilon)}$ be the regularization operator constructed above. Since $f \in C^{p-1}_{loc}(E)$, the sequence $\{g^t P(R^{(\varepsilon)} f)\}$ of $[\mathcal{E}(X_\varepsilon)]^{l'}$ converges in the weak-$*$ topology of the dual space of $[C^1_{comp}(X)]^{l'}$ as $\varepsilon \to +0$. Indeed, for every test section $\varphi \in [C^1_{comp}(X)]^{l'}$ we obviously have

$$\begin{aligned}
&\lim_{\varepsilon \to +0} \langle \varphi, g^t P(R^{(\varepsilon)} f)\rangle_X \\
&= \lim_{\varepsilon \to +0} \left\langle g\varphi, \left(\frac{1}{p} \sum_{j=1}^{n} D_j \circ P^{(j)}(x, D) + R(x, D)\right) R^{(\varepsilon)} f \right\rangle_X \\
&= -\frac{1}{p} \sum_{j=1}^{n} \left\langle D_j(g\varphi), P^{(j)}(x, D) f \right\rangle_X + \langle g\varphi, R(x, D) f\rangle_X.
\end{aligned}$$

We would like to show that $\{g^t P(R^{(\varepsilon)} f)\}$ converges to zero in the weak-$*$ topology of the dual space for $[C^1_{comp}(X)]^{l'}$. With this purpose we fix some arbitrary test section $\varphi \in [C^1_{comp}(X)]^{l'}$, and let $K \subset X$ be a compact set whose interior contains the support of φ.

By the condition imposed on $\mathcal{B}$, for every $\delta > 0$, there exist pairwise disjoint sets $\sigma_\nu \subset K$ $(\nu = 1, \ldots, N)$ of the base $\mathcal{B}$ such that: 1) diam $\sigma_\nu < \delta$; 2) meas $\partial(\cup\overline{\sigma_\nu})$ is bounded by a constant independent of δ, and 3) supp $\varphi \subset (\cup\overline{\sigma_\nu})$. We now choose points $x_\nu \in \sigma_\nu$ $(\nu = 1, \ldots, N)$. Since φ is uniformly continuous, the piecewise constant section $\sum_{\nu=1}^N \varphi(x_\nu)\chi_{\sigma_\nu}$ approximates φ uniformly on X with any prescribed precision provided δ is small.

Notice that $\sum_{\nu=1}^N \varphi(x_\nu)\chi_{\sigma_\nu}(x)$ vanishes outside K. Choosing now ε so small that K lies in X_ε, and using Stokes' Theorem and condition (3.4.11) we get

$$\begin{aligned}
&\langle \varphi, g^t P(R^{(\varepsilon)} f)\rangle_X \\
&= \left\langle \varphi - \sum_{\nu=1}^N \varphi(x_\nu)\chi_{\sigma_\nu}, g^t P(R^{(\varepsilon)} f)\right\rangle_X + \sum_{\nu=1}^N (\varphi(x_\nu))^t \left\langle g, P(R^{(\varepsilon)} f)\right\rangle_{\sigma_\nu} \\
&= \left\langle \varphi - \sum_{\nu=1}^N \varphi(x_\nu)\chi_{\sigma_\nu}, g^t P(R^{(\varepsilon)} f)\right\rangle_X + \sum_{\nu=1}^N (\varphi(x_\nu))^t \int_{\partial\sigma_\nu} G_P(g, R^{(\varepsilon)} f - f).
\end{aligned} \tag{3.4.14}$$

Denote by $\partial\sigma_\nu \cap (\partial\sigma_\mu)$ that part of the boundary of σ_ν which is common with $\partial\sigma_\mu$ (but with the orientation induced from $\partial\sigma_\nu$), and by $\partial\sigma_\nu \cap \partial(\cup\overline{\sigma_\mu})$ that part of $\partial\sigma_\nu$ which contributes to the boundary of $(\cup\overline{\sigma_\mu})$. Then $\partial\sigma_\nu$ is the union of $\cup_\mu(\partial\sigma_\nu \cap (\partial\sigma_\mu))$ and $\partial\sigma_\nu \cap \partial(\cup_\mu\overline{\sigma_\mu})$, so that

$$\begin{aligned}
&\sum_{\nu=1}^N (\varphi(x_\nu))^t \int_{\partial\sigma_\nu} G_P(g, R^{(\varepsilon)} f - f) \\
&\quad = \sum_{\substack{\nu,\mu=1\\ \nu<\mu}}^N (\varphi(x_\nu) - \varphi(x_\mu))^t \int_{\partial\sigma_\nu\cap(\partial\sigma_\mu)} G_P(g, R^{(\varepsilon)} f - f) \\
&\quad + \sum_{\nu=1}^N (\varphi(x_\nu))^t \int_{\partial\sigma_\nu\cap\partial(\cup_\mu\overline{\sigma_\mu})} G_P(g, R^{(\varepsilon)} f - f).
\end{aligned} \tag{3.4.15}$$

It follows from (3.4.14) and (3.4.15) that there are constants c_j $(j = 1, \ldots, 3)$ independent of ε and δ, such that for all sufficiently small ε and δ we have

$$\begin{aligned}
&\left|\langle \varphi, g^t P(R^{(\varepsilon)} f)\rangle_X\right| \\
&\quad \le c_1 \left(\max\left|\varphi - \sum_{\nu=1}^N \varphi(x_\nu)\chi_{\sigma_\nu}\right|\right) \|P(R^{(\varepsilon)} f)\|_{L^1(F|_K)} \\
&\quad +c_2 \ (\max|\nabla_\varphi|)\, \|R^{(\varepsilon)} f - f\|_{C^{p-1}(E|_K)} \\
&\quad +c_3 \ (\max|\varphi|)\, \|R^{(\varepsilon)} f - f\|_{C^{p-1}(E|_K)}.
\end{aligned}$$

The last two summands in the right-hand side of this estimate may be made as small as we please for small ε. And then we can make the first summand as small as we please, by choosing δ small. Hence it follows that $\lim_{\varepsilon\to+0}\langle\varphi, g^t\, P(R^{(\varepsilon)}f)\rangle = 0$.

Thus $\{g^t\, P(R^{(\varepsilon)}f)\}$ converges to zero in the weak-$*$ topology of the dual space for $[C^1_{comp}(X)]^{l'}$ as $\varepsilon \to +0$. Therefore, if $\varphi \in C^1_{comp}(F')$, we get

$$\lim_{\varepsilon\to+0}\langle\varphi, P(R^{(\varepsilon)}f)\rangle_X = \lim_{\varepsilon\to+0}\langle g^*(gg^*)^{-1}\varphi, g^t\, P(R^{(\varepsilon)}f)\rangle_X = 0,$$

i.e., $\{P(R^{(\varepsilon)}f)\}$ converges to zero in the weak-$*$ topology of $(C^1_{comp}(F'))'$.

However, it is easy to see that $\{P(R^{(\varepsilon)}f)\}$ converges to Pf in the weak-$*$ topology of $(C^1_{comp}(F'))'$. Therefore $Pf = 0$ which completes the proof.

□

If P is a differential operator with constant coefficients on an open set $X \subset \mathbb{R}^n$, the proof of Theorem 3.4.18 can be simplified because the standard regularization $R^{(\varepsilon)}$ commutes with P. Suppose that $\mathcal{B}$ is some base for the topology in X whose sets have piecewise smooth boundaries and compact closures. Moreover, $\mathcal{B}$ is assumed to be stable under small displacements of sets, i.e., $(\sigma - y) \in \mathcal{B}$ for every $\sigma \in \mathcal{B}$ and sufficiently small $y \in \mathbb{R}^n$.

Exercise 3.4.19 *Prove that under the conditions above a section $f \in C^{p-1}_{loc}(E)$ satisfies $Pf = 0$ weakly on X iff*

$$\int_{\partial\sigma} G_P(I_l, f) + \int_\sigma (P'I_l)^t f = 0 \quad \text{for all} \quad \sigma \in \mathcal{B}. \tag{3.4.16}$$

Hint. Show that $R^{(\varepsilon)}f$ satisfies condition (3.4.16) in X_ε provided $f \in C^{p-1}_{loc}(E)$ satisfies this condition. Hence it follows that $P(R^{(\varepsilon)}f) = 0$ in X_ε because $R^{(\varepsilon)}f$ is of class C^∞_{loc} on X_ε. To complete the proof, use the fact that $\{P(R^{(\varepsilon)}f)\}$ converges to Pf in the weak topology of $\mathcal{D}'(F)$ as $\varepsilon \to +0$.

□

Using a result of Harvey and Polking [99], one can strengthen Theorem 3.4.18. Namely, for a section $g \in \mathcal{E}(\oplus_1^{l'} F')$ we define $\mathcal{D}(g)$ as the set of all $x \in X$ such that $\mathrm{rank}_{\mathbb{C}}\, g(x) < l$. Clearly $\mathcal{D}(g)$ is a closed subset of X. Let $h_d(\sigma)$ be the d-dimensional Hausdorff measure of a set $\sigma \subset X$.

Corollary 3.4.20 *Theorem 3.4.18 remains true provided that g merely satisfies $h_{n-1}(K) < \infty$ for each compact set $K \subset \mathcal{D}(g)$.*

Proof. We need only prove the sufficiency. Apply Theorem 3.4.18 to the set $X \setminus \mathcal{D}(g)$ to obtain $Pf = 0$ in $X \setminus \mathcal{D}(g)$. Then the equality $Pf = 0$ on X follows from Theorem 4.3 (d) of Harvey and Polking [99].

□

3.5 Multiplication of Currents via Their Harmonic Representations

3.5.1 *A brief survey of representations of distributions*

In this section another application of the Sokhotskii-Plemelj type formulas (3.3.4) is presented. Namely, for the de Rham complex these formulas give the possibility to associate with any current f_0 on S a regularization of f_0 in some tube neighborhood of S on X (see Golubitsky and Guillemin [87, Ch.II,§7]). Using the regularizations we define an exterior product of currents on S.

Let us discuss the history of this problem.

It is well-known (see, e.g., Tillmann [237], Bremermann [31]) that for every distribution $f_0 \in \mathcal{D}'(\mathbb{R}^1)$ there exists a function f holomorphic in $C^1 \setminus (\text{supp } f_0)$ whose jump on the real axis is f_0 in the sense that for every $g \in \mathcal{D}(\mathbb{R}^1)$

$$\lim_{\varepsilon \to +0} \int_{\mathbb{R}^1} \left(f(x + \sqrt{-1}\,\varepsilon) - f(x - \sqrt{-1}\,\varepsilon) \right) g(x) dx = \langle g, f_0 \rangle. \tag{3.5.1}$$

Such a function f is called an analytic representation of f_0.

If $f_0 \in \mathcal{E}'(\mathbb{R}^1)$, one can take the Cauchy transformation of the distribution f_0 as f (see Bremermann [31]).

For distributions in $\mathcal{D}'(\mathbb{R}^n)$ $(n > 1)$ the analytic representations were studied by Martineau [159] (see also Schapira [192, Ch.IV,§3]), Vogt [249], Vladimirov [246, 247], and others. They consist of more than one function and their singularities are not contained in the support of the distribution.

Aizenberg and Kytmanov (see Aizenberg and Dautov [5, §14]) considered a representation of distributions in $\mathcal{D}'(\mathbb{R}^{2n-1})$ by $\overline{\partial}$-closed differential forms of bidegree $(n, n-1)$ with harmonic coefficients in $\mathbb{C}^n \setminus \mathbb{R}^{2n-1}$. For $f_0 \in \mathcal{E}'(\mathbb{R}^{2n-1})$, their representation is the Martinelli-Bochner transformation of f_0.

Analyzing this representation, Kytmanov [144] found a more natural representation of a distribution $f_0 \in \mathcal{D}'(\mathbb{R}^n)$ by a harmonic function in the half-space of $\mathbb{R}^{n+1}$. For $f_0 \in \mathcal{E}'(\mathbb{R}^n)$, it is the Poisson transformation of f_0.

For $n = 1$, this representation had been investigated earlier by Ivanov [116] who later replaced the Poisson kernel by an arbitrary δ-kernel (see Ivanov [117]).

The harmonic representations of distributions $f_0 \in \mathcal{D}'(\mathbb{R}^n)$ have the same properties as the analytic representations for $n = 1$.

The harmonic representations of currents $f_0 \in \mathcal{D}'(\Lambda^q T^*_{\mathbb{C}}(S))$ were studied by the author [224].

3.5.2 *Weak boundary values of harmonic forms of finite order of growth*

We limit ourselves to the local situation where X is an open set in $\mathbb{R}^n$, because the general case contains nothing new.

Suppose that S is a hypersurface of class C^∞_{loc} in X which has two sides. Thus $X \setminus S = X^- \cup X^+$ where $X^\pm$ are disjoint non-void open subsets of X, and S is oriented as the boundary of X^-. Denote by $\nu(y)$ the positive (i.e., outward) unit vector normal to S at the point $y \in S$.

There exists a positive continuous function ε_0 on S such that all the points $y + \varepsilon\nu(y)$ with $y \in S$ and $|\varepsilon| < \varepsilon_0(y)$ lie in X and form a neighborhood U of S where the data (y, ε) are coordinates. That is, every point $x \in U$ can be uniquely represented in the form $x = y + \varepsilon\nu(y)$ where $y \in S$ and $\varepsilon = \pm\text{dist }(x, S)$.

Remark 3.5.1 *Can we find a ball B of a radius so small that one could roll this ball over all S with only one point of contact? If S is insufficiently smooth, the answer is negative even locally. But for S of class C^∞_{loc} it is positive provided we limit ourselves to compact subsets of S.*

The curvilinear coordinates (y, ε) are called the geodesic coordinates of points $x \in U$. They are of the same class in U as S.

The mapping $r : U \to S$, given by $r(y + \varepsilon\nu(y)) = y$, is a retraction of U onto S, and r is of class C^s_{loc} provided $S \in C^s_{loc}$ $(s \geq 2)$.

Define the function $\rho(x)$ by $\pm\text{dist }(x, S)$ where the signs "+" or "−" are taken according to whether the point x lies in X^+ or X^-.

Lemma 3.5.2 *Suppose that S is of class C^s_{loc} $(s \geq 2)$. Then 1) $\rho \in C^s_{loc}(U)$; 2) $\nabla\rho(y + \varepsilon\nu(y)) = \nabla\rho(y)$ for all $y \in S$ and $|\varepsilon| < \varepsilon_0(y)$, and 3) $|\nabla\rho| = 1$ on S (and so everywhere in U).*

Proof. We omit the proof and mention only the model example $S = \{x \in \mathbb{R}^n : |x| = 1\}$ where we have $\rho(x) = |x| - 1$ and $\nabla\rho(x) = \frac{x}{|x|}$.

□

Let the vector field ν be given in the neighborhood U of S by $\nu(x) = \nabla\rho(x)$. It satisfies the condition that $\nu\rfloor d\rho = 1$. Notice that any two vector fields satisfying this last condition differ one from another by a vector field tangential to S. Then every current f of degree q in U can be decomposed into the sum $f = t(f) + d\rho \wedge n(f)$ where $t(f) = \nu\rfloor(d\rho \wedge f)$ is the tangential component of f, and $n(f) = \nu\rfloor f$ is the normal component of f (see Example 3.2.2). Since $\rho \in C^\infty_{loc}(U)$, the components $t(f)$ and $n(f)$ are well-defined as currents of degrees q and $(q-1)$ respectively in U.

Consider the transformation of "parallel displacement" t_ε in U given by $t_\varepsilon(y) = y + \varepsilon\nu(y)$. For every compact set $K \subset U$, there exists a positive number ε' such that $t_\varepsilon(y) \in U$ if $y \in K$ and $|\varepsilon| < \varepsilon'$. Clearly, as $\varepsilon \to 0$ the transformations t_ε together with all the derivatives converge to the identity transformation of U uniformly on compact subsets of U.

Thus, for every continuous form f given only in a one-sided neighborhood of S in $X^\pm$, its pull-back $t^\sharp_{\pm\varepsilon}(f)$ is well-defined in a neighborhood of each fixed compact set $K \subset S$ for $0 < \varepsilon < \varepsilon'$ where ε' depends on K.

Definition 3.5.3 *A form $f \in C_{loc}(\Lambda^q T^*_{\mathbb{C}}(X^\pm))$ is said to have weak limit values on S if the limit $\lim_{\varepsilon\to+0} \int_S t^\sharp_{\pm\varepsilon}(f) \wedge g$ exists for every test form $g \in \mathcal{D}(\Lambda^{n-q-1} T^*_{\mathbb{C}}(S))$.*

If the limit $\lim_{\varepsilon\to+0}\int_S t^\sharp_{\pm\varepsilon}(f)\wedge g$ exists for every test form $g\in\mathcal{D}(\Lambda^{n-q-1}T^*_{\mathbb{C}}(S))$, it defines some current in $\mathcal{D}'(\Lambda^q T^*_{\mathbb{C}}(S)) = (\mathcal{D}(\Lambda^{n-q-1}T^*_{\mathbb{C}}(S))'$ because this space is weakly complete (see de Rham [50]). We denote this current by $f|_S$.

Let us take note of two simple properties of weak limit values.

Lemma 3.5.4 *Suppose that $f\in C_{loc}(\Lambda^q T^*_{\mathbb{C}}(X^\pm))$. Then f and $t(f)$ simultaneously admit or do not admit weak limit values on S and $f|_S = t(f)|_S$.*

Proof. Indeed, we have according to Proposition 1.2.38

$$\begin{aligned} t^\sharp_{\pm\varepsilon}(f) &= t^\sharp_{\pm\varepsilon}(t(f)) + t^\sharp_{\pm\varepsilon}(d\rho)\wedge t^\sharp_{\pm\varepsilon}(n(f)) \\ &= t^\sharp_{\pm\varepsilon}(t(f)) + d(t^\sharp_{\pm\varepsilon}\rho)\wedge t^\sharp_{\pm\varepsilon}(n(f)). \end{aligned}$$

To complete the proof it suffices to note that $t^\sharp_{\pm\varepsilon}\rho = \rho(y\pm\varepsilon\nu(y)) = \pm\varepsilon$ for $y\in S$ so $t^\sharp_{\pm\varepsilon}(f) = t^\sharp_{\pm\varepsilon}(t(f))$ on S.

□

Lemma 3.5.5 *If a form $f\in C^1_{loc}(\Lambda^q T^*_{\mathbb{C}}(X^\pm))$ admits weak limit values on S, then df does also, and $(df)|_S = d(f|_S)$.*

Proof. Indeed, let $g\in\mathcal{D}(\Lambda^{n-q-2}T^*_{\mathbb{C}}(S))$ be an arbitrary test from. Using Proposition 1.2.38 and Stokes' Theorem, we get

$$\begin{aligned} \lim_{\varepsilon\to+0}\int_S t^\sharp_{\pm\varepsilon}(df)\wedge g &= \lim_{\varepsilon\to+0}\int_S dt^\sharp_{\pm\varepsilon}(f)\wedge g \\ &= (-1)^{q+1}\lim_{\varepsilon\to+0}\int_S t^\sharp_{\pm\varepsilon}(f)\wedge dg \\ &= (-1)^{q+1}\langle dg, f|_S\rangle \\ &= \langle g, d(f|_S)\rangle, \end{aligned}$$

which completes the proof.

□

Let $\Delta = dd^* + d^*d$ be the Laplace-Beltrami operator on differential forms in X (see Subsection 2.5.2). We denote by $\mathcal{E}_{\Delta_q}(X^\pm)$ the space of harmonic q-forms on $X^\pm$, i.e., the space of all $f\in\mathcal{E}(\Lambda^q T^*_{\mathbb{C}}(X^\pm))$ satisfying $\Delta_q f = 0$.

Definition 3.5.6 *We shall say that a form $f\in\mathcal{E}_{\Delta_q}(X^\pm)$ has finite order of growth near the hypersurface S if for any point $x^0\in S$ there exist a ball $B(x^0,R)$ and constants $c>0$ and $\gamma>0$ such that $|f(x)|\le c\,dist\,(x,S)^{-\gamma}$ for all $x\in B(x^0,R)\cap X^\pm$.*

Of course, the constants R, c and γ can be chosen so that they are suitable for all points x^0 on a fixed compact subset of S.

Theorem 3.5.7 *If a form $f\in\mathcal{E}_{\Delta_q}(X^\pm)$ has finite order of growth near S, then the expressions $t(f)$, $n(f)$, $t(d^*f)$ and $n(df)$ admit weak limit values on S.*

Proof. Since $\Delta_q f = 0$ in $X^{\pm}$, it follows from Proposition 2.5.2 that all the coefficients of f are harmonic functions in $X^{\pm}$. Moreover they have finite orders of growth near S. We now can apply a result of Straube (see Theorem 11.1 in Kytmanov's book [145]) or a more general result of Roitberg [190] to conclude that all the coefficients of f and their derivatives have weak limit values on S. Since the coefficients of the forms $t(f)$, $n(f)$, $t(d^*f)$ and $n(df)$ are linear combinations of the coefficients of f or their first order derivatives, we conclude that these forms as well admit weak limit values on S. This finishes the proof.

□

If S is compact, the condition of Theorem 3.5.7 is also necessary.

Corollary 3.5.8 *Suppose that S is compact. For a form $f \in \mathcal{E}_{\Delta_q}(X^{\pm})$, the expressions $t(f)$, $n(f)$, $t(d^*f)$ and $n(df)$ admit weak limit values on S if and only if f has finite order of growth near S.*

Proof. We need to prove only the necessity. For a form $f \in \mathcal{E}_{\Delta_q}(X^{\pm})$, let the expressions $t(f)$, $n(f)$, $t(d^*f)$ and $n(df)$ have weak limit values on S. Take an open set $Y \subset\subset X$ with a smooth boundary such that $Y \subset X^{\pm}$ and $\partial Y \cap S = S$. Then the restriction of f to Y satisfies $\Delta_q f = 0$, the expression $t(f)$, $n(f)$, $t(d^*f)$ and $n(df)$ have weak limit values on the part of ∂Y coinciding with S, and f is infinitely differentiable up to the rest of the boundary of Y. We write formula (2.5.11) for each of the sets $\{x \in Y : \text{dist}\,(x, S) > \varepsilon\}$, and pass to the limit as $\varepsilon \to +0$. Then we get that f is represented in Y, through its weak Cauchy data on ∂Y with respect to Δ_q, by a boundary integral whose kernel is constructed from $g(x-y)$ and its first order derivatives. Here $g(x)$ is the standard fundamental solution of the Laplace equation in $\mathbb{R}^n$. Since the weak limit values of $t(f)$, $n(f)$, $t(d^*f)$ and $n(df)$ on S necessarily have finite orders of singularity, the boundary integral can be easily estimated near S. Hence it follows that f has a finite order of growth near S.

□

For solutions of elliptic equations, the question of weak limit values on the boundary of a domain was investigated by Roitberg [190].

Corollary 3.5.9 *Suppose that a form $f \in \mathcal{E}_{\Delta_q}(X \setminus S)$ has a finite order of growth near S. If the weak limit values of $t(f)$, $n(f)$, $t(d^*f)$ and $n(df)$ from outside and from inside S coincide on S, then f is the restriction to $X \setminus S$ of a harmonic form on X.*

Proof. For $\varepsilon > 0$, we denote by U_ε the set of all points $x \in X$ such that $\text{dist}\,(x, S) < \varepsilon$.

It follows from a result of Straube (see Corollary 11.3 in Kytmanov's book [145]) that, if a form $f \in \mathcal{E}_{\Delta_q}(X^{\pm})$ has finite order of growth near S, then the limit $\lim_{\varepsilon \to +0} \int_{X^{\pm} \setminus U_\varepsilon} f \wedge g$ exists for every test form $g \in \mathcal{D}(\Lambda^{n-q} T^*_{\mathbb{C}}(X))$. Therefore this limit defines some current of $\mathcal{D}'(\Lambda^q T^*_{\mathbb{C}}(X))$.

Consider now the current $f' \in \mathcal{D}'(\Lambda^q T^*_{\mathbb{C}}(X))$ given by

$$\langle g, f' \rangle = \lim_{\varepsilon \to +0} \int_{X \setminus U_\varepsilon} f \wedge g \quad \text{for} \quad g \in \mathcal{D}(\Lambda^{n-q} T^*_{\mathbb{C}}(X)).$$

Clearly, f' coincides with f in $X \setminus S$.

Let us find the current $\Delta_q f'$ in $\mathcal{D}'(\Lambda^q T^*_{\mathbb{C}}(X))$. If $g \in \mathcal{D}(\Lambda^{n-q} T^*_{\mathbb{C}}(X))$, then we have by Stokes' Theorem

$$\langle g, \Delta_q f' \rangle = \langle \Delta'_q g, f' \rangle = \lim_{\varepsilon \to +0} \int_{X \setminus U_\varepsilon} f \wedge (dd^* + d^* d)' g$$

$$= \lim_{\varepsilon \to +0} \left\{ \int_{X \setminus U_\varepsilon} (\Delta_q f) \wedge g \right.$$

$$\left. + \int_{\partial U_\varepsilon} \left((-1)^{q+1} f \wedge (d^* g) + \star \overline{f} \wedge \star (d\overline{g}) + (d^* f) \wedge g + (-1)^{q+1} \star (d\overline{f}) \wedge \star \overline{g} \right) \right\}.$$

Since $\Delta_q f = 0$ in $X \setminus S$, we get from the last equality that

$$\begin{aligned} &\langle g, \Delta_q f' \rangle \\ &\quad = \lim_{\varepsilon \to +0} \int_S \left\{ (-1)^{q+1} \left(t^\sharp_\varepsilon(t(f)) \wedge t^\sharp_\varepsilon(t(d^\sharp g)) - t^\sharp_{-\varepsilon}(t(f)) \wedge t^\sharp_{-\varepsilon}(t(d^\sharp g)) \right) \right. \\ &\qquad + \left(t^\sharp_\varepsilon(n(f)) \wedge t^\sharp_\varepsilon(n(dg)) - t^\sharp_{-\varepsilon}(n(f)) \wedge t^\sharp_{-\varepsilon}(n(dg)) \right) \\ &\qquad + \left(t^\sharp_\varepsilon(t(d^* f)) \wedge t^\sharp_\varepsilon(t(g)) - t^\sharp_{-\varepsilon}(t(d^* f)) \wedge t^\sharp_{-\varepsilon}(t(g)) \right) \\ &\qquad \left. + (-1)^{q+1} \left(t^\sharp_\varepsilon(n(df)) \wedge t^\sharp_\varepsilon(n(g)) - t^\sharp_{-\varepsilon}(n(df)) \wedge t^\sharp_{-\varepsilon}(n(g)) \right) \right\}. \end{aligned} \tag{3.5.2}$$

For each of the forms $t(d^* g)$, $n(dg)$, $t(g)$ and $n(g)$, the difference $(t^\sharp_\varepsilon(\cdot) - t^\sharp_{-\varepsilon}(\cdot))$ converges to zero in the topology of $\mathcal{D}(\Lambda^{n-q-\cdot} T^*_{\mathbb{C}}(S))$ as $\varepsilon \to +0$, since g is of class $\mathcal{D}(\Lambda^{n-q} T^*_{\mathbb{C}}(X))$. Hence we can use Theorem 3.5.7, the condition of the theorem, and the Banach-Steinhaus Theorem (cf. Shilov [200, p.65]) to conclude that the limit in the right-hand side of (3.5.2) equals zero.

Thus $\Delta_q f' = 0$ weakly in X. Since the Laplace-Beltrami operator Δ_q is elliptic, the current f' is given by some harmonic form which coincides with f in $X \setminus S$.

□

Under the additional assumption that $f \in C^1_{loc}(\Lambda^q T^*_{\mathbb{C}}(X))$, Corollary 3.5.9 follows from (Morera's) Theorem 3.4.18 immediately.

We mention that Corollary 3.5.9 sharpens a local result of Harvey and Polking [99].

3.5.3 *Harmonic representations of currents with compact support*

Let us denote by $\mathcal{E}^{(b)}_{\Delta_q}(X \setminus S)$ the space of all harmonic q-forms f in $X \setminus S$ of finite orders of growth near S such that the weak limit values of expressions $n(f)$, $t(d^* f)$ and $n(df)$ from outside and from inside S coincide on S.

Definition 3.5.10 *A form $f \in C_{loc}(\Lambda^q T^*_{\mathbb{C}}(X \setminus S))$ is said to represent a current $f_0 \in \mathcal{D}'(\Lambda^q T^*_{\mathbb{C}}(S))$ if for every $g \in \mathcal{D}(\Lambda^{n-q-1} T^*_{\mathbb{C}}(S))$*

$$\lim_{\varepsilon \to +0} \int_S (t^\sharp_\varepsilon f - t^\sharp_{-\varepsilon} f) \wedge g = \langle g, f_0 \rangle. \tag{3.5.3}$$

Of course, for currents $f_0 \in \mathcal{D}'(\Lambda^q T^*_{\mathbb{C}}(S))$ one can seek representations from many classes. However the representations from $\mathcal{E}^{(b)}_{\Delta_q}(X \setminus S)$ seem to be more natural.

Lemma 3.5.11 *Suppose that $f \in \mathcal{E}^{(b)}_{\Delta_q}(X \setminus S)$ is a representation of a current $f_0 \in \mathcal{D}'(\Lambda^q T^*_{\mathbb{C}}(S))$. Then f is harmonic outside of the support of f_0.*

Proof. Consider the open set $X' = X \setminus \operatorname{supp} f_0$ and the hypersurface $S' = S \setminus \operatorname{supp} f_0$ in X'. We have $X' \setminus S' = X \setminus S$. By hypothesis, the form $f \in \mathcal{E}_{\Delta_q}(X' \setminus S')$ has a finite order of growth near S' and the weak limit values of $n(f)$, $t(d^* f)$ and $n(df)$ from outside and from the inside of S' coincide on S'. Moreover, it follows from (3.5.3) that the weak limit values of $t(f)$ from outside and from inside S' coincide on S'. Now we can apply Corollary 3.5.9 to conclude that f is the restriction to $X' \setminus S'$ of a harmonic form on X'. This completes the proof.

□

Thus the currents $f_0 \in \mathcal{D}'(\Lambda^q T^*_{\mathbb{C}}(S))$ measure the singularities on S of harmonic forms from $\mathcal{E}^{(b)}_{\Delta_q}(X \setminus S)$.

Lemma 3.5.12 *If $f \in \mathcal{E}^{(b)}_{\Delta_q}(X \setminus S)$ is a representation of a current $f_0 \in \mathcal{D}'(\Lambda^q T^*_{\mathbb{C}}(S))$, then df is of the class $\mathcal{E}^{(b)}_{\Delta_{q+1}}(X \setminus S)$ and represents the current df_0.*

Proof. Since $\Delta_{q+1} d = d\Delta_q$, the form df is harmonic in $X \setminus S$. Clearly, it has a finite order of growth near S. By the definition of $\mathcal{E}^{(b)}_{\Delta_q}(X \setminus S)$, the weak limit values of $n(df)$ from outside and from inside S coincide on S. We need to prove that the same is true concerning $t(d^* df)$. But $d^* df = -dd^* f$ since $\Delta_q f = 0$ in $X \setminus S$. By the hypothesis, the weak limit values of $t(d^* f)$ from outside and from inside S coincide on S. Using now Lemma 3.5.4, Proposition 1.2.38 and Stokes' Theorem, we get for every test form $g \in \mathcal{D}(\Lambda^{n-q-1} T^*_{\mathbb{C}}(S))$

$$\begin{aligned}
&\lim_{\varepsilon \to +0} \int_S \left(t^\sharp_\varepsilon(t(d^* df)) - t^\sharp_{-\varepsilon}(t(d^* df)) \right) \wedge g \\
&= - \lim_{\varepsilon \to +0} \int_S \left(t^\sharp_\varepsilon(dd^* f) - t^\sharp_{-\varepsilon}(dd^* f) \right) \wedge g \\
&= (-1)^{q-1} \lim_{\varepsilon \to +0} \int_S \left(t^\sharp_\varepsilon(t(d^* f)) - t^\sharp_{-\varepsilon}(t(d^* f)) \right) \wedge dg \\
&= 0,
\end{aligned}$$

i.e., the weak limit values of $t(d^* df)$ from outside and from inside S coincide on S. Since $d(df) = 0$ in $X \setminus S$ we may conclude that the form df is in $\mathcal{E}^{(b)}_{\Delta_{q+1}}(X \setminus S)$. Finally, the last part of the lemma follows from Theorem 3.5.7 and Lemma 3.5.5.

□

Let $\Phi(x) = d^*(-g(x)dx)$ be the fundamental solution of convolution type for the de Rham complex on X constructed in Subsection 2.5.2. The double layer potential with density $f_0 \in \mathcal{E}'(\Lambda^q T^*_{\mathbb{C}}(S))$, associated with the complex $\Lambda^\cdot T^*_{\mathbb{C}}(X)$, is by Definition 3.2.5 the current of $\mathcal{D}'(\Lambda^q T^*_{\mathbb{C}}(X))$ given by

$$\begin{aligned}
\Phi([S] \wedge f_0) &= \Phi \star ([S] \wedge f_0) \\
&= \langle \Phi_{q+1}(x, y), f_0(y) \rangle_S.
\end{aligned}$$

The following result appeared in the author's paper [224].

Theorem 3.5.13 *For $f_0 \in \mathcal{E}'(\Lambda^q T^*_{\mathbb{C}}(S))$, the potential $\Phi([S] \wedge f_0)$ is a representation of f_0 of class $\mathcal{E}^{(b)}_{\Delta_q}(X \setminus S)$, and $d\Phi([S] \wedge f_0) = \Phi([S] \wedge df_0)$.*

Proof. Clearly, the form $\Phi([S] \wedge f_0)$ is harmonic in $X \setminus (\text{supp } f_0)$ and has finite order of growth near S.

To complete the proof of the first part of the theorem, it is sufficient to show that for all test forms $g_1 \in \mathcal{D}(\Lambda^{n-q-1} T^*_{\mathbb{C}}(S))$, $g_2 \in \mathcal{D}(\Lambda^{n-q} T^*_{\mathbb{C}}(S))$, $g_3 \in \mathcal{D}(\Lambda^{n-q} T^*_{\mathbb{C}}(S))$ and $g_4 \in \mathcal{D}(\Lambda^{n-q-1} T^*_{\mathbb{C}}(S))$ the following relationship holds:

$$\begin{aligned}
\lim_{\varepsilon \to +0} \int_S \Big\{ & (-1)^{q+1} \left(t^\sharp_\varepsilon(t(\Phi([S] \wedge f_0))) - t^\sharp_{-\varepsilon}(t(\Phi([S] \wedge f_0))) \right) \wedge g_1 \\
& + \left(t^\sharp_\varepsilon(n(\Phi([S] \wedge f_0))) - t^\sharp_{-\varepsilon}(n(\Phi([S] \wedge f_0))) \right) \wedge g_2 \\
& + \left(t^\sharp_\varepsilon(t(d^*\Phi([S] \wedge f_0))) - t^\sharp_{-\varepsilon}(t(d^*\Phi([S] \wedge f_0))) \right) \wedge g_3 \\
& + (-1)^{q+1} \left(t^\sharp_\varepsilon(n(d\Phi([S] \wedge f_0))) - t^\sharp_{-\varepsilon}(n(d\Phi([S] \wedge f_0))) \right) \wedge g_4 \Big\} \\
= {} & (-1)^{q+1} \langle g_1, f_0 \rangle.
\end{aligned} \tag{3.5.4}$$

According to Lemma 3.2.9, there exists a form $g \in \mathcal{D}(\Lambda^{n-q} T^*_{\mathbb{C}}(X))$ such that

$$\left\{ \begin{array}{c} D^\alpha(\Delta_{n-q} g)|_S = 0 \quad \text{for all} \quad \alpha, \\ t(g)|_S = g_3, n(g)|_S = g_4, t(d^*g)|_S = g_1, n(dg)|_S = g_2. \end{array} \right.$$

Since as $\varepsilon \to +0$ we have

$$\begin{array}{lll}
t^\sharp_{\pm\varepsilon}(t(g)) \to g_3 & \text{in the topology of} & \mathcal{D}(\Lambda^{n-q} T^*_{\mathbb{C}}(S)), \\
t^\sharp_{\pm\varepsilon}(n(g)) \to g_4 & \text{in the topology of} & \mathcal{D}(\Lambda^{n-q-1} T^*_{\mathbb{C}}(S)), \\
t^\sharp_{\pm\varepsilon}(t(d^*g)) \to g_1 & \text{in the topology of} & \mathcal{D}(\Lambda^{n-q-1} T^*_{\mathbb{C}}(S)), \\
t^\sharp_{\pm\varepsilon}(n(dg)) \to g_2 & \text{in the topology of} & \mathcal{D}(\Lambda^{n-q} T^*_{\mathbb{C}}(S)),
\end{array}$$

the limit on the left-hand side of (3.5.4) equals

$$\begin{aligned}
& \lim_{\varepsilon \to +0} \int_{\partial U_\varepsilon} \Big\{ (-1)^{q+1} t(\Phi([S] \wedge f_0)) \wedge t(d^*g) + n(\Phi([S] \wedge f_0)) \wedge n(dg) \\
& \qquad + t(d^*\Phi([S] \wedge f_0)) \wedge t(g) + (-1)^{q+1} n(d\Phi([S] \wedge f_0)) \wedge n(g) \Big\} \\
& = \lim_{\varepsilon \to +0} \int_{\partial U_\varepsilon} \Big\{ (-1)^{q+1} \Phi([S] \wedge f_0) \wedge d^*g + \star\Phi([S] \wedge \overline{f_0}) \wedge \star(d\overline{g}) \\
& \qquad + d^*\Phi([S] \wedge f_0) \wedge g + (-1)^{q+1} \star d\Phi([S] \wedge \overline{f_0}) \wedge \star\overline{g} \Big\} \\
& = (-1)^{q+1} \lim_{\varepsilon \to +0} \int_S f_0(y) \wedge d^* \left\{ - \int_{\partial U_\varepsilon} G_{\Delta_{n-q}}(G_{n-q}(y, \cdot), g(\cdot)) \right\}.
\end{aligned}$$

Notice that the double forms $\{G_q\}$ are fundamental solutions of the Laplace-Beltrami operators $\{\Delta_q\}$. They are given by $G_q(x, y) = -\sum'_{\#I=q} g(x-y) \star dy_I \cdot dx_I$ where $g(x)$ is the standard fundamental solution of the Laplace operator in $\mathbb{R}^n$. Simple calculations show that $G_q(x, y) = (-1)^{(n-1)q} G_{n-q}(y, x)$. We have used also the formulas

$$\Phi_q(x, y) = (-1)^q d^*_y G_{q-1}(x, y) = d^*_x G_q(x, y).$$

We may apply Corollary 2.5.14 to see that

$$
\begin{aligned}
(-1)^{q+1} & \lim_{\varepsilon\to+0} \int_S f_0(y) \wedge d^* \left\{ \int_{\partial U_\varepsilon} G_{\Delta_{n-q}}(G_{n-q}(y,\cdot), g(\cdot)) \right\} \\
&= (-1)^{q+1} \left\{ \langle d^* g, f_0 \rangle_S - \lim_{\varepsilon\to+0} \int_S f_0(y) \wedge d^* \left(\int_{U_\varepsilon} (\Delta_{n-q} g(\cdot)) \wedge G_{n-q}(y,\cdot) \right) \right\} \\
&= (-1)^{q+1} \left\{ \langle g_1, f_0 \rangle_S - \lim_{\varepsilon\to+0} \langle \Phi(\chi_{U_\varepsilon}(\Delta_{n-q} g)), f_0 \rangle_S \right\}.
\end{aligned}
\tag{3.5.5}
$$

Since the form $(\Delta_{n-q} g)$ is flat on S, the sequence $\{\Phi(\chi_{U_\varepsilon}(\Delta_{n-q} g))\}$ converges to zero in the topology of $\mathcal{E}(\Lambda^{n-q-1} T^*_{\mathbb{C}}(S))$ as $\varepsilon \to +0$. Therefore the limit on the right-hand side of (3.5.5) equals zero. This finishes the proof of relationship (3.5.4).

The last assertion of Theorem 3.5.13 is quite easy. Indeed, using equality (2.5.4), we get for x off S

$$
\begin{aligned}
d\Phi([S] \wedge f_0) &= \langle d_x \Phi_{q+1}(x,y), f_0(y) \rangle_S \\
&= (-1)^{q+1} \langle d_y \Phi_{q+2}(x,y), f_0(y) \rangle_S \\
&= \langle \Phi_{q+2}(x,y), df_0(y) \rangle_S \\
&= \Phi([S] \wedge df_0).
\end{aligned}
$$

which completes the proof.

□

Thus the classical double layer potential transformation solves the problem of harmonic representation of currents with compact support on the hypersurface S.

3.5.4 *Harmonic representation of arbitrary currents*

The following statement was proved by the author [224]. However its local variant for distributions is due to Kytmanov [144].

Theorem 3.5.14 *For every current $f_0 \in \mathcal{D}'(\Lambda^q T^*_{\mathbb{C}}(S))$, there exists a harmonic representation of $\mathcal{E}^{(b)}_{\Delta_q}(X \setminus S)$.*

Proof. To pass from the harmonic representations of currents with compact supports to the harmonic representations of arbitrary currents on S, a well-known idea is used that is based on a partition of unity on S and the summation method of Mittag-Leffler (see Bremermann [31, p.76-78]).

Let $f_0 \in \mathcal{D}'(\Lambda^q T^*_{\mathbb{C}}(S))$ be an arbitrary current on S.

By Proposition 1.1.1, there exists a sequence $\{U_\nu\}$ of relatively compact open sets such that $\overline{U_\nu} \subset U_{\nu+1}$ for all ν, and $\cup_{\nu=1}^\infty U_\nu = X$. One can assume that the complement of every set U_ν has no compact components in X. Otherwise we could adjoin these components to U_ν.

Choose a sequence of functions $\{\varphi_\nu\} \subset \mathcal{D}(S)$ such that supp $\varphi_\nu \subset S \cap U_{\nu+1}$ and $\varphi_\nu \equiv 1$ on $S \cap U_\nu$. Then $(\varphi_\nu f_0) \in \mathcal{E}'(\Lambda^q T^*_{\mathbb{C}}(S))$, and $\lim_{\nu\to\infty}(\varphi_\nu f_0) = f_0$ in the topology of $\mathcal{D}'(\Lambda^q T^*_{\mathbb{C}}(S))$. According to Theorem 3.5.13 the form

$$f_\nu = \Phi([S] \wedge (\varphi_\nu f_0))$$

provides a harmonic representation of the current $(\varphi_\nu f_0)$ in $\mathcal{E}^{(b)}_{\Delta_q}(X \setminus S)$. However the sequence $\{f_\nu\}$ may be divergent as $\nu \to \infty$ so it needs to be corrected.

Since $(\varphi_{\nu+1} f_0) - (\varphi_\nu f_0) = 0$ in $S \cap U_\nu$, Lemma 3.5.11 implies that the difference $(f_{\nu+1} - f_\nu)$ is harmonic in $(X \setminus S) \cap U_\nu$. By the Runge Theorem for harmonic functions, or more generally for solutions of elliptic systems (see Malgrange [156]), we can find a harmonic form $h_\nu \in \mathcal{E}_{\Delta_q}(X)$ such that $|(f_{\nu+1} - f_\nu) - h_\nu| < \frac{1}{2^\nu}$ on $U_{\nu-1}$ provided $\nu > 1$.

Consider the series

$$f = f_1 + \sum_{\nu=1}^{\infty} (f_{\nu+1} - f_\nu - h_\nu).$$

It converges uniformly on compact subsets of X. Indeed, the remainder of this series is estimated by

$$\left| \sum_{\nu=N+1}^{\infty} (f_{\nu+1} - f_\nu - h_\nu) \right| < \frac{1}{2^N}$$

in U_N. Hence it follows that f is a harmonic q-form in $X \setminus S$.

On the other hand, the N-th partial sum of this series

$$\Sigma_N = f_1 + \sum_{\nu-1}^{N} (f_{\nu+1} - f_\nu - h_\nu)$$

equals $(f_{N+1} - \sum_{\nu=1}^{N} h_\nu)$. Therefore Σ_N is of the class $\mathcal{E}^{(b)}_{\Delta_q}(X \setminus S)$, and for every test form $g \in \mathcal{D}(\Lambda^{n-q-1} T^*_{\mathbb{C}}(S))$ we have

$$\lim_{\varepsilon \to +0} \int_S (t^\sharp_\varepsilon \Sigma_N - t^\sharp_{-\varepsilon} \Sigma_N) \wedge g = \langle g, \varphi_{N+1} f_0 \rangle_S.$$

Since the remainder $(f - \Sigma_N)$ is a harmonic q-form in $(X \setminus S) \cup U_{N+1}$, we may conclude that f is of the class $\mathcal{E}^{(b)}_{\Delta_q}(X \setminus S)$ and it represents the current f_0. Indeed, these properties need only be verified on each compact set $K \subset S$. But we can take N so large that $K \subset U_{N+1}$, and write $f = \Sigma_N + (f - \Sigma_N)$. Then Σ_N has the required properties on K, and $(f - \Sigma_N)$ does not disturb these properties because it is harmonic in a neighborhood of K.

Thus $f \in \mathcal{E}^{(b)}_{\Delta_q}(X \setminus S)$ is a representation of the current f_0 which is what was sought.

□

According to Theorem 3.5.7, for every form $f \in \mathcal{E}_{\Delta_q}(X \setminus S)$ of finite order of growth near S, we may measure its "weak jump" on S by $\delta(f) = t(f^+)|_S - t(f^-)|_S$ where $f^\pm$ are the restrictions of f to $X^\pm$.

Corollary 3.5.15 *The restriction of the mapping δ to $\mathcal{E}^{(b)}_{\Delta_q}(X \setminus S)$ determines an isomorphism*

$$\frac{\mathcal{E}^{(b)}_{\Delta_q}(X \setminus S)}{\mathcal{E}_{\Delta_q}(X)} \cong \mathcal{D}'(\Lambda^q T^*_{\mathbb{C}}(S)). \tag{3.5.6}$$

Proof. In fact, Theorem 3.5.14 asserts that the mapping

$$\delta : \mathcal{E}^{(b)}_{\Delta_q}(X \setminus S) \to \mathcal{D}'(\Lambda^q T^*_{\mathbb{C}}(S))$$

is surjective.

Besides, Corollary 3.5.9 implies that $\delta(f) = 0$ for a form $f \in \mathcal{E}^{(b)}_{\Delta_q}(X \setminus S)$ if and only if $f \in \mathcal{E}_{\Delta_q}(X)$. Thus the following sequence of vector spaces is exact

$$0 \longrightarrow \mathcal{E}_{\Delta_q}(X) \xrightarrow{i} \mathcal{E}^{(b)}_{\Delta_q}(X \setminus S) \xrightarrow{\delta} \mathcal{D}'(\Lambda^q T^*_{\mathbb{C}}(S)) \longrightarrow 0,$$

i.e., (3.5.6) holds.

□

3.5.5 *On multiplication of distributions*

The problem of multiplication of distributions has arisen in many domains of their applications. To resume the situation here one could say that each applied problem demands its own approach to multiplication of distributions.

Among them, apparently the method of Sato-Hörmander connected with the notion of a wave front of a distribution (see Hörmander [112]) has achieved the most recognition. Various developments of this approach were made by Ambrose [7], Wawak [250] and others.

Another approach is to relinquish the requirement that a product of any two distributions be again a distribution and to replace operations on distributions by operations on their representations.

For $n = 1$, the analytic representations have been used for this purpose (see Tillmann [237], Bremermann [31, p.245-249], Itano [115] and others).

This method of multiplication was taken as a principle of the theory of hyper-distributions of Ivanov [116, 117]. Within this theory one can multiply distributions with coinciding singularities, however the product is a hyperdistribution in general.

For $n > 1$, the analytic representations are badly adapted for these purposes. As Itano [115] has observed, for $n > 1$ such a method of defining the product leads to the following difficulties: 1) the product of distributions whose supports do not intersect may be non-zero, and 2) the product of the zero distribution and a non-zero one is not always zero. These difficulties are connected on the whole with the fact that the singularities of an analytic representation do not lie within the support of the distribution if $n > 1$.

Kytmanov [144] showed that one could avoid these difficulties if one uses the harmonic representations instead of the analytic ones.

For other ways of multiplying distributions, we refer the reader to Vladimirov [246, 247], Colombeau [40], Zaslavskii [258], etc.

3.5.6 *Exterior product of currents*

Currents locally are differential forms whose coefficients are distributions. Therefore each multiplication of distributions may be used to define an exterior product of currents after a suitable localization process.

Laurent-Thiebaut [146] considered the exterior product of currents in the spirit of Sato-Hörmander. The harmonic representations of currents of $\mathcal{D}'(\Lambda^q T^*_{\mathbb{C}}(S))$ described in Subsections 3.5.3, 3.5.4 allow us to outline another approach to multiplication of currents.

For a form f given in $U \setminus S$ we let $\widehat{f}(y,\varepsilon) = t^\sharp_\varepsilon f - t^\sharp_{-\varepsilon} f$ where $y \in S$ and $0 < \varepsilon < \varepsilon_0(y)$. Consider the exterior algebra $AR^\cdot = \oplus AR^q$ generated by the forms $\widehat{t(f)}$ where $f \in \mathcal{E}^{(b)}_{\Delta_q}(X \setminus S)$. Denote by RZ^q the subspace of AR^q consisting of all forms h whose weak limit values on S exist and equal zero, i.e.,

$$\lim_{\varepsilon \to +0} \int_S h(y,\varepsilon) \wedge g(y) = 0 \quad \text{for all} \quad g \in \mathcal{D}(\Lambda^{n-q-1} T^*_{\mathbb{C}}(S)).$$

Definition 3.5.16 *Elements of the quotient space* $\mathcal{D}^{*q}(S) = \frac{AR^q}{RZ^q}$ *are called hypercurrents of degree* q *on* S.

The exterior product $f_1 \circ f_2$ of currents $f_j \in \mathcal{D}'(\Lambda^{q_j} T^*_{\mathbb{C}}(S))$ $(j = 1,2)$ is defined to be the image in $\mathcal{D}^{*q_1+q_2}(S)$, under the quotient mapping, of the form $\widehat{t(F_1)} \wedge \widehat{t(F_2)} \in AR^{q_1+q_2}$ where F_j $(j = 1,2)$ is a harmonic representation of f_j in $\mathcal{E}^{(b)}_{\Delta_{q_j}}(X \setminus S)$.

Theorem 3.5.17 *The exterior product of currents $f_1 \circ f_2$ is defined uniquely.*

Proof. We have to show that the image in $\mathcal{D}^{*q_1+q_2}(S)$ of the form $\widehat{t(F_1) \wedge t(F_2)} \in AR^{q_1+q_2}$ under the quotient mapping does not depend on the choice of harmonic representations in $\mathcal{E}^{(b)}_{\Delta_{q_j}}(X \setminus S)$ of the currents f_j $(j = 1,2)$.

To prove this, it suffices to show in view of Corollary 3.5.15 that, if $F_1 \in \mathcal{E}^{(b)}_{\Delta_{q_1}}(X \setminus S)$ and $F_2 \in \mathcal{E}_{\Delta_{q_2}}(X)$, then we have for every test form $g \in \mathcal{D}(\Lambda^{n-q_1-q_2-1} T^*_{\mathbb{C}}(S))$

$$\lim_{\varepsilon \to +0} \int_S \widehat{t(F_1)}(y,\varepsilon) \wedge \widehat{t(F_2)}(y,\varepsilon) \wedge g(y) = 0. \tag{3.5.7}$$

Since $F_1 \in \mathcal{E}^{(b)}_{\Delta_{q_1}}(X \setminus S)$, the forms $\{\widehat{t(F_1)}(y,\varepsilon)\}$ converge in the topology of $\mathcal{D}'(\Lambda^{q_1} T^*_{\mathbb{C}}(S))$ to the current $\delta(F_1)$ as $\varepsilon \to +0$.

On the other hand, since $F_2 \in \mathcal{E}_{\Delta_{q_2}}(X)$, the forms $\{\widehat{t(F_2)}(y,\varepsilon) \wedge g(y)\}$ converge in the topology of $\mathcal{D}(\Lambda^{n-q_1-1} T^*_{\mathbb{C}}(S))$ to zero as $\varepsilon \to +0$. To prove this, use local coordinates $x = (x', \rho)$ near points of S so that $y = x'$ and $\varepsilon = \rho$.

Therefore, by a lemma of Shilov [200, p.65] we get

$$\lim_{\varepsilon \to +0} \int_S \widehat{t(F_1)}(y,\varepsilon) \wedge (\widehat{t(F_2)})(y,\varepsilon) \wedge g(y)) = \langle 0, \delta(F_1) \rangle_S = 0,$$

i.e., (3.5.7) holds.

□

To formulate the next property of the exterior product of currents, we define the support of a hypercurrent $[h] \in \mathcal{D}^{*q}(S)$. Namely, a point $x^0 \in S$ is said to lie outside of supp $[h]$ if there exists a ball $B(x^0, R)$ such that for every test form $g \in \mathcal{D}(\Lambda^{n-q-1} T^*_{\mathbb{C}}(S))$ with support in $S \cap B(x^0, R)$ we have $\lim_{\varepsilon \to +0} \int_S h(y,\varepsilon) \wedge g(y) = 0$, where $h \in [h]$. It is clear that this definition does not depend on a particular choice of a representation h of the hypercurrent $[h]$.

Theorem 3.5.18 *For $f_j \in \mathcal{D}(\Lambda^{q_j} T^*_{\mathbb{C}}(S))$ $(j = 1, 2)$, we have*

$$supp\ (f_1 \circ f2) \subset supp\ f_1 \cap supp\ f_2.$$

Proof. Let x^0 be a point of $S \setminus (\text{supp } f_1 \cap \text{supp } f_2)$. For example, we may assume that $x^0 \in S \setminus (\text{supp } f_2)$. Then these exists a ball $B(x^0, R)$ such that $B(x^0, R) \cap (\text{supp } f_2) = \emptyset$.

Denote by F_j $(j = 1, 2)$ some harmonic representations of the currents f_j in $\mathcal{E}^{(b)}_{\Delta_{q_j}}(X \setminus S)$. We would like to show that for every test form $\mathcal{D}(\Lambda^{n-q_1-q_2-1} T^*_{\mathbb{C}}(S))$ with support in $S \cap B(x^0, R)$ we have

$$\lim_{\varepsilon \to +0} \int_S \widehat{t(F_1)}(y, \varepsilon) \wedge \widehat{t(F_2)}(y, \varepsilon) \wedge g(y) = 0.$$

Since F_1 is a representation of f_1, the forms $\{t(F_1(y, \varepsilon)\}$ converge in the topology of $\mathcal{D}'(\Lambda^{q_1} T^*_{\mathbb{C}}(S))$ to the current f_1 as $\varepsilon \to +0$.

On the other hand, the form F_2 is harmonic in $(X \setminus S) \cup B(x^0, R)$ as follows from Lemma 3.5.11. Therefore the forms $\{\widehat{t(F_2)}(y, \varepsilon) \wedge g(y)\}$ converge to zero in the topology of $\mathcal{D}(\Lambda^{n-q_1-1} T^*_{\mathbb{C}}(S))$ as $\varepsilon \to +0$.

We may now use a lemma of Shilov [200, p.65] to obtain

$$\lim_{\varepsilon \to +0} \int_S \widehat{t(F_1)}(y, \varepsilon) \wedge (\widehat{t(F_2)})(y, \varepsilon) \wedge g(y)) = \langle 0, f_1 \rangle_S = 0.$$

Thus $x^0 \notin \text{supp } (f_1 \circ f_2)$ which finishes the proof.

□

Corollary 3.5.19 *If supp $f_1 \cap$ supp $f_2 = \emptyset$, then $f_1 \circ f_2 = 0$.*

Proof. This follows from Theorems 3.5.17 and 3.5.18.

□

Define the differential d in $\mathcal{D}^*(S)$ in the following way. For a hypercurrent $[h] \in \mathcal{D}^{*q}(S)$, we let $d[h]$ be equal to the image in $\mathcal{D}^{*q+1}(S)$ of the form $d_y h \in AR^{q+1}$ under the quotient mapping where $h \in [h]$.

Lemma 3.5.20 *The differential d is well-defined.*

Proof. We need only to show that d_y maps RZ^q into RZ^{q+1}. Indeed, if $h \in AR^q$, then $d_y h$ is easily verified to lie in AR^{q+1}. If moreover $h \in RZ^q$, then $d_y h$ is in RZ^{q+1} by Lemma 3.5.5.

□

The following theorem shows that the exterior product of currents has the same algebraic properties as the exterior product of forms.

Theorem 3.5.21 *For currents $f \in \mathcal{D}'(\Lambda^{q_j} T^*_{\mathbb{C}}(S))$ $(j = 1, 2)$, we have*

$$\begin{aligned} f_1 \circ f_2 &= (-1)^{q_1 q_2} f_2 \circ f_1, \\ d(f_1 \circ f_2) &= df_1 \circ f_2 + (-1)^{q_1} f_1 \circ df_2. \end{aligned}$$

Proof. The first equality follows immediately from the definition of the exterior product because $\widehat{t(F_1)} \wedge \widehat{t(F_2)} = (-1)^{q_1 q_2} \widehat{t(F_2)} \wedge \widehat{t(F_1)}$ where F_j $(j = 1, 2)$ are certain harmonic representations of the currents f_j in $\mathcal{E}^{(b)}_{\Delta_{q_j}}(X \setminus S)$.

To prove the second equality, we proceed from

$$\begin{aligned} d(f_1 \circ f_2) &= d[\widehat{t(F_1)} \wedge \widehat{t(F_2)}] \\ &= [d_y \widehat{t(F_1)} \wedge \widehat{t(F_2)}] + (-1)^{q_1} [\widehat{t(F_1)} \wedge d_y \widehat{t(F_2)}]. \end{aligned} \tag{3.5.8}$$

Since

$$\begin{aligned} d_y \widehat{t(F_j)} &= d_y(t^\sharp_\varepsilon(t(F_j)) - t^\sharp_{-\varepsilon}(t(F_j))) \\ &= t^\sharp_\varepsilon(d_y t(F_j)) - t^\sharp_{-\varepsilon}(d_y t(F_j)) \\ &= t^\sharp_\varepsilon(t(dF_j)) - t^\sharp_{-\varepsilon}(t(dF_j)) \\ &= \widehat{t(dF_j)} \end{aligned}$$

and dF_j $(j = 1, 2)$ is by Lemma 3.5.12 a harmonic representation of the current df_j, we derive the required result from (3.5.8).

□

The next result says that the exterior product of currents is co-ordinated with (exterior) multiplication by smooth differential forms.

Theorem 3.5.22 *If* $f_1 \in \mathcal{E}(\Lambda^{q_1} T^*_{\mathbb{C}}(S))$ *and* $f_2 \in \mathcal{D}'(\Lambda^{q_2} T^*_{\mathbb{C}}(S))$, *then* $f_1 \circ f_2 = f_1 \wedge f_2$.

Proof. Denote by F_j $(j = 1, 2)$ some harmonic representations in $\mathcal{E}^{(b)}_{\Delta_{q_j}}(X \setminus S)$ of the currents f_j. One must prove that for every test form $g \in \mathcal{D}(\Lambda^{n-q_1-q_2-1} T^*_{\mathbb{C}}(S))$ we have

$$\lim_{\varepsilon \to +0} \int_S \widehat{t(F_1)}(y, \varepsilon) \wedge \widehat{t(F_2)}(y, \varepsilon) \wedge g(y) = (-1)^{q_1 q_2} \langle f_1 \wedge g, f_2 \rangle_S.$$

Choose a function $\varphi \in \mathcal{D}(S)$ such that $\varphi \equiv 1$ in a neighborhood of supp g. Consider the difference $F_1' = F_1 - \Phi([S] \wedge (\varphi f_1))$. By Theorem 3.5.13, this is a harmonic representation in $\mathcal{E}^{(b)}_{\Delta_{q_1}}(X \setminus S)$ of the current $(1 - \varphi) f_1$. So we may use Lemma 3.5.11 to conclude that F_1' is harmonic in a neighborhood of $(X \setminus S) \cup (\text{supp } g)$.

Let us now write

$$\begin{aligned} &\lim_{\varepsilon \to +0} \int_S \widehat{t(F_1)}(y, \varepsilon) \wedge \widehat{t(F_2)}(y, \varepsilon) \wedge g(y) \\ &\quad (-1)^{q_1 q_2} \left\{ \lim_{\varepsilon \to +0} \int_S \widehat{t(F_2)}(y, \varepsilon) \wedge (\widehat{t(F_1')}(y, \varepsilon) \wedge g(y)) \right. \\ &\quad \left. + \lim_{\varepsilon \to +0} \int_S \widehat{t(F_2)}(y, \varepsilon) \wedge (t(\widehat{\Phi([S] \wedge (\varphi f_1)}))(y, \varepsilon) \wedge g(y) \right\}. \end{aligned}$$

Since F_2 is a representation of f_2, the forms $\left\{ \widehat{t(F_2)}(y, \varepsilon) \right\}$ converge in the topology of $\mathcal{D}'(\Lambda^{q_2} T^*_{\mathbb{C}}(S))$ to the current f_2 as $\varepsilon \to +0$.

The forms $\left\{(\widehat{t(F_1')}(y,\varepsilon)\wedge g(y))\right\}$ converge in the topology of $\mathcal{D}(\Lambda^{n-q_2-1}T^*_{\mathbb{C}}(S))$ to zero as $\varepsilon\to+0$ because the form F_1' is harmonic in a neighborhood of $(X\setminus S)\cup(\operatorname{supp} g)$.

Finally, since $f_1\in\mathcal{E}(\Lambda^{q_1}T^*_{\mathbb{C}}(S))$, it follows from Theorem 3.2.6 that the forms

$$\left\{(t(\Phi(\widehat{[S]\wedge(\varphi f_1)}))(y,\varepsilon)\wedge g(y)\right\}$$

converge in the topology of $\mathcal{D}(\Lambda^{n-q_2-1}T^*_{\mathbb{C}}(S))$ to the form $(\varphi f_1)\wedge g=f_1\wedge g$ as $\varepsilon\to+0$.

Thus, using a lemma of Shilov [200, p.65], we obtain

$$\begin{aligned}\lim_{\varepsilon\to+0}\int_S \widehat{t(F_1)}(y,\varepsilon)\wedge\widehat{t(F_2)}(y,\varepsilon)\wedge g(y)\\ &=(-1)^{q_1q_2}\left\{\langle 0,f_2\rangle_S+\langle f_1\wedge g,f_2\rangle_S\right\}\\ &=(-1)^{q_1q_2}\langle f_1\wedge g,f_2\rangle_S\end{aligned}$$

which completes the proof.

□

We introduce, in addition, the notion of singular support of a hypercurrent $[h]\in\mathcal{D}^{*q}(S)$ as a subset of S. Namely, a point $x^0\in S$ is said to lie outside of sing supp $[h]$ if there exist a ball $B(x^0,R)$ and a form $f_0\in\mathcal{E}(\Lambda^qT^*_{\mathbb{C}}(S\cap B(x^0,R)))$ such that for all test forms $g\in\mathcal{D}(\Lambda^{n-q-1}T^*_{\mathbb{C}}(S))$ with supports in $S\cap B(x^0,R)$ we have $\lim_{\varepsilon\to+0}\int_S h(y,\varepsilon)\wedge g(y)=\langle g,f_0\rangle_S$, where $h\in[h]$. Clearly, this definition does not depend on a particular choice of a representation h of the hypercurrent $[h]$.

Exercise 3.5.23 *Prove that, for currents* $f_j\in\mathcal{D}'(\Lambda^{q_j}T^*_{\mathbb{C}}(S))$ $(j=1,2)$, *we have*

$$\text{sing supp }(f_1\circ f_2)\subset(\text{sing supp }f_1)\cup(\text{sing supp }f_2).$$

Hint. Follow the proof of Theorem 12.2 of Kytmanov [145].

□

Exercise 3.5.24 *Define the pull-back of a current under a smooth mapping.*

Hint. Use the harmonic representations of currents in Subsections 3.5.3, 3.5.4. See the author's thesis [225].

□

In conclusion we consider a curious example that has however been standard for applications of the results of Subsection 3.5.6.

3.5.7 *Index of intersection of currents*

The notion of intersection index is introduced in the following way (see de Rham [50, §20]):

Let $f_j\in\mathcal{D}'(\Lambda^{q_j}T^*_{\mathbb{C}}(S))$ $(j=1,2)$ be two currents on S such that $q_1+q_2=n-1$. If, for any regularizations $R^{(\varepsilon')}$ and $R^{(\varepsilon'')}$ on S, the expression $(R^{(\varepsilon')}f_1)\wedge(R^{(\varepsilon'')}f_2)(1)$

has a certain limit as $\varepsilon', \varepsilon'' \to 0$, then this limit is called the index of intersection of the currents f_1 and f_2 and denoted by $f_1 \wedge f_2(1)$. For properties of the index of intersection of currents see de Rham [50, §20].

If F_j $(j = 1, 2)$ are harmonic representations of the currents f_j in $\mathcal{E}^{(b)}_{\Delta_{q_j}}(X \setminus S)$, then the forms $\left\{\widehat{t(F_j)}(y, \varepsilon)\right\}$ may also be considered as regularizations of f_j. Therefore it would be possible to define the generalized index of intersection of f_1 and f_2 as the limit

$$f_1 \circ f_2\,(1) = \lim_{\varepsilon \to +0} \int_S \widehat{t(F_1)}(y, \varepsilon) \wedge \widehat{t(F_2)}(y, \varepsilon)$$

provided it exists. However this definition could depend on the particular choice of the representations F_j of the currents f_j.

When the intersection $\cap_{j=1}^{2}(\mathrm{supp}\, f_j)$ is compact, we may correct the definition by introducing some cut-off function.

Definition 3.5.25 *Let $\cap_{j=1}^{2}(\mathrm{supp}\, f_j)$ be compact, and let a function $\varphi \in \mathcal{D}(S)$ equal 1 in a neighborhood of $\cap_{j=1}^{2}(\mathrm{supp}\, f_j)$. Then the generalized index of intersection of f_1 and f_2 is the limit*

$$f_1 \circ f_2\,(1) = \lim_{\varepsilon \to +0} \int_S \varphi(y)\, \widehat{t(F_1)}(y, \varepsilon) \wedge \widehat{t(F_2)}(y, \varepsilon)$$

provided it exists.

Exercise 3.5.26 *Prove that Definition 3.5.25 does not depend on a particular choice of the representations F_j $(j = 1, 2)$ of f_j and the function φ.*

Hint. Use a partition of unity on $S \setminus \cap_{j=1}^{2}(\mathrm{supp}\, f_j)$ subordinated to the covering $\{S \setminus \mathrm{supp}\, f_j\}_{j=1,2}$ of this set.

□

It is very likely that, if $\cap_{j=1}^{2}(\mathrm{supp}\, f_j)$ is compact and one of the symbols $f_1 \wedge f_2(1)$ or $f_1 \circ f_2(1)$ makes sense, then the other symbol has a sense also and $f_1 \wedge f_2(1) = f_1 \circ f_2(1)$.

If the hypersurface S behaves as a hyperplane at infinity, i.e.,

$$\mathrm{meas}\,(S \cap B(0, R)) = 0(R^{n-1}),$$

then there is a convenient formula for the index.

Exercise 3.5.27 *Suppose that the intersection $\cap_{j=1}^{2}(supp\ f_j)$ is compact and a function $\varphi \in \mathcal{D}(S)$ equals 1 in a neighborhood of this intersection. Prove that*

$$f_1 \circ f_2(1) = \lim_{\varepsilon \to +0} \int_S t(\widehat{\Phi([S] \wedge (\varphi f_1))})(y, \varepsilon) \wedge t(\widehat{\Phi([S] \wedge (\varphi f_2))})(y, \varepsilon). \tag{3.5.9}$$

Hint. Use Exercise 3.5.26. Take into account that the difference $F_j - \Phi([S] \wedge (\varphi f_j))$ is a harmonic representation of the current $(1-\varphi)f_j$ for $j = 1,2$, so it is harmonic in a neighborhood of $(X \setminus S) \cup \cap_{j=1}^2(\text{supp } f_j)$ by Lemma 3.5.11. Finally observe that $\Phi([S] \wedge (\varphi f_j)) = 0(\frac{1}{|x|^{n-1}})$ as $|x| \to \infty$, as is easily seen from Proposition 2.5.4.

□

Let us rewrite formula (3.5.9) for the local situation where S is a hyperplane.

Example 3.5.28 Suppose that the hypersurface S is given by $\{x \in X : x_n = 0\}$ so that $\rho(x) = x_n$.

We use the coordinates $y = (y_1, \dots, y_{n-1})$ (and $z = (z_1, \dots, z_{n-1})$) for points of S so that $x = (y, \varepsilon)$.

It easily follows from Proposition 2.5.4 that, if $f(y) = \sum_{\#I=q} f_I(y) dy_I$ is a current in $\mathcal{E}'(\Lambda^q T^*_{\mathbb{C}}(S))$, then

$$t(\Phi(\widehat{[S] \wedge f}))(y,\varepsilon) = {\sum}'_{\#I=q} \left\langle \frac{2}{\sigma_n} \frac{\varepsilon}{|y-z|^2 + \varepsilon^2)^{n/2}}, f_I(z) \right\rangle_S dy_I,$$

i.e., $t(\Phi(\widehat{[S] \wedge f_1}))$ is the Poisson transformation applied to each coefficient of the current f.

Let now f_j $(j = 1,2)$ be two currents in $\mathcal{E}'(\Lambda^{q_j} T^*_{\mathbb{C}}(S))$ such that $q_1 + q_2 = n-1$. Using Propositions 2.5.4 and 2.5.5, we obtain

$$\begin{aligned} t(\Phi(\widehat{[S] \wedge f_1}))(y,\varepsilon) &= (-1)^{n-1} {\sum}'_{\#I=q_1} \left\langle \frac{2}{\sigma_n} \frac{\varepsilon}{|y-z|^2 + \varepsilon^2)^{n/2}} \star dz_I, f_1(z) \right\rangle_S dy_I, \\ t(\Phi(\widehat{[S] \wedge f_2}))(y,\varepsilon) &= (-1)^{q_1 q_2 + (n-1)} {\sum}'_{\#I=q_1} \left\langle \frac{2}{\sigma_n} \frac{\varepsilon}{|y-z|^2 + \varepsilon^2)^{n/2}} dz_I, f_2(z) \right\rangle_S \star dy_I. \end{aligned}$$

According to Exercise 3.5.27, the generalized index of intersection of the currents f_1 and f_2 is given by

$$\begin{aligned} f_1 \circ f_2(1) = \lim_{\varepsilon \to +0} (-1)^{q_1 q_2} \int_S \Bigg({\sum}'_{\#I=q_1} \\ \left\langle \frac{2}{\sigma_n} \frac{\varepsilon}{|y-z|^2 + \varepsilon^2)^{n/2}} \star dz_I, f_1(z) \right\rangle_S \left\langle \frac{2}{\sigma_n} \frac{\varepsilon}{|y-z|^2 + \varepsilon^2)^{n/2}} dz_I, f_2(z) \right\rangle_S \Bigg) dy. \end{aligned}$$

In particular, this formula is of use when the currents f_j $(j = 1,2)$ are given by chains c_j of dimensions $(n - q_j - 1)$ respectively. It gives an obvious expression for the Kroneker index (see de Rham [50, §33]).

□

We shall return to this question in Section 6.1.

Chapter 4

Boundary Problems for Differential Complexes

4.1 The Neumann-Spencer Problem

4.1.1 *Representation of cohomologies of differential complexes on manifolds with boundary*

Suppose that X is an oriented differentiable manifold of dimension n countable at infinity and E is a (cochain) differential complex on X.

Let Y be a relatively compact open subset of X whose boundary $S = \partial Y$ is a differentiable submanifold of X

We endow S with the orientation induced from the orientation of Y. Then $\overline{Y}$ is a compact oriented differentiable manifold with (the) boundary (S).

For a section $f \in \mathcal{E}(E^i|_{\overline{Y}})$, we shall write $n(f) = 0$ if f has zero Cauchy data on S with respect to the differential operator P^*_{i-1}.

We are interested in the following problem of global solvability on $\overline{Y}$.

Problem 4.1.1 (the P-problem) *Given section $f \in \mathcal{E}(E^i|_{\overline{Y}})$, under what conditions does these exist a section $u \in \mathcal{E}(E^{i-1}|_{\overline{Y}})$ satisfying $Pu = f$, and how does u depend on f?*

A necessary condition for solvability of this problem is almost obvious.

Lemma 4.1.2 *In order that Problem 4.1.1 may be solvable, it is necessary that*

$$f \perp \left\{g \in \mathcal{E}(E^i|_{\overline{Y}}) : \ P^*g = n(g) = 0\right\}. \tag{4.1.1}$$

Proof. Indeed, if $u \in \mathcal{E}(E^{i-1}|_{\overline{Y}})$ and $g \in \mathcal{E}(E^i|_{\overline{Y}})$, then we have by Stokes' formula

$$(Pu, g)_Y = -\overline{\int_S G_{P^*}(\star u, g)} + (u, P^*g)_Y.$$

The required statement follows immediately from this equality.

□

It is easy to see that condition (4.1.1) implies in particular that $Pf = 0$, i.e., $f \in Z^{\imath}(\mathcal{E}(E^{\cdot}|_{\overline{Y}}))$.

Denote by $\mathcal{E}_{n_\imath}(\overline{Y})$ the subspace of $\mathcal{E}(E^{\imath}|_{\overline{Y}})$ consisting of all sections with zero Cauchy data on S with respect to the differential operator $P^*_{\imath-1}$. It follows from Lemma 3.1.16 that the differential P^* maps $\mathcal{E}_{n_\imath}(\overline{Y})$ into $\mathcal{E}_{n_{\imath-1}}(\overline{Y})$ for all i. Thus we get a chain complex $\mathcal{E}_n^{\cdot}(\overline{Y})$ as a subcomplex of $\{\mathcal{E}(E^{\cdot}|_{\overline{Y}}), P^*\}$. Then condition (4.1.1) means that $f \perp Z_\imath(\mathcal{E}_n^{\cdot}(\overline{Y}))$. The homologies $H_\imath(\mathcal{E}_n^{\cdot}(\overline{Y}))$ could be a measure of deviation of condition (4.1.1) from the condition $f \in Z^{\imath}(\mathcal{E}(E^{\cdot}|_{\overline{Y}}))$.

As in Section 3.4, we let

$$\mathcal{H}^{\imath}(E^{\cdot}|_{\overline{Y}}) = \left\{ f \in \mathcal{E}(E^{\imath}|_{\overline{Y}}) : \ Pf = P^*f = n(f) = 0 \right\}.$$

These vector spaces are respectively endowed with the topologies induced from the $\mathcal{E}(E^{\imath}|_{\overline{Y}})$. Sections $f \in \mathcal{H}^{\imath}(E^{\cdot}|_{\overline{Y}})$ are called harmonic sections of $E^{\imath}|_{\overline{Y}}$. By analogy with Proposition 2.2.1 we have

Proposition 4.1.3 *The natural mappings* $i : \mathcal{H}^{\imath}(E^{\cdot}|_{\overline{Y}}) \to H^{\imath}(\mathcal{E}(E^{\cdot}|_{\overline{Y}}))$ *are continuous monomorphisms.*

Proof. Let $f \in \mathcal{H}^{\imath}(E^{\cdot}|_{\overline{Y}})$ and $f = Pu$ for some section $u \in \mathcal{E}(E^{\imath-1}|_{\overline{Y}})$. Then we get by condition (4.1.1) that $\|f\|^2 = (f, f)_Y = 0$, i.e., $f = 0$. And the property of continuity of i is obvious.

□

In connection with the problem of global solvability of overdetermined nonhomogeneous systems of differential equations Spencer [205] suggested an approach that in certain cases allowed one to prove the surjectivity of the natural mapping $\mathcal{H}^{\imath}(E^{\cdot}|_{\overline{Y}}) \to H^{\imath}(\mathcal{E}(E^{\cdot}|_{\overline{Y}}))$. It may be considered as an attempt to extend Hodge's theory for elliptic complexes from compact manifolds (see Section 2.2) to the case of compact manifolds with boundary. We now turn to this approach.

Lemma 4.1.4 *For every elliptic complex* $E^{\cdot}$ *on* X, *there exists an associated complex* $\tilde{E}^{\cdot} = \{E^{\imath}, \tilde{P}_\imath\}$ *such that the differential* $\tilde{P}$ *may be divided by* P *from the left and from the right, and moreover* $\tilde{P}\mathcal{E}(E^{\imath})$ *is dense in* $P\mathcal{E}(E^{\imath})$ *for all* i.

Proof. Of course, if all the differential operators $P_\imath$ in the complex $E^{\cdot}$ have the same order p, one can take the complex $E^{\cdot}$ itself as $\tilde{E}^{\cdot}$.

In general we proceed as in Subsection 3.4.2. Namely, fix some integer p with the property that $p_\imath \le p/2$ for all i. Given any $\imath$, we choose a differential operator $D_\imath \in \mathrm{do}_{p-2p_{\imath-1}}(E^{\imath} \to F^{\imath})$ with injective symbol where $F^{\imath}$ is some vector bundle over X. Finally, we let $\tilde{P}_\imath = P_\imath((D_{\imath+1}P_\imath)^*(D_{\imath+1}P_\imath)+1)$ so that $\tilde{P}_\imath$ is a differential operator of type $E^{\imath} \to E^{\imath+1}$ and order $(2p - p_\imath)$.

According to Lemma 3.4.1, the complex $\tilde{E}^{\cdot} = \{E^{\imath}, \tilde{P}_\imath\}$ is an associated complex for $E^{\cdot}$. Moreover, we have $\tilde{P}_\imath = P_\imath Q_\imath = \tilde{Q}_{\imath+1} P_\imath$ for suitable differential operators $Q_\imath$ and $\tilde{Q}_\imath$ of type $E^{\imath} \to E^{\imath}$ on X.

To complete the proof, it remains to show that the subspace $\tilde{P}\mathcal{E}(E^{\imath})$ is dense in $P\mathcal{E}(E^{\imath})$ for all i.

With this purpose in mind we perform some additional constructions. For every $\imath$, we choose a differential operator $D_\imath \in \mathrm{do}_{p-p_\imath-p_{\imath+1}}(E^\imath \to \tilde{F}^\imath)$ with injective symbol where $\tilde{F}^\imath$ is some vector bundle over X. Given any $\imath$, we consider the differential operator

$$L_\imath = (P_{\imath-1}\widetilde{D}^*_{\imath-1})(P_{\imath-1}\widetilde{D}^*_{\imath-1})^* + (D_{\imath+1}P_\imath)^*(D_{\imath+1}P_\imath) + 1 \quad \text{on} \quad X.$$

By analogy with Lemma 3.4.1 we get that $L_\imath$ is an elliptic differential operator of type $E^\imath \to E^\imath$ and order $(2p - 2p_\imath)$ on X.

We claim that $L_\imath\mathcal{E}(E^\imath)$ is dense in $\mathcal{E}(E^\imath)$ for every $\imath$. To prove this we use two properties of the (formal) adjoint operator $L_\imath{}^*$, namely, that $L_\imath{}^*$ is hypoelliptic and positive on $\mathcal{D}(E^\imath)$.

Indeed, let g be a continuous linear functional on $\mathcal{E}(E^\imath)$ vanishing on $L_\imath\mathcal{E}(E^\imath)$. Then g can be identified with a section of $\mathcal{E}'(E^{\imath\prime})$ satisfying $L_\imath' g = 0$.

Since the differential operator $L_\imath'$ is elliptic, we may conclude that g is smooth, i.e., $g \in \mathcal{D}(E^{\imath\prime})$. We now get

$$\begin{aligned} 0 &= \langle g, \star^{-1}(L_\imath' g)\rangle_X \\ &= (L_\imath{}^*(\star^{-1}g), (\star^{-1}g))_X \\ &= (\widetilde{D}P^*(\star^{-1}g), \widetilde{D}P^*(\star^{-1}g))_X + (DP(\star^{-1}g), DP(\star^{-1}g))_X + ((\star^{-1}g), (\star^{-1}g))_X, \end{aligned}$$

so $g = 0$.

By the Hahn-Banach Theorem, the subspace $L_\imath\mathcal{E}(E^\imath)$ is dense in $\mathcal{E}(E^\imath)$ as required.

We can now finish the proof of the lemma. Suppose that f is an arbitrary section of $P\mathcal{E}(E^\imath)$, i.e., $f = Pu$ for some $u \in \mathcal{E}(E^\imath)$. There exists a sequence $\{u_\nu\} \subset \mathcal{E}(E^\imath)$ such that $\{L_\imath u_\nu\}$ converges to u in the topology of $\mathcal{E}(E^\imath)$. Since $P_\imath L_\imath = \tilde{P}_\imath$, the sequence $\{\tilde{P}u_\nu\}$ converges to f in the topology of $\mathcal{E}(E^{\imath+1})$. Therefore $\tilde{P}\mathcal{E}(E^\imath)$ is dense in $P\mathcal{E}(E^\imath)$ for all $\imath$.

□

As one can see from the proof, the differential operators $D_\imath$ may be chosen so that every differential operator $\Delta_\imath$ is selfadjoint on the symbol level, and even formally selfadjoint when E is a complex of differential operators with constant coefficients.

We fix some associated complex $\tilde{E}$ for E with the properties listed in Lemma 4.1.4 and consider the Laplacians $\Delta_\imath = P_{\imath-1}\tilde{P}^*_{\imath-1} + \tilde{P}^*_\imath P_\imath$.

The vector spaces

$$\mathcal{N}^\imath(E\,|_{\overline{Y}}) = \left\{ f \in \mathcal{E}(E^\imath|_{\overline{Y}}) : \; n(f) = n(\tilde{P}f) = 0 \right\}$$

are called the Neumann spaces. They are endowed with the respective topologies induced from the $\mathcal{E}(E^\imath|_{\overline{Y}})$.

Problem 4.1.5 (the $\tilde{P}$-Neumann problem) *Given a section $f \in \mathcal{E}(E^\imath|_{\overline{Y}})$, under what conditions does there exist a section $u \in \mathcal{N}^\imath(E\,|_{\overline{Y}})$ satisfying $\Delta^* u = f$, and how does u depend on f?*

Let us find a necessary condition for the solvability of this problem.

Lemma 4.1.6

$$\mathcal{H}^{\iota}(E\,|_{\overline{Y}}) = \left\{f \in \mathcal{E}(E^{\iota}|_{\overline{Y}}) :\ \Delta f = c_{\widetilde{P}^*}(f) = c_{\widetilde{P}^*}(Pf) = 0\right\}.$$

Proof. Suppose first that $f \in \mathcal{H}^{\iota}(E^{\cdot}|_{\overline{Y}})$. Since $\widetilde{P} = PQ$ for a suitable differential operator Q, we obtain just as in proving the sufficiency in Lemma 3.4.10 that $\Delta f = c_{\widetilde{P}^*}(f) = c_{\widetilde{P}^*}(Pf) = 0$.

On the other hand, let a section $f \in \mathcal{E}(E^{\iota}|_{\overline{Y}})$ satisfy $\Delta f = c_{\widetilde{P}^*}(f) = c_{\widetilde{P}^*}(Pf) = 0$. We can no longer follow the proof of Lemma 3.4.10 because we lack the appropriate tools.

Since $\Delta f = 0$, we have $\widetilde{P}^*Pf = -P\widetilde{P}^*f$. Using now the condition $c_{\widetilde{P}^*}(Pf) = 0$ and the equality $\widetilde{P} \circ P = 0$, we get $(\widetilde{P}^*Pf, \widetilde{P}^*Pf)_Y = 0$, i.e., $\widetilde{P}^*Pf = 0$. Hence it follows that also $P\widetilde{P}^*f = 0$.

Since $\widetilde{P} = \widetilde{Q}P$ for a suitable differential operator $\widetilde{Q}$, we may conclude that $\widetilde{P}\widetilde{P}^*f = 0$. Therefore $(\widetilde{P}^*f, \widetilde{P}^*f)_Y = 0$ because $c_{\widetilde{P}^*}(f) = 0$. We have shown that $\widetilde{P}^*f = 0$.

Let us use the equalities $\widetilde{P}^*f = c_{\widetilde{P}^*}(f) = 0$ to obtain that $(f, \widetilde{P}v)_Y = 0$ for all sections $v \in \mathcal{E}(E^{\iota-1}|_{\overline{Y}})$. Since the subspace $\widetilde{P}\mathcal{E}(E^{\iota-1})$ is dense in $P\mathcal{E}(E^{\iota-1})$ and $\overline{Y}$ is compact, the subspace $\widetilde{P}\mathcal{E}(E^{\iota-1})|_{\overline{Y}})$ is dense in $P\mathcal{E}(E^{\iota-1})|_{\overline{Y}})$ also. Hence it follows that $(f, Pv)_Y = 0$ for all sections $v \in \mathcal{E}(E^{\iota-1})|_{\overline{Y}})$. Since $\mathcal{D}(E^{\iota-1})|_{\overline{Y}}) \subset \mathcal{E}(E^{\iota-1})|_{\overline{Y}})$, we easily get that $P^*f = n(f) = 0$.

At last, since $\widetilde{P}^*(Pf) = c_{\widetilde{P}^*}(Pf) = 0$, we can repeat the same reasoning to obtain that $P^*(Pf) = n(Pf) = 0$. Hence it follows easily that $Pf = 0$.

□

Lemma 4.1.7 *For Problem 4.1.5 to be solvable, it is necessary that*

$$f \perp \mathcal{H}^{\iota}(E\,|_{\overline{Y}}). \tag{4.1.2}$$

Proof. Indeed, the general observation is that, for Problem 4.1.5 to be solvable, it is necessary that f should be orthogonal to the space of solutions of the homogeneous (formal) adjoint boundary problem to Problem 4.1.5. However Theorem 3.3.16 implies that the (formal) adjoint boundary problem to Problem 4.1.5 is

$$\left\{\begin{array}{c} \Delta v = g, \\ c_{\widetilde{P}^*}(v) = c_{\widetilde{P}^*}(Pv) = 0. \end{array}\right. \tag{4.1.3}$$

Therefore, to complete the proof it suffices to invoke Lemma 4.1.6.

□

Proposition 4.1.8 *If Problem 4.1.5 is solvable under condition (4.1.2), then a necessary and sufficient condition of solvability of Problem 4.1.1 is*

$$\left\{\begin{array}{l} Pf = 0 \quad \textit{and} \\ f \perp \mathcal{H}^{\iota}(E\,|_{\overline{Y}}). \end{array}\right. \tag{4.1.4}$$

Proof. We need prove the sufficiency only. Let a section $f \in \mathcal{E}(E^{\cdot}|_{\overline{Y}})$ satisfy condition (4.1.2). By hypothesis, there exists a section $u \in \mathcal{N}^{\cdot}(E|_{\overline{Y}})$ such that $\Delta^{\cdot} u = f$. If we have in addition $Pf = 0$, then the equality $\Delta^{\cdot} u = f$ implies $PP^*\tilde{P}u = 0$ because $P\tilde{P} = 0$. Since $n(\tilde{P}u) = 0$, we derive from here that $P^*\tilde{P}u = 0$. Using now the equalities $\tilde{P} = PQ$ and again $n(\tilde{P}u) = 0$ we obtain $(\tilde{P}u, \tilde{P}u)_Y = (P^*\tilde{P}u, Qu)_Y = 0$, i.e., $\tilde{P}u = 0$. Hence it follows that $f = \tilde{P}P^*u = P(QP^*u)$, i.e., $(PQ^*)^*u$ is a solution of Problem 4.1.1 as desired.

□

Under the conditions of this proposition, Problems 4.1.1 and 4.1.5 are normally solvable. In particular, by the Open Mapping Theorem the operator $P : \mathcal{E}(E^{\cdot-1}|_{\overline{Y}}) \to \mathcal{E}(E^{\cdot}|_{\overline{Y}})$ is a (topological) homomorphism.

Proposition 4.1.9 *If Problem 4.1.5 is solvable under condition (4.1.2) and the space $\mathcal{H}^{\cdot}(E^{\cdot}|_{\overline{Y}})$ is finite-dimensional, then*

$$H^{\cdot}(\mathcal{E}(E^{\cdot}|_{\overline{Y}})) \overset{top.}{\cong} \mathcal{H}^{\cdot}(E^{\cdot}|_{\overline{Y}}).$$

Proof. First we notice that the space $\mathcal{H}^{\cdot}(E^{\cdot}|_{\overline{Y}})$ is finite-dimensional if and only if it is closed in $L^2(E^{\cdot}|_{\overline{Y}})$. Indeed, as far as the sufficiency is concerned, we may apply the Open Mapping Theorem to conclude that the intrinsic topology of $\mathcal{H}^{\cdot}(E|_{\overline{Y}})$ coincides with the topology induced from $L^2(E^{\cdot}|_{\overline{Y}})$, and then use the Arzela-Ascoli Theorem to derive that the identity mapping of $\mathcal{H}^{\cdot}(E|_{\overline{Y}})$ is compact.

Let now f be some section of $\mathcal{E}(E^{\cdot}|_{\overline{Y}})$. Denote by Hf the L^2-orthogonal projection onto the space $\mathcal{H}^{\cdot}(E^{\cdot}|_{\overline{Y}})$. Then the difference $(f - Hf)$ satisfies condition (4.1.2), so there exists a section $u \in \mathcal{N}^{\cdot}(E|_{\overline{Y}})$ such that $\Delta^{\cdot} u = f - Hf$. Therefore we have

$$f = Hf + \tilde{P}P^*u + P^*\tilde{P}u.$$

If $Pf = 0$, then $PP^*\tilde{P}u = 0$. This implies just as in proving Proposition 4.1.8 that $\tilde{P}u = 0$. It follows that $f = Hf + P(QP^*u)$, so the natural mapping $\imath : \mathcal{H}^{\cdot}(E^{\cdot}|_{\overline{Y}}) \to H^{\cdot}(\mathcal{E}(E^{\cdot}|_{\overline{Y}}))$ is surjective.

Now use Proposition 4.1.3 to see that $\imath$ is an isomorphism of vector spaces (without topologies). Since the space $H^{\cdot}(\mathcal{E}(E|_{\overline{Y}}))$ is separated in view of Proposition 4.1.8, $\imath$ is a homeomorphism by the Open Mapping Theorem. This finishes the proof.

□

4.1.2 *L^2-version of the Neumann problem*

In this subsection we would like to sketch the treatment of the $\tilde{P}$-Neumann problem via the L^2-approach. Of course, other approaches using Hilbert spaces are also possible.

Definition 4.1.10 *The functional $D(f) = (\|\tilde{P}f\|^2 + \|P^*f\|^2 + \|f\|^2)^{\frac{1}{2}}$ is called the Dirichlet norm on $\mathcal{E}(E^{\cdot}|_{\overline{Y}})$.*

Let us denote by $\mathcal{D}^\iota$ the completion of the space $\mathcal{E}_{n_\iota}(\overline{Y})$ under the norm $D(f)$. Since $D(f) \geq \|f\|$, we can identify $\mathcal{D}^\iota$ with a subspace of $L^2(E^\iota|_{\overline{Y}})$. At the same time, for $f \in \mathcal{D}^\iota$ the values $\tilde{P}f$ and P^*f are well-defined as elements of $L^2(E^{\iota+1}|_{\overline{Y}})$ and $L^2(E^{\iota-1}|_{\overline{Y}})$ respectively.

Remark 4.1.11 *For $f \in \mathcal{D}^\iota$, the elements $\tilde{P}f$ and P^*f are easily seen to be the differentials of f in the weak sense on Y.*

Smooth elements of $\mathcal{D}^\iota$ inherit the property $n(\cdot) = 0$ from $\mathcal{E}_{n_\iota}(\overline{Y})$.

Lemma 4.1.12

$$\mathcal{D}^\iota \cap \mathcal{E}(E^\iota|_{\overline{Y}}) = \mathcal{E}_{n_\iota}(\overline{Y}).$$

Proof. Indeed, let f be some section of $\mathcal{E}_{n_\iota}(\overline{Y})$. This means that $(P^*f, g)_Y = (f, Pg)_Y$ for all sections $g \in \mathcal{E}(E^{\iota-1}|_{\overline{Y}})$. Since

$$|(P^*f, g)_Y - (f, Pg)_Y| \leq D(f)\ (\|Pg\|^2 + \|g\|^2)^{1/2},$$

this equality remains valid for all elements $f \in \mathcal{D}^\iota$. In particular, if $f \in \mathcal{D}^\iota \cap \mathcal{E}(E^\iota|_{\overline{Y}})$, then $n(f) = 0$ which completes the proof.

□

Consider the following problem.

Problem 4.1.13 *Given $f \in L^2(E^\iota|_{\overline{Y}})$, find a section $u \in \mathcal{D}^\iota$ satisfying*

$$(\tilde{P}u, Pg)_Y + (P^*u, \tilde{P}^*g)_Y = (f, g)_Y \quad \textit{for all} \quad g \in \mathcal{E}(E^\iota|_{\overline{Y}}) \ \textit{such that}\ c_{\tilde{P}^*}(g) = 0. \tag{4.1.5}$$

Lemmas 4.1.14 and 4.1.16 below show the relevance of this problem to the $\tilde{P}$-Neumann problem.

Lemma 4.1.14 *If $u \in \mathcal{D}^\iota$ satisfies equation (4.1.5), then $\Delta^*u = f$ weakly on Y.*

Proof. Indeed, let g be any test section of $\mathcal{D}(E^\iota|_{\overline{Y}})$. Since $g \in \mathcal{E}(E^\iota|_{\overline{Y}})$ and $c_{\tilde{P}^*}(g) = 0$, then in view of Remark 4.1.11 we get

$$(u, \Delta g)_Y = (\tilde{P}u, Pg)_Y + (P^*u, \tilde{P}^*g)_Y = (f, g)_Y$$

so $\Delta^*u = f$ weakly on Y.

□

The next lemma is a purely technical result.

Lemma 4.1.15 *If $f \in \mathcal{E}(E^{\iota+1}|_{\overline{Y}})$ satisfies $(P^*f, g)_Y = (f, Pg)_Y$ for all sections $g \in \mathcal{E}(E^\iota|_{\overline{Y}})$ with $c_{\tilde{P}^*}(g) = 0$, then $n(f) = 0$.*

Proof. Indeed, let g be an arbitrary section of $\mathcal{E}(E^{\cdot}|_{\overline{Y}})$. It follows from Corollary 3.3.12 that g can be written in the form $g = g_1 + g_2$ where $g_j \in \mathcal{E}(E^{\cdot}|_{\overline{Y}})$ $(j = 1, 2)$ satisfy $c_{\tilde{P}^*}(g_1) = 0$ and $c_P(g_2) = 0$. Therefore we have

$$\begin{aligned} (P^* f, g)_Y &= (P^* f, g_1)_Y + (P^* f, g_2)_Y \\ &= (f, P g_1)_Y + (f, P g_2)_Y \\ &= (f, P g)_Y, \end{aligned}$$

i.e., $n(f) = 0$. This finishes the proof.

□

Lemma 4.1.16 *If* $u \in \mathcal{D}^{\cdot} \cap \mathcal{E}(E^{\cdot}|_{\overline{Y}})$ *satisfies equation (4.1.5), then* $u \in \mathcal{N}^{\cdot}(E^{\cdot}|_{\overline{Y}})$.

Proof. First by Lemma 4.1.12 we conclude that $n(u) = 0$.

Further we consider the section $\tilde{P}u \in \mathcal{E}(E^{\cdot+1}|_{\overline{Y}})$. According to Lemma 4.1.14 we have $\Delta^* u = f$. So, for all sections $g \in \mathcal{E}(E^{\cdot}|_{\overline{Y}})$ satisfying $c_{\tilde{P}^*}(g) = 0$, we get from (4.1.5)

$$\begin{aligned} (P^*(\tilde{P}u), g)_Y &= (\Delta^* u, g)_Y - (\tilde{P} P^* u, g)_Y \\ &= (f, g)_Y - (P^* u, \tilde{P}^* g)_Y \\ &= (\tilde{P}u, P g)_Y. \end{aligned}$$

We now invoke Lemma 4.1.15 to conclude that $n(\tilde{P}u) = 0$, i.e., $u \in \mathcal{N}^{\cdot}(E^{\cdot}|_{\overline{Y}})$.

□

Thus, if $f \in \mathcal{E}(E^{\cdot}|_{\overline{Y}})$ and for a solution $u \in \mathcal{D}^{\cdot}$ of (4.1.5) we could prove the regularity up to the boundary, i.e., $u \in \mathcal{E}(E^{\cdot}|_{\overline{Y}})$, then u would be a solution of Problem 4.1.5.

Problem 4.1.13 is said to be (globally) hypoelliptic if, for every section $f \in \mathcal{E}(E^{\cdot}|_{\overline{Y}})$, all the solutions $u \in \mathcal{D}^{\cdot}$ of (4.1.5) are infinitely differentiable up to the boundary in $\overline{Y}$.

Lemma 4.1.17 *If Problem 4.1.13 is (globally) hypoelliptic, then the space of solutions of the homogeneous equation, corresponding to (4.1.5), is finite-dimensional.*

Proof. Let us denote the space in question by L, i.e., L consists of all sections $u \in \mathcal{D}^{\cdot}$ satisfying (4.1.5) with $f = 0$. By hypothesis, L may be identified with a (vector) subspace of $\mathcal{E}(E^{\cdot}|_{\overline{Y}})$.

Consider the following two topologies on L.

The topology t' is the topology on L induced from the $\mathcal{E}(E^{\cdot}|_{\overline{Y}})$ topology. It follows from Proposition 1.1.7 that the topological vector space $\{L, t'\}$ is a Fréchet-Schwartz space.

The topology t'' is the topology on L given by means of the norm $D(u)$. The topological vector space $\{L, t''\}$ is easily seen to be a Banach space because

$$|(\tilde{P}u, Pg)_Y + (P^* u, \tilde{P}^* g)_Y| \le D(u)(\|Pg\|^2 + \|\tilde{P}^* g\|^2)^{1/2}$$

for all $u \in \mathcal{D}^{\cdot}$ and $g \in \mathcal{E}(E^{\cdot}|_{\overline{Y}})$.

Since the topology t' is stronger than the topology t'', it follows from the Open Mapping Theorem that these topologies on L coincide. Therefore, as each normed Fréchet-Schwartz space is finite-dimensional, the space L is too.

□

To investigate Problem 4.1.13 the dual equations have usually been considered along with (4.1.5).

We introduce the dual Dirichlet norm $D'(g) = (\|Pg\|^2 + \|\tilde{P}^* g\|^2 + \|g\|^2)^{\frac{1}{2}}$ on $\mathcal{E}(E^{\cdot}|_{\overline{Y}})$, and denote by $\mathcal{D}^{\cdot\prime}$ the completion of the space of all $g \in \mathcal{E}(E^{\cdot}|_{\overline{Y}})$ satisfying $c_{\tilde{P}^*}(g) = 0$ under the norm $D'(g)$.

Lemma 4.1.18 *If $g \in \mathcal{D}^{\cdot\prime} \cap \mathcal{E}(E^{\cdot}|_{\overline{Y}})$, then $c_{\tilde{P}^*}(g) = 0$.*

Proof. This is quite similar to the proof of Lemma 4.1.12.

□

Clearly, if $u \in \mathcal{D}^{\cdot}$ is a solution of Problem 4.1.13, then (4.1.5) is fulfilled also for all sections $g \in \mathcal{D}^{\cdot\prime}$.

Problem 4.1.19 *Given $g \in L^2(E^{\cdot}|_{\overline{Y}})$, find a section $v \in \mathcal{D}$ satisfying*

$$(\tilde{P} f, Pv)_Y + (P^* f, \tilde{P}^* v)_Y = (f, g)_Y \quad \text{for all} \quad f \in \mathcal{D}^{\cdot}. \tag{4.1.6}$$

The next lemmas reveals the "strong" boundary problem which hides behind ("weak") Problem 4.1.19.

Lemma 4.1.20 *If $v \in \mathcal{D}^{\cdot\prime}$ satisfies equation (4.1.6), then $\Delta v = g$ weakly on Y.*

Proof. Follow the proof of Lemma 4.1.14.

□

Lemma 4.1.21 *If $g \in \mathcal{E}(E^{\cdot+1}|_{\overline{Y}})$ satisfies $(f, \tilde{P}^* g)_Y = (\tilde{P} f, g)_Y$ for all sections $f \in \mathcal{E}(E^{\cdot}|_{\overline{Y}})$ with $n(f) = 0$, then $c_{\tilde{P}^*}(g) = 0$.*

Proof. The proof is similar to the proof of Lemma 4.1.15 with the exception that one must use Corollary 3.3.15 instead of Corollary 3.3.12.

□

Lemma 4.1.22 *If $v \in \mathcal{D}^{\cdot\prime} \cap \mathcal{E}(E^{\cdot}|_{\overline{Y}})$ satisfies equation (4.1.6), then $c_{\tilde{P}^*}(v) = c_{\tilde{P}^*}(Pv) = 0$.*

Proof. Follow the proof of Lemma 4.1.16 using Lemmas 4.1.18, 4.1.20 and 4.1.21.

□

Thus, if $g \in \mathcal{E}(E^{\cdot}|_{\overline{Y}})$ and for a solution $v \in \mathcal{D}^{\cdot\prime}$ of (4.1.6) we could prove the smoothness up to the boundary, i.e., $v \in \mathcal{E}(E^{\cdot}|_{\overline{Y}})$, then v would be a solution of the P-Neumann problem (4.1.3).

Corollary 4.1.23 *If Problem 4.1.19 is (globally) hypoelliptic, then the space of solutions of the homogeneous equation, corresponding to (4.1.6), coincides with* $\mathcal{H}^i(E^\cdot|_{\overline{Y}})$.

Proof. It follows from Lemmas 4.1.20 and 4.1.22 that, if $v \in \mathcal{D}^{i\prime}$ satisfies the homogeneous equation corresponding to (4.1.6), then $v \in \mathcal{E}(E^i|_{\overline{Y}})$ and $\Delta v = c_{\widetilde{P}^*}(v) = c_{\widetilde{P}^*}(Pv) = 0$. To complete the proof it suffices to apply Lemma 4.1.6.

□

Under the hypothesis of Corollary 4.1.23, the space $\mathcal{H}^i(E^\cdot|_{\overline{Y}})$ is finite-dimensional. To prove this it suffices to repeat the proof of Lemma 4.1.17.

Lemma 4.1.24 *For Problem 4.1.13 to be solvable, it is necessary that f should be orthogonal to all solutions of the homogeneous equation corresponding to (4.1.6).*

Proof. This is obvious. We have presented this lemma only to emphasize the dual nature of Problems 4.1.13 and 4.1.19.

□

Let us denote by K^i the space of all sections $u \in \mathcal{D}^i$ satisfying (4.1.5) with $f = 0$, and by R^i the space of all sections $f \in L^2(E^i|_{\overline{Y}})$ such that equation (4.1.5) has some solution in $\mathcal{D}^i$.

Recall that Problem 4.1.13 is said to be *normally solvable* if the subspace R^i is closed in $L^2(E^i|_{\overline{Y}})$, and noetherian if in addition the spaces K^i and $L^2(E^i|_{\overline{Y}})/R^i$ are finite-dimensional.

Definition 4.1.25 *The $\widetilde{P}$-Neumann problem for the complex $E^\cdot|_{\overline{Y}}$ is said to be solvable at degree i if: 1) the space $\mathcal{H}^i(E^\cdot|_{\overline{Y}})$ is finite-dimensional; 2) Problem 4.1.13 is solvable under condition (4.1.2), and 3) Problem 4.1.13 is hypoelliptic.*

Besides the assumption on noetherity of Problem 4.1.13, this involves the assumption on hypoellipticity of this problem.

Proposition 4.1.26 *If the $\widetilde{P}$-Neumann problem for the complex $E^\cdot|_{\overline{Y}}$ is solvable at degree i, then there exist bounded operators H_i and N_i in $L^2(E^i|_{\overline{Y}})$ such that*

1) H_i is the L^2-orthogonal projection onto the subspace $\mathcal{H}^i(E^\cdot|_{\overline{Y}})$;

2) N_i is a bounded operator from $L^2(E^i|_{\overline{Y}})$ to $\mathcal{D}^i$ mapping $\mathcal{E}(E^i|_{\overline{Y}})$ to $\mathcal{N}^i(E^\cdot|_{\overline{Y}})$ and satisfying $(\mathrm{proj}_{K^i}) \circ N_i = N_i \circ H_i = 0$, -

and for every section $f \in L^2(E^i|_{\overline{Y}})$

$$f = Hf + P^*\widetilde{P}Nf + \widetilde{P}P^*Nf \quad \textit{weakly on} \quad Y. \tag{4.1.7}$$

Proof. Let us denote by H_i the L^2-orthogonal projection onto the closed subspace $\mathcal{H}^i(E^\cdot|_{\overline{Y}})$ in $L^2(E^i|_{\overline{Y}})$.

Given a section $f \in L^2(E^i|_{\overline{Y}})$, the difference $(f - Hf)$ is L^2-orthogonal to $\mathcal{H}^i(E^\cdot|_{\overline{Y}})$. Therefore there exists a section $u \in \mathcal{D}^i$ satisfying equation (4.1.5) with $(f - Hf)$ instead of f.

Since Problem 4.1.13 is assumed to be hypoelliptic, the space K^i is finite-dimensional by Lemma 4.1.17. We set $N_i f = u - \mathrm{proj}_{K^i}(u)$. This operator is well-defined because, if u' is another solution of (4.1.5) with $(f - Hf)$ instead of f, then the difference $(u - u')$ is in $K^{\cdot}$ so

$$(u - \mathrm{proj}_{K^{\cdot}}(u)) - (u' - \mathrm{proj}_{K^{\cdot}}(u')) = (u - u') - \mathrm{proj}_{K^{\cdot}}(u - u') = 0.$$

Clearly, $Nf \in \mathcal{D}^{\cdot}$ satisfies equation (4.1.5) with $(f-Hf)$ instead of f. Moreover we have $(\mathrm{proj}_{K^{\cdot}}) \circ N_{\cdot} = N_{\cdot} \circ H_{\cdot} = 0$ by the construction.

Consider now two Hilbert spaces $L^1 = \{R^{\cdot}, \|\cdot\|\}$ and $L^2 = \{\mathcal{D}^{\cdot}, D(\cdot)\}$, and the mapping $M : L^1 \to L^2$ given by $M(f) = u$ where $u \in \mathcal{D}^{\cdot}$ is the solution of (4.1.5) orthogonal to $K^{\cdot}$. The graph of M is easily verified to be closed in $L^1 \times L^2$. By the Closed Graph Theorem M is a continuous mapping, i.e., there exists a constant $c > 0$ such that $D(Mf) \leq c\|f\|$ for all $f \in R^{\cdot}$.

Applying this estimate to the section $(f - Hf)$, we get

$$D(Nf) \leq c\|f - Hf\| \leq c\|f\|,$$

i.e., $N_{\cdot}$ is a bounded operator from $L^2(E^{\cdot}|_{\overline{Y}})$ to $\mathcal{D}^{\cdot}$.

Thus Nf is the unique solution in $\mathcal{D}^{\cdot} \cap (K^{\cdot})^{\perp}$ of the equation

$$(\tilde{P}u, Pg)_Y + (P^*u, \tilde{P}^*g)_Y = (f - Hf, g)_Y \quad \text{for all} \quad g \in \mathcal{D}^{\cdot\prime}.$$

If $f \in \mathcal{E}(E^{\cdot}|_{\overline{Y}})$, then the difference $(f - Hf)$ is in $\mathcal{E}(E^{\cdot}|_{\overline{Y}})$. Hence it follows by hypothesis that Nf is also in $\mathcal{E}(E^{\cdot}|_{\overline{Y}})$. We now use Lemma 4.1.16 to conclude that $N_{\cdot}$ maps $\mathcal{E}(E^{\cdot}|_{\overline{Y}})$ to $\mathcal{N}^{\cdot}(E^{\cdot}|_{\overline{Y}})$.

Finally equality (4.1.7) follows from Lemma 4.1.14 which finishes the proof.

□

We put off studying further properties of the operator N, called *Neumann's operator*, to the next section. Meanwhile, we notice that decomposition (4.1.7) allows one to get a continuous $(\tilde{P}-)$ homotopy between the identity and H endomorphisms of the complex $\mathcal{E}(E^{\cdot}|_{\overline{Y}})$. Such a homotopy is given by the operator P^*N, at least, on the cocycles of $\mathcal{E}(E^{\cdot}|_{\overline{Y}})$ because $Pf = 0$ implies $\tilde{P}Nf = 0$ for $f \in \mathcal{E}(E^{\cdot}|_{\overline{Y}})$. The relationship $\tilde{P}N = N\tilde{P}$ would follow from an affirmative answer to the next question.

Exercise 4.1.27 *Check whether* $\Delta^* f = n(f) = n(\tilde{P}f) = 0$ *is valid for* $f \in \mathcal{E}(E^{\cdot}|_{\overline{Y}})$ *if and only if* $\tilde{P}f = P^*f = n(f) = 0$.

Hint. The proofs of Lemmas 2.2.3 and 4.1.6 may be of use.

□

4.1.3 *Examples*

We shall limit ourselves to three well-known examples, around which the Neumann problem has been mainly studied.

Example 4.1.28 (the d-Neumann problem) The Neumann problem for the de Rham complex on $\overline{Y}$ was considered by Duff and Spencer [59], Conner [41], Friedrichs [68], Morrey and Eells [164], and Morrey [163]. In this form the Neumann problem is quite classical. In particular, the d-Neumann problem at degree 0 is the usual Neumann problem for harmonic functions, and the d-Neumann problem at degree n is the classical Dirichlet problem. Simple calculations show that the d-Neumann problem at an arbitrary degree $\imath$ is reduced locally to the Dirichlet and Neumann problems for coefficients of a differential form f. Thus the d-Neumann problem is a regular elliptic boundary value problem and it may be investigated by standard techniques (see Triebel [241, Ch.5], for example). In particular, this problem is solvable at each degree (in the L^2-sense) and for each compact manifold $\overline{Y} \subset X$ with smooth boundary (see Triebel [241, §5.4.4]).

□

Example 4.1.29 (the $\overline{\partial}$-Neumann problem) The Neumann problem for the Dolbeault complex with coefficients in $\Lambda^{p,0}$ on X is much more difficult because it is non-coercive with the exception of the case $\imath = n$. The study of the $\overline{\partial}$-Neumann problem turned out to be very useful and its solution had led to important discoveries not only in complex analysis but also in the theory of partial differential equations. The first formulation of the $\overline{\partial}$-Neumann problem was given by Spencer in his Paris lectures of 1955 (unpublished). It may be found also in the paper of Kohn and Spencer [134] where an analogous problem was considered for a differential operator which is a linear combination of ∂ and $\overline{\partial}$. Using a priori estimates in Hilbert spaces, Kohn [125, 126] proved that the $\overline{\partial}$-Neumann problem was solvable at every degree $\imath > 0$ for each compact complex manifold $\overline{Y} \subset X$ with strongly pseudo-convex boundary Hörmander [109] further simplified and developed the $\overline{\partial}$-Neumann problem. Describing compact complex manifolds with boundaries, on which the $\overline{\partial}$-Neumann problem was subelliptic, was one of the central problems of complex analysis in the 70-s. We refer to the paper of Kohn [128], his lectures [130] and the survey [129]. We mention also the papers of Lieb and Range [151] and of Range [187], where an integral representation for the principal part of the $\overline{\partial}$-Neumann operator on a strongly pseudo-convex complex manifold was obtained.

□

Example 4.1.30 (the D-Neumann problem) Let P be a formally integrable differential operator with injective symbol on X, and S be the second Spencer sequence for P whose differential is denoted by D (see Definition 1.3.49). Spencer [205, 206] announced that the D-Neumann problem for the complex $S\,|_{\overline{Y}}$ was solvable (in the L^2 - sense) and the harmonic spaces $\mathcal{H}^{\cdot}(S\,|_{\overline{Y}})$ were trivial at positive degrees provided Y was a sufficiently small convex domain in X. But the proof had not been published. A quite unexpected example was found by Sweeney [210] who constructed a first order differential operator P, with injective symbol and constant coefficients in $\mathbb{R}^3$, for which the D-Neumann problem in the unit ball is not solvable at the degree 1 (see Example 1.3.55). Sweeney [211] derived also some algebraic condition for P, under which the D-Neumann problem is solvable at positive degrees. However this condition is far from being necessary because it is not fulfilled even for the Cauchy-Riemann operator on a strongly pseudo-convex manifold. MacKichan [154, 155] and Sweeney [212] found a proof of the above formulated hypothesis of Spencer under the condition that P satisfies the δ-estimate. See Subsection 1.3.13 for more detailed information. Cenkl [37] rearranged the Dirichlet integral in the D-Neumann problem to single out a leading term. It appears as a semilinear form expressed through the coefficients of P and the geometry of the bundles. A condition of "sufficient" positivity of this form ensures that the harmonic spaces $\mathcal{H}^{\cdot}(S\,|_{\overline{Y}})$ vanish at positive degrees, i.e., for $i > 0$. Also the case of compact Y is not excluded. For the Cauchy-Riemann operator this statement becomes the well-known theorem of Kodaira [123]. The subellipticity of the D-Neumann problem was considered by Sweeney [214, 215, 216, 217]. The part of proving regularity up to the boundary in all the mentioned results was based on the fundamental work of Kohn and Nirenberg [131], where a wider class of non-coercive boundary value problems had been investigated. Further advances in the Neumann problem for arbitrary complexes of first order differential operators were exposed in the lecture of Kohn [127].

□

4.1.4 *Reducing the question of solvability to a subelliptic estimate*

Suppose that $E^{\cdot}$ is an elliptic complex of first order differential operators on X. We are allowed to take the complex $E^{\cdot}$ itself as an associated complex for $E^{\cdot}$.

In this case the Dirichlet norm on $\mathcal{E}(E^{\cdot}|_{\overline{Y}})$ is $D(f) = (\|Pf\|^2+\|P^*f\|^2+\|f\|^2)^{1/2}$. As before, we denote by $\mathcal{D}^{\cdot}$ the completion of $\mathcal{E}_{n,}(\overline{Y})$ in the Dirichlet norm.

Now Problem 4.1.5 is formally selfadjoint, i.e., Problems 4.1.13 and 4.1.19 coincide. Kohn and Nirenberg [131] proved the following fundamental result concerning these problems.

Theorem 4.1.31 *If the inclusion $\mathcal{D}^{\cdot} \hookrightarrow L^2(E^{\cdot}|_{\overline{Y}})$ is compact, then the P-Neumann problem for the complex $E^{\cdot}|_{\overline{Y}}$ is solvable at degree i.*

Proof. The proof of this theorem is based on studying commutators of pseudodifferential operators of fractional orders in Sobolev spaces. It deviates from the framework of this book. See Kohn and Nirenberg [131, Theorem III] for details.

□

(Theorem 4.1.31 is carried over to elliptic complexes of higher order by Sweeney [213].)

According to the well-known Rellich Theorem (see, for instance, Taylor [236, Ch.I]) the inclusion $W^{s'',2}(E^{\cdot}|_{\overline{Y}}) \hookrightarrow W^{s',2}(E^{\cdot}|_{\overline{Y}})$ is compact provided $0 \le s' < s''$. It follows from here that the inclusion $\mathcal{D}^{\cdot} \hookrightarrow L^2(E^{\cdot}|_{\overline{Y}})$ is compact if there exists a real $s > 0$ such that

$$\|f\|_{W^{s,2}(E^{\cdot}|_{\overline{Y}})} \le c\, D(f) \quad \text{for all} \quad f \in \mathcal{E}_{n,}(\overline{Y}). \tag{4.1.8}$$

Remark 4.1.32 *A priori estimates of the type (4.1.8) are called subelliptic estimates because one loses* $(1-s)$ *in the order of the estimated derivatives.*

Hörmander [111] characterized very general norms $D(f)$, for which the estimate (4.1.8) is fulfilled with $s = \frac{1}{2}$. Further results were presented in the book of Egorov [63, Ch.VIII]. But it is very difficult to interpret what these results say concerning the initial complex $E\,|_{\overline{Y}}$. This question was discussed by Kohn [127].

Remark 4.1.33 *Kohn and Nirenberg [131, Theorem IV] observed that the estimate (4.1.8) with* $s = \frac{1}{2}$ *follows from the following estimate:*

$$\int_{\partial Y} |f|^2 ds \le c\, D(f) \quad \text{for all} \quad f \in \mathcal{E}_{n,i}(\overline{Y}). \tag{4.1.9}$$

We shall refer to (4.1.9) as the Kohn-Nirenberg estimate at degree i for the complex $E\,|_{\overline{Y}}$.

Corollary 4.1.34 *If the Kohn-Nirenberg estimate for the complex* $E\,|_{\overline{Y}}$ *is satisfied at degree* i*, then the P-Neumann problem for this complex is solvable at degree* i*, and the Neumann operator* N_i *has the following property of pseudo-locality:*

for each open set $U \subset \overline{Y}$*, compact set* $K \subset U$ *and integer* $s \ge 0$*, there exists a constant* $c > 0$*, depending on* U, K *and* s*, such that*

$$\|Nf\|_{W^{s+1,2}(E^i|_K)} \le c\, \left(\|f\|_{W^{s,2}(E^i|_{\overline{U}})} + \|Nf\|_{L^2(E^i|_{\overline{U}})}\right) \quad \text{for all } f \in \mathcal{E}(E^i|_{\overline{Y}}). \tag{4.1.10}$$

Proof. We must justify estimate (4.1.10) only. But it was noted by Kohn and Nirenberg [131].

□

It is unlikely that the condition of Theorem 4.1.31 is right for the general Problem 4.1.13.

Exercise 4.1.35 *Given an arbitrary elliptic complex* E *on* X*, find out whether the condition, that the inclusion* $\mathcal{D}^i \hookrightarrow L^2(E^i|_{\overline{Y}})$ *is compact, is sufficient for the* $\tilde{P}$*-Neumann problem for* $E\,|_{\overline{Y}}$ *to be solvable at degree* i.

Hint. The results of Sweeney [213] could be of use.

□

4.1.5 *A geometrical condition for solvability of the Neumann problem*

Let $d_0^{(E)}$ be a sufficiently regular differential operator with injective symbol on X, and E be a compatibility complex for $d_0^{(E)}$. According to Theorem 1.3.24, $d_0^{(E)}$ may be transformed by a finite number of steps into an equivalent normalized differential operator $d_0^{(F)}$ with injective symbol. According to Theorem 1.3.37, one can construct by a finite number of steps a compatibility complex $F^{\cdot}$ for $d_0^{(F)}$, consisting of normalized differential operators. By Proposition 1.3.38, the complex $F^{\cdot}$ is elliptic. Finally, it follows from Proposition 1.2.8 that $H^{\cdot}(E^{\cdot}|_{\overline{Y}}) \cong H^{\cdot}(F^{\cdot}|_{\overline{Y}})$.

Thus, by representing cohomologies of formally exact elliptic complexes, the Neumann problem for elliptic complexes $E^{\cdot}$ of first order differential operators is of primary interest.

We start with historical remarks concerning the Dolbeault complex which is a building block for the general theory. Making the result of Kohn [125] more precise, Hörmander [109, Theorem 3.2.5] showed that, if $\overline{Y} \subset X$ is a compact complex manifold with boundary and at each point of ∂Y the Levi form has either $(n-i)$ positive eigenvalues or $(i+1)$ negative ones, then estimate (4.1.9) for the complex $\Lambda^{p,\cdot} T^*_{\mathbb{C}}(\overline{Y})$ is fulfilled at degree i.

Later on Hörmander [111, Theorem 1.2.7] extended this assertion to Koszul complexes of first order differential operators (see Subsection 1.2.7).

For general elliptic complexes $E^{\cdot}$ that satisfy some restriction of the type of the δ-estimate for the second Spencer resolvent, there is also a geometrical condition on the boundary ∂Y of $\overline{Y}$ that is sufficient for the Kohn-Nirenberg estimate for $E^{\cdot}|_{\overline{Y}}$ to be fulfilled at degree i. It consists in a certain semilinear form the "Levi form" at degree i for the complex being "sufficiently" positive on ∂Y.

We limit ourselves to the situation when $\overline{Y}$ lies in a coordinate neighborhood U on X over which the bundles E^{i-1}, E^{i} and E^{i+1} are trivial, i.e., $E^{i}|_U \cong U \times \mathbb{C}^{k_i}$ etc.

Under this assumption each of the differential operators P_{i-1} and P_i may be represented in U in the form $P(x,D) = \sum_{j=1}^n P^{(j)}(x) D_j$ where $P^{(j)}(x)$ are suitable matrices of differentiable functions on U.

Exercise 4.1.36 *Can $E^{\cdot}$ be transformed into a homotopically equivalent elliptic complex of first order differential operators satisfying the condition:*

there exists a constant $c > 0$ such that, for each point $x \in \overline{Y}$ and vectors v and v_j $(j = 1, \ldots, n)$ of E^{i}_x,

$$-\sum_{j,k=1}^n \left(P_{i-1}^{(j)*} v_k + \partial_k P_{i-1}^{(j)*} v\right)^* \left(P_{i-1}^{(k)*} v_j + \partial_j P_{i-1}^{(k)*} v\right) \leq c\left(\left|\sum_{j=1}^n P_i^{(j)} v_j\right|^2 + \left|\sum_{j=1}^n P_{i-1}^{(j)*} v_j\right|^2 + |v|^2\right) \ ? \tag{4.1.11}$$

Hint. This is an unsolved problem. Although in general the answer seems to be negative, a reasonable condition for $E^{\cdot}$ may be formal exactness. If so, we would obtain an affirmative answer to the old question of Spencer [208] : whether each formally exact elliptic complex is sheaf exact at positive degrees.

□

Let us consider a few examples.

Example 4.1.37 Suppose that $E^{\cdot} = \Lambda^{\cdot} T^*_{\mathbb{C}}(X)$ is the de Rham complex on X. It is easy to check that $\left(P_{i-1}^{(j)} P_{i-1}^{(k)*} + P_i^{(k)*} P_i^{(j)}\right)$ equals 0, if $j \neq k$, and I_{k_i}, if $j = k$.

Hence it follows that, for all vectors v_i $(j = 1, \ldots, n)$ of E_x^i, we have

$$-\sum_{j,k=1}^{n} \left(P_{i-1}^{(j)*} v_k\right)^* \left(P_{i-1}^{(k)*} v_j\right) = \left|\sum_{j=1}^{n} P_i^{(j)} v_j\right|^2 - \sum_{j=1}^{n} |v_j|^2,$$

so condition (4.1.11) is fulfilled with each constant $c \geq 1$.

□

Example 4.1.38 Suppose that $E^\cdot = \Lambda^{p,\cdot} T_{\mathbb{C}}^*(X)$ is the Dolbeault complex with coefficients in $\Lambda^{p,0} T_{\mathbb{C}}^*(X)$ on a complex manifold X of dimension n. We fix the complex structure in U by $z_j = x_j + \sqrt{-1} x_{n+j}$ $(j = 1, \ldots, n)$. An easy calculation shows that, for all vectors v_j $(j = 1, \ldots, 2n)$ of E_x^i, we have

$$-\sum_{j,k=1}^{n} \left(P_{i-1}^{(j)*} v_k\right)^* \left(P_{i-1}^{(k)*} v_j\right) = \left|\sum_{j=1}^{n} P_i^{(j)} v_j\right|^2 - \sum_{j=1}^{n} \left|\frac{1}{2}(v_j + \sqrt{-1} v_{n+j})\right|^2,$$

so condition (4.1.11) is fulfilled again with each constant $c \geq 1$.

□

Example 4.1.39 Suppose that $E^\cdot = S^\cdot$ is the second Spencer sequence for a formally integrable differential operator P with injective symbol on X. In this case the coefficients of the principal parts of the differential operators P_{i-1} are also constant (see the remark after Proposition 1.3.48). So condition (4.1.11) becomes

$$-\sum_{j,k=1}^{n} \left(P_{i-1}^{(j)*} v_k\right)^* \left(P_{i-1}^{(k)*} v_j\right) \leq c \left(\left|\sum_{j=1}^{n} P_i^{(j)} v_j\right|^2 + \left|\sum_{j=1}^{n} P_i^{(j)*} v_j\right|^2 \right),$$

for all vectors v_j $(j = 1, \ldots, n)$ of E_x^i. I.M. Singer showed that, if P satisfies the δ-estimate, then condition (4.1.11) is fulfilled for $i = 1$. Later on MacKichan [154] proved that the δ-estimate implies estimate (4.1.11) at every degree i. This is the way through which the δ-estimate enters into the D-Neumann problem.

□

Example 4.1.40 Suppose that $E^\cdot$ is the elliptic Hilbert complex in $\mathbb{R}^3$ given by Example 1.3.55 (Sweeney's [210] example). We take $i = 1$, and the vectors v_j $(j = 1, \ldots, 3)$ of $\mathbb{C}^4$ in the following way:

$$v_1 = -\sqrt{-1} v_2, \quad v_2 = \begin{pmatrix} 0 \\ v_2^{(2)} \\ 0 \\ 0 \end{pmatrix} \quad \text{and} \quad v_3 = \begin{pmatrix} 0 \\ 0 \\ 0 \\ v_3^{(4)} \end{pmatrix}.$$

An easy calculation shows that

$$-\sum_{j,k=1}^{3} \left(P_{i-1}^{(j)*} v_k\right)^* \left(P_{i-1}^{(k)*} v_j\right) = 2 \operatorname{Re} \overline{v_2^{(2)}} \cdot v_3^{(4)} - \left|v_3^{(4)}\right|^2,$$

$$\left|\sum_{j=1}^{3} P_i^{(j)} v_j\right| = 0, \quad \text{and} \quad \left|\sum_{j=1}^{3} P_{i-1}^{(j)\,*} v_j\right| = \left|v_3^{(4)}\right|.$$

Clearly the components $v_2^{(2)}$ and $v_3^{(4)}$ may be chosen so that estimate (4.1.11) fails whatever constant $c > 0$ has been fixed. Thus condition (4.1.11) is not fulfilled for the complex $E^{\cdot}$ at degree 1.

□

The "Levi form" at degree i appeared from the analysis of the relationship $n(f) = 0$ for a section $f \in \mathcal{E}(E^i|_{\overline{Y}})$. Let $\rho \in C^\infty_{loc}(X)$ be a defining function for Y, i.e., $Y = \{x \in X : \rho(x) < 0\}$ and $d\rho|_{\partial Y} \neq 0$. For $1 \leq j \leq n$, the restriction of the form $(\star dx_j)$ to ∂Y equals $\frac{\partial_j \rho}{|d\rho|} ds$ where ds is the volume form on ∂Y induced by the volume form dx in U. Since the Green operator for P^*_{i-1} is given by the formula $\frac{1}{\sqrt{-1}} g\left(\sum_{j=1}^n \left(P_{i-1}^{(j)\,*} f\right) \star dx_j\right)$ in view of (2.4.13), it follows that $n(f) = 0$ if and only if $\sum_{j=1}^n \partial_j \rho \left(P_{i-1}^{(j)\,*} f\right) = 0$ on ∂Y.

Definition 4.1.41 *The "Levi form" at degree i for the complex $E^{\cdot}$ is the mapping associating with every point $x \in U$ the following Hermitian form on E^i_x:*

$$\mathcal{L}^{(i)}(x)(v,w) = \sum_{j,k=1}^{n} \frac{\partial^2 \rho(x)}{\partial x_j \partial x_k} \left(P_{i-1}^{(j)}(x)^* w\right)^* \left(P_{i-1}^{(k)}(x)^* v\right) \quad (v, w \in E^i_x).$$

The geometrical condition, mentioned above, on the boundary ∂Y of $\overline{Y}$ consists in the following:

$$\begin{array}{l} \text{given any point } x \in \partial Y, \text{ we have } \mathcal{L}^{(i)}(x)(v,v) > 0 \text{ for all} \\ \text{non-zero } v \in E^i_x \text{ satisfying } \sum_{j=1}^n \frac{\partial \rho(x)}{\partial x_j}\left(P_{i-1}^{(j)}(x)^* v\right) = 0. \end{array} \tag{4.1.12}$$

Exercise 4.1.42 *Prove that condition (4.1.12) does not depend on the particular choice of a defining function ρ for Y.*

Hint. If $\rho' \in C^\infty_{loc}(X)$ is another defining function for Y, then the quotient $\frac{\rho'}{\rho}$ is a positive smooth function on X. To prove this, use local coordinates near the points of ∂Y.

□

Exercise 4.1.43 *Prove that condition (4.1.12) is equivalent to the following one:*

$$\begin{array}{l} \textit{there exists a constant } c > 0 \textit{ such that } \mathcal{L}^{(i)}(x)(v,v) \geq c\,|v|^2 \textit{ for all} \\ x \in \partial Y \textit{ and all vectors } v \in E^i_x \textit{ satisfying } \sum_{j=1}^n \frac{\partial \rho(x)}{\partial x_j}\left(P_{i-1}^{(j)}(x)^* v\right) = 0. \end{array} \tag{4.1.13}$$

Hint. Consider the set σ in $\partial Y \times E^i$ formed by the pairs (x, v) where $x \in \partial Y$, and v lies on the unit sphere of E^i_x and satisfies $\sum_{j=1}^n \frac{\partial \rho(x)}{\partial x_j}\left(P_{i-1}^{(j)}(x)^* v\right) = 0$. Show that σ is compact. Denote by c the infimum of $\mathcal{L}^{(i)}(x)(v,v)$ over all $(x,v) \in \sigma$.

Under condition (4.1.12), we have $c > 0$. Check that c is an admissible constant for (4.1.13).

□

The following result was noted by the author [227]. However the most important cases of it were well-known (see Kohn [125], Hörmander [108, 109, 111], MacKichan [154], Andreotti [8], and others).

Theorem 4.1.44 *Let $E^{\cdot}$ be an elliptic complex of first order differential operators on X satisfying (4.1.11), and Y be a relatively compact open subset of X having a boundary of class C^2_{loc} satisfying (4.1.12). Then the Kohn-Nirenberg estimate for $E|_{\overline{Y}}$ is fulfilled at degree i.*

Proof. Fix some section $f \in \mathcal{E}(E^i|_{\overline{Y}})$ which satisfies $n(f) = 0$.

Let us rewrite the integral $\left\|\sum_{j=1}^n D_j \left({P_{i-1}^{(j)}}^* f\right)\right\|^2$, twice integrating by parts. We get

$$\begin{aligned}\left\|\sum_{j=1}^n D_j \left({P_{i-1}^{(j)}}^* f\right)\right\|^2 &= \int_Y \sum_{j,k=1}^n \partial_j \left({P_{i-1}^{(j)}}^* f\right)^* \partial_k \left({P_{i-1}^{(k)}}^* f\right) dx \\ &= \int_{\partial Y} \sum_{j,k=1}^n \left(\left({P_{i-1}^{(j)}}^* f\right)^* \partial_k \left({P_{i-1}^{(k)}}^* f\right) \star dx_j - \left({P_{i-1}^{(j)}}^* f\right)^* \partial_j \left({P_{i-1}^{(k)}}^* f\right) \star dx_k\right) \\ &\quad + \int_Y \sum_{j,k=1}^n \partial_k \left({P_{i-1}^{(j)}}^* f\right)^* \partial_j \left({P_{i-1}^{(k)}}^* f\right) dx.\end{aligned}$$

Use now the equalities $\star dx_j|_S = \frac{\partial_j \rho}{|d\rho|} ds \quad (j = 1, \ldots, n)$ to rewrite

$$\begin{aligned}&\left\|\sum_{j=1}^n D_j \left({P_{i-1}^{(j)}}^* f\right)\right\|^2 \\ &\quad = \int_{\partial Y} \left(\left(\sum_{j=1}^n \partial_j \rho \left({P_{i-1}^{(j)}}^* f\right)\right)^* \left(\sum_{k=1}^n \partial_k \left({P_{i-1}^{(k)}}^* f\right)\right)\right. \\ &\quad\quad \left. - \sum_{j=1}^n \left({P_{i-1}^{(j)}}^* f\right)^* \left(\sum_{k=1}^n \partial_k \rho \, \partial_j \left({P_{i-1}^{(k)}}^* f\right)\right)\right) \frac{ds}{|d\rho|} \\ &\quad\quad + \int_Y \sum_{j,k=1}^n \partial_k \left({P_{i-1}^{(j)}}^* f\right)^* \partial_j \left({P_{i-1}^{(k)}}^* f\right) dx. \end{aligned} \qquad (4.1.14)$$

Since $n(f) = 0$, we have $\sum_{j=1}^n \partial_j \rho \left({P_{i-1}^{(j)}}^* f\right) = \rho s$ for some $s \in C^1(E^{i-1}|_{\overline{Y}})$. Hence it follows that

$$\sum_{j=1}^n \left({P_{i-1}^{(j)}}^* f\right)^* \partial_j \left(\sum_{k=1}^n \partial_k \rho \left({P_{i-1}^{(k)}}^* f\right)\right) = \rho \left(\sum_{j=1}^n \left({P_{i-1}^{(j)}}^* f\right)^* \partial_j s + s^* s\right),$$

so

$$-\sum_{j=1}^{n}\left(P_{i-1}^{(j)*}f\right)^{*}\left(\sum_{k=1}^{n}\partial_k\rho\,\partial_j\left(P_{i-1}^{(k)*}f\right)\right)$$
$$=\mathcal{L}^{(i)}(x)(f,f)-\rho\left(\sum_{j=1}^{n}\left(P_{i-1}^{(j)*}f\right)^{*}\partial_j s+s^{*}s\right).$$

Thus the integrand in the boundary integral of the right-hand side of (4.1.14) is equal to

$$\mathcal{L}^{(i)}(x)(f,f)+\rho\left(s^{*}\sum_{k=1}^{n}\partial_k\left(P_{i-1}^{(k)*}f\right)-\sum_{j=1}^{n}\left(P_{i-1}^{(j)*}f\right)^{*}\partial_j s-s^{*}s\right),$$

and we get

$$\int_{\partial Y}\mathcal{L}^{(i)}(x)(f,f)\frac{ds}{|d\rho|}$$
$$=\left\|\sum_{j=1}^{n}D_j\left(P_{i-1}^{(j)*}f\right)\right\|^{2}-\int_{Y}\sum_{j,k=1}^{n}\partial_k\left(P_{i-1}^{(j)*}f\right)^{*}\partial_j\left(P_{i-1}^{(k)*}f\right)\,dx. \tag{4.1.15}$$

It follows from condition (4.1.11) that there exists a constant $c>0$ such that at each point of Y we have

$$-\sum_{j,k=1}^{n}\partial_k\left(P_{i-1}^{(j)*}f\right)^{*}\partial_j\left(P_{i-1}^{(k)*}f\right)\le c\left(\left|\sum_{j=1}^{n}P_i^{(j)}\partial_j f\right|^{2}+\left|\sum_{j=1}^{n}P_{i-1}^{(j)*}\partial_j f\right|^{2}+|f|^{2}\right).$$

Therefore (4.1.15) implies

$$\int_{\partial Y}\mathcal{L}^{(i)}(x)(f,f)\frac{ds}{|d\rho|}$$
$$\le\left\|\sum_{j=1}^{n}D_j\left(P_{i-1}^{(j)*}f\right)\right\|^{2}+c\left(\left\|\sum_{j=1}^{n}P^{(j)}D_j f\right\|^{2}+\left\|\sum_{j=1}^{n}P_{i-1}^{(j)*}D_j f\right\|^{2}+\|f\|^{2}\right).$$

Use now the inequality $\|v_1+v_2\|^2\le 2(\|v_1\|^2+\|v_2\|^2)$ to obtain

$$\int_{\partial Y}\mathcal{L}^{(i)}(x)(f,f)\frac{ds}{|d\rho|}\le c\,D(f),$$

with some constant $c>0$ independent of f.

To complete the proof, it suffices to apply condition (4.1.12) and Exercise 4.1.43. □

We shall not attempt any discussion of the geometrical meaning of the "Levi condition" (4.1.12). We note however that, if Y is strictly convex, then this condition holds.

Theorem 4.1.45 *If the symbol mapping $\sigma(P_{i-1}) : E^{i-1} \otimes T^*(U) \to E^i$ is surjective, then condition (4.1.12) is fulfilled for each strictly convex domain $Y \subset\subset U$.*

Proof. Indeed, let $Y \subset\subset U$ be a strictly convex domain, and $x \in \partial Y$. Then

$$\sum_{j,k=1}^{n} \frac{\partial^2 \rho(x)}{\partial x_j \partial x_k} \overline{w_j} w_k > 0$$

for all non-zero vectors $w = (w_1, \ldots, w_n)$ of $\mathbb{R}^n \cong T_x(U)$ satisfying $\sum_{j=1}^{n} \frac{\partial \rho(x)}{\partial x_j} w_j = 0$.

Since

$$\sum_{j,k=1}^{n} \frac{\partial^2 \rho(x)}{\partial x_j \partial x_k} \overline{w_j} w_k = \sum_{j,k=1}^{n} \frac{\partial^2 \rho(x)}{\partial x_j \partial x_k} (\operatorname{Re} w_j \operatorname{Re} w_k + \operatorname{Im} w_j \operatorname{Im} w_k),$$

this remains valid for all non-zero vectors $w \in \mathbb{C}^n$ also.

Fix now some non-zero vector $v \in E^i_x$ which satisfies $\sum_{j=1}^{n} \frac{\partial \rho(x)}{\partial x_j} \left(P^{(j)}_{i-1}(x)^* v\right) = 0$.

If $P^{(i)}_{i-1}(x)^* v = 0$ for all $j = 1, \ldots, n$, then $(\sigma(P_{i-1})(x,z))^* v = 0$ for all $z \in T^*_x(U)$. Hence it follows that $v = 0$, because the symbol mapping $\sigma(P_{i-1}) : E^{i-1}_x \otimes T^*_x(U) \to E^i_x$ is assumed to be surjective. This contradiction shows that $P^{(i)}_{i-1}(x)^* v \neq 0$ for some j with $1 \le j \le n$.

We let

$$P^{(i)}_{i-1}(x)^* v = \begin{pmatrix} w_j^{(1)} \\ w_j^{(2)} \\ \cdots \\ w_j^{(k_{i-1})} \end{pmatrix} \quad (j = 1, \ldots, n),$$

and consider the following vectors of $\mathbb{C}^n$:

$$w^{(\nu)} = (w_1^{(\nu)}, \ldots, w_n^{(\nu)}) \quad (\nu = 1, \ldots, k_{i-1}).$$

As we have seen, at least one of these vectors is non-zero, and

$$\sum_{i=1}^{n} \frac{\partial \rho(x)}{\partial x_j} w_j^{(\nu)} = 0 \quad \nu = 1, \ldots, k_{i-1}.$$

Therefore

$$\begin{aligned} \mathcal{L}^{(i)}(x)(v,v) &= \sum_{j,k=1}^{n} \frac{\partial^2 \rho(x)}{\partial x_j \partial x_k} \left(\overline{w_j^{(1)}} \ldots \overline{w_k^{(k_{i-1})}}\right) \cdot \begin{pmatrix} w_k^{(1)} \\ \cdots \\ w_k^{(k_{i-1})} \end{pmatrix} \\ &= \sum_{\nu=1}^{k_{i-1}} \left(\sum_{j,k=1}^{n} \frac{\partial^2 \rho(x)}{\partial x_j \partial x_k} \overline{w_j^{(\nu)}} w_k^{(\nu)} \right) \\ &> 0, \end{aligned}$$

which completes the proof.

□

We mention that the surjectivity of the symbol mapping is one of the properties of normalized differential operators. See Definition 1.3.23 and the remark after it for details.

Remark 4.1.46 *As Sweeney's [210] example suggests, condition (4.1.12) is not likely in general to be the right "Levi criterion" for determining existence domains of solutions of the overdetermined system $P_{i-1}u = f$. See Nacinovich [166] for further information.*

To formulate a few exercises we withdraw ourselves from the particular situation under consideration.

Suppose we are given a family $M^{(j)} : B^1 \to B^2 \quad (j = 1, \ldots, n)$ of homomorphisms of smooth vector bundles over U. Supposing that B^2 is endowed with some Hermitian structure, we consider the "Levi form"

$$\mathcal{L}^{M^{(\,)}}(x)(v,w) = \sum_{j,k=1}^{n} \frac{\partial^2 \rho(x)}{\partial x_j \partial x_k} \left(M^{(j)}v, M^{(k)}w\right)_x$$

on B^1. The set Y is said to be $M^{(\,)}$-convex (respectively strictly $M^{(\,)}$-convex) at a point $x \in \partial Y$ if $\mathcal{L}^{M^{(\,)}}(x)(v,v) \geq 0$ (respectively $\mathcal{L}^{M^{(\,)}}(x)(v,v) > 0$) for all vectors $v \in B^1_x$ satisfying $\sum_{j=1}^{n} \frac{\partial \rho(x)}{\partial x_j}(M^{(j)}v) = 0$. This definition is easily verified to be independent of the particular choice of a defining function ρ for Y.

Exercise 4.1.47 *Study the geometrical content of $M^{(\,)}$-convexity.*

Hint. Follow two important particular cases which are convexity and (complex) pseudo-convexity.

□

For the classical complexes, condition (4.1.12) implies both the solvability of the P-Neumann problem and the uniqueness of solutions.

Exercise 4.1.48 *Let $E = \Lambda\, T^*_{\mathbb{C}}(X)$ be the de Rham complex on X. Prove that if a domain $Y \subset\subset U$ is ${P^{(\,)}_{i-1}}^*$- convex at each point of ∂Y and there is a point $x \in \partial Y$ such that Y is strictly ${P^{(\,)}_{i-1}}^*$-convex at x, then $\mathcal{H}^i(E\,|_{\overline{Y}}) = 0$.*

Hint. Derive from (4.1.15) and Example 4.1.37 that for all sections $f \in \mathcal{E}_{n_i}(\overline{Y})$

$$\int_{\partial Y} \mathcal{L}^{(i)}(x)(f,f)\frac{ds}{|d\rho|} + \int_Y \sum_{j=1}^{n} |\partial_j f|^2 dx = \|df\|^2 + \|d^*f\|^2.$$

Therefore, if $df = d^*f = 0$ and Y is ${P^{(\,)}_{i-1}}^*$-convex at each point of ∂Y, then every coefficient of f is constant, and $\int_{\partial Y} \mathcal{L}^{(i)}(x)(f,f)\frac{ds}{|d\rho|} = 0$. Since $\mathcal{L}^{(i)}(x)(f,f)$ is continuous, we may conclude that $\mathcal{L}^{(i)}(x)(f,f) \equiv 0$ on ∂Y.

If Y is strictly ${P^{(\,)}_{i-1}}^*$-convex at some point $x \in \partial Y$, then $f(x) = 0$ since otherwise we would have $\mathcal{L}^{(i)}(x)(f,f) > 0$.

Use now the fact, that Y is connected, to conclude that $f \equiv 0$ in Y.

□

Exercise 4.1.49 *Let $E = \Lambda^{p,\cdot}\, T^*_{\mathbb{C}}(X)$ be the Dolbeault complex with coefficients in $\Lambda^{p,0}T^*_{\mathbb{C}}(X)$. Prove that, if a domain $Y \subset\subset U$ is ${P^{(\,)}_{i-1}}^*$-convex at each point of ∂Y and strictly ${P^{(\,)}_{i-1}}^*$-convex on a subset of ∂Y of positive measure, then $\mathcal{H}^i(E\,|_{\overline{Y}}) = 0$.*

Hint. Use the relationship

$$\int_{\partial Y} \mathcal{L}^{(\imath)}(x)(f,f)\frac{ds}{|d\rho|} + \int_Y \sum_{j=1}^{n} \left|\frac{\partial f}{\partial \bar{z}_j}\right|^2 dx = \|\overline{\partial} f\|^2 + \|\overline{\partial}^* f\|^2 \quad \text{for all} \quad f \in \mathcal{E}_{n,\imath}(\overline{Y}),$$

which follows from (4.1.15) and Example 4.1.38.

Check that, if $\overline{\partial} f = \overline{\partial}^* f = n(f) = 0$ and Y is $P^{(\,)\,*}_{\imath-1}$-convex at each point of ∂Y, then every coefficient of f is a holomorphic function in Y, and moreover $\mathcal{L}^{(\imath)}(x)(f,f) \equiv 0$ on ∂Y.

Therefore, if Y is strictly $P^{(\cdot)\,*}_{\imath-1}$-convex at a point $x \in \partial Y$, then we have $f(x) = 0$. Finally, apply the well-known uniqueness theorem for holomorphic functions.

□

4.2 The L^2-Cohomologies of Differential Complexes and the Bergman Projector

4.2.1 L^2-*cohomologies*

It was noted in Section 4.1, that the proof of regularity of Neumann's operator near the boundary is the most difficult part of solving the P-Neumann problem. In this section we consider a "weak version" of this problem finessing the question of boundary regularity. We follow on the whole the paper of Kohn [125] using functional analysis. Applications will be given.

Assume that $E^{\cdot}$ is a differential complex on X and Y is an open subset of X (maybe, $Y = X$). We will use the notations $\mathcal{L}^{\imath} = L^2(E^{\imath}|_Y)$ and $\mathcal{E}^{\imath} = \mathcal{E}(E^{\imath}|_Y)$. We mention that $\mathcal{L}^{\imath}$ is the completion of the vector space $\{f \in \mathcal{E}^{\imath} : \|f\| < \infty\}$ in the norm $\|f\| = (f,f)_Y^{1/2}$.

The reader may see details concerning the following definition in the book of Desin [52, Ch.II, §2].

Definition 4.2.1 *Let $\mathcal{D}^{\imath}_T$ be the set of all sections $f \in \mathcal{L}^{\imath}$, for which there exists a sequence $\{f_\nu\}$ with the following properties: 1) $f_\nu \in \mathcal{L}^{\imath} \cap \mathcal{E}^{\imath}$; 2) $f = \lim f_\nu$ in $\mathcal{L}^{i}$; and 3) $Pf_\nu \in \mathcal{L}^{\imath+1}$ and $\{Pf_\nu\}$ is a Cauchy sequence in $\mathcal{L}^{\imath+1}$. The mapping $T : \mathcal{D}^{\imath}_T \to \mathcal{L}^{\imath+1}$ defined by $Tf = \lim Pf_\nu$, where $\{f_\nu\}$ is a sequence with properties 1)–3), is called the maximal operator generated by P.*

We note that T is well-defined. Indeed, if $\{f'_\nu\}$ is another sequence with the properties 1)–3) and $\psi = \lim Pf'_\nu$, then for each section $g \in \mathcal{D}(E^{\imath+1}|_Y)$

$$\begin{aligned} (Tf - \psi, g)_Y &= \lim\,(Pf_\nu - Pf'_\nu, g)_Y \\ &= \lim\,(f_\nu - f'_\nu, P^* g)_Y \\ &= 0. \end{aligned}$$

Consequently $Tf = \psi$.

We will consider T as an (unbounded) operator from $\mathcal{L}^\imath$ to $\mathcal{L}^{\imath+1}$, with domain $\mathcal{D}_T^\imath$. Since $\mathcal{D}_T^\imath \supset \mathcal{D}(E^\imath|_Y)$, then the domain of T is dense in $\mathcal{L}^\imath$.

We note that from the lemma of Du Bois-Reymond and the uniqueness of a weak limit, it follows that if $f \in \mathcal{D}_T^\imath \cap \mathcal{E}^\imath$ then $Tf = Pf$.

Lemma 4.2.2 *$T\mathcal{D}_T^\imath \subset \mathcal{D}_T^{\imath+1}$ and $T^2 = 0$.*

Proof. Let $f \in \mathcal{D}_T^\imath$ and $\{f_\nu\}$ be a sequence from Definition 4.2.1. We set $\psi_\nu = Pf_\nu$. Then $Tf = \lim \psi_\nu$. And since $P\psi_\nu = 0$, we obtain that $Tf \in \mathcal{D}_T^{\imath+1}$ and $T(Tf) = 0$.

□

Thus we have the following complex of Hilbert spaces and (closed) linear mappings :

$$\mathcal{L} \,:\, 0 \longrightarrow \mathcal{L}^0 \xrightarrow{T} \mathcal{L}^1 \xrightarrow{T} \cdots. \qquad (4.2.1)$$

The L^2-cohomologies of the complex $E\,|_Y$ are defined to be the cohomologies of the complex (4.2.1), that is

$$H^\imath(\mathcal{L}\,) = \frac{\ker\,\{T:\ \mathcal{D}_T^\imath \to \mathcal{L}^{\imath+1}\}}{T\mathcal{D}_T^{\imath-1}}.$$

4.2.2 *Adjoint complex*

We now define T^*, the adjoint of T, as usual for unbounded operators.

Definition 4.2.3 *Let $\mathcal{D}_{T^*}^\imath$ be the set of all $g \in \mathcal{L}^\imath$ such that there exists $\psi \in \mathcal{L}^{\imath-1}$ satisfying the equation $(Tf, g)_Y = (f, \psi)_Y$ for all $f \in \mathcal{D}_T^{\imath-1}$. We define $T^* : \mathcal{D}_{T^*}^\imath \to \mathcal{L}^{\imath-1}$ by $T^*g = \psi$.*

The operator T^* is well-defined because $\mathcal{D}_T^{\imath-1}$ is dense in $\mathcal{L}^{\imath-1}$. It is easy to see that if $g \in \mathcal{D}_{T^*}^\imath \cap \mathcal{E}^\imath$ then $T^*g = P^*g$.

By analogy with Lemma 4.2.2 we have the following simple result.

Lemma 4.2.4 *$T^*\mathcal{D}_{T^*}^\imath \subset \mathcal{D}_{T^*}^{\imath-1}$ and $T^{*2} = 0$.*

Proof. Indeed if $g \in \mathcal{D}_{T^*}^\imath$, and $f \in \mathcal{D}_T^{\imath-2}$ then by Definition 4.2.3 and Lemma 4.2.2 we get $(Tf, T^*g)_Y = (T(Tf), g)_Y = 0$. Therefore $T^*g \in D_{T^*}^{\imath-1}$ and $T^*(T^*f) = 0$, which completes the proof.

□

Thus we obtain the following (chain) complex of Hilbert spaces and "closed" linear mappings:

$$\mathcal{L}^* : 0 \longleftarrow \mathcal{L}^0 \xleftarrow{T^*} \mathcal{L}^1 \xleftarrow{T^*} \cdots. \qquad (4.2.2)$$

The complex (4.2.2) is called the adjoint complex for (4.2.1), and its homologies are denoted by

$$H_\imath(\mathcal{L}^*) = \frac{\ker\,\{T^*:\ \mathcal{D}_{T^*}^\imath \to \mathcal{L}^{\imath-1}\}}{T^*\mathcal{D}_{T^*}^{\imath-1}}.$$

4.2.3 Weak version of the Neumann problem

We shall introduce an operator L on $\mathcal{L}^{\cdot}$ with a domain $\mathcal{D}^{\cdot}_L$ which has the property that, if $f \in \mathcal{D}^{\cdot}_L \cap \mathcal{E}^{\cdot}$, then $Lf = \Delta f$ where $\Delta = P^*P + PP^*$ is the Laplacian of the complex E.

Definition 4.2.5 *Let $\mathcal{D}^{\cdot}_L$ be the set of all $f \in \mathcal{D}^{\cdot}_T \cap \mathcal{D}^{\cdot}_{T^*}$ such that $Tf \in \mathcal{D}^{\cdot+1}_{T^*}$ and $T^*f \in \mathcal{D}^{\cdot-1}_T$. Then $L: \mathcal{D}^{\cdot}_L \to \mathcal{L}^{\cdot}$ is defined by $Lf = T^*Tf + TT^*f$.*

The weak (L^2-) version of Problem 4.1.5 consists in the following :

given a section $f \in \mathcal{L}^{\cdot}$, when is there $u \in \mathcal{D}^{\cdot}_L$
such that $Lu = f$, and how does u depend on f?

4.2.4 Weak orthogonal decomposition

The weak orthogonal decomposition is the first step in solving the Neumann problem.

Let us denote

$$\mathcal{H}^{\cdot} = \{f \in \mathcal{D}^{\cdot}_T \cap \mathcal{D}^{\cdot}_{T^*} : \ Tf = T^*f = 0\}.$$

Since the operators T and T^* are closed, $\mathcal{H}^{\cdot}$ is a closed subset of $\mathcal{L}^{\cdot}$. Let $H: \mathcal{L}^{\cdot} \to \mathcal{H}^{\cdot}$ be the orthogonal projection of $\mathcal{L}^{\cdot}$ onto $\mathcal{H}^{\cdot}$.

Lemma 4.2.6 *$f \in \mathcal{H}^{\cdot}$ if and only if $f \in \mathcal{D}^{\cdot}_L$ and $Lf = 0$.*

Proof. If $f \in \mathcal{H}^{\cdot}$, then obviously $f \in \mathcal{D}^{\cdot}_L$ and $Lf = 0$. If $Lf = 0$, then $(Lf, f)_Y = 0$, and since $(Lf, f)_Y = \|Tf\|^2 + \|T^*f\|^2$, we have $f \in \mathcal{H}^{\cdot}$.

□

Proposition 4.2.7 *The operator L is selfadjoint, and $(L+1)^{-1}$ exists, is bounded, and is everywhere in $\mathcal{L}^{\cdot}$ defined.*

Proof. Since T is a closed operator and the domain of T is dense, the same is also true for T^*, and $(T^*)^* = T$ (see Riesz and Sz.-Nagy [189, Subsection 117]).

It follows that the operators $(TT^* + 1)^{-1}$ and $(T^*T + 1)$ exist, are bounded, selfadjoint and defined everywhere in $\mathcal{L}^{\cdot}$ (see Riesz and Sz.-Nagy [189, Subsection 118]).

We will show that $(L + 1)^{-1}$ exists, is bounded, is everywhere defined, and is given by the formula

$$(L+1)^{-1} = (TT^* + 1)^{-1} + (T^*T + 1)^{-1} - 1. \qquad (4.2.3)$$

First of all we note that

$$\begin{aligned} (TT^* + 1)^{-1} - 1 &= (1 - (TT^* + 1))(TT^* + 1)^{-1} \\ &= -TT^*(TT^* + 1)^{-1}, \end{aligned}$$

and by analogy

$$(T^*T+1)^{-1}-1=-T^*T(T^*T+1)^{-1}.$$

Denote by R the right-hand side of (4.2.3). It follows from above and Lemma 4.2.4 that the range of R is contained in the domain of TT^*, and

$$\begin{aligned} TT^*R &= TT^*(TT^*+1)^{-1}+TT^*((T^*T+1)^{-1}-1) \\ &= TT^*(TT^*+1)^{-1}. \end{aligned}$$

Similarly by using Lemma 4.2.2 we obtain $T^*TR = T^*T(T^*T+1)^{-1}$. Hence $(L+1)R = 1$ which proves (4.2.3).

Thus the operator $(L+1)^{-1}$ is selfadjoint because it is the sum of bounded selfadjoint operators. Therefore $(L+1)$ and L are selfadjoint (see Riesz and Sz.-Nagy [189, Subsection 119]), which completes the proof.

□

Corollary 4.2.8 (weak orthogonal decomposition) *The range of L is orthogonal to $\mathcal{H}^\iota$, and*

$$\mathcal{L}^\iota = \mathcal{H}^\iota \oplus \overline{L\mathcal{D}^\iota_L}, \tag{4.2.4}$$

where $\overline{L\mathcal{D}^\iota_L}$ denotes the closure of $L\mathcal{D}^\iota_L$ in $\mathcal{L}^\iota$.

Proof. This follows immediately from the selfadjointness of L and Lemma 4.2.6.

□

In particular, if $L\mathcal{D}^\iota_L$ is closed then we get the "strong orthogonal decomposition"

$$\mathcal{L}^\iota = \mathcal{H}^\iota \oplus T^*T\mathcal{D}^\iota_L \oplus TT^*\mathcal{D}^\iota_L. \tag{4.2.5}$$

4.2.5 *The Neumann operator*

The Neumann operator was encountered in the proof of Proposition 4.1.26.

Definition 4.2.9 *Let $L\mathcal{D}^\iota_L$ be closed and $f \in \mathcal{L}^\iota$, then $f = Hf + L\psi$ where $\psi \in \mathcal{D}^\iota_L$. The (Neumann) operator $N: \mathcal{L}^\iota \to \mathcal{D}^\iota_L$ is defined by $Nf = \psi - H\psi$.*

Note that N is well-defined. Indeed, if also $f = Hf + L\psi'$, where $\psi' \in \mathcal{D}^\iota_L$ then $L(\psi - \psi') = 0$, therefore $(\psi - H\psi) - (\psi' - H\psi') = (\psi - \psi') - H(\psi - \psi') = 0$.

We summarize the properties of the Neumann operator. They generalize properties of the Green operator from Hodge theory (see Corollary 2.2.6).

Proposition 4.2.10 *Let $L\mathcal{D}^\iota_L$ be closed. Then the Neumann operator N has the following properties:*

1) N is bounded, selfadjoint, $HN = NH = 0$, and we have the (L^2-) orthogonal decomposition

$$f = Hf + T^*TNf + TT^*Nf \quad \text{for all} \quad f \in \mathcal{L}^\iota; \tag{4.2.6}$$

2) if $f \in \mathcal{D}_T^i$ and $Tf = 0$, then $TNf = 0$, and moreover, if $L\mathcal{D}_L^{i+1}$ is closed, then $TNf = NTf$;

3) if $f \in \mathcal{D}_{T^*}^i$ and $T^*f = 0$, then $T^*Nf = 0$, and moreover, if $L\mathcal{D}_L^{i-1}$ is closed, then $T^*Nf = NT^*f$;

4) if the differential operator Δ_i is hypoelliptic, then $\mathcal{H}^i \subset \mathcal{E}^i$, and also, $f \in \mathcal{L}^i \cap \mathcal{E}^i$ implies $Nf \in \mathcal{E}^i$. That is, N preserves the interior regularity.

Proof.

1) The correlations $HN = NH = 0$ and formula (4.2.6) follow immediately from the definition of N.

Further, by the Closed Graph Theorem there exists a constant $c > 0$ such that, if $f \in \mathcal{D}_L^i$ and $f \perp \mathcal{H}^i$, then we have $\|Lf\| \geq c\|f\|$. Applying this to Nf, we obtain

$$\|Nf\| \leq \frac{1}{c}\ \|LNf\| = \frac{1}{c}\ \|f - Hf\| \leq \frac{1}{c}\ \|f\|.$$

Hence N is bounded.

Finally, the selfadjointness of N follows immediately from Proposition 4.2.7 because

$$\begin{aligned}(Nf, g)_Y &= (Nf, Hg + LNg)_Y = (Nf, LNg)_Y \\ &= (LNf, Ng)_Y = (f, Ng)_Y.\end{aligned}$$

2) Let $f \in \mathcal{D}_L^i$. Then from formula (4.2.6) and Lemma 4.2.2 we get $T^*TNf \in \mathcal{D}_T^i$, and $Tf = 0$ implies $TT^*TNf = 0$. Hence it easily follows that $TNf = 0$.

If also $L\mathcal{D}_L^{i+1}$ is closed, then for a section $f \in \mathcal{D}_T^i$, we have $Tf = TT^*TNf$ on the one hand, and $Tf = TT^*NTf$ on the other hand. Hence it follows that $L(TNf - NTf) = 0$, and since $(TNf - NTf) \perp \mathcal{H}^{i+1}$ then $TNf - NTf = 0$ as required.

3) Part 3) is proved by analogy with part 2).

4) Assume that Δ^i is a hypoelliptic operator. If $f \in \mathcal{H}^i$, then for each section $g \in \mathcal{D}(E^i|_Y)$ we have $(Lf, g)_Y = (f, \Delta g)_Y = 0$. That is, $\Delta f = 0$ in the sense of distributions. Therefore, $f \in \mathcal{E}^i$. Thus $\mathcal{H}^i \subset \mathcal{E}^i$.

If $f \in \mathcal{L}^i \cap \mathcal{E}^i$, then for all $g \in \mathcal{D}(E^i|_Y)$ we get $(Nf, \Delta g)_Y = (LNf, g)_Y = (f - Hf, g)_Y$. That is, $\Delta(Nf) = f - Hf$ in the sense of distributions. Since $f - Hf \in \mathcal{E}^i$, then $Nf \in \mathcal{E}^i$ because the operator Δ_i is hypoelliptic.

This completes the proof.

□

4.2.6 *The Dirichlet norm*

Beginning with its classical forms, the Dirichlet norm has been an important technical tool in studying the Neumann problem.

Definition 4.2.11 *Let $f, g \in \mathcal{D}_T^i \cap \mathcal{D}_{T^*}^i$. The Dirichlet inner product of f and g is defined by $D(f, g) = (Tf, Tg)_Y + (T^*f, T^*g)_Y + (f, g)_Y$, and the Dirichlet norm is $D(f) = D(f, f)^{1/2}$.*

The space $\mathcal{D}^\iota_T \cap \mathcal{D}^\iota_{T^*}$ with the Dirichlet norm is a complete (Hilbert) space. It is denoted by $\mathcal{D}^\iota$.

Since $D(f) \geq \|f\|$ for all $f \in \mathcal{D}^\iota$, there exists only one selfadjoint operator M with a domain $\mathcal{D}^\iota_M \subset \mathcal{D}^\iota$ such that, if $f \in \mathcal{D}^\iota_M$ and $g \in \mathcal{D}^\iota$, then

$$D(f,g) = (Mf,g)_Y \tag{4.2.7}$$

(see Riesz and Sz.-Nagy [189, Subsection 124]).

The following lemma gives a useful description of the operator L because our estimates will be in the norm $D(f)$.

Lemma 4.2.12 $\mathcal{D}^\iota_L = \mathcal{D}^\iota_M$ *and* $L = M - 1$, *where the operator* M *is defined above.*

Proof. If $f \in \mathcal{D}^\iota_L$ and $g \in \mathcal{D}^\iota$, then clearly $D(f,g) = ((L+1)f,g)_Y$. Hence by the uniqueness of M, we have $M = L + 1$.

□

Definition 4.2.13 *Let* $\|\cdot\|_1$ *and* $\|\cdot\|_2$ *be two norms on a vector space* V. *We will say that the norm* $\|\cdot\|_1$ *is completely continuous with respect to the norm* $\|\cdot\|_2$ *if every sequence which is bounded in the norm* $\|\cdot\|_1$ *has a convergent subsequence in the norm* $\|\cdot\|_2$.

Proposition 4.2.14 *If the norm* D *on* $\mathcal{D}^\iota$ *is completely continuous with respect to* $\|\cdot\|$ *then* $\mathcal{H}^\iota$ *is finite-dimensional.*

Proof. Observe that if $f, g \in \mathcal{H}^\iota$ then $D(f,g) = (f,g)_Y$. Suppose that the dimension of $\mathcal{H}^\iota$ is infinite. Then there exists an infinite sequence $\{f_\nu\}$ of orthonormal elements in $\mathcal{H}^\iota$. Since $D(f_\nu) = \|f_\nu\| = 1$, the sequence $\{f_\nu\}$ contains a convergent subsequence. But this is at variance with the fact that if $\nu \neq \mu$ then $\|f_\nu - f_\mu\| = \sqrt{2}$.

□

Lemma 4.2.15 *If the norm* D *on* $\mathcal{D}^\iota$ *is completely continuous with respect to* $\|\cdot\|$ *then there exists a constant* $c > 0$ *such that for all* $f \in \mathcal{D}^\iota$ *with* f *orthogonal to* $\mathcal{H}^\iota$, *we have*

$$\|Tf\|^2 + \|T^*f\|^2 \geq c\,\|f\|^2.$$

Proof. Consider the Hilbert space $\mathcal{L}^{\iota+1} \times \mathcal{L}^{\iota-1}$ with the norm $\|\{v_1, v_2\}\| = (\|v_1\|^2 + \|v_2\|^2)^{1/2}$. Let $M : \mathcal{D}^\iota \to \mathcal{L}^{\iota+1} \times \mathcal{L}^{\iota-1}$ be the mapping defined by $Mf = \{Tf, T^*f\}$. We note that M is a closed operator.

We will prove that the range of M is closed. Suppose that $M\mathcal{D}^\iota$ is not closed. Then there exists a sequence $\{f_\nu\} \in \mathcal{D}^\iota$ such that $\lim\ Mf_\nu = \{v_1, v_2\}$ and $\{v_1, v_2\} \notin M\mathcal{D}^\iota$.

Set $\psi_\nu = f_\nu - Hf_\nu$, then ψ_ν are orthogonal to $\mathcal{H}^\iota$ and $\lim\ M\psi_\nu = \{v_1, v_2\}$. If $\|\psi_\nu\|$ are bounded, then $D(\psi_\nu) = (\|M\psi_\nu\|^2 + \|\psi_\nu\|^2)^{1/2}$ are bounded too. Then by hypothesis $\{\psi_\nu\}$ has a convergent subsequence with a limit ψ, and since M is closed then $M\psi = \{v_1, v_2\}$ which contradicts the assumption that $\{v_1, v_2\} \notin M\mathcal{D}^\imath$. Thus by choosing a subsequence, if necessary, we may assume that $\lim \|\psi_\nu\| = \infty$.

Now set $\Theta_\nu = \psi_\nu / \|\psi_\nu\|$. Then $\lim \|M\Theta_\nu\| = 0$ and $D(\Theta_\nu)$ are bounded. Therefore $\{\Theta_\nu\}$ has a convergent subsequence $\{\Theta_{\nu_k}\}$ such that

$$\lim \Theta_{\nu_k} = \Theta \quad \text{and} \quad \lim M\Theta_{\nu_k} = \{0, 0\}.$$

Hence $M\Theta = 0$ so that $\Theta \in \mathcal{H}^\imath$. Since $\Theta_\nu \perp \mathcal{H}^\imath$, we have $\Theta = 0$, but $\|\Theta_\nu\| = 1$. This contradiction proves that $M\mathcal{D}^\imath$ is closed in $\mathcal{L}^{\imath+1} \times \mathcal{L}^{\imath-1}$.

Let M_1 be the restriction of M to the orthogonal complement of $\mathcal{H}^\imath$ in $\mathcal{D}^\imath$. Then M^1 is injective and has a closed range. By the Closed Graph Theorem, the operator M_1^{-1} is bounded. Thus there is a constant $c > 0$ such that $\|M_1 f\|^2 \geq c\, \|f\|^2$. The lemma has been proved.

□

Theorem 4.2.16 *If the norm D on $\mathcal{D}^\imath$ is completely continuous with respect to the norm $\|\cdot\|$, then $L\mathcal{D}^\imath$ is closed.*

Proof. By Lemma 4.2.15 there exists $c > 0$ such that for all $f \in \mathcal{D}^\imath_L$ which are orthogonal to $\mathcal{H}^\imath$ we have $(Lf, f)_Y \geq c\, \|f\|^2$, so that $\|Lf\| \geq c\, \|f\|$.

Let $\psi = \lim Lf_\nu$. We may suppose that f_ν are orthogonal to $\mathcal{H}^\imath$, and then $\|f_\nu\|$ are uniformly bounded. Therefore $\{f_\nu\}$ has a subsequence whose arithmetic means converge (see Riesz and Sz.-Nagy [189, Subsections 32,38]). Denoting this limit by f, we get $\psi = Lf$ which completes the proof.

□

Usually for open subsets $Y \subset X$ the subspace $\mathcal{H}^0$ is infinite-dimensional, so by Proposition 4.2.14 D may be completely continuous with respect to the norm $\|\cdot\|$ on $\mathcal{D}^0$. But the following result holds.

Theorem 4.2.17 *If the norm D on $\mathcal{D}^1$ is completely continuous with respect to the norm $\|\cdot\|$ then $L\mathcal{D}^0_L$ is closed.*

Proof. It suffices to prove that there exists a constant $c > 0$ such that $\|Lf\| \geq c\, \|f\|$ for all $f \in \mathcal{D}^0_L$ which are orthogonal to $\mathcal{H}^0$.

First, if $\psi \in \mathcal{D}^0_L$, then $T\psi \in \mathcal{D}^1$ and $T\psi \perp \mathcal{H}^1$. Thus by Lemma 4.2.15 we obtain $\|T^*T\psi\|^2 = \|Lf\|^2 \geq c\, \|T\psi\|^2$.

Further, since $f \perp \mathcal{H}^0$, then, by the weak orthogonal decomposition (4.2.4), $f \in \overline{L\mathcal{D}^0_L}$. Hence, for each $\varepsilon > 0$ there exists $\psi \in \mathcal{D}^0_L$ such that $\|f - L\psi\| < \varepsilon$. Thus,

$$\begin{aligned} \|f\|^2 &\leq |(L\psi, f)_Y| + \varepsilon\, \|f\| \\ &\leq \|T\psi\|\, \|Tf\| + \varepsilon\, \|f\| \\ &\leq \frac{1}{c}\, \|L\psi\|\, \|Lf\| + \varepsilon\, \|f\| \\ &\leq \frac{1}{c}\, \|f\|\, \|Lf\| + \varepsilon \left(\frac{1}{c}\, \|Lf\| + \|f\|\right). \end{aligned}$$

Since ε can be made arbitrarily small by choosing $L\psi$ close enough to f, we obtain $\|Lf\| \geq c\, \|f\|$ which concludes the proof.

□

4.2.7 A representation of L^2-cohomologies of differential complexes

The question of when the norm D on $\mathcal{D}^i$ is completely continuous with respect to the norm $\|\cdot\|$, is very difficult in the general case and it requires special consideration. It was touched upon partially in Subsections 4.1.4, 4.1.5. We present some consequences here.

Corollary 4.2.18 *Let the norm D on $\mathcal{D}^i$ be completely continuous with respect to the norm $\|\cdot\|$. Then for the complex $E^{\cdot}|_Y$ the weak Neumann problem is solvable at degree i in the sense that there exist operators H and N in $\mathcal{L}^i$ with properties 1) – 4) of Proposition 4.2.10.*

Proof. This follows immediately from Proposition 4.2.10 and Theorem 4.2.16. □

Corollary 4.2.19 *If the norm D on $\mathcal{D}^i$ is completely continuous with respect to the norm $\|\cdot\|$, then*

$$Z^i(\mathcal{L}^{\cdot}) = \mathcal{H}^i \overset{orth.}{\oplus} T\mathcal{D}_T^{i-1}. \tag{4.2.8}$$

Proof. This follows from 1) and 2) of Proposition 4.2.10, and Theorem 4.2.16. □

In particular, (4.2.8) implies that the natural mapping $i:\ \mathcal{H}^i \to H^i(\mathcal{L}^{\cdot})$ is a topological isomorphism.

Corollary 4.2.20 *If the norm D on $\mathcal{D}^i$ is completely continuous with respect to the norm $\|\cdot\|$, then*

$$Z^i(\mathcal{L}^{\cdot *}) = \mathcal{H}^i \overset{orth.}{\oplus} T^*\mathcal{D}_{T^*}^{i+1}. \tag{4.2.9}$$

Proof. This follows from 1) and 3) of Proposition 4.2.10, and Theorem 4.2.16. □

In particular, (4.2.9) implies that the natural mapping $i:\ \mathcal{H}^i \to H_i(\mathcal{L}^{\cdot *})$ is a topological isomorphism.

Taking this into account together with the remark following Corollary 4.2.19, we obtain that, under the conditions of the corollary, $H^i(\mathcal{L}^{\cdot}) \cong H_i(\mathcal{L}^{\cdot *})$. However this is a special case of the following assertion (cf. (5.1.2)).

Theorem 4.2.21 *If the space $H^{i+1}(\mathcal{L}^{\cdot})$ is separated, then*

$$H^i(\mathcal{L}^{\cdot})' \overset{top.}{\cong} H_i(\mathcal{L}^{\cdot *}). \tag{4.2.10}$$

Proof. This is fairly straightforward and can be obtained from Riesz's description of the dual space of a Hilbert space and the weak orthogonal decomposition for T_i^*, namely

$$\mathcal{L}^i = Z^i(\mathcal{L}^{\cdot}) \oplus \overline{T^*\mathcal{D}_{T^*}^{i+1}}.$$

The hypothesis of the theorem means that $T^*\mathcal{D}^{i+1}_{T^*}$ is closed. The reader can easily fill in the details.

□

Some theorems on general complexes of Hilbert spaces and their densely defined mappings can be found in the articles of Hörmander [109] and Andreotti and Hill [10].

Nacinovich [167] outlined the extension to general complexes of an alternative method, also initially developed for the study of $\overline{\partial}$, but not solving the $\overline{\partial}$-Neumann problem. It consists in the use of a priori estimates involving certain weight functions, that are related to a method developed in the 30-s by Carleman to prove uniqueness for solutions of the Cauchy problem. The idea of using this method was suggested by the observation that the existence problem $Pu = f$ (where $Pf = 0$) was easily dealt with in the case of compact manifolds without boundary, and then a next reasonable step was to investigate manifolds endowed with a complete metric. The weight functions played an essential role for the completeness of the metric. For the use of weight functions for $\overline{\partial}$, see also Hörmander [109].

Though the two methods give equivalent results for $\overline{\partial}$, it turns out that the first, having stronger implications (regularity up to the boundary) requires a priori estimates which are more difficult to establish, while it cannot be applied directly on domains which are either unbounded or have non-smooth boundary.

4.2.8 *The Bergman projector and a generalization of a theorem of Bungart*

The next result follows from Proposition 4.2.10 and Theorem 4.2.16.

Corollary 4.2.22 *Let the norm D on $\mathcal{D}^1$ be completely continuous with respect to the norm $\|\cdot\|$. Then, for each section $f \in \mathcal{D}^0_T$, we have $f = Hf + T^*NTf$, where $H : \mathcal{L}^0 \to \mathcal{H}^0 = ker\, T_0$ is the L^2-orthogonal projection.*

If also the differential operator P_0 has injective symbol, then the Laplacian $\Delta_0 = P_0^* P_0$ is elliptic so according to Proposition 4.2.10

$$\mathcal{H}^0 = \{f \in \mathcal{L}^0 \cap \mathcal{E}^0 : \ Pf = 0\}.$$

Thus the operator H_0 is a generalization of the Bergman projector from complex analysis (see, for instance, Aizenberg and Yuzhakov [6, §6]), and we would like to present the corresponding results. As a matter of fact, this is a part of the theory of function spaces with reproducing kernels of Aronszajn [17].

First, we need a well-known a priori estimate.

Theorem 4.2.23 *Let $P \in do_p(E \to F)$ be a differential operator with injective symbol on X, and let non-negative integers $s_j \quad (j = 1,2)$ and a real $q > 1$ satisfy $\frac{1}{q} < \frac{s_1 - s_2}{n}$. Then for each open set $U \subset\subset X$ and compact set $K \subset U$ there exists a constant $c > 0$ depending on P, s_1, s_2, q, U and K, such that*

$$\|f\|_{C^{s_2}(E|_K)} \leq c \left(\|Pf\|_{W^{s_1-p,q}(F|_U)} + \|f\|_{L^q(E|_U)} \right) \quad \textit{for all} \quad f \in \mathcal{E}(E). \tag{4.2.11}$$

Proof. This follows from the well-known a priori (L^q-) estimates for solutions of elliptic systems (see Agmon, Douglis and Nirenberg [2, §15]), and the Sobolev Embedding Theorem 1.1.8.

□

Estimate (4.2.11) implies in particular that, for each point $x \in X$ the "evaluation functional" $\delta_x(f) = f(x)$ is bounded on $\mathcal{H}^0$. Therefore $\mathcal{H}^0$ is a space with reproducing kernel (see Aronszajn [17]).

We have not stipulated the choice of a density v on Y, but we will suppose that the space $\mathcal{L}^0$ is separated. For example, this is the case when Y is a bounded domain in $\mathbb{R}^n$ and $v = \omega dx$, where $\omega \in C(\overline{Y})$ is a positive function on Y.

Let $\{h^{(\nu)}\}_{\nu \in \mathbb{Z}_+}$ be some complete orthonormal system in $\mathcal{H}^0$. If $f \in \mathcal{H}^0$, then this section decomposes into the Fourier series $f = \sum c_\nu h^{(\nu)}$, which converges in the norm of the space $\mathcal{L}^0$, and hence in the topology of the space $\mathcal{E}^0$ because of estimate (4.2.11).

Further, we consider the series

$$H(x,y) = \sum_{\nu=0}^{\infty} h^{(\nu)}(x) \otimes \star h^{(\nu)}(y). \tag{4.2.12}$$

Theorem 4.2.24 *Series (4.2.12) converges in the topology of the space $\mathcal{E}(E^0 \boxtimes E^{0\prime}|_{Y\times Y})$, so that $H \in \mathcal{E}(E^0 \boxtimes E^{0\prime}|_{Y\times Y})$. If $x \in Y$ is fixed, then this series converges in the norm of the space $L^2(E^0_x \otimes E^{0\prime}|_Y)$.*

Proof. It is sufficient to consider the case when Y is an open set in $\mathbb{R}^n$.
Let

$$h^{(\nu)}(x) = \begin{pmatrix} h_1^{(\nu)}(x) \\ \dots \\ h_{k_0}^{(\nu)}(x) \end{pmatrix}$$

be representations of the sections $h^{(\nu)}$ where k_0 is the rank of E^0.

Assume that K_1 and K_2 are compact subsets of Y, and $K = K_1 \times K_2$. If $x \in K_1$ is a fixed point, then in view of the orthonormality of the system $\{h^{(\nu)}\}$ and estimate (4.2.11) (with $P = P_0(y)$) we obtain for $j = 1, \dots, k_0$

$$\begin{aligned} \left(\sum_{\nu=0}^{N} |h_j^{(\nu)}(x)|^2\right)^2 &\le \left|\sum_{\nu=0}^{N} \overline{h_j^{(\nu)}(x) h^{(\nu)}(x)}\right|^2 \\ &\le c_1 \int_Y \left|\sum_{\nu=0}^{N} \overline{h_j^{(\nu)}(x) h^{(\nu)}(y)}\right|^2 v(y) = c_1 \left(\sum_{\nu=0}^{N} |h_j^{(\nu)}(x)|^2\right), \end{aligned}$$

where the constant $c_1 > 0$ depends on P_0 and K_1 only.

Hence

$$\sum_{\nu=0}^{N} |h^{(\nu)}(x)|^2 \le c_1 k_0 \quad \text{for all} \quad x \in K_1. \tag{4.2.13}$$

Therefore, denoting by c_2 the constant obtained by analogy for compact K_2, we have for $(x,y) \in K$

$$\sum_{\nu=0}^{N} |h^{(\nu)}(x) \otimes \star h^{(\nu)}(y)| = \sum_{\nu=0}^{N} |h^{(\nu)}(x)| \cdot |h^{(\nu)}(y)| \le k_0\sqrt{c_1 c_2}.$$

This proves the absolute and uniform convergence on compact subsets of $X \times Y$ of the series (4.2.12). Then the convergence of this series in the topology of $\mathcal{E}(E^0 \boxtimes E^{0\prime}|_{Y\times Y})$ follows from Theorem 4.2.23.

Finally, (4.2.13) implies that, for fixed $y \in Y$, equality (4.2.12) gives the expansion of $H(x,y)$ in the complete orthonormal system $\{h^{(\nu)}\}$. To finish the proof, it is sufficient to observe that x and y enter into $H(x,y)$ in a symmetric way.

□

Theorem 4.2.25 (Bergman formula) *If $f \in \mathcal{H}^0$ then*

$$f(x) = \int_Y H(x,y)f(y) \quad \textit{for} \quad x \in Y. \tag{4.2.14}$$

Proof. Let $f = \sum c_\nu h^{(\nu)}$. Then by the previous theorem we get for fixed $x \in Y$

$$\begin{aligned} \int_Y H(x,y)f(y) &= \sum_{\mu,\nu} c_\mu h^{(\nu)}(x)\, \langle \star h^{(\nu)}, h^{(\mu)} \rangle_Y \\ &= \sum_{\mu,\nu} c_\mu h^{(\nu)}(x)\, \left(h^{(\mu)}, h^{(\nu)}\right)_Y \\ &= f(x), \end{aligned}$$

and the proof is complete.

□

For the Dolbeault complex in $\mathbb{C}^n$ Theorems 4.2.24 and 4.2.25 were proved by S. Bergman (see Aizenberg and Yuzhakov [6, §6]). Applications of Bergman's kernel function $H(x,y)$ in complex analysis are elucidated in the survey of Diederich [53]]. An asymptotic expansion of $H(x,y)$ near the diagonal of $(\partial Y \times \partial Y)$ for strongly pseudo-convex domains $Y \subset \mathbb{C}^n$ can be found in the survey of Henkin [103, (5.1)].

Combining formula (4.2.14) with Green's formula (2.5.9), as in the book of Aizenberg and Yuzhakov [6, §6], we can get the following generalization of a theorem of Bungart [34]. This theorem guarantees the existence for holomorphic functions of several variables of a representation by a boundary integral with a holomorphic kernel.

Theorem 4.2.26 *Let $P \in do_p(E \to F)$ be a differential operator with injective symbol on X which has a left fundamental solution. Then for each open set $Y \subset\subset X$ with smooth boundary there exists a kernel $K(x,y) \in C^{p-1}_{loc}(E \boxtimes F'|_{Y\times\partial Y})$ such that $P(x)K(x,y) = 0$ for $x \in Y$, and for all $f \in \mathcal{E}_P(Y) \cap C^{p-1}(E|_{\overline{Y}})$ we have*

$$f(x) = -\int_{\partial Y} G_P(K(x,y), f(y)) \quad (x \in Y). \tag{4.2.15}$$

Formula (4.2.15) is interesting from the standpoint of analysis rather than from the standpoint of partial differential equations. It seems that the information on the kernel is not sufficient to find useful applications in the theory of boundary value problems.

Bungart [33] constructed also a boundary kernel function (the Szegö kernel) for the closure in $L^2(\partial Y)$ of the set of functions holomorphic on $\overline{Y}$.

4.3 The Mayer-Vietoris sequence

4.3.1 *Basic example*

The Cauchy type integral in the theory of functions of a single complex variable provides a solution of the so-called "additive Riemann-Hilbert problem" (see Muskhelishvili [165, §37]).

Problem 4.3.1 *Let Y be a relatively compact domain in $\mathbb{C}^1$ with smooth boundary, and $f \in C^{0,\lambda}_{loc}(\partial Y)$ $(\lambda > 0)$. It is required to find two functions $f^\pm \in C_{loc}(\overline{\mathbb{C}^{1\pm}})$ homotopic in $\mathbb{C}^{1+} = \mathbb{C}^1 \setminus \overline{Y}$ and $\mathbb{C}^{1-} = Y$ respectively such that $f^+ - f^- = f$ on ∂Y.*

This problem is well-posed if f^+ is assumed to be regular at infinity.

In this section a straightforward generalization of the "additive Riemann-Hilbert problem" is considered, namely an analytic version of the Mayer-Vietoris sequence for cohomologies of a differential complex. The idea of such a generalization is due to Andreotti and Hill [9].

4.3.2 *Long cohomological sequence*

Let us be given some (cochain) complexes of topological vector spaces $K^\cdot$, $L^\cdot$ and $\Sigma^\cdot$ whose differentials are d_K, d_L and d_Σ respectively.

Suppose that these complexes are included in an exact sequence $0 \to \Sigma^\cdot \overset{M}{\to} L^\cdot \overset{M}{\to} K^\cdot \to 0$. This means that there is the following commutative diagram with exact rows:

$$\begin{array}{ccccccccc}
 & & \dots & & \dots & & \dots & & \\
 & & \downarrow & & \downarrow & & \downarrow & & \\
0 & \longrightarrow & \Sigma^0 & \overset{M}{\longrightarrow} & L^0 & \overset{M}{\longrightarrow} & K^0 & \longrightarrow & 0 \\
 & & \downarrow d_\Sigma & & \downarrow d_L & & \downarrow d_K & & \\
0 & \longrightarrow & \Sigma^1 & \overset{M}{\longrightarrow} & L^1 & \overset{M}{\longrightarrow} & K^1 & \longrightarrow & 0 \\
 & & \downarrow d_\Sigma & & \downarrow d_L & & \downarrow d_K & & \\
0 & \longrightarrow & \Sigma^2 & \overset{M}{\longrightarrow} & L^2 & \overset{M}{\longrightarrow} & K^2 & \longrightarrow & 0 \\
 & & \downarrow d_\Sigma & & \downarrow d_L & & \downarrow d_K & & \\
 & & \dots & & \dots & & \dots & &
\end{array} \qquad (4.3.1)$$

The mapping M induces homomorphisms of the cohomology spaces $H^i(\Sigma^\cdot) \stackrel{M}{\to} H^i(L^\cdot) \stackrel{M}{\to} H^i(K^\cdot)$ $(i \in \mathbb{Z})$ which are denoted by M again. To join these fragments we define the so-called binding homomorphisms $M_b : H^i(K^\cdot) \to H^{i+1}(\Sigma^\cdot)$ in the following way.

Let $[k_i]$ be some class of $H^i(K^\cdot)$, and $k_i \in Z^i(K^\cdot)$ be a representative of this class. Since the i-th row of (4.3.1) is exact, there exists an element $l_i \in L^i$ such that $Ml_i = k_i$. Consider the element $d_L l_i$ of L^{i+1}. We have $M(d_L l_i) = d_K k_i = 0$ because (4.3.1) is commutative. So there exists an element $\sigma_{i+1} \in \Sigma^{i+1}$ satisfying $M\sigma_{i+1} = d_L l_i$. We have $d_\Sigma \sigma_{i+1} = 0$ because $M(d_\Sigma \sigma_{i+1}) = (d_L)^2 l_i = 0$ and the mapping $M : \Sigma^{i+2} \to L^{i+2}$ is injective. We now let $M_b([k_i]) = [\sigma_{i+1}]$ where $[\sigma_{i+1}]$ is the class of the element σ_{i+1} in $H^{i+1}(\Sigma^\cdot)$.

Exercise 4.3.2 *Prove that M_b is thus well-defined, i.e., $M_b([k_i])$ does not depend on the particular choice of a representative of the class $[k_i]$ in $Z^i(K^\cdot)$.*

Hint. If $k_i = d_K k_{i-1}$ for some $k_{i-1} \in K^{i-1}$ and $k_{i-1} = Ml_{i-1}$ for some $l_{i-1} \in L^{i-1}$, then $l_i - d_L l_{i-1} = M\sigma_i$ for some $\sigma_i \in \Sigma^i$ and $\sigma_{i+1} = d_\Sigma \sigma_i$.

□

In homological algebra the homomorphism M_b is called the Bockstein homomorphism (see Dold [56, Ch.II,§2]).

Theorem 4.3.3 *The following long cohomological sequence is exact:*

$$\begin{array}{cccccccc} \ldots & \longrightarrow & H^0(\Sigma^\cdot) & \stackrel{M}{\longrightarrow} & H^0(L^\cdot) & \stackrel{M}{\longrightarrow} & H^0(K^\cdot) & \stackrel{M_b}{\longrightarrow} \\ & \longrightarrow & H^1(\Sigma^\cdot) & \stackrel{M}{\longrightarrow} & H^1(L^\cdot) & \stackrel{M}{\longrightarrow} & H^1(K^\cdot) & \stackrel{M_b}{\longrightarrow} \\ & \longrightarrow & H^2(\Sigma^\cdot) & \stackrel{M}{\longrightarrow} & H^2(L^\cdot) & \stackrel{M}{\longrightarrow} & H^2(K^\cdot) & \stackrel{M_b}{\longrightarrow} \ \ldots . \end{array} \qquad (4.3.2)$$

Proof. See Dold [56, Ch.II, §2].

□

As Andreotti and Kas [12] proved, the binding homomorphism M_b is a continuous mapping provided the spaces $K^\cdot$, $L^\cdot$ and $\Sigma^\cdot$ are not too "bad".

4.3.3 *The Mayer-Vietoris sequence for open sets (the case of arbitrary supports)*

Suppose that X is a differentiable manifold of dimension n countable at infinity and $E^\cdot$ is a (cochain) differential complex on X.

The Mayer-Vietoris sequence for a pair of open sets $U = \{U_\nu\}_{\nu=1,2}$ is an easy consequence of the algebraic construction of a long cohomological sequence (see Subsection 4.3.2), and partition of unity on $U_1 \cup U_2$ subordinate to the covering U.

Theorem 4.3.4 *There exists an exact sequence of cohomology spaces:*

$$\begin{array}{llllll} 0 \to & Z^0(\mathcal{E}^\cdot(U_1 \cup U_2)) & \stackrel{\delta}{\to} & Z^0(\mathcal{E}^\cdot(U_1)) \oplus Z^0(\mathcal{E}^\cdot(U_2)) & \stackrel{\delta}{\to} & Z^0(\mathcal{E}^\cdot(U_1 \cap U_2)) \stackrel{\delta_b}{\to} \\ \to & H^1(\mathcal{E}^\cdot(U_1 \cup U_2)) & \to & \ldots & & \\ \to & H^i(\mathcal{E}^\cdot(U_1 \cup U_2)) & \stackrel{\delta}{\to} & H^i(\mathcal{E}^\cdot(U_1)) \oplus H^i(\mathcal{E}^\cdot(U_2)) & \stackrel{\delta}{\to} & H^i(\mathcal{E}^\cdot(U_1 \cap U_2)) \stackrel{\delta_b}{\to} \\ \to & H^{i+1}(\mathcal{E}^\cdot(U_1 \cup U_2)) & \to & \ldots . & & \end{array}$$

(4.3.3)

Proof. Using a partition of unity on $U_1 \cup U_2$ subordinate to the covering U (see Corollary 1.1.6), it is easy to show that the following sequence of cochain complexes is exact:

$$0 \dashrightarrow \mathcal{E}\,(U_1 \cup U_2) \xrightarrow{\delta} \mathcal{E}\,(U_1) \oplus \mathcal{E}\,(U_2) \xrightarrow{\delta} \mathcal{E}\,(U_1 \cap U_2) \longrightarrow 0. \qquad (4\,3.4)$$

Here the mapping δ in the first fragment is given by the restriction from $U_1 \cup U_2$ to U_1 and U_2, and the mapping δ in the second fragment measures the "jump" when crossing $U_1 \cap U_2$, i.e., $\delta(f_1, f_2) = f_2 - f_1$.

To obtain (4.3.3) it remains to write the long cohomological sequence corresponding to (4.3.4).

□

The construction of the binding homomorphism δ_b in sequence (4.3.3) is the following.

Remark 4.3.5 *Let $[f]$ be some class of $H^\imath(\mathcal{E}\,(U_1 \cup U_2))$, and $f \in Z^\imath(\mathcal{E}\,(U_1 \cup U_2))$ be a representative of $[f]$. Choose a partition of unity $\{\varphi_\nu\}_{\nu=1,2}$ of the class C^∞_{loc} on $U_1 \cup U_2$ subordinate to the covering U. Then $\delta_b([f])$ is the image in $H^{\imath+1}(\mathcal{E}\,(U_1 \cup U_2))$ of the section from $Z^{\imath+1}(\mathcal{E}\,(U_1 \cup U_2))$ given by $P(-\varphi_2 f)$ on U_1 and by $P(\varphi_1 f)$ on U_2. Actually the class $\delta_b([f])$ does not depend on the particular choice of $\{\varphi_\nu\}$.*

Let us consider two instructive examples.

Example 4.3.6 Suppose that $E = \Lambda\, T^*_{\mathbb{C}}(X)$ is the de Rham complex on X. By Proposition 1.2.42, we have isomorphisms $H^\imath(\mathcal{E}\,(U)) \cong H^\imath(U, \mathbb{C})$ $(\imath \in \mathbb{Z})$. Using now Theorem 4.3.4 we get the following exact sequence of cohomology spaces

$$\begin{array}{cccccccc}
0 & \to & \mathbb{C}(U_1 \cup U_2) & \xrightarrow{\delta} & \mathbb{C}(U_1) \oplus \mathbb{C}(U_2) & \xrightarrow{\delta} & \mathbb{C}(U_1 \cap U_2) & \xrightarrow{\delta} \\
 & \to & H^1(U_1 \cup U_2, \mathbb{C}) & \to & \dots & & & \\
 & \to & H^\imath(U_1 \cup U_2, \mathbb{C}) & \xrightarrow{\delta} & H^\imath(U_1, \mathbb{C}) \oplus H^\imath(U_2, \mathbb{C}) & \xrightarrow{\delta} & H^\imath(U_1 \cap U_2, \mathbb{C}) & \xrightarrow{\delta_b} \\
 & \to & H^{\imath+1}(U_1 \cup U_2, \mathbb{C}) & \to & \dots & & & \\
 & \to & H^n(U_1 \cup U_2, \mathbb{C}) & \xrightarrow{\delta} & H^n(U_1, \mathbb{C}) \oplus H^n(U_2, \mathbb{C}) & \xrightarrow{\delta} & H^n(U_1 \cap U_2, \mathbb{C}) & \to 0.
\end{array} \qquad (4.3.5)$$

This is just the classical Mayer-Vietoris sequence for cohomologies (see Dold [56, Ch.III, §8], Massey [160, §8.6], etc.).

□

Example 4.3.7 Suppose that $E = \Lambda^{p,\cdot}\, T^*_{\mathbb{C}}(X)$ is the Dolbeaut complex with coefficients in the bundle $\Lambda^{p,0}T^*_{\mathbb{C}}(X)$ on a complex manifold X of dimension n. Then by Proposition 1.2.49 we have isomorphisms $H^\imath(\mathcal{E}\,(U)) \cong H^\imath(U, \mathcal{O}^{(p)})$ $(\imath \in \mathbb{Z})$. Using now Theorem 4.3.4 we obtain the following exact sequence of cohomology spaces

$$
\begin{array}{llllll}
 & & & & 0 & \to \\
\to & \mathcal{O}^{(p)}(U_1 \cup U_2) & \xrightarrow{\delta} & \mathcal{O}^{(p)}(U_1) \oplus \mathcal{O}^{(p)}(U_2) & \xrightarrow{\delta} \mathcal{O}^{(p)}(U_1 \cap U_2) & \xrightarrow{\delta_b} \\
\to & H^1(U_1 \cup U_2, \mathcal{O}^{(p)}) & \to & \dots & & \\
\to & H^i(U_1 \cup U_2, \mathcal{O}^{(p)}) & \xrightarrow{\delta} & H^i(U_1, \mathcal{O}^{(p)}) \oplus H^i(U_2, \mathcal{O}^{(p)}) & \xrightarrow{\delta} H^i(U_1 \cap U_2, \mathcal{O}^{(p)}) & \xrightarrow{\delta_b} \\
\to & H^{i+1}(U_1 \cup U_2, \mathcal{O}^{(p)}) & \to & \dots & & \\
\to & H^n(U_1 \cup U_2, \mathcal{O}^{(p)}) & \xrightarrow{\delta} & H^n(U_1, \mathcal{O}^{(p)}) \oplus H^n(U_2, \mathcal{O}^{(p)}) & \xrightarrow{\delta} H^n(U_1 \cap U_2, \mathcal{O}^{(p)}) & \to \\
\to & 0. & & & &
\end{array}
\tag{4.3.6}
$$

This is a particular case of the Mayer-Vietoris sequence for cohomologies with coefficients in a sheaf (see Godement [72, Ch.II, §5.6]).

□

Notice that (4.3.4) is simply the Cěch bicomplex $C\,(U, \mathcal{E}\,)$ corresponding to the covering U of $U_1 \cup U_2$ (see diagram (5.2.2) below). Then, the long cohomological sequence (4.3.3) is an expression of the fact that the δ-cohomology of this bicomplex is trivial. From this point of view it is clear how to extend the Mayer-Vietoris principle to the case of any finite or countable coverings $U = \{U_\nu\}_{\nu \in N}$ of the manifold X. Namely, by the bicomplex $C\,(U, \mathcal{E}\,)$ one can construct the complex L with the spaces $L^p = \oplus_{q+i=p} C^q(U, \mathcal{E}^i)$ and the differential d_L of degree 1 defined by $d_L = \delta + (-1)^q P$ on $C^q(U, \mathcal{E}^i)$. Then the generalized Mayer-Vietoris principle states that the mapping $\delta : \mathcal{E}\,(X) \to L$, given by restriction, induces the isomorphism of cohomologies $H\,(\mathcal{E}\,(X)) \cong H\,(L\,)$.

Exercise 4.3.8 *Prove that the mapping $\delta : \mathcal{E}\,(X) \to L$ is homotopically invertible.*

Hint. Follow Bott and Tu [25] who proved this fact for the de Rham complex.

□

It is pertinent to note that the analogy between the theory of complexes of differential operators and complex analysis is not superficial.

Remark 4.3.9 *Because of (4.3.6), the Dolbeault cohomology of the union of two 1-convex domains in $\mathbb{C}^n$ (= domains of holomorphy) is supported at the degrees 0 and 1. However, the de Rham cohomology of the union of two 1-convex domains in $\mathbb{R}^n$ may be much more complicated (say, this cohomology may be non-trivial at degree 2).*

4.3.4 *The Mayer-Vietoris sequence for open sets (the case of compact supports)*

For complexes of spaces of compactly supported sections, the sequence (4.3.4) fails. The point is that the correspondence $U \to \mathcal{E}'\,(U)$ is a covariant functor with respect to inclusions of open sets (i.e., cosheaf) whereas the correspondence $U \to \mathcal{E}\,(U)$ is a contravariant functor (i.e., sheaf). Hence, the sequence (4.3.3) needs to be modified.

Theorem 4.3.10 *There exists an exact sequence of homology spaces*

$$\begin{array}{llllll} 0 & \leftarrow & Z_0(\mathcal{E}_\cdot'(U_1 \cup U_2)) & \stackrel{\partial}{\leftarrow} & Z_0(\mathcal{E}_\cdot'(U_1)) \oplus Z_0(\mathcal{E}_\cdot'(U_2)) & \stackrel{\partial}{\leftarrow} & Z_0(\mathcal{E}_\cdot'(U_1 \cap U_2)) & \stackrel{\partial_b}{\leftarrow} \\ & \leftarrow & H_1(\mathcal{E}_\cdot'(U_1 \cup U_2)) & \leftarrow & \ldots & & & \\ & \leftarrow & H_\imath(\mathcal{E}_\cdot'(U_1 \cup U_2)) & \stackrel{\partial}{\leftarrow} & H_\imath(\mathcal{E}_\cdot'(U_1)) \oplus H_\imath(\mathcal{E}_\cdot'(U_2)) & \stackrel{\partial}{\leftarrow} & H_\imath(\mathcal{E}_\cdot'(U_1 \cap U_2)) & \stackrel{\partial_b}{\leftarrow} \\ & \leftarrow & H_{\imath+1}(\mathcal{E}_\cdot'(U_1 \cup U_2)) & \leftarrow & \ldots . & & & \end{array} \qquad (4.3.7)$$

Proof. We use a partition of unity on $U_1 \cup U_2$ subordinate to the covering U (see Corollary 1.1.6) to conclude that the following sequence of chain complexes is exact:

$$0 \leftarrow \mathcal{E}_\cdot'(U_1 \cup U_2) \stackrel{\partial}{\leftarrow} \mathcal{E}_\cdot'(U_1) \oplus \mathcal{E}_\cdot'(U_2) \stackrel{\partial}{\leftarrow} \mathcal{E}_\cdot'(U_1 \cap U_2) \leftarrow 0. \qquad (4.3.8)$$

Here the mapping ∂ in the first fragment is given by $(g_1, g_2) \to g_1 + g_2$, and the mapping ∂ in the second fragment is given by $g \to (-g|_{U_1}, g|_{U_2})$.

To finish the proof it remains to write out the long homological sequence corresponding to (4.3.8).

□

The construction of the binding homomorphism ∂_b in the sequence (4.3.7) is the following

Remark 4.3.11 *Let $[g]$ be some class in $H_{\imath+1}(\mathcal{E}_\cdot'(U_1 \cup U_2))$, and $g \in Z_{\imath+1}(\mathcal{E}_\cdot'(U_1 \cup U_2))$ be a representative of $[g]$. Choose a partition of unity $\{\varphi_\nu\}_{\nu=1,2}$ of the class C^∞_{loc} on $U_1 \cup U_2$ subordinate to the covering U. Then $\partial_b([g])$ is the image in $H_\imath(\mathcal{E}_\cdot'(U_1 \cap U_2))$ of the section from $Z_\imath(\mathcal{E}_\cdot'(U_1 \cap U_2))$ given by $P'(\varphi_\nu g)$ on U_ν $(\nu = 1, 2)$. This class does not depend on the particular choice of $\{\varphi_\nu\}$.*

For the de Rham complex on X, the Mayer-Vietoris sequence (4.3.7) is well-known (see Bott and Tu [25, p.35]).

4.3.5 *The Mayer-Vietoris sequence for closed sets*

We now pass to a more complicated version of the Mayer-Vietoris sequence.

Suppose that S is a smooth (i.e., C^∞_{loc}) hypersurface in X which has two sides in X. This means that $X \setminus S = X^- \cup X^+$ where $X^\pm$ are disjoint non-void open subsets of X, and S is oriented as the boundary of $\overline{X^-}$.

The following result is due to Andreotti et al. [8] who developed the earlier approach of Andreotti and Hill [9] for the Dolbeault complex.

Theorem 4.3.12 *Let $\mathfrak{C}$ be $\mathcal{E}$ or $\mathcal{D}$, and S be formally non-characteristic for the complex $\mathfrak{C}^\cdot(X)$. Then there exists an exact sequence of cohomology spaces*

$$\begin{array}{llllll} 0 & \to & Z^0(\mathfrak{C}^\cdot(X)) & \stackrel{\delta}{\to} & Z^0(\mathfrak{C}^\cdot(\overline{X^-})) \oplus Z^0(\mathfrak{C}^\cdot(\overline{X^+})) & \stackrel{\delta}{\to} & Z^0(\mathfrak{C}^\cdot_{[t]}(S)) & \stackrel{\delta_b}{\to} \\ & \to & H^1(\mathfrak{C}^\cdot(X)) & \to & \ldots & & & \\ & \to & H^\imath(\mathfrak{C}^\cdot(X)) & \stackrel{\delta}{\to} & H^\imath(\mathfrak{C}^\cdot(\overline{X^-})) \oplus H^\imath(\mathfrak{C}^\cdot(\overline{X^+})) & \stackrel{\delta}{\to} & H^\imath(\mathfrak{C}^\cdot_{[t]}(S)) & \stackrel{\delta_b}{\to} \\ & \to & H^{\imath+1}(\mathfrak{C}^\cdot(X)) & \to & \ldots . & & & \end{array} \qquad (4.3.9)$$

Proof. Using the exactness of the third sequence in (3.1.4), it is easy to see that the following sequence of cochain complexes is exact:

$$0 \longrightarrow \mathfrak{C}^{\cdot}(X) \xrightarrow{\delta} \mathfrak{C}^{\cdot}(\overline{X^-}) \oplus \mathfrak{C}^{\cdot}(\overline{X^+}) \xrightarrow{\delta} \mathfrak{C}^{\cdot}(S) \longrightarrow 0.$$

Here the mapping δ in the first fragment is given by the restriction from X to $\overline{X^-}$ and $\overline{X^+}$, and the mapping δ in the second fragment is determined by $\delta(f_1, f_2) = j_S^\infty(f_2) - j_S^\infty(f_1)$. In other words it measures the jumps of all normal derivatives when crossing S.

Applying the long cohomological sequence of Theorem 4.3.3 we get

$$\begin{array}{llllll} 0 & \to & Z^0(\mathfrak{C}^{\cdot}(X)) & \xrightarrow{\delta} & Z^0(\mathfrak{C}^{\cdot}(\overline{X^-})) \oplus Z^0(\mathfrak{C}^{\cdot}(\overline{X^+})) & \xrightarrow{\delta} & Z^0(\mathfrak{C}^{\cdot}(S)) & \xrightarrow{\delta_b} \\ & \to & H^1(\mathfrak{C}^{\cdot}(X)) & \to & \ldots & & & \\ & \to & H^i(\mathfrak{C}^{\cdot}(X)) & \xrightarrow{\delta} & H^i(\mathfrak{C}^{\cdot}(\overline{X^-})) \oplus H^i(\mathfrak{C}^{\cdot}(\overline{X^+})) & \xrightarrow{\delta} & H^i(\mathfrak{C}^{\cdot}(S)) & \xrightarrow{\delta_b} \\ & \to & H^{i+1}(\mathfrak{C}^{\cdot}(X)) & \to & \ldots . & & & \end{array} \tag{4.3.10}$$

We now write out the first exact sequence of (3.1.10)

$$0 \longrightarrow \mathfrak{C}^{\cdot}_t(S) \xrightarrow{\iota} \mathfrak{C}^{\cdot}(S) \xrightarrow{t} \mathfrak{C}^{\cdot}_{[t]}(S) \longrightarrow 0.$$

From Theorem 4.3.3 we obtain the exact sequence of cohomology spaces

$$\begin{array}{llllllll} 0 & \longrightarrow & Z^0(\mathfrak{C}^{\cdot}_t(S)) & \longrightarrow & Z^0(\mathfrak{C}^{\cdot}(S)) & \longrightarrow & Z^0(\mathfrak{C}^{\cdot}_{[t]}(S)) & \longrightarrow \\ & \longrightarrow & H^1(\mathfrak{C}^{\cdot}_t(S)) & \longrightarrow & \ldots & & & \\ & \longrightarrow & H^i(\mathfrak{C}^{\cdot}_t(S)) & \longrightarrow & H^i(\mathfrak{C}^{\cdot}(S)) & \longrightarrow & H^i(\mathfrak{C}^{\cdot}_{[t]}(S)) & \longrightarrow \\ & \longrightarrow & H^{i+1}(\mathfrak{C}^{\cdot}_t(S)) & \longrightarrow & \ldots . & & & \end{array} \tag{4.3.11}$$

Since the hypersurface S is formally non-characteristic for the complex $\mathfrak{C}^{\cdot}(X)$, we have $H^i(\mathfrak{C}^{\cdot}_t(S)) = H^{i+1}(\mathfrak{C}^{\cdot}_t(S)) = 0$. Therefore (4.3.11) implies $H^i(\mathfrak{C}^{\cdot}(S)) \cong H^i(\mathfrak{C}^{\cdot}_{[t]}(S))$ for all $i \in \mathbb{Z}$. Substituting these equalities in (4.3.10), we get the exact sequence (4.3.9) as required.

□

Let us now describe the homomorphisms for the sequence (4.3.9).

Remark 4.3.13 *The homomorphism* $H^i(\mathfrak{C}^{\cdot}(X)) \xrightarrow{\delta} H^i(\mathfrak{C}^{\cdot}(\overline{X^-})) \oplus H^i(\mathfrak{C}^{\cdot}(\overline{X^+}))$ *is induced by the restriction mapping* $f \to (f|_{\overline{X^-}}, f|_{\overline{X^+}})$. *The homomorphism*

$$H^i(\mathfrak{C}^{\cdot}(\overline{X^-})) \oplus H^i(\mathfrak{C}^{\cdot}(\overline{X^+})) \xrightarrow{\delta} H^i(\mathfrak{C}^{\cdot}_{[t]}(S))$$

is induced by the mapping $(f_1, f_2) \to t(f_2) - t(f_1)$ *where* $t : \mathfrak{C}^i(\overline{X^\pm}) \to \mathfrak{C}_{[t_i]}(S)$ *is the quotient mapping (see Subsection 3.1.4). To describe the homomorphism* δ_b, *we note that each class* $[f] \in H^i(\mathfrak{C}^{\cdot}_{[t]}(S))$ *has a (distinguished) representative* $f_0 \in \mathfrak{C}^i(X)$ *such that* Pf_0 *is flat on* S. *In fact, we have* $Pf \in \mathfrak{C}_{t_{i+1}}(X)$ *for any representative* f *of* $[f]$. *Since* $P(Pf) = 0$, *it follows from the equality* $H^{i+1}\left(\frac{\mathfrak{C}_t(X)}{\mathfrak{C}_j(X)}\right) = 0$ *that there exists a section* $f' \in \mathfrak{C}_{t_i}(X)$ *such that* $(Pf - Pf')$ *is flat on* S. *Therefore* $f_0 = f - f'$ *is a distinguished representative of* $[f]$ *because* $[f - f'] = [f]$ *in* $H^i(\mathfrak{C}^{\cdot}_{[t]}(S))$. *We now let* $f_1 = Pf_0$ *on* $\overline{X^-}$ *and* $f_1 = -Pf_0$ *on* $\overline{X^+}$, *where* $f_0 \in \mathfrak{C}^i(X)$ *is a distinguished representative of the class* $[f] \in H^i(\mathfrak{C}^{\cdot}_{[t]}(S))$. *Then* $f_1 \in Z^{i+1}(\mathfrak{C}^{\cdot}(X))$ *and* δ_b *is induced by the mapping* $f \to f_1$. *As to* δ_b *being well-defined, see Andreotti and Hill [9].*

4.3.6 *The Mayer-Vietoris sequence for elliptic complexes*

For elliptic complexes $E^\cdot$, one can find a more satisfactory variant of the Mayer-Vietoris sequence.

Let Π be the properly supported parametrix of the complex $E^\cdot$ constructed in Theorem 2.1.12, so that $\Pi P f + P\Pi f = f - Rf$ for all $f \in \mathcal{D}'(E^\cdot)$.

Theorem 4.3.14 *For $\mathfrak{C} = \mathcal{E}$ or $\mathcal{D}$, the following sequence of cohomology spaces is exact:*

$$\begin{array}{lllllll} 0 & \to & Z^0(\mathfrak{C}^\cdot(X)) & \overset{\delta}{\to} & Z^0(\mathfrak{C}^\cdot(\overline{X^-})) \oplus Z^0(\mathfrak{C}^\cdot(\overline{X^+})) & \overset{\delta}{\to} & Z^0(\mathfrak{C}_{[t\,]}(S)) \overset{R\circ[S]^P}{\to} \\ & \to & H^1(\mathfrak{C}^\cdot(X)) & \to & \cdots & & \\ & \to & H^\cdot(\mathfrak{C}^\cdot(X)) & \overset{\delta}{\to} & H^\cdot(\mathfrak{C}^\cdot(\overline{X^-})) \oplus H^\cdot(\mathfrak{C}^\cdot(\overline{X^+})) & \overset{\delta}{\to} & H^\cdot(\mathfrak{C}_{[t\,]}(S)) \overset{R\circ[S]^P}{\to} \\ & \to & H^{\cdot+1}(\mathfrak{C}^\cdot(X)) & \to & \cdots . & & \end{array} \tag{4.3.12}$$

Proof.

1) The exactness of (4.3.12) at $Z^0(\mathfrak{C}\,(X))$ is obvious.

2) Now we check the exactness of the sequence (4.3.12) at the term $H^\cdot(\mathfrak{C}^\cdot(\overline{X^-}))\oplus H^\cdot(\mathfrak{C}^\cdot(\overline{X^+}))$.

Let $f \in Z^\cdot(\mathfrak{C}^\cdot(X))$. Then $\delta^2 f = t(f^+) - t(f^-) = 0$ where $f^\pm$ means the restriction of f to $X^\pm$. Thus we have $\delta^2 = 0$.

Suppose further that $f_1 \in Z^\cdot(\mathfrak{C}^\cdot(\overline{X^-}))$, $f_2 \in Z^\cdot(\mathfrak{C}^\cdot(\overline{X^+}))$ and $t(f_2) - t(f_1) = P_b u_0$ where $u_0 \in \mathfrak{C}_{[t_\cdot -1]}(S)$. We extend f_ν $(\nu = 1,2)$ to sections from $\mathfrak{C}^\cdot(X)$ and we denote these extensions by f_ν again.

We let $u = \Pi([S]^P u_0) - \Pi(\chi^-(f_2 - f_1))$. By Proposition 3.3.7, $u^\pm \in \mathfrak{C}^{\cdot-1}(\overline{X^\pm})$. Using the first equalities of (3.3.4) and (3.3.24), we conclude also that $t(u^+) - t(u^-) = u_0$.

Applying Theorems 2.1.12, 3.1.26 and 2.4.8, we obtain

$$\begin{aligned} Pu &= \left\{[S]^P u_0 - R([S]^P u_0) - \Pi P([S]^P u_0)\right\} - P\Pi(\chi^-(f_2 - f_1)) \\ &= [S]^P u_0 - R([S]^P u_0) - \left\{-\Pi([S]^P(f_2 - f_1)) + \Pi P(\chi^-(f_2 - f_1))\right\} \\ &= [S]^P u_0 - \chi^-(f_2 - f_1) \\ &\qquad + \left\{R(-[S]^P u_0 - \chi^-(f_2 - f_1)) + \Pi(\chi^- P(f_2 - f_1))\right\}. \end{aligned} \tag{4.3.13}$$

Since the section $P(f_2 - f_1)$ is flat on S, we see that $\chi^- P(f_2 - f_1)$ in $\mathfrak{C}^{\cdot+1}(X)$. Therefore (4.3.13) implies that $Pu = [S]^P u_0 - \chi^-(f_2 - f_1)$ modulo $\mathfrak{C}^\cdot(X)$.

Hence it follows that the section $(Pu^+ - Pu^-) - (f_2 - f_1)$ is flat on S.

Consider now the section $f = \chi^- f_1 + \chi^+ f_2 - Pu$. As we saw, $f \in \mathfrak{C}^\cdot(X)$ so $f \in Z^\cdot(\mathfrak{C}^\cdot(X))$. Besides that, the classes of the sections δf and (f_1, f_2) in $H^\cdot(\mathfrak{C}^\cdot(\overline{X^-})) \oplus H^\cdot(\mathfrak{C}^\cdot(\overline{X^+}))$ coincide.

This proves the desired equality im δ = ker δ.

3) Now we show that the sequence (4.3.12) is exact at $H^\cdot(\mathfrak{C}_{[t\,]}(S))$.

Let $f \in \mathfrak{C}_{[t,]}(S)$ and $f = t(f_2) - t(f_1)$, where $f_1 \in Z^{\cdot}(\mathfrak{C}^{\cdot}(\overline{X^-}))$ and $f_2 \in Z^{\cdot}(\mathfrak{C}^{\cdot}(\overline{X^+}))$. Using Theorem 2.1.12 and formula (2.4.23), we get

$$\begin{aligned} R([S]^P f) &= R([S]^P f_2) - R([S]^P f_1) = RP(\chi^+ f_2) + RP(\chi^- f_1) \\ &= PR(\chi^- f_1 + \chi^+ f_2) \end{aligned}$$

where $R(\chi^- f_1 + \chi^+ f_2) \in \mathfrak{C}^{\cdot}(X)$. Thus the class in $H^{\cdot+1}(\mathfrak{C}^{\cdot}(X))$ of the section $R([S]^P f)$ is zero which proves semiexactness.

Suppose now that $f \in Z^{\cdot}(\mathfrak{C}_{[t\,]}(S))$ and $R([S]^P f) = Pu$, where $u \in \mathfrak{C}^{\cdot}(X)$. We write $F = \Pi([S]^P f) + u$. According to Theorem 3.3.9, $F^{\pm} \in \mathfrak{C}^{\cdot}(\overline{X^{\pm}})$ and $t(F^+) - t(F^-) = f$.

Apply now Theorems 2.1.12 and 3.1.26 to obtain

$$PF = \left\{[S]^P f - R([S]^P f) - \Pi P([S]^P f)\right\} + Pu = [S]^P f.$$

Thus $F^{\pm} \in Z^{\cdot}(\mathfrak{C}^{\cdot}(\overline{X^{\pm}}))$ and $\delta(F^-, F^+) = f$, which proves the equality im $\delta =$ ker $R \circ [S]^P$ as desired.

4) It remains to check the exactness at the term $H^{\cdot+1}(\mathfrak{C}^{\cdot}(X))$.

Let $f \in \mathfrak{C}^{\cdot+1}(X)$ and $f = R([S]^P u)$ for some jet $u \in Z^{\cdot}(\mathfrak{C}_{[t\,]}(S))$. Using Theorems 2.1.12 and 3.1.26, we obtain

$$f = [S]^P u - \Pi P([S]^P u) - P\Pi([S]^P u) = [S]^P u - P\Pi([S]^P u),$$

so $\delta f = (P\Pi(-[S]^P u)^-, P\Pi(-[S]^P u)^+)$. Since the restrictions $\Pi(-[S]^P u)^{\pm}$ are of class $\mathfrak{C}^{\cdot}(\overline{X^{\pm}})$ in view of Corollary 3.3.8, we conclude that the class of δf in $H^{\cdot+1}(\mathfrak{C}^{\cdot}(\overline{X^-})) \oplus H^{\cdot+1}(\mathfrak{C}^{\cdot}(\overline{X^+}))$ is zero. This proves semiexactness.

Suppose now that $f \in Z^{\cdot+1}(\mathfrak{C}^{\cdot}(X))$ and $f = Pu_1$ on X^-, where $u_1 \in \mathfrak{C}^{\cdot}(\overline{X^-})$, while $f = Pu_2$ on X^+, where $u_2 \in \mathfrak{C}^{\cdot}(\overline{X^+})$.

Let $u = t(u_1) - t(u_2)$. Obviously $u \in Z^{\cdot}(\mathfrak{C}_{[t\,]}(S))$, and it suffices to prove that $R([S]^P u) = f$ modulo $B^{\cdot+1}(\mathfrak{C}^{\cdot}(X))$.

To do this we again use Theorem 2.1.12 and formula (2.4.23), according to which

$$\begin{aligned} R([S]^P u) &= R([S]^P u_1) - R([S]^P u_2) \\ &= R(\chi^- Pu_1 - P(\chi^- u_1)) + R(\chi^+ Pu_2 - P(\chi^+ u_2)) \\ &= Rf - PR(\chi^- u_1 + \chi^+ u_2) \\ &= f - P\left\{R(\chi^- u_1 + \chi^+ u_2) + \Pi f\right\}, \end{aligned}$$

and the proof is completed by observing that the expression in braces on the right-hand side lies in $\mathfrak{C}^{\cdot}(X)$.

Therefore im $R \circ [S]^P =$ ker δ, and the theorem is proved.

□

Remark 4.3.15 *As one sees from the proof of the theorem, the inverse for the mapping* $H^{\cdot}(\mathfrak{C}^{\cdot}(\overline{X^-})) \oplus H^{\cdot}(\mathfrak{C}^{\cdot}(\overline{X^+})) \xrightarrow{\delta} H^{\cdot}(\mathfrak{C}_{[t\,]}(S))$ *in (4.3.12) is* $\delta \circ (\Pi([S]^P \cdot) + P^{-1}R([S]^P \cdot))$, *where* Π *is any properly supported pseudo-differential parametrix of the complex* $E^{\cdot}$.

It follows from (the proof of) Theorem 4.3.14 that if $E^\cdot$ has a fundamental solution at degree $(i+1)$, say, Φ, then the mapping $H^i(\mathcal{E}^\cdot(\overline{X^-})) \oplus H^i(\mathcal{E}^\cdot(\overline{X^+})) \xrightarrow{\delta} H^i(\mathcal{E}_{[t]}(S))$ is surjective provided S is compact. The inverse mapping is given by $\delta \circ \Phi([S]^P \cdot)$. For the Dolbeault complex with coefficients in $\Lambda^{p,0}T^*_{\mathbb{C}}(\mathbb{C}^n)$ this result was proved by Dautov (see Theorem 2.11 in Aizenberg and Dautov [5]). For the de Rham complex in $\mathbb{R}^n$, it was proved by the author [219], and for arbitrary elliptic complexes of differential operators with constant coefficients in $\mathbb{R}^n$, it was also proved by the author [226]. In full generality Theorem 4.3.14 was proved by the author [231], [233]. The tools of integral representations allow one to construct versions of the Mayer-Vietoris sequence for cohomologies of the complex $E^\cdot$ in the classes of Hölder, Sobolev, generalized and other sections also. The weaker the restrictions on a class of sections the stronger the restrictions on the smoothness of S. This techniques has been well-known in complex analysis (see Chirka [38] for instance).

4.3.7 *Examples*

We consider three interesting examples.

Example 4.3.16 Suppose that $E^\cdot$ is some elliptic complex on X. According to Theorem 3.1.22, there exists a differential complex $E^\cdot_t$ on S such that $\mathfrak{C}_{[t]}(S) \cong \mathfrak{C}(E^\cdot_t)$. For complexes of first order differential operators, a construction of $E^\cdot_t$ is given in Corollary 3.2.14. Then, Theorem 4.3.14 shows that, for $\mathfrak{C} = \mathcal{E}$ and $\mathcal{D}$, the following sequence of cohomology spaces is exact:

$$
\begin{array}{rlcccl}
0 \to & Z^0(\mathfrak{C}^\cdot(X)) & \xrightarrow{\delta} & Z^0(\mathfrak{C}^\cdot(\overline{X^-})) \oplus Z^0(\mathfrak{C}^\cdot(\overline{X^+})) & \xrightarrow{\delta} & Z^0(\mathfrak{C}(E^\cdot_t)) \overset{R\circ[S]^P}{\to} \\
\to & H^1(\mathfrak{C}^\cdot(X)) & \to & \cdots & & \\
\to & H^i(\mathfrak{C}^\cdot(X)) & \xrightarrow{\delta} & H^i(\mathfrak{C}^\cdot(\overline{X^-})) \oplus H^i(\mathfrak{C}^\cdot(\overline{X^+})) & \xrightarrow{\delta} & H^i(\mathfrak{C}(E^\cdot_t)) \overset{R\circ[S]^P}{\to} \\
\to & H^{i+1}(\mathfrak{C}^\cdot(X)) & \to & \cdots . & &
\end{array}
$$

□

Example 4.3.17 Let $E^\cdot = \Lambda^\cdot T^*_{\mathbb{C}}(X)$ be the de Rham complex on a starlike open set $X \subset \mathbb{R}^n$, and S be a smooth hypersurface in X. Because of Example 3.1.24, the tangential complex $E^\cdot_t$ on S is just the de Rham complex $\Lambda^\cdot T^*_{\mathbb{C}}(S)$. Since the complex $\mathcal{E}(\Lambda^\cdot T^*_{\mathbb{C}}(X))$ is exact at positive degrees, sequence (4.3.12) reduces to $(n-1)$ topological isomorphisms

$$H^i(\mathcal{E}(\Lambda^\cdot T^*_{\mathbb{C}}(\overline{X^-}))) \oplus H^i(\mathcal{E}(\Lambda^\cdot T^*_{\mathbb{C}}(\overline{X^+}))) \xrightarrow{\delta} H^i(\mathcal{E}(\Lambda^\cdot T^*_{\mathbb{C}}(S))) \quad (i > 0).$$

Remark 4.3.15 gives the inverse mapping for δ provided S is compact. Namely,

$$\delta^{-1}([f]) = ([\Phi([S] \wedge f)^-], [\Phi([S] \wedge f)^+]),$$

where Φ is the fundamental solution of the convolution type of the de Rham complex in $\mathbb{R}^n$ constructed in Subsection 2.5.2. This result was published by the author [219].

□

Example 4.3.18 Suppose that $E = \Lambda^{p,\cdot} T^*_{\mathbb{C}}(X)$ is the Dolbeault complex with coefficients in $\Lambda^{p,\cdot} T^*_{\mathbb{C}}(X)$ in a domain of holomorphy $X \subset \mathbb{C}^n$, and S is a smooth hypersurface in X. In view of Example 3.1.25, the tangential complex E_t on S is just the tangential Cauchy-Riemann complex. Since the complex $\mathcal{E}(\Lambda^{p,\cdot} T^*_{\mathbb{C}}(X))$ is exact at positive degrees, sequence (4.3.12) reduces to $(n-1)$ topological isomorphisms

$$H^{\imath}(\mathcal{E}(\Lambda^{p,\cdot} T^*_{\mathbb{C}}(\overline{X^-}))) \oplus H^{\imath}(\mathcal{E}(\Lambda^{p,\cdot} T^*_{\mathbb{C}}(\overline{X^+}))) \xrightarrow{\delta} H^{\imath}(\mathcal{E}(\Lambda^{p,\cdot} T^*_{\mathbb{C}}(S)_t)) \quad (\imath > 0).$$

Using the fundamental solution of the convolution type of the complex $\Lambda^{p,\cdot} T^*_{\mathbb{C}}(\mathbb{C}^n)$ constructed in Subsection 2.5.3, one can give the inverse mapping for δ when S is compact. Namely,

$$\delta^{-1}([f]) = ([\Phi([S] \wedge t(f))^-], [\Phi([S] \wedge t(f))^+]),$$

where Φ is the mentioned fundamental solution. This result is due to Dautov (see Theorem 2.11 in the book of Aizenberg and Dautov [5].

□

4.4 The Cauchy problem for cohomology classes of differential complexes

4.4.1 *The Cauchy problem for sections*

As before, we suppose that E is some differential complex on X, and S is a smooth hypersurface in X. It is convenient to denote X^- by Y. Consider the following problem.

Problem 4.4.1 (the Cauchy problem) *Given a section $f \in \mathcal{E}^{\imath}(\overline{Y})$ and jet $u_0 \in \mathcal{E}^{(p_{\imath-1}-1)}(E^{\imath-1}|_S)$, it is required to find a section $u \in \mathcal{E}^{\imath-1}(\overline{Y})$ such that $Pu = f$ and $j_S^{p_{\imath-1}-1}(u) = u_0$.*

Lemma 4.4.2 *In order that Problem 4.4.1 may be solvable, it is necessary that*

$$\begin{cases} Pf = 0 \quad \text{and} \\ P_b t(u_0) = t(f). \end{cases} \tag{4.4.1}$$

Proof. Let $\tilde{u}_0 \in \mathcal{E}^{\imath-1}(\overline{Y})$ be some representative of the jet u_0, and u be a solution of Problem 4.4.1. Then $j_S^{p_{\imath-1}-1}(u - \tilde{u}_0) = 0$, so $t(u - \tilde{u}_0) = 0$. Hence it follows that $P_b t(u_0) = t(P\tilde{u}_0) = t(Pu) - t(Pu - P\tilde{u}_0) = t(f)$. The condition $Pf = 0$ is obvious. This complete the proof.

□

Our purpose is to find an appropriate reformulation of Problem 4.4.1 whose solvability conditions may be given in terms of the cohomologies of E defined in Section 3.1.

Problem 4.4.3 *Given a section $f \in \mathcal{E}_{t_\imath}(\overline{Y})$ satisfying $Pf = 0$, it is required to find a section $u \in \mathcal{E}_{t_{\imath-1}}(\overline{Y})$ such that $Pu = f$.*

As in Subsection 3.1.6, we denote by $\sigma_\imath(x)$ the symbol of the differential operator $P_\imath$ evaluated at a point $\left(x, \frac{d\rho(x)}{|d\rho(x)|}\right)$ of $T^*(X)$.

Proposition 4.4.4 *Suppose that the sequence $E_x^{\imath-2} \xrightarrow{\sigma} E_x^{\imath-1} \xrightarrow{\sigma} E_x^\imath$ is exact for each point $x \in S$. Then Problem 4.4.1 is solvable under the condition of Lemma 4.4.2 if and only if Problem 4.4.3 is solvable.*

Proof. We prove this proposition under the stronger assumption that the complex E is elliptic. Namely, we shall use the formal version of the Cartan-Kähler Theorem of Subsection 3.3.7. To prove Proposition 4 4.4 in full, one needs additional technical reasonings like those used in proving Theorem 3.1.30. Thus, the necessity is obvious, and we have to prove the sufficiency only. Suppose that Problem 4.4.3 is always solvable, and suppose we are given some section $f \in \mathcal{E}^\imath(\overline{Y})$ and jet $u_0 \in \mathcal{E}^{(p_{\imath-1}-1)}(E^{\imath-1}|_S)$ satisfying $Pf = 0$ and $P_b t(u_0) = t(f)$. Choose some representative $\tilde{u}_0 \in \mathcal{E}^{\imath-1}(\overline{Y})$ of the jet u_0, i.e., $\jmath_S^{p_{\imath-1}-1}(\tilde{u}_0) = u_0$. Then we obtain $f - P\tilde{u}_0 \in \mathcal{E}_{t_\imath}(\overline{Y})$ and $P(f - P\tilde{u}_0) = 0$. By hypothesis there exists a section $\tilde{u} \in \mathcal{E}_{t_{\imath-1}}(\overline{Y})$ such that $P\tilde{u} = f - P\tilde{u}_0$. Using Corollary 3.3.18, we can find a section $v \in \mathcal{E}_{t_{\imath-2}}(\overline{Y})$ such that the difference $(Pv - \tilde{u})$ is $(p_{\imath-1} - 1)$-flat on S. We now let $u = \tilde{u}_0 - (Pv - \tilde{u})$. Then $u \in \mathcal{E}^{\imath-1}(\overline{Y})$ and $\jmath_S^{p_{\imath-1}-1}(u) = u_0$. Besides that, we have $Pu = f$, i.e., u is a solution of Problem 4.4.1. This completes the proof.

□

Corollary 4.4.5 *Assume that the sequence $E_x^{\imath-2} \xrightarrow{\sigma} E_x^{\imath-1} \xrightarrow{\sigma} E_x^\imath$ is exact for all $x \in S$. Then Problem 4.4.1 is solvable under the condition of Lemma 4.4.2 if and only if $H^\imath(\mathcal{E}_t(\overline{Y})) = 0$.*

Proof. This follows immediately from Proposition 4.4.4.

□

For the Dolbeault complex, the Cauchy problem was studied by Andreotti and Hill [9] who further developed the earlier approach of Kohn and Rossi [133]. For arbitrary complexes of first order differential operators, the Cauchy problem was considered by MacKichan [155]. Corollary 4.4.5 is in that case easy. MacKichan [155] showed that the Cauchy problem was solvable for complexes satisfying the hyperbolicity condition. This implies the isomorphism of cohomologies of the complexes $\mathcal{E}(\overline{Y})$ and $\mathcal{E}_{[t]}(S)$ (see Subsection 4.4.4).

4.4.2 *Reducing to boundary differential complexes*

Proposition 4.4.4 is of interest because Problem 4.4.3 has a useful homological interpretation.

First of all we notice that the quotient mapping $t : \mathcal{E}(\overline{Y}) \to \mathcal{E}_{[t]}(S)$ is a cochain mapping of degree 0 of the complexes. To see this, use Definition 3.1.18 of the tangential complex.

Problem 4.4.6 *Given sections $f \in \mathcal{E}^{i}(\overline{Y})$ and $u_0 \in \mathcal{E}_{[t_i-1]}(S)$ satisfying $-Pf = 0$ and $t(f) + P_b u_0 = 0$, it is required to find sections $u \in \mathcal{E}^{i-1}(\overline{Y})$ and $v \in \mathcal{E}_{[t_i-2]}(S)$ such that $-Pu = f$ and $t(u) + P_b v = u_0$.*

Proposition 4.4.7 *Problem 4.4.3 is solvable if and only if Problem 4.4.6 is solvable.*

Proof. *Necessity.* Suppose that (every) Problem 4.4.3 is solvable and let $f \in \mathcal{E}^{i}(\overline{Y})$ and $u_0 \in \mathcal{E}_{[t_i-1]}(S)$ satisfy $-Pf = 0$ and $t(f) + P_b u_0 = 0$. Choose some representative $\tilde{u}_0 \in \mathcal{E}^{i-1}(\overline{Y})$ of the class u_0, so that $t(\tilde{u}_0) = u_0$ and $t(P\tilde{u}_0) = P_b u_0$. Then we obtain $P(f + P\tilde{u}_0) = 0$ and $t(f + P\tilde{u}_0) = 0$. By hypothesis, there exists a section $\tilde{u} \in \mathcal{E}_{t_i-1}(\overline{Y})$ satisfying $P\tilde{u} = f + P\tilde{u}_0$. We let $u = \tilde{u}_0 - \tilde{u}$ and $v = 0$. Obviously, (u, v) is a solution of Problem 4.4.6, as desired.

Sufficiency. We now assume that (every) Problem 4.4.6 is solvable. Let some section $f \in \mathcal{E}_{t_i}(\overline{Y})$ satisfying $Pf = 0$ be given. Then the pair $(f, 0)$ satisfies the necessary condition of Problem 4.4.6, so there exist sections $\tilde{u} \in \mathcal{E}^{i-1}(\overline{Y})$ and $v \in \mathcal{E}_{[t_i-2]}(S)$ such that $-P\tilde{u} = f$ and $t(\tilde{u}) + P_b v = 0$. Choose some representative $\tilde{v} \in \mathcal{E}^{i-2}(\overline{Y})$ of the class v, so that $t(\tilde{v}) = v$ and $t(P\tilde{v}) = P_b v$. Then the section $u = -(\tilde{u} + P\tilde{v})$ is easily verified to be a solution of Problem 4.4.3, as desired.

□

Consider the complex of topological vector spaces L whose spaces are $L^{i} = \mathcal{E}^{i}(\overline{Y}) \oplus \mathcal{E}_{[t_i-1]}(S)$ $(i \in \mathbb{Z})$ and whose differential d is defined by

$$d(f, f_0) = (-Pf, t(f) + P_b f_0).$$

Namely,

$$\begin{matrix} \mathcal{E}\ (\overline{Y}) \\ \oplus \\ \mathcal{E}_{[t\ -1]}(S) \end{matrix} : 0 \longrightarrow \begin{matrix} \mathcal{E}^{0}(\overline{Y}) \\ \oplus \\ 0 \end{matrix} \overset{\begin{pmatrix} -P & 0 \\ t & P_b \end{pmatrix}}{\longrightarrow} \begin{matrix} \mathcal{E}^{1}(\overline{Y}) \\ \oplus \\ \mathcal{E}_{[t_0]}(S) \end{matrix} \overset{\begin{pmatrix} -P & 0 \\ t & P_b \end{pmatrix}}{\longrightarrow} \dots . \qquad (4.4.2)$$

Remark 4.4.8 *Complex (4.4.2) is the cone of the cochain mapping $t : \mathcal{E}\ (\overline{Y}) \to \mathcal{E}_{[t\ ]}(S)$ (see Dold [56]).*

Complexes of the form (4.4.2) are said to be boundary differential complexes. They have arisen as compatibility complexes for overdetermined boundary value problems. These subjects go beyond the range of this book, and we refer to the papers of Dynin [60], Pillat and Schulze [181], Rempel and Schulze [188, (3.2.3)], Samborskii [191], Dudnikov and Samborskii [57, 58] etc.

Corollary 4.4.9 *Problem 4.4.3 is solvable if and only if $H^{i}(\mathcal{E}\ (\overline{Y}) \oplus \mathcal{E}_{[t\ -1]}(S)) = 0$.*

Proof. This follows immediately from Proposition 4.4.7.

□

Let us consider an example.

Example 4.4.10 Let $E^\cdot = \Lambda\, T^*_{\mathbb{C}}(X)$ be the de Rham complex on X. Then (4.4.2) is the relative de Rham complex of the pair $(\overline{Y}, S)$ (see Dynin [60]). Its cohomology is isomorphic to the relative cohomology of the pair $(\overline{Y}, S)$ with complex coefficients, i.e., $H^\cdot((\overline{Y}, S), \mathbb{C})$. Thus, Proposition 4.4.4 and Corollary 4.4.9 imply that Problem 4.4.1 is solvable under the condition of Lemma 4.4.2 if and only if $H^\imath((\overline{Y}, S), \mathbb{C}) = 0$.

□

4.4.3 *The Cauchy problem for cohomology classes*

Consider the homogeneous version of Problem 4.4.6. Namely, for a given section $u_0 \in Z^{\imath-1}(\mathcal{E}_{[t\,]}(S))$, it is required to find sections $u \in Z^{\imath-1}(\mathcal{E}^\cdot(\overline{Y}))$ and $v \in \mathcal{E}_{[t,-2]}(S)$ such that $t(u) + P_b v = u_0$.

An admissible Cauchy datum u_0 determines a unique cohomology class $[u_0] \in H^{\imath-1}\mathcal{E}_{[t\,]}(S))$. And any solution u determines a unique cohomology class $[u] \in H^{\imath-1}\mathcal{E}^\cdot(\overline{Y}))$. Denote by Ht (or simply t) the mapping $H^\cdot(\mathcal{E}^\cdot(\overline{Y})) \to H^\cdot(\mathcal{E}_{[t\,]}(S))$ induced by the mapping t on sections. Namely, let $[u] \in H^{\imath-1}(\mathcal{E}^\cdot(\overline{Y}))$ and let $u \in Z^{\imath-1}(\mathcal{E}^\cdot(\overline{Y}))$ be a representative of the class $[u]$. Then $P_b t(u) = t(Pu) = 0$, and we set $t([u])$ equal to the class of $t(u)$ in $H^{\imath-1}(\mathcal{E}_{[t\,]}(S))$. Now the condition $t(u) + P_b v = u_o$ means that $t([u]) = [u_0]$. And we have arrived at the following problem.

Problem 4.4.11 *Given a boundary class* $[u_0] \in H^{\imath-1}(\mathcal{E}_{[t\,]}(S))$*, it is required to find a class* $[u] \in H^{\imath-1}(\mathcal{E}^\cdot(\overline{Y}))$ *satisfying* $t([u]) = [u_0]$.

It is clear that, if Problem 4.4.6 is solvable, then Problem 4.4.11 is solvable too. The converse fails unless $H^\imath(\mathcal{E}^\cdot(\overline{Y})) = 0$. The uniqueness or the existence of solutions of Problem 4.4.11 is equivalent to the injectivity or the surjectivity of the homomorphism $H^{\imath-1}(\mathcal{E}^\cdot(\overline{Y})) \xrightarrow{t} H^{\imath-1}(\mathcal{E}_{[t\,]}(S))$. In the next theorem these questions are reduced to the vanishing of some cohomology spaces of the complex $E^\cdot$. For the Dolbeault complex it was proved by Andreotti and Hill [9], and for arbitrary differential complexes it is due to Andreotti et al. [11].

Theorem 4.4.12 *Let* $\mathfrak{C} = \mathcal{E}$ *or* $\mathcal{D}$*, and* S *be formally non-characteristic for the complex* $\mathfrak{C}^\cdot(X)$*. Then there is the following commutative diagram with exact rows:*

$$\begin{array}{ccccccccc}
0 \to & Z^0(\mathfrak{C}_t(\overline{X^\pm})) & \xrightarrow{\imath} & Z^0(\mathfrak{C}^\cdot(X)) & \xrightarrow{\delta^\mp} & Z^0(\mathfrak{C}^\cdot(\overline{X^\mp})) & \xrightarrow{P} & H^1(\mathfrak{C}_t(\overline{X^\pm})) & \to \ldots \\
& \downarrow \wr & & \downarrow \delta^\pm & & \downarrow t & & \downarrow & \\
0 \to & Z^0(\mathfrak{C}_t(\overline{X^\pm})) & \xrightarrow{\imath} & Z^0(\mathfrak{C}^\cdot(\overline{X^\pm})) & \xrightarrow{t} & Z^0(\mathfrak{C}_{[t\,]}(S)) & \xrightarrow{P} & H^1(\mathfrak{C}_t(\overline{X^\pm})) & \to \ldots \\
& & & & & & & & \\
\to & H^\imath(\mathfrak{C}_t(\overline{X^\pm})) & \xrightarrow{\imath} & H^\imath(\mathfrak{C}^\cdot(X)) & \xrightarrow{\delta^\mp} & H^\imath(\mathfrak{C}^\cdot(\overline{X^\mp})) & \xrightarrow{P} & H^{\imath+1}(\mathfrak{C}_t(\overline{X^\pm})) & \to \ldots \\
& \downarrow \wr & & \downarrow \delta^\pm & & \downarrow t & & \downarrow & \\
\to & H^i(\mathfrak{C}_t(\overline{X^\pm})) & \xrightarrow{\imath} & H^i(\mathfrak{C}^\cdot(\overline{X^\pm})) & \xrightarrow{t} & H^\imath(\mathfrak{C}_{[t\,]}(S)) & \xrightarrow{P} & H^{\imath+1}(\mathfrak{C}_t(\overline{X^\pm})) & \to \ldots .
\end{array} \tag{4.4.3}$$

Proof. According to Proposition 3.1.11, we have the following commutative diagram with exact rows

$$\begin{array}{ccccccccc} 0 & \longrightarrow & \mathfrak{C}_{\overline{X^{\pm}}}(E) & \xrightarrow{\imath} & \mathfrak{C}(X) & \xrightarrow{\delta^{\mp}} & \mathfrak{C}(\overline{X^{\mp}}) & \longrightarrow & 0 \\ & & \downarrow{\scriptstyle \delta^{\pm}} & & \downarrow{\scriptstyle \delta^{\pm}} & & \downarrow{\scriptstyle \jmath} & & \\ 0 & \longrightarrow & \mathfrak{C}_{\jmath}(\overline{X^{\pm}}) & \xrightarrow{\imath} & \mathfrak{C}(\overline{X^{\pm}}) & \xrightarrow{\jmath} & \mathfrak{C}(S) & \longrightarrow & 0. \end{array}$$

Writing the corresponding long cohomological sequence of Theorem 4.3.3, we get the commutative diagram with exact rows

$$\begin{array}{ccccccccccc} 0 & \to & Z^0(\mathfrak{C}_{\overline{X^{\pm}}}(E)) & \xrightarrow{\imath} & Z^0(\mathfrak{C}(X)) & \xrightarrow{\delta^{\mp}} & Z^0(\mathfrak{C}(\overline{X^{\mp}})) & \xrightarrow{h_b} & H^1(\mathfrak{C}_{\overline{X^{\pm}}}(E)) & \to & \dots \\ & & \downarrow{\scriptstyle \delta^{\pm}} & & \downarrow{\scriptstyle \delta^{\pm}} & & \downarrow{\scriptstyle \jmath} & & \downarrow & & \\ 0 & \to & Z^0(\mathfrak{C}_{\jmath}(\overline{X^{\pm}})) & \xrightarrow{\imath} & Z^0(\mathfrak{C}(\overline{X^{\pm}})) & \xrightarrow{\jmath} & Z^0(\mathfrak{C}(S)) & \xrightarrow{h_b} & H^1(\mathfrak{C}_{\jmath}(\overline{X^{\pm}})) & \to & \dots \\ & & & & & & & & & & \\ & \to & H^{\imath}(\mathfrak{C}_{\overline{X^{\pm}}}(E)) & \xrightarrow{\imath} & H^{\imath}(\mathfrak{C}(X)) & \xrightarrow{\delta^{\mp}} & H^{\imath}(\mathfrak{C}(\overline{X^{\mp}})) & \xrightarrow{h_b} & H^{\imath+1}(\mathfrak{C}_{\overline{X^{\pm}}}(E)) & \to & \dots \\ & & \downarrow{\scriptstyle \delta^{\pm}} & & \downarrow{\scriptstyle \delta^{\pm}} & & \downarrow{\scriptstyle \jmath} & & \downarrow & & \\ & \to & H^{\imath}(\mathfrak{C}_{\jmath}(\overline{X^{\pm}})) & \xrightarrow{\imath} & H^{\imath}(\mathfrak{C}(\overline{X^{\pm}})) & \xrightarrow{\jmath} & H^{\imath}(\mathfrak{C}(S)) & \xrightarrow{h_b} & H^{\imath+1}(\mathfrak{C}_{\jmath}(\overline{X^{\pm}})) & \to & \dots, \end{array} \tag{4.4.4}$$

where h_b is the binding homomorphism.

Since the hypersurface S is formally non-characteristic for the complex $\mathfrak{C}(X)$, the cohomologies of the complexes $\mathfrak{C}(S)$ and $\mathfrak{C}_{[t]}(S)$ are the same, i.e., $H(\mathfrak{C}(S)) \cong H(\mathfrak{C}_{[t]}(S))$ (see (4.3.11). Besides that, we obviously have

$$H^{\imath}(\mathfrak{C}_{\overline{X^{\pm}}}(E)) \cong H^{\imath}(\mathfrak{C}_{\jmath}(\overline{X^{\pm}}))$$

for all $\imath \in \mathbb{Z}$.

Now we use Proposition 3 1 27 to obtain the exact sequence of cochain complexes

$$0 \longrightarrow \mathfrak{C}_{\jmath}(\overline{X^{\pm}}) \xrightarrow{\imath} \mathfrak{C}_{t}(\overline{X^{\pm}}) \xrightarrow{\jmath} \mathfrak{C}_{t}(S) \longrightarrow 0.$$

Applying the long cohomological sequence of Theorem 4.3.3, we get

$$\begin{array}{cccccccc} 0 & \longrightarrow & Z^0(\mathfrak{C}_{\jmath}(\overline{X^{\pm}})) & \longrightarrow & Z^0(\mathfrak{C}_{t}(\overline{X^{\pm}})) & \longrightarrow & Z^0(\mathfrak{C}_{t}(S)) & \longrightarrow \\ & \longrightarrow & H^1(\mathfrak{C}_{\jmath}(\overline{X^{\pm}})) & \longrightarrow & \dots & & & \\ & \longrightarrow & H^{\imath}(\mathfrak{C}_{\jmath}(\overline{X^{\pm}})) & \longrightarrow & H^{\imath}(\mathfrak{C}_{t}(\overline{X^{\pm}})) & \longrightarrow & H^{\imath}(\mathfrak{C}_{t}(S)) & \longrightarrow \\ & \longrightarrow & H^{\imath+1}(\mathfrak{C}_{\jmath}(\overline{X^{\pm}})) & \longrightarrow & \dots . & & & \end{array} \tag{4.4.5}$$

Since S is formally non-characteristic for the complex $\mathfrak{C}(X)$, we have

$$H^{\imath-1}(\mathfrak{C}_{t}(S)) = H^{\imath}(\mathfrak{C}_{t}(S)) = 0.$$

Therefore (4.4.5) implies $H^{\imath}(\mathfrak{C}_{\jmath}(\overline{X^{\pm}})) \cong H^{\imath}(\mathfrak{C}_{t}(\overline{X^{\pm}}))$ for all $\imath \in \mathbb{Z}$. Substituting these equalities in (4.4.4) we obtain (4.4.3), as desired.

□

Let us describe the homomorphisms of the sequence (4.4.3)

Remark 4.4.13 *To describe the homomorphism i at the upper row of (4.4.3), we notice that each class $[f] \in H^\iota(\mathfrak{C}_t(\overline{X^\pm}))$ has a (distinguished) representative $f_0 \in Z^\iota(\mathfrak{C}_t(\overline{X^\pm}))$ such that $j_S^\infty(f_0) = 0$. In fact, we have $Pf = 0$ for any representative $f \in Z^\iota(\mathfrak{C}_t(\overline{X^\pm}))$ of $[f]$. Since $H^{\iota+1}\left(\frac{\mathfrak{C}_t(X)}{\mathfrak{C}_J(X)}\right) = 0$, there exists a section $u \in \mathfrak{C}_{t_{\iota-1}}(\overline{X^\pm})$ such that the difference $(f - Pu)$ is flat on S. Then $f_0 = f - Pu$ is a distinguished representative of $[f]$ because $[f - Pu] = [f]$ in $H^\iota(\mathfrak{C}_t(\overline{X^\pm}))$. Now we let $f_1 = f_0$ on $\overline{X^\pm}$ and $f_1 = 0$ on $X^\mp$, where f_0 is a distinguished representative of the class $[f]$. Then $f_1 \in Z^\iota(\mathfrak{C}(X))$, and the homomorphism i in the upper row of (4.4.3) is defined by $i([f]) = [f_1] \in H^\iota(\mathfrak{C}^\cdot(X))$. The homomorphism i in the lower row of (4.4.3) is induced by inclusion of the complexes. The homomorphisms $\delta^\pm$ are induced by restrictions of sections from $\mathfrak{C}^\cdot(X)$ to $X^\pm$ respectively. The homomorphism t was described above. The homomorphism P in the upper row of (4.4.3) is defined in the following way. Let $[f] \in H^\iota(\mathfrak{C}^\cdot(\overline{X^\mp})$, and $f \in Z^\iota(\mathfrak{C}^\cdot(\overline{X^\mp}))$ be some representative of the class $[f]$. We extend f to a section $\tilde{f} \in \mathfrak{C}^\iota(X)$. Then $P\tilde{f}^\pm$ lies in $Z^{\iota+1}(\mathfrak{C}_t(\overline{X^\pm}))$, and $P([f])$ is defined to be equal to the class of the section $P\tilde{f}^\pm$ in $H^{\iota+1}(\mathfrak{C}_t(\overline{X^\pm}))$. Finally, to describe the homomorphism P in the lower row of (4.4.3), we notice that every class $[f] \in H^\iota(\mathfrak{C}_{[t]}(S))$ is represented by a section $f \in \mathfrak{C}^\iota(X)$ such that $t(Pf) = 0$. Then $P([f])$ is defined to be equal to the class of the section $Pf^\pm$ in $H^{\iota+1}(\mathfrak{C}_t(\overline{X^\pm}))$. To see that these homomorphisms are well-defined, see Andreotti and Hill [9].*

When the hypersurface S is formally non-characteristic for the complex $\mathfrak{C}^\cdot(X)$, Theorem 4.4.12 provides a full analysis of Problem 4.4.11.

Corollary 4.4.14 (uniqueness) *In order that the mapping*

$$H^{\iota-1}(\mathfrak{C}^\cdot(\overline{X^\pm})) \xrightarrow{t} H^{\iota-1}(\mathfrak{C}_{[t]}(S))$$

may be injective, each of the conditions $H^{\iota-1}(\mathfrak{C}_t(\overline{X^\pm})) = 0$ or $H^{\iota-1}(\mathfrak{C}^\cdot(X)) = 0$ is sufficient.

Proof. As is seen from the commutative diagram (4.4.3), under the corollary's condition the range of the mapping $H^{\iota-1}(\mathfrak{C}_t(\overline{X^\pm})) \xrightarrow{i} H^{\iota-1}(\mathfrak{C}^\cdot(X))$ in the upper row is zero. Hence it follows that the range of the mapping $H^{\iota-1}(\mathfrak{C}_t(\overline{X^\pm})) \xrightarrow{i} H^{\iota-1}(\mathfrak{C}(\overline{X^\pm}))$ in the lower row is zero too. To complete the proof, it remains only to use the exactness of the lower row of (4.4.3).

□

Corollary 4.4.15 (existence) *In order that the mapping*

$$H^{\iota-1}(\mathfrak{C}^\cdot(\overline{X^\pm})) \xrightarrow{t} H^{\iota-1}(\mathfrak{C}_{[t]}(S))$$

may be surjective,

1) $H^\iota(\mathfrak{C}_t(\overline{X^\pm})) = 0$ is sufficient;

2) $H^\iota(\mathfrak{C}_t(\overline{X^\pm})) = 0$ is necessary provided $H^\iota(\mathfrak{C}^\cdot(X)) = 0$ or $H^\iota(\mathfrak{C}^\cdot(\overline{X^\pm})) = 0$;

3) $H^{\iota-1}(\mathfrak{C}^\cdot(\overline{X^\mp})) = 0$ is sufficient provided $H^\iota(\mathfrak{C}^\cdot(X)) = 0$.

Proof. 1) is obvious.

To prove 2), we notice that if the mapping $H^{i-1}(\mathfrak{C}^\cdot(\overline{X^\pm})) \xrightarrow{t} H^{i-1}(\mathfrak{C}_{[t\,]}(S))$ is surjective, then the mappings $H^{i-1}(\mathfrak{C}\,(\overline{X^\mp}) \xrightarrow{P} H^i(\mathfrak{C}_t\,(\overline{X^\pm}))$ in the upper row and $H^{i-1}(\mathfrak{C}_{[t\,]}(S)) \xrightarrow{P} H^i(\mathfrak{C}_t\,(\overline{X^\pm}))$ in the lower row of (4.4.3) have zero ranges. Then the equality $H^i(\mathfrak{C}^\cdot(X)) = 0$ implies $H^i(\mathfrak{C}_t\,(\overline{X^\pm})) = 0$ because the upper row is exact, while the equality $H^i(\mathfrak{C}^\cdot(\overline{X^\pm})) = 0$ implies $H^i(\mathfrak{C}_t\,(\overline{X^\pm})) = 0$ because the lower row is exact. Finally, 3) follows from 1) because the upper row of (4.4.3) is exact. This completes the proof.

□

4.4.4 *Representation of cohomologies of a differential complex by means of cohomologies of the tangential complex*

Versions of the following theorem may be found in the papers of Andreotti and Hill [9], Kuranishi [141], MacKichan [155], Nacinovich [168, 169] etc.

Theorem 4.4.16 *Let $\mathfrak{C} = \mathcal{E}$ or $\mathcal{D}$, and S be formally non-characteristic for the complex $\mathfrak{C}^\cdot(X)$. If $H^i(\mathfrak{C}_t\,(\overline{Y})) = H^{i+1}(\mathfrak{C}_t\,(\overline{Y})) = 0$, then there is the isomorphism*

$$H^i(\mathfrak{C}^\cdot(\overline{Y})) \stackrel{t}{\cong} H^i(\mathfrak{C}_{[t\,]}(S)). \tag{4.4.6}$$

Proof. This follows directly from the exactness of the lower row of (4.4.3).

□

As to other consequences of Theorem 4.4.12, see Andreotti and Hill [9, §4].

Apparently, Theorem 4.4.16 is the most important consequence of all because repeated application of the procedure allows one to reduce calculation of the cohomology of the complex $\mathfrak{C}^\cdot(\overline{Y})$ to calculation of cohomologies of complexes on manifolds of smaller dimensions.

Remark 4.4.17 *If $H^i(\mathfrak{C}^\cdot(X)) = H^{i+1}(\mathfrak{C}^\cdot(X)) = 0$ and $H^i(\mathfrak{C}^\cdot(X \setminus Y)) = 0$, then the isomorphism (4.4.6) holds. To prove this, use Corollaries 4.4.14 and 4.4.15. Another way is to apply the Mayer-Vietoris sequence (4.3.9).*

A favourable situation for applying Remark 4.4.17 will be considered in the next subsection.

4.4.5 *Removable singularities of solutions of overdetermined systems and a generalization of Bochner's theorem*

Equality (4.4.6) for $i = 0$ is related to automatic extension of solutions of the overdetermined system $P_0 f = 0$ from a neighborhood of S to Y, whose uniqueness is guaranteed by the condition $H^0(\mathfrak{C}_t\,(\overline{Y})) = 0$. Bochner [24] and Martinelli [159] independently of each other found a rigorous proof of the so-called Hartogs theorem asserting that, if a function f is holomorphic in a domain $Y \subset \mathbb{C}^n$ $(n > 1)$ except for a compact set $K \subset Y$ which does not divide Y, then f may be holomorphically extended to the whole domain Y. Moreover, it follows from the proof of Bochner [24] that, if Y is a bounded domain in $\mathbb{C}^n$ $(n > 1)$ with a smooth connected boundary, then each function $f_0 \in C^1(\partial Y)$ satisfying the tangential Cauchy-Riemann equations on ∂Y, may be extended to a function $f \in C^1(\overline{Y})$ holomorphic in Y.

Ehrenpreis [66] found a proof of the Hartogs theorem valid in the context of overdetermined systems of differential equations with constant coefficients. Let $E^{\cdot}$ be a Hilbert complex on an open set $X \subset \mathbb{R}^n$ associated with a $\mathcal{P}$-module $M =$ coker $P_0(z)^t$. In the book of Palamodov [175, Ch.VIII, §14] a $\mathcal{P}$-module M (and any differential operator P_0 with the property $M \cong \mathcal{P}^k / P_0(z)^t \mathcal{P}^l$ is said to be determined if $\mathrm{Ext}^0(M, \mathcal{P}) = 0$, and overdetermined if $\mathrm{Ext}^0(M, \mathcal{P}) = \mathrm{Ext}^1(M, \mathcal{P}) = 0$. Determined $\mathcal{P}$-modules M (and differential operators P_0) are characterized by the condition $\dim \mathfrak{M}(M) < n$, and overdetermined $\mathcal{P}$-modules M (and differential operators P_0) are characterized by the condition $\dim \mathfrak{M}(M) < n-1$ where $\mathfrak{M}(M)$ is the algebraic variety associated to M (see Subsection 2.3.5). Ehrenpreis [66] showed that, if the differential operators P_0 is overdetermined, then each (differentiable or generalized) solution of the system $P_0 f = 0$ in the complement of a complex compact set $K \subset X$ has a unique extension to the whole set X also satisfying the system. The condition for P_0 to be overdetermined is also necessary (see Palamodov [175, Ch.VIII, §14]). Zecca [259] sharpened the result of Ehrenpreis [66] by proving that, if the differential operator P_0 has injective symbol, then the convexity condition for K may be replaced by the condition that the complement of K in X is connected. But he considered systems $P_0 f = 0$ of a particular form with one unknown function. Remark 4.4.17 allows one to strengthen the theorem of Ehrenpreis [66] in the spirit of Bochner's [24] theorem.

Theorem 4.4.18 *Let $E^{\cdot}$ be a Hilbert complex on X associated with an overdetermined $\mathcal{P}$-module, and $Y \subset\subset X$ be a convex domain with a smooth boundary formally non-characteristic for $\mathcal{E}^{\cdot}(X)$. Then $Z^0(\mathcal{E}^{\cdot}(\overline{Y})) \xrightarrow{t} Z^0(\mathcal{E}_{[t]}(\partial Y))$ is an isomorphism.*

Proof. We are going to apply Remark 4.4.17 for $\mathfrak{C} = \mathcal{D}$ and $i = 0$. Let us check that all the conditions are satisfied. Since Y is convex, one can assume that X is convex also.

The equality $H^0(\mathcal{D}^{\cdot}(X)) = 0$ follows from Corollary 2.3.25 because the differential operator P_0 is assumed to be (over) determined.

As to the condition $H^1(\mathcal{D}^{\cdot}(X)) = 0$, we take some $f \in Z^1(\mathcal{D}^{\cdot}(X))$. According to Theorem 1.2.21, there exists a section $u \in \mathcal{E}^0(X)$ such that $P_0 u = f$. Another way to prove this is to use Corollary 2.3.25. In particular we have $P_0 u = 0$ outside the support of f. Choose a convex compact set $K \subset X$ such that $\operatorname{supp} f \subset K$. Since the operator P_0 is overdetermined, the Hartogs theorem implies existence of a solution $\tilde{u} \in Z^0(\mathcal{E}^{\cdot}(X))$ coinciding with u in $X \setminus K$. We let $u' = u - \tilde{u}$. Then $u' \in \mathcal{D}^0(X)$ and $Pu' = Pu = f$, i.e., $f \in B^1(\mathcal{D}^{\cdot}(X))$ as desired.

Finally we check the condition $H^0(\mathcal{D}^{\cdot}(X \setminus Y)) = 0$. Since P_0 is overdetermined, we can apply the Hartogs theorem to conclude that $H^0(\mathcal{D}^{\cdot}(X \setminus Y)) \cong H^0(\mathcal{D}^{\cdot}(X))$.

It remains only to use the equality $H^0(\mathcal{D}(X)) = 0$ proved above.

Hence by Remark 4.4.17 we get the isomorphism $Z^0(\mathcal{D}(\overline{Y})) \overset{t}{\cong} Z^0(\mathcal{D}_{[t]}(\partial Y))$, and the proof is completed by observing that $\mathcal{D}(\overline{Y}) = \mathcal{E}(\overline{Y})$ and $\mathcal{D}_{[t]}(\partial Y) = \mathcal{E}_{[t]}(\partial Y)$ because of the compactness of $\overline{Y}$.

□

For elliptic complexes E, one can sharpen Theorem 4.4.18. We assume that E is an elliptic complex on a real analytic manifold X.

Theorem 4.4.19 *Suppose that the differential operator P_0 has real analytic coefficients, and $H^1(\mathcal{D}(X)) = 0$. If $Y \subset\subset X$ is an open set with boundary of class $C^{p_0}_{loc}$ and $X \setminus Y$ has no compact connected components, then for every jet $f_0 \in C^{p_0-1}(E^0|_{\partial Y})$ satisfying $P_b f_0 = 0$ there exists a unique section $f \in C^{p_0-1}(E^0|_{\overline{Y}})$ such that $Pf = 0$ in Y and $j^{p_0-1}_{\partial Y}(f) = f_0$.*

Remark 4.4.20 *The condition $P_b f_0 = 0$ is understood in the generalized sense, i.e., as $P([\partial Y]^P f_0) = 0$.*

Proof. Let Π be the properly supported parametrix of the complex E given by Theorem 2.1.12. Since the hypersurface $S = \partial Y$ is compact, we have $[S]^P f_0 \in \mathcal{E}'^1(X)$. Therefore $R([S]^P f_0) \in \mathcal{D}^1(X)$. Besides that, this section satisfies

$$PR([S]^P f) = 0$$

on X because R is an endomorphism of the complex $\mathcal{E}'(X)$ and $P_b f_0 = 0$. Using the condition $H^1(\mathcal{D}(X)) = 0$, we now find a section $u \in \mathcal{D}^0(X)$ such that $Pu = R([S]^P f_0)$. Consider the section $f = -(\Pi([S]^P f_0) + u)$ which obviously belongs to $\mathcal{E}'(X)$. It follows from the homotopy formula (2.1 7) and the condition $P_b f_0 = 0$ that $Pf = -[S]^P f_0$ on X. In particular, f satisfies $Pf = 0$ in the complement of S so this section is real analytic outside S. Since the complement of the set Y has no compact connected components in X, we may conclude that the restriction of f to $X \setminus \overline{Y}$ is zero. Hence it follows that the restriction of $-\Pi([S]^P f_0)$ to $X \setminus \overline{Y}$ may be extended to a differentiable section of E^0 in the closure $X \setminus Y$ because $-\Pi([S]^P f_0) = u$ in $X \setminus \overline{Y}$. Applying Corollary 3.3.23 we see that the derivatives up to order $(p_0 - 1)$ of the potential $-\Pi([S]^P f_0)$ in Y can be extended continuously to the closure of Y. By Theorem 3.3.22, the $(p_0 - 1)$-jet on S of this extension coincides with $f_0 + j^{p_0-1}_S(u)$. Therefore the restriction of f to Y belongs to the class $C^{p_0-1}(E^0|_{\overline{Y}})$, and $j^{p_0-1}_S(f|_Y) = f_0$. Thus $f|_Y$ provides the required continuation of the jet f_0.

This continuation is unique because all solutions of the homogeneous equation $P_0 f = 0$ are real analytic (see Subsection 2.1.6). In fact, if $f \in C^{p_0-1}(E^0|_{\overline{Y}})$ satisfies $Pf = 0$ in Y and $j^{p_0-1}_S(f) = 0$, then the section $\chi_Y f \in \mathcal{E}'^0(X)$ satisfies $P(\chi_Y f) = 0$ everywhere in X. So $\chi_Y f$ vanishes identically on non-compact connected components of X which intersect Y. We get $f \equiv 0$, as desired.

□

4.4.6 The Dirichlet problem and representation of cohomologies of the complex of spaces of sections with zero Cauchy data

In this subsection we sketch how the results of solvability of the P-Neumann problem (see Section 4.1) may be used to study (the Cauchy) Problem 4.4.11.

For the Dolbeault complex, this relation was discovered by Kohn and Rossi [133]. Sweeney [211] developed their idea for the case of the second Spencer sequence.

Let Y be a relatively compact open subset of X whose boundary $S = \partial Y$ is a differentiable submanifold of X.

Let us introduce the spaces $\widetilde{\mathcal{H}}^{\cdot}(E^{\cdot}|_{\overline{Y}})$ consisting of all sections $f \in \mathcal{E}(E^{\cdot}|_{\overline{Y}})$ such that $Pf = P^{*}f = t(f) = 0$. We endow these spaces with the topologies induced from $\mathcal{E}(E^{\cdot}|_{\overline{Y}})$. Then it turns out that the natural mapping $\widetilde{\mathcal{H}}^{\cdot}(E^{\cdot}|_{\overline{Y}}) \to H^{\cdot}(\mathcal{E}_t(\overline{Y}))$ is a continuous monomorphism.

The P-Dirichlet problem is formulated by analogy with the P-Neumann problem where the P-Neumann data $\{n(f), n(Pf)\}$ are replaced by the P-Dirichlet data $\{t(f), t(P^{*}f)\}$.

We recall the relevant notation. Consider the Laplacians $\Delta_i = P_{i-1}P_{i-1}^{*} + P_i^{*}P_i$ of the complex $E^{\cdot}$. The vector spaces $\mathfrak{D}^{\cdot}(E^{\cdot}|_{\overline{Y}})$, consisting of all sections $f \in \mathcal{E}(E^{\cdot}|_{\overline{Y}})$ satisfying $t(f) = t(P^{*}f) = 0$, are called the Dirichlet spaces. They are endowed with the induced topologies. The spaces $\mathfrak{D}^{\cdot}(E^{\cdot}|_{\overline{Y}})$ have thus been defined in such a way that $\widetilde{\mathcal{H}}^{\cdot}(E^{\cdot}|_{\overline{Y}}) = \{f \in \mathfrak{D}^{\cdot}(E^{\cdot}|_{\overline{Y}}) : \Delta f = 0\}$. Hence it follows that the condition $f \perp \widetilde{\mathcal{H}}^{\cdot}(E^{\cdot}|_{\overline{Y}})$ is necessary for solvability of the following (formally) selfadjoint boundary value problem.

Problem 4.4.21 (the P-Dirichlet problem) *Given a section $f \in \mathcal{E}(E^{\cdot}|_{\overline{Y}})$ satisfying $f \perp \widetilde{\mathcal{H}}^{\cdot}(E^{\cdot}|_{\overline{Y}})$, it is required to find a section $u \in \mathfrak{D}^{\cdot}(E^{\cdot}|_{\overline{Y}})$ such that $\Delta u = f$.*

As is it easy to see, the P-Dirichlet problem is the same as the P^{*}-Neumann problem so that all the results of Section 4.1 are applicable.

Proposition 4.4.22 *If (each) Problem 4.4.21 is solvable and the space $\widetilde{\mathcal{H}}^{\cdot}(E^{\cdot}|_{\overline{Y}})$ is finite-dimensional, then*

$$H^{\cdot}(\mathcal{E}_t(\overline{Y})) \overset{top}{\cong} \widetilde{\mathcal{H}}^{\cdot}(E^{\cdot}|_{\overline{Y}}). \tag{4.4.7}$$

Proof. Notice that the subspace $\widetilde{\mathcal{H}}^{\cdot}(E^{\cdot}|_{\overline{Y}})$ of $L^2(E^{\cdot}|_{\overline{Y}})$ is finite-dimensional if and only if it is closed.

Let now f be some section of $\mathcal{E}(E^{\cdot}|_{\overline{Y}})$. Denote by $\widetilde{H}f$ the L^2-orthogonal projection of f onto the subspace $\widetilde{\mathcal{H}}^{\cdot}(E^{\cdot}|_{\overline{Y}})$. Then the difference $(f - \widetilde{H}f)$ belongs to $\mathcal{E}(E^{\cdot}|_{\overline{Y}})$ and it is orthogonal to $\widetilde{\mathcal{H}}^{\cdot}(E^{\cdot}|_{\overline{Y}})$. By hypothesis, there exists a section $u \in \mathfrak{D}^{\cdot}(E^{\cdot}|_{\overline{Y}})$ such that $f - \widetilde{H}f = \Delta u$.

Denote by $\mathfrak{D}(f)$ the unique section $u \in \mathfrak{D}^{\cdot}(E\,|_{\overline{Y}})$ with the property $f - \widetilde{H}f = \Delta u$ which is orthogonal to $\widetilde{\mathcal{H}}^{\cdot}(E^{\cdot}|_{\overline{Y}})$. It follows from the Closed Graph Theorem that the mapping $\mathfrak{D}: \mathcal{E}(E^{\cdot}|_{\overline{Y}}) \to \mathfrak{D}^{\cdot}(E\,|_{\overline{Y}})$ is a (topological) homomorphism. Thus each section $f \in \mathcal{E}(E^{\cdot}|_{\overline{Y}})$ may be decomposed into the sum

$$f = \widetilde{H}f + P^*P\mathfrak{D}f + PP^*\mathfrak{D}f. \tag{4.4.8}$$

Suppose that $f \in \mathcal{E}_{t,}(\overline{Y})$ and $Pf = 0$. Since $t(P^*\mathfrak{D}f) = 0$, (4.4.8) implies $t(P^*P\mathfrak{D}f) = 0$. Besides that, we get $PP^*P\mathfrak{D}f = 0$. Hence it follows that $(P^*P\mathfrak{D}f, P^*P\mathfrak{D}f)_Y = 0$, i.e., $P^*P\mathfrak{D}f = 0$. Using the condition $t(\mathfrak{D}f) = 0$ one can see that $P\mathfrak{D}f = 0$, but we do not need this equality.

Therefore we obtain $f = \widetilde{H}f + Pu$ where $u = P^*\mathfrak{D}f$ is an element of $\mathcal{E}_{t,-1}(\overline{Y})$. This proves the surjectivity of the natural mapping $\widetilde{\mathcal{H}}^{\cdot}(E\,|_{\overline{Y}}) \to H^{\cdot}(\mathcal{E}_t\,(\overline{Y}))$, as desired.

□

Corollary 4.4.23 *Under the conditions of Proposition 4.4.22, for a section* $f \in Z^{\cdot}(\mathcal{E}_t\,(\overline{Y}))$ *there exists a section* $u \in \mathcal{E}_{t,-1}(\overline{Y})$ *with* $Pu = f$ *if and only if* $f \perp \widetilde{\mathcal{H}}^{\cdot}(E^{\cdot}|_{\overline{Y}})$.

Proof. This follows from the decomposition (4.4.8).

□

4.4.7 *A generalization of the theorem of Kohn and Rossi*

Proposition 4.4.22 and Corollary 4.4.15 make clear the obstacles to the solvability of Problem 4.4.11. These are the non-zero elements of the space $\widetilde{\mathcal{H}}^{\cdot}(E\,|_{\overline{Y}})$. A mechanism for getting over these obstacles is found in Theorem 4.4.24 below. For the Dolbeault complex on strongly pseudo-convex manifolds it was proved by Kohn and Rossi [133] who used the results of Kohn [125, 126] on solvability of the $\overline{\partial}$-Neumann problem.

Theorem 4.4.24 *Suppose that (each) Problem 4.4.21 is solvable and the space* $\widetilde{\mathcal{H}}^{\cdot}(E^{\cdot}|_{\overline{Y}})$ *is finite-dimensional. Then for* $u_0 \in Z^{\cdot-1}(\mathcal{E}_{[t\,]}(S))$ *there exists a section* $u \in Z^{\cdot-1}(\mathcal{E}^{\cdot}(\overline{Y}))$ *such that* $t(u) = u_0$ *if and only if*

$$\int_S G_{P_{\cdot-1}}(\star f, u_0) = 0 \quad \textit{for all} \quad f \in \widetilde{\mathcal{H}}^{\cdot}(E\,|_{\overline{Y}}). \tag{4.4.9}$$

Proof. *Necessity.* Let $u \in \mathcal{E}^{\cdot-1}(\overline{Y})$ and $t(u) = u_0$. Then for each section $f \in \mathcal{E}^{\cdot}(\overline{Y})$ we obtain by Stokes' Theorem

$$\int_S G_{P_{\cdot-1}}(\star f, u_0) = \int_S G_{P_{\cdot-1}}(\star f, u) = (Pu, f)_Y - (u, P^*f)_Y, \tag{4.4.10}$$

wherefrom the desired assertion follows.

Sufficiency. We now suppose that a jet $u_0 \in Z^{\cdot-1}(\mathcal{E}_{[t\,]}(S))$ satisfies condition (4.4.9). Choose some representative $\tilde{u}_0 \in \mathcal{E}^{\cdot-1}(\overline{Y})$ of this jet, that is $t(\tilde{u}_0) = u_0$.

Then $P\tilde{u}_0 \in Z^\cdot(\mathcal{E}_t(\overline{Y}))$. Using (4.4.10) we get $P\tilde{u}_0 \perp \widetilde{\mathcal{H}}^\cdot(E\,|_{\overline{Y}})$. According to Corollary 4.4.23 there exists a section $\tilde{u} \in \mathcal{E}_{t,-1}(\overline{Y})$ such that $P\tilde{u} = P\tilde{u}_0$. We let $u = \tilde{u}_0 - \tilde{u}$. Then $Pu = 0$ and $t(u) = u_0$ which completes the proof.

□

We present also a condition independent of the Hermitian metric.

Corollary 4.4.25 *Let the conditions of Theorem 4.4.24 hold. Then for a jet $u_0 \in \mathcal{E}_{[t,-1]}(S)$ there exists a section $u \in Z^{\cdot-1}(\mathcal{E}(\overline{Y}))$ such that $t(u) = u_0$ if and only if*

$$\int_S G_{P_{\cdot-1}}(g, u_0) = 0 \quad \text{for all} \quad g \in Z_\cdot(\mathcal{E}(E'|_{\overline{Y}})). \tag{4.4.11}$$

Proof. Necessity follows from Stokes' Theorem. Let us prove sufficiency. Since $P^* = \star^{-1}P'\star$ (see Subsection 2.1.3), condition (4.4.11) is stronger than condition (4.4.9). Therefore we need check only that condition (4.4.11) implies in addition $P_b u_0 = 0$. Let $\tilde{u}_0 \in \mathcal{E}^{\cdot-1}(\overline{Y})$ be some representative of the jet u_0, i.e., $t(\tilde{u}_0) = u_0$. Using Proposition 2.4.6 we obtain for each section $v \in \mathcal{E}(E^{\cdot+1\prime})$

$$\int_S G_{P_\cdot}(v, P\tilde{u}_0) = -\int_S G_{P_{\cdot-1}}(P'v, \tilde{u}_0) = -\int_S G_{P_{\cdot-1}}(P'v, u_0) = 0 \tag{4.4.12}$$

because $P'v \in Z_\cdot(\mathcal{E}(E^{\cdot\prime}|_{\overline{Y}}))$. So $t(P\tilde{u}_0) = 0$, i.e., $P_b u_0 = 0$ which completes the proof.

□

As to further results on the Cauchy problem for the Dolbeault complex, we refer to the survey of Henkin [104] and the book of Aizenberg and Yuzhakov [6, §26].

Remark 4.4.26 *We mention the role of Theorem 3.3.9 in studying Problem 4.4.11. Suppose that $E^\cdot$ has a fundamental solution at degree i (say, Φ), and S is compact. If $u_0 \in Z^{\cdot-1}(\mathcal{E}_{[t\,]}(S))$, then $\Phi([S]^P u_0)^\pm \in Z^{\cdot-1}(\mathcal{E}(\overline{X^\pm}))$ and*

$$t(\Phi([S]^P u_0)^+) - t(\Phi([S]^P u_0)^-) = u_0.$$

Thus in order that there be a section $u \in Z^{\cdot-1}(\mathcal{E}(\overline{X^\pm}))$ satisfying $t(u) = u_0$ it is necessary and sufficient that there be a section $\tilde{u} \in Z^{\cdot-1}(\mathcal{E}(\overline{X^\pm}))$ satisfying $t(\tilde{u}) = t(\Phi([S]^P u_0)^\mp)$. This approach was developed in the author's paper [226].

4.5 The Kernel Approach to Solving the Equation $Pu = f$

4.5.1 *General remark*

Good integral representations for solutions of the non-homogeneous Cauchy-Riemann system in strongly pseudo-convex domains and analytic polyhedra, found in the 70's, resulted in the constructive solution of the main problems of complex analysis (see the survey of Henkin [104]).

From the standpoint of the algebraic properties of the Dolbeault complex, this approach is provided by the fact that the spaces of sections $\mathcal{D}'(\Lambda^{0,q}T^*_{\mathbb{C}}(X))$ have the structure of $\mathcal{O}^{(p)}(X)$-modules and the differential $\overline{\partial}^{(0)}$ is an $\mathcal{O}^{(0)}(X)$-homomorphism. The method of integral representations works very fruitfully if the ring $\mathcal{O}^{(0)}(X)$ is rich enough. For example, this is the case on Stein manifolds X.

For an arbitrary differential complex $E^{\cdot}$, one can not count on such a picture.

Example 4.5.1 Let us call a function $\varphi \in \mathcal{E}(X)$ holomorphic with respect to the complex $E^{\cdot}$ if $P(\varphi f) = \varphi(Pf)$ for each section $f \in \mathcal{E}(E^{\cdot})$. Denote by $\mathcal{O}^{(E^{\cdot})}(X)$ the set of all the functions $\varphi \in \mathcal{E}(X)$ holomorphic with respect to $E^{\cdot}$. Obviously, $\mathcal{O}^{(E^{\cdot})}(X)$ is a ring. It is easy to see that a function $\varphi \in \mathcal{E}(X)$ belongs to $\mathcal{O}^{(E^{\cdot})}(X)$ if and only if φ satisfies some system of differential equations on X. However the system is too overdetermined, so that the ring $\mathcal{O}^{(E^{\cdot})}(X)$ is trivial in most cases. i.e., $\mathcal{O}^{(E^{\cdot})}(X) = \mathbb{C}$.

□

When the ring $\mathcal{O}^{(E^{\cdot})}(X)$ is rich enough, Naruki [170] found an approach to the equation $Pu = f$ based on the idea of Oka's mapping and called the localization principle. Apparently, this is the case where the algebraic structure of the complex allows one still to construct "explicit" formulas for solutions of $Pu = f$.

For general complexes of differential operators, the localization principle is of no use.

We derive some rough information on kernels for solutions of $Pu = f$ by using Theorem 1.4.9.

4.5.2 *Decomposition of the fundamental solution*

Let $E^{\cdot}$ be a differential complex on X which is assumed to be either elliptic or homotopically equivalent to an elliptic one.

Suppose that $E^{\cdot}$ has a fundamental solution at degree q. According to Proposition 2.3.9, $E^{\cdot}$ has also a pseudo-differential fundamental solution at degree q. We fix one of them, say, Φ.

Remark 4.5.2 *As formally compared with the q-convexity (see Subsection 2.3.3), one can call X q-concave with respect to the (chain) complex $E^{\cdot\prime}$ if $H_i(\mathcal{E}(E^{\cdot\prime})) = 0$ for all $i \leq q$.*

For the de Rham complex, the following lemma was proved in the author's paper [219].

Lemma 4.5.3 *Let Y be an open subset of X such that $X \setminus \overline{Y}$ is q-concave with respect to $E^{\cdot\prime}|_{X\setminus\overline{Y}}$. Then there exist operators $K^{(-1)}_{q+1} \in \mathcal{L}(\mathcal{E}'^{q+1}(X \setminus \overline{Y}) \to \mathcal{E}^q(Y))$ and $K^{(-2)}_{q+1} \in \mathcal{L}(\mathcal{E}'^{q+1}(X \setminus \overline{Y}) \to \mathcal{E}^{q-1}(Y))$ such that: 1) $K^{(-1)}_{q+1} P_q = 0$, and 2) $\Phi_{q+1} = K^{(-1)}_{q+1} + P_{q-1} K^{(-2)}_{q+1}$ on $\mathcal{E}'^{q+1}(X \setminus \overline{Y})$.*

Proof. Using the relation between kernels and operators given by Theorem 1.5.1, we have to construct kernels $K_{q+1}^{(-1)} \in \mathcal{E}(E^q \boxtimes E^{q+1\prime}|_{Y\times(X\setminus\overline{Y})})$ and $K_{q+1}^{(-2)} \in \mathcal{E}(E^{q-1} \boxtimes E^{q+1\prime}|_{Y\times(X\setminus\overline{Y})})$ such that: 1) $P'_q(y)K_{q+1}^{(-1)} = 0$, and 2) $\Phi_{q+1} = K_{q+1}^{(-1)} + P_{q-1}(x)K_{q+1}^{(-2)}$ in the product $Y \times (X \setminus \overline{Y})$.

With this purpose we write equation (2.1.3) for the components of the fundamental solution Φ, which are of interest for us:

$$P'_\imath(y)\Phi_{\imath+1}(x,y) + P_{\imath-1}(x)\Phi_\imath(x,y) = \text{tr}\ \Delta^* - S_\imath(x,y) \quad (\imath = 0,\dots,q). \tag{4.5.1}$$

Here the kernels $S_\imath \in \mathcal{E}(E^\imath \boxtimes E^{\imath\prime})$ satisfy the system (2.3.7), and $S_q = 0$.

First we would like to find a correction $M \in \mathcal{E}((E \boxtimes E')^{-1}|_{X\times(X\setminus\overline{Y})})$ supported at the degrees $\leq q$ (i.e., $M_\imath = 0$ for $\imath > q$), such that the kernel $\Phi^{(q)} = \Phi + M$ satisfies system (4.5.1) on $X \times (X \setminus \overline{Y})$ with the zero errors $S_\imath$. For this it is required that the components of the kernel M satisfy the following system of differential equations considered on the product $X \times (X \setminus \overline{Y})$:

$$P'_\imath(y)M_{\imath+1}(x,y) + P_{\imath-1}(x)M_\imath(x,y) = S_\imath(x,y) \quad (\imath = 0,\dots,q). \tag{4.5.2}$$

The system (4.5.2) is solved by the chain rule. Namely, using the condition $H_0(\mathcal{E}(E'|_{X\setminus\overline{Y}})) = 0$ and Theorem 1.4.9 we may find $M_1 \in \mathcal{E}(E^0 \boxtimes E^{1\prime}|_{X\times(X\setminus\overline{Y})})$ from the first equation (4.5.2) (i.e., for $\imath = 0$). This gives automatically the compatibility condition for the second equation (4.5.2) (i.e., for $\imath = 1$). Using the condition $H_1(\mathcal{E}(E'|_{X\setminus\overline{Y}})) = 0$ and Theorem 1.4.9 we find then $M_2 \in \mathcal{E}(E^1 \boxtimes E^{2\prime}|_{X\times(X\setminus\overline{Y})})$ from the second equation (4.5.2). This gives the compatibility condition for the third equation (4.5.2), then M_3, etc. Such considerations have already been used in the proof of Theorem 2.3.12.

But in the last step we meet a new difficulty consisting in that we have to find a solution $M_q \in \mathcal{E}(E^{q-1} \boxtimes E^{q\prime}|_{X\times(X\setminus\overline{Y})})$ of the equation $P'_{q-1}(y)M_q(x,y) = S_{q-1}(x,y) - P_{q-2}(x)M_{q-1}(x,y)$, which also satisfies $P_{q-1}(x)M_q(x,y) = 0$ on $X \times (X \setminus \overline{Y})$. In order to do this, Theorem 1.4.9 is used in full force.

Consider the (chain) complex L on $X \times (X \setminus \overline{Y})$ endowed with the differential $P'(y)$, where $L_\imath$ is the subspace of $\mathcal{E}(E^{q-1} \boxtimes E^{\imath\prime}|_{X\times(X\setminus\overline{Y})})$ formed by all sections f satisfying $P_{q-1}(x)f = 0$. Since $H_{q-1}(\mathcal{E}(E'|_{X\setminus\overline{Y}})) = 0$, we get by Theorem 1.4.9 that $H_{q-1}(L) = 0$. Therefore, for the kernel $(S_{q-1} - P_{q-2}M_{q-1}) \in Z_{q-1}(L)$ there exists a kernel $M_q \in L_q$ such that $P'_{q-1}(y)M_q = S_{q-1} - P_{q-2}M_{q-1}$ on $X \times (X \setminus \overline{Y})$, which we wanted to prove.

Thus the components of the kernel $\Phi^{(q)} = \Phi + M$ satisfy the following system of differential equations on $X \times (X \setminus \overline{Y})$:

$$P'_\imath(y)\Phi_{\imath+1}^{(q)}(x,y) + P_{\imath-1}(x)\Phi_\imath^{(q)}(x,y) = \text{tr}\ \Delta^* \quad (\imath = 0,\dots,q). \tag{4.5.3}$$

This system is solved analogously to system (4.5.2).

Consider the complex $\mathcal{E}(E^0 \boxtimes E'|_{Y\times(X\setminus\overline{Y})})$ endowed with the differential $P'(y)$. Since $H_1(\mathcal{E}(E'|_{X\setminus\overline{Y}})) = 0$, Theorem 1.4.9 implies that $H_1(\mathcal{E}(E^0 \boxtimes E'|_{Y\times(X\setminus\overline{Y})})) = 0$. It follows from the first equation of (4.5.3) (i.e., for $\imath = 0$) that $\Phi_1^{(q)} \in Z_1(\mathcal{E}(E^0 \boxtimes$

$E^{\cdot\prime}|_{Y\times(X\setminus\overline{Y})}))$. So for the kernel $\Phi_1^{(q)}$ we can find a kernel $K_2^{(-2)} \in \mathcal{E}(E^0 \boxtimes E^{2\prime}|_{Y\times(X\setminus\overline{Y})})$ such that $P_1'(y)K_2^{(-2)} = -\Phi_1^{(q)}$ on $Y \times (X \setminus \overline{Y})$.

Define now the kernel $K_2^{(-1)} \in \mathcal{E}(E^1 \boxtimes E^{2\prime}|_{Y\times(X\setminus\overline{Y})})$ from the equality $\Phi_2^{(q)} = K_2^{(-1)} + P_0(x)K_2^{(-2)}$ on $Y \times (X \setminus \overline{Y})$. Then, using the second equality (4.5.3) (i.e., for $i = 1$), we obtain

$$P_1'(y)K_2^{(-1)} = P_1'(y)\Phi_2^{(q)} - P_0(x)(P_1'(y)K_2^{(-2)}) = 0 \quad \text{on} \quad Y \times (X \setminus \overline{Y}).$$

Consider the complex $\mathcal{E}(E^1 \boxtimes E^{\cdot\prime}|_{Y\times(X\setminus\overline{Y})})$ endowed with the differential $P'(y)$. Since $H_2(\mathcal{E}(E^{\cdot\prime}|_{X\setminus\overline{Y}})) = 0$, it follows from Theorem 1.4.9 that $H_2(\mathcal{E}(E^1 \boxtimes E^{\cdot\prime}|_{Y\times(X\setminus\overline{Y})})) = 0$. Therefore, for the kernel $K_2^{(-1)} \in Z_2(\mathcal{E}(E^1 \boxtimes E^{2\prime}|_{Y\times(X\setminus\overline{Y})}))$ we can find a kernel $K_3^{(-2)} \in \mathcal{E}(E^1 \boxtimes E^{3\prime}|_{Y\times(X\setminus\overline{Y})})$ such that $P_2'(y)K_3^{(-2)} = -K_2^{(-1)}$ on $Y \times (X \setminus \overline{Y})$.

Define the kernel $K_3^{(-1)} \in \mathcal{E}(E^2 \boxtimes E^{3\prime}|_{Y\times(X\setminus\overline{Y})})$ from the equality $\Phi_3^{(q)} = K_3^{(-1)} + P_1(x)K_3^{(-2)}$ on $Y \times (X \setminus \overline{Y})$. Then, using the third equality (4.5.3), we obtain $P_2'(y)K_3^{(-1)} = 0$ on $Y \times (X \setminus \overline{Y})$.

Continuing this process as for as the q-concavity of $X \setminus \overline{Y}$ allows, we construct kernels $K_i^{(-2)} \in \mathcal{E}(E^{i-2} \boxtimes E^{i\prime}|_{Y\times(X\setminus\overline{Y})})$ $(i = 2, \ldots, q+1)$ such that $\Phi_i^{(q)} = -P_i'(y)K_{i+1}^{(-2)} + P_{i-2}(x)K_i^{(-2)}$ $(i = 1, \ldots, q)$ on $Y \times (X \setminus \overline{Y})$.

Define the kernel $K_{q+1}^{(-1)} \in \mathcal{E}(E^q \boxtimes E^{q+1\prime}|_{Y\times(X\setminus\overline{Y})})$ from the equality $\Phi_{q+1}^{(q)} = K_{q+1}^{(-1)} + P_{q-1}(x)K_{q+1}^{(-2)}$ on $Y \times (X \setminus \overline{Y})$. Then, using the last equality (4.5.3), we get on the set $Y \times (X \setminus \overline{Y})$:

$$\begin{aligned}
P_q'(y)K_{q+1}^{(-1)} &= P_q'(y)\Phi_{q+1}^{(q)} - P_{q-1}(x)\left(P_q'(y)K_{q+1}^{(-2)}\right) \\
&= P_q'(y)\Phi_{q+1}^{(q)} + P_{q-1}(x)\left(\Phi_q^{(q)} - P_{q-2}(x)K_q^{(-2)}\right) \\
&= 0.
\end{aligned}$$

To complete the proof of the lemma, it remains only to note that $\Phi_{q+1}^{(q)} = \Phi_{q+1}$ on the set $X \times (X \setminus \overline{Y})$.

□

A decomposition of the fundamental solution Φ, like the one guaranteed by Lemma 4.5.3, appeared first for the Dolbeault complex. However, there it was based on the existence of some "support function" for Y. Such a decomposition provides a generalization of the Cauchy-Fantappiè formula to differential forms (see Henkin [104, §8.2]). For the de Rham complex, the approach was developed by the author [219] (see also Section 6.1).

4.5.3 *Special homotopy formulas on manifolds with boundary*

Let Y be an open subset of X such that $X \setminus \overline{Y}$ is q-concave with respect to the complex $E^{\cdot\prime}|_{X\setminus\overline{Y}}$.

Theorem 4.5.4 *If U is a neighborhood of $\overline{Y}$ with piecewise smooth boundary, then for each section $f \in C^{p_q-1}_{comp}(E^q|_{\overline{U}})$ satisfying $Pf = 0$ in U the following formula holds:*

$$f(x) = -K^{(-1)}_{q+1}([\partial U]^P f) + P(-K^{(-2)}_{q+1}([\partial U]^P f) + \Phi(\chi_U f)) \quad (x \in Y). \tag{4.5.4}$$

Proof. In fact, using Corollary 2.4.9, we obtain

$$f(x) = -\Phi_{q+1}([\partial U]^P f) + P\Phi(\chi_U f) \quad (x \in U).$$

Since $[\partial U]^P f \in \mathcal{E}'^{q+1}(X \setminus \overline{Y})$, we may apply the decomposition of Lemma 4.5.3 to the integral $\Phi_{q+1}([\partial U]^P f)$ for $x \in Y$. This gives precisely formula (4.5.4), as desired.

□

Notice that, under the hypothesis of Theorem 4.5.4, we have $f(x) = hPf + Phf$ $(x \in Y)$ for each section $f \in C^{p_i+p_{i+1}-1}_{comp}(E^i|_{\overline{U}})$ $(0 \le i \le q-1)$ where the homotopy operator is given by $h(f) = -K^{(-2)}_{i+1}([\partial U]^P f) + \Phi(\chi_U f)$ in Y.

4.5.4 *Kernels for solutions of the equation $Pu = f$*

We start with the following lemma.

Lemma 4.5.5 *Let U be a neighborhood of $\overline{Y}$ with piecewise smooth boundary. In order that for a section $f \in C^{p_q-1}_{comp}(E^q|_{\overline{U}})$ there exist a section $u \in C^{p_{q-1}+p_q-2}_{comp}(E^{q-1}|_{\overline{U}})$ satisfying $Pu = f$ in U, it is necessary that*

$$\int_{\partial U} G_{P_q}(g, f) = 0 \quad \text{for all} \quad g \in Z_{q+1}(\mathcal{E}(E'|_{X\setminus\overline{Y}})). \tag{4.5.5}$$

Proof. In fact, using Proposition 2.4.6 and Stokes' Theorem, we get

$$\int_{\partial U} G_{P_q}(g, f) = \int_{\partial U} G_{P_q}(g, Pu) = -\int_{\partial U} G_{P_{q-1}}(P'g, u) = 0,$$

as desired.

□

If $X \setminus \overline{Y}$ is q-concave with respect to the complex $E'|_{X\setminus\overline{Y}}$, then condition (4.5.5) for a cocycle f is also sufficient in order that the equation $Pu = f$ be solvable, at least, in Y.

Corollary 4.5.6 *Suppose that U is a neighborhood of $\overline{Y}$ with piecewise smooth boundary, and $f \in C^{p_q-1}_{comp}(E^q|_{\overline{U}})$ satisfies $Pf = 0$ in U as well as condition (4.5.5). Then $f = Ph(f)$ in Y where $h(f) = -K^{(-2)}_{q+1}([\partial U]^P f) + \Phi(\chi_U f)$ in Y.*

Proof. This follows immediately from Theorem 4.5.4 because

$$K^{(-1)}_{q+1}([\partial U]^P f)(x) = \int_{\partial U} G_{P_q}(K^{(-1)}_{q+1}(x, \cdot), f) \quad (x \in Y),$$

and $P'(y)K_{q+1}^{(-1)}(x,y) = 0$ for (x,y) in the product $Y \times (X \setminus \overline{Y})$.

$\square$

What can be said of the solution $u = h(f)$ is only that u inherits the local smoothness of f in Y. The information on the kernel $K_{q+1}^{(-2)}$ is too scanty. What should further be known, is the behavior of $K_{q+1}^{(-2)}(x,y)$ near the distinquished boundary $\partial Y \times \partial Y$. However, this requires much more delicate methods.

Chapter 5

Duality Theory for Cohomologies of Differential Complexes

5.1 The Poincaré Duality and the Alexander-Pontryagin Duality

5.1.1 *Duality Lemma*

The results of this section are essentially statements from functional analysis and in this context are well-known (see Serre [198]).

We recall that with each topological vector space L one can associate a separated topological vector space $L_{sep} = L/\overline{\{0\}}$, which is called the separated topological vector space associated to L (see Edwards [62, (1.8.5)]). In particular if Σ is a (generally speaking, unclosed) subspace of L then $(L/\Sigma)_{sep} = L/\overline{\Sigma}$. . We emphasize that always we endow a subspace with the induced topology, a quotient space with the canonical topology, and a dual space with the strong topology. The following "Duality Lemma", originating with the work of Serre [198], is used in all papers on duality theory in complex analytic geometry (see in particular Andreotti and Kas [12]). Following Golovin [86, p.161], we present it in a more general form.

Lemma 5.1.1 *Let $L^1 \overset{M_1}{\to} L^2 \overset{M_2}{\to} L^3$ be a complex of separated locally convex topological vector spaces and continuous linear mappings thereof, where L^2 is semireflexive. Then there is a natural isomorphism of vector spaces (without topologies)*

$$\left(\frac{\ker M_2}{\overline{im\ M_1}}\right)' \cong \frac{\ker M_1'}{\overline{im\ M_2'}}. \tag{5.1.1}$$

Proof. Denote $V^1 = \ker M_2/\overline{\mathrm{im}\ M_1}$ and $V^2 = \ker M_1'/\overline{\mathrm{im}\ M_2'}$.

Define the mapping $m : V^2 \to (V^1)'$ as follows. Assume that $[f] \in V^2$ and let $f \in \ker M_1'$ be a representative of the class $[f]$, then $\langle m([f]), [x]\rangle = f(x)$, where $x \in \ker M_2$ is a representative of the class $[x] \in V^1$. It easy to see this is well-defined, that is m does not depend on the choice of representatives $f \in [f]$ and $x \in [x]$, and $m([f])$ is a continuous linear functional on V^1.

The mapping m is always surjective. Indeed, assume that $F \in (V^1)'$. Then F can be considered as an element of $(\ker M_2)'$ vanishing on $\overline{\operatorname{im} M_1}$. By the Hahn-Banach Theorem, we may extend F to a functional $\widetilde{F} \in (L^2)'$. For $x \in L^1$, we have $(M_1'\widetilde{F})(x) = \widetilde{F}(M_1, x) = F(M_1 x) = 0$; hence $\widetilde{F} \in \ker M_1'$. Obviously $m([\widetilde{F}]) = F$ as required.

Finally, since L^2 is semireflexive, the mapping m is injective. Indeed, one must show that if $f \in \ker M_1'$ and $f(x) = 0$ for all $x \in \ker M_2$ then $f \in \overline{\operatorname{im} M_2'}$. By the Hahn-Banach Theorem, it is sufficient to show that, if $F \in (L^2)''$ vanishes on $\operatorname{im} M_2'$ then $F(f) = 0$. Since the space L^2 semireflexive, there exists a vector $x \in L^2$ such that $F(\varphi) = \varphi(x)$ for all $\varphi \in (L^2)'$. As $F = 0$ on $\operatorname{im} M_2'$, then for all $\psi \in (L^3)'$ we have $\psi(M_2 x) = (M_2'\psi)(x) = F(M_2'\psi) = 0$. Therefore $M_2 x = 0$, that is $x \in \ker M_2$. It follows that $F(f) = f(x) = 0$ and the proof of the lemma is complete. □

Remark 5.1.2 *Of course,* $\left(\frac{\ker M_2}{\overline{im\ M_1}}\right)' = \left(\frac{\ker M_2}{im M_1}\right)'$.

Remark 5.1.3 *The condition of semireflexivity of L^2 can be replaced by the condition that M_2 be a homomorphism. Then $(im\,M_2')$ is closed (see Proposition 8 of Palamodov [175, Ch. V, §1]).*

The importance of the class of the Fréchet-Schwartz spaces, introduced by Grothendieck [91], is clear from the following statement.

Lemma 5.1.4 *If in addition to the conditions of Lemma 5.1.1 L^i $(i = 1, \ldots, 3)$ are Fréchet-Schwartz spaces, then the algebraic isomorphism (5.1.1) is topological. It is also true if L^i $(i = 1, \ldots, 3)$ are dual to Fréchet-Schwartz spaces.*

Proof. This follows from variants of the Open Mapping Theorem and the Closed Graph Theorem (see Grothendieck [91]). Since the Fréchet-Schwartz spaces are reflexive and barreled, the algebraic isomorphism $m: V^2 \to (V^1)'$ is continuous (see Edwards [62, Exercise 8.43]). On the other hand, since the Fréchet-Schwartz spaces are quite complete (see Edwards [62, Theorem 8.10.5]) and the dual spaces for them are barreled (see Theorem 8.4.3 therein) the mapping m is open (see Theorem 8.9.3(1) therein). Therefore m is a topological isomorphism. The second part of the lemma follows easily from first part.

□

The conditions, under which M_2 is a homomorphism were earlier found by Dieudonné and Schwartz [54] (see also Proposition 7 of Palamodov [175, Ch.V, §1]).

Lemma 5.1.5 *Let $L \xrightarrow{M} K$ be a continuous mapping of Fréchet spaces. Then the following conditions are equivalent: 1) M is a homomorphism; 2) $(im\ M)$ is closed in K, and 3) $(im\ M')$ is closed in L'.*

Remark 5.1.6 *Palamodov (op. cit.) showed that if L and K are Fréchet-Schwartz spaces, than the conditions 1)–3) of the lemma are also equivalent to: 4) M' is a homomorphism.*

Serre [198, p.21] discovered a simple condition for $\operatorname{im} M_2$ to be closed.

Lemma 5.1.7 *Assume that $L \overset{M}{\to} K$ is a continuous linear mapping of Fréchet spaces. If the range of M has finite codimension then it is closed.*

Proof. Let us denote by C the algebraic complement of $M(L)$ in K. Consider the mapping $m: L \times C \to K$ given by $m(x,y) = M(x) + y$ for $x \in L$, $y \in C$. It is easy to see that this mapping is linear and continuous. Since C is a separated topological vector space of a finite dimension, we conclude that $L \times C$ is a Fréchet space. By the Open Mapping Theorem, m is a (topological) homomorphism. Hence it follows that M is a homomorphism also.

□

5.1.2 *Poincaré duality*

Assume now that X is an n-dimensional differentiable manifold countable at infinity and E is a (cochain) differential complex on X.

The following result was observed in the work of Tarkhanov [231].

Theorem 5.1.8 *Let $\mathfrak{C}$ be $\mathcal{E}$ or $\mathcal{E}'$. If the cohomology space $H^{q+1}(\mathfrak{C}(E))$ is a separated then there is the topological isomorphism*

$$H^q(\mathfrak{C}(E))' \overset{tr}{\cong} H_q(\mathfrak{C}'(E')). \tag{5.1.2}$$

Proof. Consider the complex of topological vector spaces and continuous linear mappings thereof

$$\mathfrak{C}(E^{q-1}) \overset{P_{q-1}}{\to} \mathfrak{C}(E^q) \overset{P_q}{\to} \mathfrak{C}(E^{q+1}). \tag{5.1.3}$$

The linear form tr $= \langle , \rangle_X$ determines the pairings $\mathfrak{C}'(E^{\cdot\prime}) \otimes \mathfrak{C}(E^{\cdot}) \to \mathbb{C}$, where $\mathfrak{C}'(E^{\cdot\prime}) = (\mathfrak{C}(E^{\cdot}))'$. According to Proposition 1.1.25, the complex transposed to the complex (5.1.3), is

$$\mathfrak{C}(E^{q-1\prime}) \overset{P'_{q-1}}{\leftarrow} \mathfrak{C}'(E^{q\prime}) \overset{P'_q}{\leftarrow} \mathfrak{C}'(E^{q+1\prime}).$$

As the manifold X is countable at infinity, $\mathcal{E}(E^{\cdot})$ are the Fréchet-Schwartz spaces (see Proposition 1.1.7) and by Lemma 5.1.4 we obtain the topological isomorphism

$$H^q(\mathfrak{C}(E))' \overset{tr}{\cong} H_q(\mathfrak{C}'(E'))_{sep}. \tag{5.1.4}$$

Since the space $H^{q+1}(\mathfrak{C}(E))$ is separated, by Proposition 1.1.12 the range $P_q\mathfrak{C}(E^q)$ is closed in $\mathfrak{C}(E^{q+1})$. By Lemma 5.1.5 we deduce that the range $P'_q\mathfrak{C}'(E^{q+1\prime})$ is closed in $\mathfrak{C}'(E^{q\prime})$. That is, the space $H_q(\mathfrak{C}'(E'))$ is separated.

Thus $H_q(\mathfrak{C}'(E'))_{sep} = H_q(\mathfrak{C}'(E'))$, and (5.1.2) implies (5.1.2) which concludes the proof.

□

Remark 5.1.9 *Taking into account the classical aspects of Theorem 5 1.8 (see Example 5.1.12), we call (5.1.2) the Poincaré duality. Here is a more general form of it, valid for* $\mathfrak{C} = \mathcal{E}, \mathcal{D}, \mathcal{D}'$, *and* $\mathcal{E}'$: $H(\mathfrak{C}(E))' \overset{alg}{\cong} H(\mathfrak{C}'(E'))_{sep}$. *This follows from Lemma 5.1.1 and Remark 5.1.2 because the spaces* $\mathfrak{C}(E^{\cdot})$ *are Montel spaces and therefore reflexive.*

Corollary 5.1.10 *If the cohomology spaces* $H^q(\mathcal{E}(E))$ *and* $H^{q+1}(\mathcal{E}(E))$ *are finite-dimensional, then* $H^q(\mathcal{E}(E))$ *is a Fréchet-Schwartz space and there is the topological isomorphism*

$$H^q(\mathcal{E}(E))' \overset{tr}{\cong} H_q(\mathcal{E}'(E')).$$

Proof. By Lemma 5.1.7 we get that the spaces $H^q(\mathcal{E}(E))$ and $H^{q+1}(\mathcal{E}(E))$ are separated and consequently they are Fréchet-Schwartz spaces. It remains only to invoke Theorem 5.1.8.

□

Let us consider some examples.

Example 5.1.11 Assume that E is an elliptic complex on X. Then from Theorem 5.1.8 and Corollary 2.1.15 we obtain that, if the space $H^{q+1}(\mathcal{E}(E))$ is separated, then $H^q(\mathcal{E}(E))' \overset{top}{\cong} H_q(\mathcal{D}(E'))$. Similarly, if the space $H^{q+1}(\mathcal{D}(E))$ is separated, then $H^q(\mathcal{D}(E))' \overset{top}{\cong} H_q(\mathcal{E}(E'))$. Moreover in that case it is easy to indicate the inverse mapping for $H_q(\mathfrak{C}'(E')) \ni [g] \to \mathrm{tr}\,(g \otimes \cdot) \in H^q(\mathfrak{C}(E'))$. Namely, let Π be a properly supported parametrix of the complex E from Theorem 2.1.12, and R be the corresponding smooth endomorphism of E Then the inverse mapping is $H^q(\mathfrak{C}(E))' \ni g \to [R'(\tilde{g})] \in H_q(\mathfrak{C}'(E'))$ where $\tilde{g} \in \mathfrak{C}'(E^{q'})$ is a suitable extension of g. The reader can enter into details without any assistance. In particular, if the manifold X is compact, then by Corollary 2.2.4 the spaces $H(\mathcal{E}(E))$ are finite-dimensional. Thus, according to Corollary 5.1 10 there is a topological isomorphism $H(\mathcal{E}(E))' \overset{tr}{\cong} H(\mathcal{E}(E'))$. This result was announced by Atiyah and Bott [19]

□

Example 5.1.12 Let $E = \Lambda\, T^*_{\mathbb{C}}(X)$ be the de Rham complex on an orientable manifold X. Then $E' \cong E$ where the isomorphism is obtained by identifying $E^{i'}$ with E^{n-i}. Therefore $H^i(\mathcal{E}(E)) \cong H^i(X,\mathbb{C})$ and $H_i(\mathcal{D}(E')) \cong H^{n-i}_{comp}(X,\mathbb{C})$ where the symbol "comp" indicates cohomologies with compact supports. We endow the spaces $H^i(X,\mathbb{C})$ and $H^{n-i}_{comp}(X,\mathbb{C})$ with the topologies induced by these isomorphisms. Then by Theorem 5.1.8 we have $H^q(X,\mathbb{C})' \overset{top}{\cong} H^{n-q}_{comp}(X,\mathbb{C})$ provided the space $H^{q+1}(X,\mathbb{C})$ is separated. Similarly, if $H^{q+1}_{comp}(X,\mathbb{C})$ is separated, then $H^q_{comp}(X,\mathbb{C})' \overset{top}{\cong} H^{n-q}(X,\mathbb{C})$. Assume that $H^q(X,\mathbb{C})$ is finite-dimensional. Then $H^q(X,\mathbb{C})' = \mathrm{Hom}\,(H^q(X,\mathbb{C}) \to \mathbb{C}) = H^{comp}_q(X,\mathbb{C})$, where $H^{comp}_q(X,\mathbb{C})$ is the q-th group of compact homologies with complex coefficients of a manifold X. Similarly, if $H^q_{comp}(X,\mathbb{C})$ is finite-dimensional, then $H^q_{comp}(X,\mathbb{C})' = H_q(X,\mathbb{C})$. Thus we get the isomorphisms

$$\begin{aligned} H_q^{comp}(X,\mathbb{C}) &\cong H^{n-q}_{comp}(X,\mathbb{C}), \\ H_q(X,\mathbb{C}) &\cong H^{n-q}(X,\mathbb{C}). \end{aligned}$$

This is the classic Poincaré duality in algebraic topology (see Massey [160, §11.3] and de Rham [50, Theorem 17]).

□

Certain versions of the Poincaré duality for the de Rham complex were discussed by Gol'dshtein, Kuz'minov, and Shvedov [80].

Example 5.1.13 Let E be a holomorphic vector bundle over an n-dimensional complex manifold X and $E = \{E\otimes\Lambda^{0,\cdot}T^*_{\mathbb{C}}(X), \overline{\partial}^{(E)} = 1\otimes\overline{\partial}^{(0)}\}$ be the Dolbeault complex with coefficients in E on X (see Wells [254]). Then $H^{\cdot}(\mathcal{E}(E^{\cdot})) \cong H^{\cdot}(X,\mathcal{O}^{(E)})$, where $\mathcal{O}^{(E)}$ is the sheaf of germs of holomorphic sections of E on X. See Serre [198]) or Wells [254] for this generalization of the Dolbeault Theorem. Further $E^{\cdot\prime} \cong (E^*\otimes\Lambda^{n,0}T^*_{\mathbb{C}}(X))\otimes\Lambda^{0,\cdot}T^*_{\mathbb{C}}(X)$, where the isomorphism is obtained by identifying $E^{i\prime}$ with $E^*\otimes\Lambda^{n,n-i}T^*_{\mathbb{C}}(X)$. Hence $H_{\cdot}(\mathcal{D}(E^{\cdot\prime})) \cong H^{n-\cdot}_{comp}(X,\mathcal{O}^{(E^*\otimes\Lambda^{n\,0}T^*_{\mathbb{C}}(X))})$. We endow the spaces $H^{\cdot}(X,\mathcal{O}^{(E)})$ and $H^{n-\cdot}_{comp}(X,\mathcal{O}^{(E^*\otimes\Lambda^{n\,0}T^*_{\mathbb{C}}(X))})$ with the topologies induced by the isomorphisms above. By Theorem 5.1.8, if the space $H^{q+1}(X,\mathcal{O}^{(E)})$ is separated, then $H^q(X,\mathcal{O}^{(E)})' \overset{top}{\cong} H^{n-q}_{comp}(X,\mathcal{O}^{(E^*\otimes\Lambda^{n\,0}T^*_{\mathbb{C}}(X))})$. In particular, this is valid provided $H^q(X,\mathcal{O}^{(E)})$ and $H^{q+1}(X,\mathcal{O}^{(E)})$ are finite-dimensional (Corollary 5.1.10). This is the Serre Duality Theorem (see Serre [198, p.20]). A more general statement is due to Golovin [86, p.31]). Similarly, if $H^{q+1}_{comp}(X,\mathcal{O}^{(E)})$ is separated, then $H^q_{comp}(X,\mathcal{O}^{(E)})' \overset{top}{\cong} H^{n-q}(X,\mathcal{O}^{(E^*\otimes\Lambda^{n\,0}T^*_{\mathbb{C}}(X))})$. Thus, by suitable conditions we obtain the isomorphisms

$$\begin{aligned} H_q^{comp}(X,\mathcal{O}^{(E)}) &\cong H^{n-q}_{comp}(X,\mathcal{O}^{(E^*\otimes\Lambda^{n\,0}T^*_{\mathbb{C}}(X))}), \\ H_q(X,\mathcal{O}^{(E)}) &\cong H^{n-q}(X,\mathcal{O}^{(E^*\otimes\Lambda^{n\,0}T^*_{\mathbb{C}}(X))}). \end{aligned}$$

This is a version of the Poincaré duality in complex analytic geometry (see Golovin [86, p.79]).

□

Example 5.1.14 Assume that the manifold X is 1-convex with respect to $E^{\cdot}$ that is, the complex $\mathcal{E}(E^{\cdot})$ is exact at positive degrees. Then by Corollary 5.1.10 the complex $\mathcal{E}'(E^{\cdot\prime})$ is exact at positive dimensions, and $Z^0(\mathcal{E}(E^{\cdot}))' \overset{top}{\cong} H_0(\mathcal{E}'(E^{\cdot\prime}))$. In particular, this is valid for the Dolbeault complex on a Stein manifold (see Serre [198, p.21]).

□

Example 5.1.15 Let $\{U\}$ be an open covering of X such that each open covering of X has a refinement consisting of elements of $\{U\}$. Suppose further that each set $U \in \{U\}$ is 1-convex with respect to the complex E. This implies first of all that the sequence of sheaves $\mathcal{E}$ over X is exact at positive degrees. Thus, by the Abstract de Rham Theorem (see Subsection 5.2.4) we have $H(\mathcal{E}(E)) \cong H(X,S)$, where $S = \mathcal{E}_{P_0}$ is the sheaf of germs of differentiable solutions of the equation $P_0 f = 0$. Further, according to the statement of Example 5.1.14 the complex $\mathcal{E}'(E'|_U)$ is exact at positive dimensions and $(S(U))' \overset{top}{\cong} H_0(\mathcal{E}'(E'|_U))$ for each set $U \in \{U\}$. Consequently, we get (see Subsection 5.2.7) that $H(\mathcal{E}'(E')) \cong H(X,S')$, where $H(X,S')$ is the space of homologies of X with values in the cosheaf S' defined by $S'(\sigma) = H_0(\mathcal{E}'(E'|_\sigma))$. Define the topologies in $H(X,S)$ and $H(X,S')$ by means of the isomorphisms above. Then by Theorem 5.1.8, if the space $H^{q+1}(X,S)$ is separated, then there is the topological isomorphism $H^q(X,S)' \overset{tr}{\cong} H_q(X,S')$. This was observed by Andreotti, Hill, Łojasiewicz and MacKichan [11, §13].

□

5.1.3 *Alexander-Pontryagin duality*

The Alexander-Pontryagin duality is an interesting application of the Poincaré duality (5.1.2). Let us consider the corresponding notions.

Let Y be an open subset of X, so that $X \setminus Y$ is closed.

Denote by $\vec{\mathcal{E}}(E^{i\prime}|_{X\setminus Y})$ the vector space of differentiable sections of the bundle $E^{i\prime}$ in a neighborhood of $X \setminus Y$ depending on the section. Introduce in this space an equivalence relation by letting two sections be equivalent if they coincide in some neighborhood of $X \setminus Y$. For the set of equivalence classes we will keep the notation $\vec{\mathcal{E}}(E^{i\prime}|_{X\setminus Y})$.

With the inductive limit topology induced by the Fréchet-Schwartz spaces $\mathcal{E}(E^{i\prime}|_U)$ which are (partly) ordered through inclusions of neighborhoods U of the set $X \setminus Y$, the space $\vec{\mathcal{E}}(E^{i\prime}|_{X\setminus Y})$ is locally convex and complete [1].

These topological vector spaces are united into the complex $\vec{\mathcal{E}}(E'|_{X\setminus Y})$ which is endowed with the differential P'.

For elements $g \in \vec{\mathcal{E}}(E^{i\prime}|_{X\setminus Y})$, the values are well-defined on the set $X \setminus Y$ only. Hence the notation $\vec{\mathcal{D}}(E^{i\prime}|_{X\setminus Y})$ seems to be appropriate for the (vector) subspace of $\vec{\mathcal{E}}(E^{i\prime}|_{X\setminus Y})$ formed by elements which are equal to zero outside some compact subset of $X \setminus Y$. The topology in $\vec{\mathcal{D}}(E^{i\prime}|_{X\setminus Y})$ is stronger than in $\vec{\mathcal{E}}(E^{i\prime}|_{X\setminus Y})$, namely it is determined by the condition that convergent sequences do not "wash away" the supports. Thus we arrive at the subcomplex $\vec{\mathcal{D}}(E'|_{X\setminus Y})$ of the complex $\vec{\mathcal{E}}(E'|_{X\setminus Y})$.

If the complement of the set Y is compact, then $\vec{\mathcal{D}}(E'|_{X\setminus Y}) = \vec{\mathcal{E}}(E'|_{X\setminus Y})$.

The complex $\vec{\mathcal{D}'}(E'|_{X\setminus Y})$ and its subcomplex $\vec{\mathcal{E}'}(E'|_{X\setminus Y})$ are defined by analogy.

Then by the Alexander-Pontryagin duality we mean the following proposition presented in the author's paper [231].

[1] See the notion of inductive (or direct) limit, for example, in the book of Palamodov [175, Pt 2, §1]

Theorem 5.1.16 *Let $H^q(\mathcal{E}(E^\cdot)) = H^{q+1}(\mathcal{E}(E^\cdot)) = 0$. If the spaces $H^{q+2}(\mathcal{E}(E^\cdot))$ and $H^{q+1}(\mathcal{E}(E^\cdot|_Y))$ are separated then there is a topological isomorphism*

$$H^q(\mathcal{E}(E^\cdot|_Y))' \cong H_{q+1}(\vec{\mathcal{E}'}\,(E^{\cdot\prime}|_{X\setminus Y})). \tag{5.1.5}$$

Proof. First of all, under the conditions of the theorem it follows from Poincaré duality (5.1.2) that

$$H_q(\mathcal{E}'(E^{\cdot\prime})) = H_{q+1}(\mathcal{E}'(E^{\cdot\prime})) = 0, \tag{5.1.6}$$

and

$$H^q(\mathcal{E}(E^\cdot|_Y))' \overset{top.}{\cong} H_q(\mathcal{E}'(E^{\cdot\prime}|_Y)). \tag{5.1.7}$$

To prove the theorem, it is sufficient, by (5.1.7), to show that under conditions (5.1.6) we have

$$H_q(\mathcal{E}'(E^{\cdot\prime}|_Y)) \overset{top.}{\cong} H_{q+1}(\vec{\mathcal{E}'}\,(E^{\cdot\prime}|_{X\setminus Y})). \tag{5.1.8}$$

Define a mapping $m: H_q(\mathcal{E}'(E^{\cdot\prime}|_Y)) \to H_{q+1}(\vec{\mathcal{E}'}\,(E^{\cdot\prime}|_{X\setminus Y}))$ in the following way. Suppose that $[g] \in H_q(\mathcal{E}'(E^{\cdot\prime}|_Y))$ and let $g \in Z_q(\mathcal{E}'(E^{\cdot\prime}|_Y))$ be a representative of the class $[g]$. Then since $g \in Z_q(\mathcal{E}'(E^{\cdot\prime}))$, by the first equation (5.1.6) there exists a section $v \in \mathcal{E}'(E^{q+1\prime})$ such that $g = P'_q v$ on X. Since supp $g \subset Y$, we have $P'_q = 0$ in some neighborhood of $X \setminus Y$. Consequently, v determines some homology class $[[v]] \in H_{q+1}(\vec{\mathcal{E}'}\,(E^{\cdot\prime}|_{X\setminus Y}))$. We set $m([g]) = [[v]]$.

We show that m is well-defined. It is necessary to show, if $g = P'_q\tilde{v}$ for some section $\tilde{v} \in \mathcal{E}'(E^{q+1\prime}|_Y)$, then $m([g]) = 0$. Indeed, let $g = P'_q v$ for $v \in \mathcal{E}'(E^{q+1\prime})$, then $P'_q(v - \tilde{v}) = 0$ on X. Hence, by the second equality (5.1.6) there exists a section $w \in \mathcal{E}'(E^{q+2\prime})$ such that $v - \tilde{v} = P'_{q+1}w$ on X. In particular, v coincides with $P'w$ in some neighborhood of $X \setminus Y$, so $[[v]] = 0$, as required.

We will now prove the injectivity of the mapping m. It is necessary to show that if $g \in \mathcal{E}'(E^{q\prime}|_Y)$ and $g = P'_q v$ on X where $v \in \mathcal{E}'(E^{q+1\prime})$ and $v = P'_{q+1}w$ in a neighborhood of $X \setminus Y$ for some section $w \in \mathcal{E}'(E^{q+2\prime})$, then $[g] = 0$. We set $\tilde{v} = v - P'_{q+1}w$. Then $\tilde{v} \in \mathcal{E}'(E^{q+1\prime}|_Y)$ and $P'_q\tilde{v} = g$ in Y, as required.

Finally we will show the surjectivity of m. Assume that $[[v]] \in H_{q+1}(\vec{\mathcal{E}'}\,(E^{\cdot\prime}|_{X\setminus Y}))$ and let v be a representative of the class $[[v]]$. This means that $v \in \mathcal{E}'(E^{q+1\prime})$ and $P'_q v = 0$ in some neighborhood of $X \setminus Y$. Set $g = P'v$. Then $g \in Z_q(\mathcal{E}'(E^{\cdot\prime}|_Y))$ and by construction we have $m([g]) = [[v]]$, as required.

It is easy to show that m and m^{-1} are continuous. That is, m is a homeomorphism.

□

Example 5.1.17 Let $E^\cdot$ be an elliptic complex on X. Then from Theorem 5.1.16 we get that if $H^q(\mathcal{E}(E^\cdot)) = H^{q+1}(\mathcal{E}(E^\cdot)) = 0$ and the spaces $H^{q+2}(\mathcal{E}(E^\cdot))$ and $H^{q+1}(\mathcal{E}(E^\cdot|_Y))$ are separated then $H^q(\mathcal{E}(E^\cdot|_Y))' \overset{top.}{\cong} H_{q+1}(\vec{\mathcal{D}}(E^{\cdot\prime}|_{X\setminus Y}))$. In particular, if the manifold X is compact, then the condition of being separated for $H^{q+2}(\mathcal{E}(E^\cdot))$ is automatic (see Corollaries 2.2.4 and 5.1.10), and the isomorphism can be rewritten as $H^q(\mathcal{E}(E^\cdot|_Y))' \overset{top.}{\cong} H_{q+1}(\vec{\mathcal{E}}(E^{\cdot\prime}|_{X\setminus Y}))$. The mechanism of this duality will be considered in Subsection 5.5.

□

Example 5.1.18 For the de Rham complex $E^\cdot = \Lambda^\cdot T^*_{\mathbb{C}}(X)$ on an orientable manifold X Theorem 5.1.16 can be reformulated in the following way. Suppose that $H^q(X, \mathbb{C}) = H^{q+1}(X, \mathbb{C}) = 0$. If the spaces $H^{q+2}(X, \mathbb{C})$ and $H^{q+1}(Y, \mathbb{C})$ are separated, then there is a topological isomorphism $H^q(Y, \mathbb{C})' \cong H^{n-q-1}_{comp}(X \backslash Y, \mathbb{C})$. In particular, if $H^q(Y, \mathbb{C})$ is finite-dimensional, then we have the isomorphism $H^q(Y, \mathbb{C}) \cong H_{n-q-1}(X \setminus Y, \mathbb{C})$. This is the classical Alexander-Pontryagin duality in algebraic topology (see Massey [160, §11.5]).

□

Example 5.1.19 For the Dolbeault complex $E^\cdot = E \otimes \Lambda^{0,\cdot} T^*_{\mathbb{C}}(X)$ with the notations of Example 5.1.13, we obtain the following statement. If $H^q(X, \mathcal{O}^{(E)}) = H^{q+1}(X, \mathcal{O}^{(E)}) = 0$ and the cohomology spaces $H^{q+2}(X, \mathcal{O}^{(E)})$ and $H^{q+1}(Y, \mathcal{O}^{(E)})$ are separated, then $H^q(Y, \mathcal{O}^{(E)})' \overset{top.}{\cong} H^{n-q-1}_{comp}(X \setminus Y, \mathcal{O}^{(E^* \otimes \Lambda^{n,0} T^*_{\mathbb{C}}(X))})$. This is a special case of a result of Golovin [86, p.174-175] for arbitrary coherent analytic sheaves on X.

□

Example 5.1.20 Under the assumptions of Example 5.1.15, Theorem 5.1.16 gives the following result. Let $H^q(X, S) = H^{q+1}(X, S) = 0$. If the spaces $H^{q+2}(X, S)$ and $H^{q+1}(Y, S)$ are separated, then $H^q(Y, S)' \overset{top.}{\cong} H_{q+1}(X \setminus Y, S')$.

□

In conclusion we remark that for $q = 0$ the conditions of Theorem 5.1.16 are not fulfilled as a rule.

Exercise 5.1.21 *Find a suitable modification of isomorphism (5.1.5) with $q = 0$ which is valid without the condition $H^0(\mathcal{E}(E^\cdot)) = 0$.*

Hint. See Massey [160, §11.5] and Golovin [86, p.174–175].

□

5.2 The Weil Homomorphism

5.2.1 *Preliminaries*

The Čech cohomology is one of the methods of calculating the cohomology groups with coefficients in sheaves (see Godement [72, Ch.II,§5]). They play an important role in complex analysis and arise naturally in, for example, the Cousin problems. Weil [252] considered this approach in the general theory of differential complexes. The main ideas of Weil [252] are given and developed in the book of Bott and Tu [25] which was published not long ago. Palamodov [175, §9 and §10 of Ch.VII] used the Čech cohomology to investigate Hilbert complexes. We aim at developing this approach for differential complexes on manifolds. As a matter of fact this is a fragment of Čech cohomology theory (see Godement [72, Ch.II, §5]).

5.2.2 *The Weil homomorphism*

Suppose that X is an n-dimensional differentiable manifold countable at infinity and E is a (cochain) differential complex on X.

Denote by $\mathcal{E}^{\cdot}$ $(= \widetilde{E}^{\cdot})$ a sheaf of germs of differentiable sections of the bundle $E^{\cdot}$ and let $S^{\cdot}$ be a subsheaf of $\mathcal{E}^{\cdot}$ of germs of (differentiable) solutions of the equation $P_{\cdot} f = 0$ so that $\mathcal{E}(E^{\cdot}) = \mathcal{E}^{\cdot}(X)$ and $Z^{\cdot}(\mathcal{E}(E^{\cdot})) = S^{\cdot}(X)$.

Fix an open covering $U = \{U_\nu\}_{\nu\in\mathcal{N}}$ of the manifold X. Elements $\mathcal{I} = (\imath_0, \dots, \imath_q)$ of $\mathcal{N}^{q+1}$ are (still) said to be multi-indices. We let $\#\mathcal{I} = q+1$ and abbreviate $U_{\imath_0} \cap \dots \cap U_{\imath_q}$ by $U_{\mathcal{I}}$ for $\mathcal{I} \in \mathcal{N}^{q+1}$.

First we define the Čech cohomology groups $H^q(U, S)$, where S is a sheaf of complex vector spaces over X.

A q-cochain f for U with values in S is a mapping which assigns to every multi-index $\mathcal{I} \in \mathcal{N}^{q+1}$ a section $f_{\mathcal{I}} \in S(U_{\mathcal{I}})$ so that $f_{\mathcal{I}}$ is an alternating function of $\mathcal{I}$. Let $C^q(U, S)$ be the vector space of all q-cochains for U with values in S. We set $C^q(U, S) = 0$ for $q < 0$.

Further $\delta : \; C^q(U, S) \to C^{q+1}(U, S)$ is the usual coboundary operator defined by

$$(\delta f)_{\mathcal{I}} = \sum_{k=0}^{q+1} (-1)^k f_{\imath_0 \dots \widehat{\imath_k} \dots \imath_{q+1}},$$

where the sign "$\hat{}$" over the index indicates deletion of that index.

Then there is the Čech complex $C^{\cdot}(U, S) = \{C^q(U, S), \delta\}$ whose cohomology groups are denoted by $H^q(U, S)$ and called the Čech cohomology groups for the covering U with values in S.

It will be convenient to consider the complex $C^{\cdot}(U, S)$ together with the "bordered" complex $\widetilde{C}^{\cdot}(U, S)$ which is defined as follows. Set $\widetilde{C}^q(U, S) = 0$ for $q < -1$, $\widetilde{C}^{-1}(U, S) = S(X)$ and $\widetilde{C}^q(U, S) = C^q(U, S)$ for $q \geq 0$. We define the mapping $\delta : \; \widetilde{C}^{-1}(U, S) \to \widetilde{C}^0(U, S)$ by $(\delta f)_\nu = f|_{U_\nu}$ for $f \in S(X)$. Then the new complex $\widetilde{C}^{\cdot}(U, S)$ is exact at degrees ≤ 0, and $H^q(\widetilde{C}^{\cdot}(U, S)) = H^q(U, S)$ for $q > 0$.

To prove that the complex $\widetilde{C}^{\cdot}(U, \mathcal{E}^{\cdot})$ is exact, the standard way is to construct a cochain homotopy between the identity endomorphism of this complex and the zero endomorphism.

Lemma 5.2.1 *There exists a family $\{\varphi_\nu\}_{\nu\in\mathcal{N}}$ of functions $\varphi_\nu \in \mathcal{E}(X)$ such that*
1) $\operatorname{supp} \varphi_\nu \subset U_\nu$, that is, $\varphi_\nu \equiv 0$ in a neighborhood of $X \setminus U_\nu$;
2) the family $\{\operatorname{supp} \varphi_\nu\}$ is locally finite; and
3) $\sum \varphi_\nu \equiv 1$ on X.

Proof. See Corollary 1.1.6

□

The family $\{\varphi_\nu\}_{\nu\in\mathcal{N}}$ can be used to construct a cochain homotopy above. Namely, define the mapping $h : \; \widetilde{C}^q(U, \mathcal{E}^{\cdot}) \to \widetilde{C}^{q-1}(U, \mathcal{E}^{\cdot})$ by

$$(hf)_{\mathcal{I}} = \sum_{\nu} \varphi_\nu f_{\nu\, \mathcal{I}} \quad (\mathcal{I} \in \mathcal{N}^q) \tag{5.2.1}$$

for $f \in \widetilde{C}^q(U, \mathcal{E}^{\cdot})$.

Since φ_ν vanishes on a neighborhood of $X \setminus U_\nu$ and $\mathcal{E}^\cdot(U_I)$ is an $\mathcal{E}(U_I)$-module, $\varphi_\nu f_I$ uniquely determines an element $\mathcal{E}^\cdot(U_I)$ which we also denote by $\varphi_\nu f_I$. Hence $(hf)_I \in \mathcal{E}^\cdot(U_I)$ so that the mapping h is well-defined.

Lemma 5.2.2 *$h\delta + \delta h = id$ on $\widetilde{C}^q(U, \mathcal{E}^\cdot)$.*

Proof. Let f be in $\widetilde{C}^q(U, \mathcal{E}^\cdot)$. Then

$$\begin{aligned}(h\delta f)_I &= \sum_\nu \varphi_\nu (\delta f)_{\nu I} \\ &= \sum_\nu \varphi_\nu \left(f_I - \sum_{k=0}^{q} (-1)^k f_{\nu\imath_0 \ldots \widehat{\imath_k} \ldots \imath_q} \right) \\ &= f_I - \sum_\nu \sum_{k=0}^{q} (-1)^k \varphi_\nu f_{\nu\imath_0 \ldots \widehat{\imath_k} \ldots \imath_q},\end{aligned}$$

and

$$\begin{aligned}(\delta h f)_I &= \sum_{k=0}^{q} (-1)^k (hf)_{\imath_0 \ldots \widehat{\imath_k} \ldots \imath_q} \\ &= \sum_{k=0}^{q} (-1)^k \sum_\nu \varphi_\nu f_{\nu\imath_0 \ldots \widehat{\imath_k} \ldots \imath_q} \\ &= \sum_\nu \sum_{k=0}^{q} (-1)^k \varphi_\nu f_{\nu\imath_0 \ldots \widehat{\imath_k} \ldots \imath_q}.\end{aligned}$$

Therefore $h\delta f + \delta h f = f$.

□

Corollary 5.2.3 *The complex $\widetilde{C}^\cdot(U, \mathcal{E}^\cdot)$ is exact. In particular, $H^0(U, \mathcal{E}^\cdot) \cong \mathcal{E}^\cdot(X)$ and $H^q(U, \mathcal{E}^\cdot) = 0$ for $q > 0$.*

In the terminology of the sheaf theory Lemma 5.2.1 means that the sheaves $\mathcal{E}^\cdot$ are fine and so Corollary 5.2.3 is well-known (see Theorem 4.4.3 of Godement [72, Ch.II]).

Thus we have the Čech bicomplex $C^\cdot(U, \mathcal{E}^\cdot)$ with commuting differentials δ and P of bidegrees $(1,0)$ and $(0,1)$, given respectively by the Čech coboundary and the differential of $E^{\cdot}$, and the corresponding "bordered" complex with exact rows:

$$\begin{array}{ccccccccc}
 & & 0 & & 0 & & 0 & & \\
 & & \downarrow & & \downarrow & & \downarrow & & \\
0 & \to & S^0(X) & \stackrel{\delta}{\to} & C^0(U,S^0) & \stackrel{\delta}{\to} & C^1(U,S^0) & \stackrel{\delta}{\to} & \dots \\
 & & \downarrow i & & \downarrow i & & \downarrow i & & \\
0 & \to & \mathcal{E}^0(X) & \underset{h}{\overset{\delta}{\rightleftarrows}} & C^0(U,\mathcal{E}^0) & \underset{h}{\overset{\delta}{\rightleftarrows}} & C^1(U,\mathcal{E}^0) & \underset{h}{\overset{\delta}{\rightleftarrows}} & \dots \\
 & & \downarrow P & \swarrow M & \downarrow P & \swarrow M & \downarrow P & \swarrow M & \\
0 & \to & \mathcal{E}^1(X) & \underset{h}{\overset{\delta}{\rightleftarrows}} & C^0(U,\mathcal{E}^1) & \underset{h}{\overset{\delta}{\rightleftarrows}} & C^1(U,\mathcal{E}^1) & \underset{h}{\overset{\delta}{\rightleftarrows}} & \dots \\
 & & \downarrow P & \swarrow M & \downarrow P & \swarrow M & \downarrow P & \swarrow M & \\
0 & \to & \mathcal{E}^2(X) & \underset{h}{\overset{\delta}{\rightleftarrows}} & C^0(U,\mathcal{E}^2) & \underset{h}{\overset{\delta}{\rightleftarrows}} & C^1(U,\mathcal{E}^2) & \underset{h}{\overset{\delta}{\rightleftarrows}} & \dots \\
 & & \downarrow P & \swarrow M & \downarrow P & \swarrow M & \downarrow P & \swarrow M & \\
 & & \dots & & \dots & & \dots & &
\end{array} \qquad (5.2.2)$$

Consider a "skew" mapping $M_q : \tilde{C}^q(U,\mathcal{E}^{\cdot}) \to \tilde{C}^{q-1}(U,\mathcal{E}^{\cdot+1})$ defined by the formula $M = [P,h] = Ph - hP$. More precisely, given $f \in \tilde{C}^q(U,\mathcal{E}^{\cdot})$, let

$$(Mf)_{\mathcal{I}} = \sum_{\nu} (P(\varphi_\nu f_{\nu\mathcal{I}}) - \varphi_\nu(Pf_{\nu\mathcal{I}})) \qquad (\mathcal{I} \in \mathcal{N}^{q+1}). \qquad (5.2.3)$$

Lemma 5.2.4 *The operations M and P, and M and δ anticommute. Hence M induces a mapping HM from $H^q(U,S^{\cdot})$ into $H^{q-1}(U,S^{\cdot+1})$ for all $q > 1$.*

Proof. Indeed, by formula (5.2.3) we get

$$MP + PM = (Ph - hP)P + P(Ph - hP) = 0,$$

and by Lemma 5.2.2

$$\begin{aligned}
M\delta + \delta M &= (Ph - hP)\delta + \delta(Ph - hP) \\
&= (Ph\delta + \delta Ph) - (hP\delta + \delta hP) \\
&= P(h\delta + \delta h) - (h\delta + \delta h)P \\
&= P - P = 0
\end{aligned}$$

which proves the first part of the lemma.

It follows that M maps $Z^q(\tilde{C}^{\cdot}(U,S^{\iota}))$ into $Z^{q-1}(\tilde{C}^{\cdot}(U,S^{\iota+1}))$, and $\delta\tilde{C}^{q-1}(U,S^{\iota})$ into $\delta\tilde{C}^{q-2}(U,S^{\iota+1})$. So M induced a homomorphism HM from $H^q(\tilde{C}^{\cdot}(U,S^{\iota}))$ into $H^{q-1}(\tilde{C}^{\cdot}(U,S^{\iota+1}))$ and in particular the homomorphism HM from $H^q(U,S^{\iota})$ into $H^{q-1}(U,S^{\iota+1})$ for $q>1$.

□

Lemma 5.2.5 *The composition $h\circ M: C^1(U,\mathcal{E}^{\iota})\to\mathcal{E}^{\iota+1}(X)$ induces the mapping $H(hM)$ from $H^1(U,S^{\iota})$ into $H^{\iota+1}(\mathcal{E}^{\cdot}(X))$.*

Proof. Indeed, if $f\in Z^1(C^{\cdot}(U,\mathcal{E}^{\iota}))$, then by Lemmas 5.2.4 and 5.2.2 we have

$$\delta(PhM(f)) = P(\delta h)\; M(f) = PM(f) = -M(Pf).$$

So, (hM) maps $Z^1(C\,(U,S^{\iota}))$ into $Z^{\iota+1}(\mathcal{E}\,(X))$.

On the other hand, if $u\in C^0(U,S^{\iota})$, then the same Lemmas 5.2.4 and 5.2.2 give

$$hM(\delta u) = -(h\delta)Mu = -Mu = P(-hu),$$

where $hu\in\mathcal{E}^{\iota}(X)$. Therefore, (hM) maps $B^1(C^{\cdot}(U,S^{\iota}))$ into $B^{\iota+1}(\mathcal{E}^{\cdot}(X))$ which finishes the proof.

□

The homomorphism HM ($H(hM)$ if $q=1$) which was constructed in Lemmas 5.2.4 and 5.2.5, is called Weil's homomorphism:

$$H^q(U,S^{\iota}) \stackrel{HM}{\to} H^{q-1}(U,S^{\iota+1}) \stackrel{HM}{\to} \dots \stackrel{HM}{\to} H^1(U,S^{\iota+q-1}) \stackrel{H(hM)}{\to} H^{q+\iota}(\mathcal{E}\,(X)). \quad (5.2.4)$$

Weil's homomorphism can be equivalently described as follows.

Definition 5.2.6 *The class $[f^{q-1}]\in H^{q-1}(U,S^{\iota+1})$ is the image of the class $[f^q]\in H^q(U,S^{\iota})$ under Weil's homomorphism if there exists a cochain $u^{q-1}\in C^{q-1}(U,\mathcal{E}^{\iota})$ such that: 1) $f^q=\delta u^{q-1}$ and 2) $[Pu^{q-1}]=[f^{q-1}]$. If $q=1$, then $[f^{q-1}]\in H^{q-1}(U,S^{\iota+1})$ and 2) are replaced by $[f]\in H^{\iota+1}(\mathcal{E}^{\cdot}(X))$ and 3) $Pu^0=\delta f$, respectively.*

To prove that this definition is equivalent to the above one, it is sufficient to show that $u\in Z^{q-1}(C^{\cdot}(U,\mathcal{E}^{\iota}))$ implies $[Pu]=0$. But we have $u=\delta hu$ by Lemma 5.2.2. Hence $Pu=\delta(Phu)$ with $Phu\in C^{q-2}(U,\mathcal{E}^{\iota+1})$; that is, $[Pu]=0$. If $q=1$, then $hu\in\mathcal{E}^{\iota}(X)$ and so $Pu\in\delta B^{\iota+1}(\mathcal{E}^{\cdot}(X))$, as required.

5.2.3 *Properties of the Weil homomorphism*

We can see from Definition 5.2.6 that Weil's homomorphism is independent of the particular mapping M, i.e, of the choice of $\{\varphi_\nu\}$. We will describe some of its properties.

Lemma 5.2.7 *If $H^{\iota+1}(\mathcal{E}^{\cdot}(U_{\mathcal{I}}))=0$ for all $\mathcal{I}\in\mathcal{N}^{q-1}$, then Weil's homomorphism $HM: H^q(U,S^{\iota})\to H^{q-1}(U,S^{\iota+1})$ for $q>1$ or $H(hM): H^1(U,S^{\iota})\to H^{\iota+1}(\mathcal{E}^{\cdot}(X))$ for $q=1$ is injective.*

Proof. First let $q > 1$, $[f] \in H^q(U, S^\cdot)$ and $HM[f] = 0$ in $H^{q-1}(U, S^{\cdot+1})$. Choose a representative $f \in Z^q(C^\cdot(U, S^\cdot))$ of the class $[f]$ and a cochain $f_1 \in C^{q-2}(U, S^{\cdot+1})$ such that $Phf = \delta f_1$. Since $H^{\cdot+1}(\mathcal{E}^\cdot(U_\mathcal{I})) = 0$ for all $\mathcal{I} \in \mathcal{N}^{q-1}$, there exists a cochain $u_1 \in C^{q-2}(U, \mathcal{E}^\cdot)$ such that $Pu_1 = f_1$. We set $u = hf - \delta u_1 \in C^{q-1}(U, \mathcal{E}^\cdot)$. Since $Pu = Phf - P\delta u_1 = \delta f_1 - \delta f_1 = 0$, we get $u \in C^{q-1}(U, S^\cdot)$. Moreover by Lemma 5.2.2 $\delta u = \delta hf = f - h\delta f = f$; hence $[f] = 0$ in $H^q(U, S^\cdot)$.

Now let $q = 1$. Then there are no conditions for the covering U. Assume that $[f] \in H^1(U, S^\cdot)$ and $H(hM)[f] = 0$ in $H^{\cdot+1}(\mathcal{E}^\cdot(X))$. Choose a representative $f \in Z^1(C^\cdot(U, S^\cdot))$ of the cohomology class $[f]$ and a section $u_1 \in \mathcal{E}^\cdot(X)$ such that $hPhf = Pu_1$. Set $u = hf - \delta u_1 \in C^0(U, \mathcal{E}^\cdot)$. Since $Pu = Phf - \delta h(Phf) = 0$, we get $u \in C^0(U, S^\cdot)$. By Lemma 5.2.2 $\delta u = \delta hf = f$, so $[f] = 0$ in $H^1(U, S^\cdot)$ as required.

□

Lemma 5.2.8 *If $H^{\cdot+1}(\mathcal{E}^\cdot(U_\mathcal{I})) = 0$ for all $\mathcal{I} \in \mathcal{N}^q$, then Weil's homomorphism, $HM : H^q(U, S^\cdot) \to H^{q-1}(U, S^{\cdot+1})$ for $q > 1$ or $H(hM) : H^1(U, S^\cdot) \to H^{\cdot+1}(\mathcal{E}^\cdot(X))$ for $q = 1$, is surjective.*

Proof. Let $q > 1$, $[f] \in H^{q-1}(U, S^{\cdot+1})$ and $f \in Z^{q-1}(C^\cdot(U, S^{\cdot+1}))$ be a representative of the cohomology class $[f]$. Since $H^{\cdot+1}(\mathcal{E}^\cdot(U_\mathcal{I})) = 0$ for all $\mathcal{I} \in \mathcal{N}^q$, there exists a cochain $u \in C^{q-1}(U, \mathcal{E}^\cdot)$ such that $Pu = f$. Consider $u_1 = \delta u \in Z^q(C^\cdot(U, \mathcal{E}^\cdot))$. We have $Pu_1 = P\delta u = \delta Pu = \delta f = 0$ and therefore $u_1 \in Z^q(C^\cdot(U, S^\cdot))$. Denote by $[u_1]$ the cohomology class of u_1 in $H^q(U, S^\cdot)$. By Lemma 5.2.2,

$$\begin{aligned} Phu_1 &= Ph\delta u \\ &= P(u - \delta hu) \\ &= Pu - P\delta hu \\ &= f - \delta(Phu), \end{aligned}$$

and $Phu \in C^{q-2}(U, S^{\cdot+1})$. Hence it follows that $HM[u_1] = [f]$.

Assume now $q = 1$, $[f] \in H^{\cdot+1}(\mathcal{E}^\cdot(X))$ and $f \in S^{\cdot+1}(X)$ is a representative of $[f]$.

Since $H^{\cdot+1}(\mathcal{E}^\cdot(U_\nu)) = 0$ for all $\nu \in \mathcal{N}$, there exists a cochain $u \in C^0(U, \mathcal{E}^\cdot)$ such that $Pu = \delta f$. Denote $u_1 = \delta u \in Z^1(C^\cdot(U, \mathcal{E}^\cdot))$. Then $Pu_1 = \delta Pu = 0$; that is $u_1 \in Z^1(C^\cdot(U, S^\cdot))$. Let $[u_1] \in H^1(U, S^\cdot)$ be the cohomology class containing u_1. By Lemma 5.2.2,

$$hPhu_1 = hPh\delta u = hP(u - \delta hu) = f - P(hu)$$

and $hu \in \mathcal{E}^\cdot(X)$. Hence $H(hM)[u_1] = [f]$ as required.

□

Let $V = \{V_\mu\}_{\mu \in \mathcal{M}}$ be a refinement of the covering U. Then there is a mapping $T : \mathcal{M} \to \mathcal{N}$ such that $V_\mu \subset U_{T(\mu)}$. This mapping induces a mapping $T^\sharp : C^q(U, \mathcal{E}^\cdot) \to C^q(V, \mathcal{E}^\cdot)$ by the formula $(T^\sharp f)_\mathcal{J} = f_{T(\mathcal{J})}|_{V_\mathcal{J}}$. It is easy to see that $T^\sharp$ commutes with δ and P, and hence it induces a mapping $HT^\sharp : H^q(U, S^\cdot) \to H^q(V, S^\cdot)$. This mapping $HT^\sharp$ can be shown to be transitive and independent of the particular mapping T (see Godement [72, Ch.II,§5.7]).

Lemma 5.2.9 *Weil's homomorphism is natural in the sense that, if V is a refinement of the covering U, then there is the commutative diagram*

$$\begin{array}{ccccc}
H^q(U,S^\imath) & \xrightarrow{HM} & H^{q-1}(U,S^{\imath+1}) & \xrightarrow{HM} & \dots \\
\downarrow{\scriptstyle HT^\imath} & & \downarrow{\scriptstyle HT^\imath} & & \\
H^q(V,S^\imath) & \xrightarrow{HM} & H^{q-1}(V,S^{\imath+1}) & \xrightarrow{HM} & \dots
\end{array}$$

$$\begin{array}{ccccc}
\xrightarrow{HM} & H^1(U,S^{\imath+q-1}) & \xrightarrow{H(hM)} & H^{q+\imath}(\mathcal{E}(X)) \\
 & \downarrow{\scriptstyle HT^\imath} & & \downarrow{\scriptstyle 1} \\
\xrightarrow{HM} & H^1(V,S^{\imath+q-1}) & \xrightarrow{H(hM)} & H^{q+\imath}(\mathcal{E}(X))
\end{array} \qquad (5.2.5)$$

Proof. Let $\{\psi_\mu\}$ be a partition of unity of Lemma 5 2 1 subordinate to the covering V. Then $\{\sum_{T(\mu)=\nu}\psi_\mu\}$ is a partition of unity subordinate to the covering U. To prove that diagram (5.2.5) commutes, simple note that Weil's homomorphism does not depend on the choice of a partition of unity.

□

Define $W_q = (-1)^{\frac{q(q+1)}{2}} h \circ M^q$ where $M^q = M_1 \circ \dots \circ M_q$ is the q-th power of the mapping M and $W_0 = h$. This is a well-defined mapping from $C^q(U,\mathcal{E}^\imath)$ into $\mathcal{E}^{q+\imath}(X)$.

Lemma 5.2.10 *For each cochain $f \in C^q(U,\mathcal{E}^\imath)$, we have*

$$PW_q(f) = W_{q+1}(\delta f) + (-1)^q W_q(Pf). \qquad (5.2.6)$$

Proof. According to Lemmas 5.2.4 and 5.2.2

$$\begin{aligned}
W_{q+1}\circ\delta &= (-1)^{\frac{(q+1)(q+2)}{2}} h\circ(M_1\circ\dots\circ M_{q+1})\circ\delta \\
&= (-1)^{\frac{(q+1)(q+4)}{2}} (h\circ\delta)\circ M_0\circ M^q \\
&= (-1)^{\frac{(q+1)(q+4)}{2}} M_0\circ M^q,
\end{aligned}$$

and

$$\begin{aligned}
W_q\circ P &= (-1)^{\frac{q(q+1)}{2}} h\circ(M_1\circ\dots\circ M_q)\circ P \\
&= (-1)^{\frac{q(q+3)}{2}} h\circ P\circ M^q.
\end{aligned}$$

Hence it follows that

$$\begin{aligned}
W_{q+1}\circ\delta + (-1)^q W_q\circ P &= (-1)^{\frac{q(q+1)}{2}} (M_0 + hP)\circ M^q \\
&= (-1)^{\frac{q(q+1)}{2}} (Ph)\circ M^q \\
&= P\circ W_q,
\end{aligned}$$

as required.

□

Theorem 5.2.11 *$W : C(U, S^0) \to \mathcal{E}(X)$ is a cochain mapping of complexes (of degree zero) which induces the homomorphism $HW : H(U, S^0) \to H(\mathcal{E}(X))$ independent of the choice of a partition of unity $\{\varphi_\nu\}_{\nu \in \mathcal{N}}$ and natural in the sense that, for each refinement V of the covering U, the following diagram is commutative:*

$$\begin{array}{ccc} H(U, S^0) & \xrightarrow{HM} & H(\mathcal{E}(X)) \\ \downarrow{\scriptstyle HT^{\cdot}} & & \downarrow{\scriptstyle 1} \\ H(V, S^0) & \xrightarrow{HM} & H(\mathcal{E}(X)). \end{array}$$

Proof. This follows immediately from Lemmas 5 2.10 and 5.2.9.

□

The above construction of the homomorphism HW for the de Rham complex is due to Weil [252] who proceeded from the mappings

$$(-1)^{\frac{q(q+1)}{2}} (Ph)^q : Z^q(C(U, S^{\cdot})) \to Z^0(C(U, S^{\cdot+q}))$$

on cochains with values in the sheaves $S^{\cdot}$. Gleason [71] improved Weil's construction and applied it to integral formulas. Harvey [98] developed Gleason's ideas for the Dolbeault complex.

Theorem 5.2.12 *1) If $H^i(\mathcal{E}(U_\mathcal{I})) = 0$ $(\mathcal{I} \in \mathcal{N}^{q-i})$ for every $i = 1, \dots, q$, then $HW : H^q(U, S^0) \to H^q(\mathcal{E}(X))$ is a monomorphism.*

2) If $H^i(\mathcal{E}(U_\mathcal{I})) = 0$ $(\mathcal{I} \in \mathcal{N}^{q-i+1})$ for every $i = 1, \dots, q$, then $HW : H^q(U, S^0) \to H^q(\mathcal{E}(X))$ is an epimorphism.

Proof. Since

$$HW = (-1)^{\frac{q(q+1)}{2}} H(hM_1) \circ HM_2 \circ \dots \circ HM_q,$$

the desired statements follow immediately from Lemmas 5.2.7 and 5.2.8. This completes the proof.

□

The Čech cohomologies $H^q(U, S^0)$ arose as the obstructions to the solvability of some problem of separating the singularities for solutions of the system $P_0 f = 0$. It is an analog of the additive Cousin problem (see Gunning and Rossi [96, Ch.VIII, B]). From this point of view, Theorem 5.2.12 part 1) gives a criterion of solvability of such a problem, namely the image in $H^q(\mathcal{E}(X))$ of the cocycle of data should be zero under the homomorphism W_q And part 2) of Theorem 5.2.12 gives the necessary condition for solvability of all the problems, namely $H^q(\mathcal{E}(X)) = 0$.

5.2.4 *A localized version of the Weil homomorphism*

We discuss a localized version of sequence (5.2.4).

Thanks to Lemma 5.2.9, we are allowed in (5.2.4) to pass to the inductive limit over refined coverings. By Theorem 5.10.1 of Godement [72, Ch II], Weil's homomorphism HM is lowered to a homomorphism HM_{loc} of cohomology groups of the manifold X,

$$\begin{array}{rl} H^q(X,S^i) \xrightarrow{HM_{loc}} H^{q-1}(X,S^{i+1}) \xrightarrow{HM_{loc}} & \dots \\ \xrightarrow{HM_{loc}} H^1(X,S^{i+q-1}) \xrightarrow{H(hM)_{loc}} & H^{q+i}(\mathcal{E}^{\cdot}(X)). \end{array} \tag{5.2.7}$$

Denote by HW_{loc} the composition of the homomorphisms in sequence (5.2.7) multiplied by $(-1)^{\frac{q(q+1)}{2}}$ and let $\mathcal{E}^{\cdot}: 0 \to \mathcal{E}^0 \xrightarrow{P} \mathcal{E}^1 \xrightarrow{P} \dots$ be the complex of sheaves associated to $E^{\cdot}$. The following facts are well-known in sheaf theory (see Theorem 4.7.1 and the corollary to Theorem 5.4.1 of Godement [72, Ch.II]).

Theorem 5.2.13 (Abstract de Rham Theorem) *Suppose that* $H^1(\mathcal{E}^{\cdot}) = \dots = H^q(\mathcal{E}^{\cdot}) = 0$. *Then*

$$HW_{loc}: \ H^q(X,S^0) \to H^q(\mathcal{E}^{\cdot}(X))$$

is an isomorphism.

Proof. From Lemmas 5.2.7 and 5.2.8 by passing to the inductive limit over refined coverings we get that, if $H^{i+1}(\mathcal{E}^{\cdot}) = 0$, then the mappings

$$\begin{array}{rl} HM_{loc} & : \ H^q(X,S^i) \to H^{q-1}(X,S^{i+1}) \qquad (q>1), \\ H(hM)_{loc} & : \ H^1(X,S^i) \to H^{i+1}(\mathcal{E}^{\cdot}(X)) \end{array}$$

in sequence (5.2.7) are isomorphisms. Hence the theorem follows.

□

Theorem 5.2.14 (Leray's Theorem on Acyclic Coverings) *If* $H^1(\mathcal{E}^{\cdot}) = \dots = H^q(\mathcal{E}^{\cdot}) = 0$ *and* $H^i(\mathcal{E}^{\cdot}(U_{\mathcal{I}})) = 0 \quad (\mathcal{I} \in \mathcal{N}^{q-i} \cup \mathcal{N}^{q-i+1})$ *for every* $i = 1, \dots, q$, *then*

$$(HW_{loc})^{-1} \circ HW: \ H^q(U,S^0) \to H^q(X,S^0)$$

is an isomorphism.

Proof. This follows from Theorems 5.2.12 and 5.2.13.

□

Let us consider two well-known examples.

Example 5.2.15 Let $E^{\cdot} = E \otimes \Lambda^{0,\cdot}T^*_{\mathbb{C}}(X)$ be the Dolbeault complex with coefficients in a holomorphic vector bundle E on a complex manifold X. Assume that $U = \{U_\nu\}$ is a Stein covering of X, that is, all U_ν are Stein manifolds. Then by Theorem B of Cartan (see Gunning and Rossi [96, Ch.VIII, A]) all the conditions of the Theorem 5.2.14 (and 5.2.13) are fulfilled. Therefore

$$\begin{array}{rl} H^q(\mathcal{E}^{\cdot}(X)) & \cong \ H^q(U,\mathcal{O}^{(E)}) \\ & \cong \ H^q(X,\mathcal{O}^{(E)}) \end{array}$$

for all q.

□

Example 5.2.16 Suppose that E is a Hilbert complex on an open set $X \subset \mathbb{R}^n$. Let $U = \{U_\nu\}$ be an open covering of X by convex sets. Then by Theorem 1.2.21 all the conditions of Theorems 5.2.14 (and 5.2.13) are fulfilled. So we have

$$\begin{aligned} H^q(\mathcal{E}\,(X)) &\cong H^q(U, S^0) \\ &\cong H^q(X, S^0) \end{aligned}$$

for all q (cf. Palamodov [175, Ch.VII,§9]).

□

5.2.5 *The adjoint Weil homomorphism*

Now we wish to dualize Weil's homomorphism.

Denote by $\mathcal{E}'_{\cdot}$ the cosheaf [2] dual to the sheaf $\mathcal{E}^{\cdot}$, i.e., $\mathcal{E}'_{\cdot}(\mathcal{O}) = \mathcal{E}^{\cdot}(\mathcal{O})' = \mathcal{E}'(E^{\cdot\prime}|_{\mathcal{O}})$ for each open set $\mathcal{O} \subset X$.

The covering U is always assumed to be locally finite in the sequel. Since X is countable at infinity, such a covering U has to be countable.

Let $\mathfrak{C}$ be a cosheaf of complex vector spaces over X. A q-chain g for U with values in $\mathfrak{C}$ is a mapping which assigns to every multi-index $\mathcal{I} \in \mathcal{N}^{q+1}$ a section $g_{\mathcal{I}} \in \mathfrak{C}(U_{\mathcal{I}})$ so that $g_{\mathcal{I}}$ is an alternating function of $\mathcal{I}$, and so that $g_{\mathcal{I}} = 0$ except for a finite number of multi-indices $\mathcal{I}$. The vector space of all $q-$ chains for U with values in $\mathfrak{C}$ is denoted by $C_q(U, \mathfrak{C})$.

Defining the boundary mapping $\partial:\ C_q(U, \mathfrak{C}) \to C_{q-1}(U, \mathfrak{C})$ by the equality

$$(\partial g)_{\mathcal{I}} = \sum_{\nu} g_{\nu\mathcal{I}} \qquad (\mathcal{I} \in \mathcal{N}^q),$$

we get a (chain) complex $C\,(U, \mathfrak{C})$. Its homology groups are denoted by $H_q(U, \mathfrak{C})$, and they are called the Čech homology groups for the covering U with values in the cosheaf $\mathfrak{C}$.

Together with the complex $C\,(U, \mathfrak{C})$ we will consider also the "bordered" complex $\tilde{C}\,(U, \mathfrak{C})$ such that $\tilde{C}_q(U, \mathfrak{C}) = C_q(U, \mathfrak{C})$ for $q \neq -1$ and $\tilde{C}_{-1}(U, \mathfrak{C}) = \mathfrak{C}(X)$, with the mapping $\partial:\ \tilde{C}_0(U, \mathfrak{C}) \to \tilde{C}_{-1}(U, \mathfrak{C})$ given by $\partial g = \Sigma g_\nu$. For this new complex we obviously have $H_q(\tilde{C}\,(U, \mathfrak{C})) = H_q(U, \mathfrak{C})$ if $q > 0$.

Using the mapping h' defined below (5.2.9) and the definition of ∂ we could now prove that $(h'\partial + \partial h')$ is the identity on $\tilde{C}_q(U, \mathcal{E}'_{\cdot})$ by a direct calculation. Instead we deduce this fact from Lemma 5.2.2 and the fact that ∂ and h' are the transposed operators of δ and h respectively.

The vector space $\tilde{C}^q(U, \mathcal{E}^{\cdot})$ with the relative topology induced from the Fréchet space $\prod_{\mathcal{I} \in \mathcal{N}^{q+1}} \mathcal{E}^{\cdot}(U_{\mathcal{I}})$ is obviously a Fréchet space.

For $g \in \tilde{C}_q(U, \mathcal{E}'_{\cdot})$ and $f \in \tilde{C}^q(U, \mathcal{E}^{\cdot})$ we let

$$\langle g, f\rangle = \frac{1}{(q+1)!} \sum\nolimits_{\mathcal{I} \in \mathcal{N}^{q+1}} \langle g_{\mathcal{I}}, f_{\mathcal{I}}\rangle = \sum\nolimits'_{\mathcal{I} \in \mathcal{N}^{q+1}} \langle g_{\mathcal{I}}, f_{\mathcal{I}}\rangle, \tag{5.2.8}$$

where $\sum'$ denotes summation over strictly increasing multi-indices.

[2] See the notion of a cosheaf, for example, in the book of Palamodov [175, Ch V,§2(9)]

Now $\widetilde{C}_q(U, \mathcal{E}'_c)$ is easily seen to be isomorphic to the dual space of the Fréchet space $C^q(U, \mathcal{E}^\cdot)$ under the pairing (5.2.8).

Lemma 5.2.17 *The coboundary mapping* $\delta : \widetilde{C}^q(U, \mathcal{E}^\cdot) \to \widetilde{C}^{q+1}(U, \mathcal{E}^\cdot)$ *is a continuous linear mapping with transposed mapping* $\partial : \widetilde{C}_{q+1}(U, \mathcal{E}'_c) \to \widetilde{C}_q(U, \mathcal{E}'_c)$.

Proof. Assume that $g \in \widetilde{C}_{q+1}(U, \mathcal{E}'_c)$ and $f \in \widetilde{C}^q(U, \mathcal{E}^\cdot)$. Then

$$
\begin{aligned}
\langle g, \delta f\rangle &= \frac{1}{(q+2)!}\sum_{\mathcal{I}\in\mathcal{N}^{q+2}}\sum_{k=0}^{q+1}(-1)^k\langle g_{\mathcal{I}}, f_{\imath_0\ldots\hat{\imath}_k\ldots\imath_{q+1}}\rangle \\
&= \frac{1}{(q+1)!}\sum_{\mathcal{J}\in\mathcal{N}^{q+1}}\sum_{\nu}\langle g_{\nu\mathcal{J}}, f_{\mathcal{J}}\rangle \\
&= \langle \partial g, f\rangle,
\end{aligned}
$$

where the middle equality follows by letting $\mathcal{J} = (\imath_0, \ldots, \hat{\imath}_k, \ldots, \imath_{q+1})$ and $\nu = \imath_k$. The lemma is proved.

□

A family of functions $\{\varphi_\nu\}_{\nu\in\mathcal{N}}$ satisfying the conditions of Lemma 5.2.1 can be used to construct a (chain) homotopy h' between the identity and zero endomorphisms of the complex $\widetilde{C}\,(U, \mathcal{E}'_c)$. Given $g \in \widetilde{C}_{q-1}(U, \mathcal{E}'_c)$, let

$$
(h'g)_{\mathcal{I}} = \sum_{k=0}^{q}(-1)^k\varphi_{\imath_k} g_{\imath_0\ldots\hat{\imath}_k\ldots\imath_q} \qquad (\mathcal{I}\in\mathcal{N}^{q+1}). \tag{5.2.9}
$$

Since $\varphi_{\imath_k}$ vanishes in a neighborhood of $X \setminus U_{\imath_k}$ and $g_{\imath_0\ldots\hat{\imath}_k\ldots\imath_q}$ is compactly supported in $U_{\imath_0\ldots\hat{\imath}_k\ldots\imath_q}$, their product is compactly supported in $U_{\mathcal{I}}$. Hence, $(h'g)_{\mathcal{I}} \in \mathcal{E}'_c(U_{\mathcal{I}})$ for each $\mathcal{I} \in \mathcal{N}^{q+1}$, so that h' is a well-defined mapping from $\widetilde{C}_{q-1}(U, \mathcal{E}'_c)$ into $\widetilde{C}_q(U, \mathcal{E}'_c)$.

Lemma 5.2.18 *The mapping* $h : \widetilde{C}^q(U, \mathcal{E}^\cdot) \to \widetilde{C}^{q-1}(U, \mathcal{E}^\cdot)$ *is a continuous linear mapping with transposed mapping* $h' : \widetilde{C}_{q-1}(U, \mathcal{E}'_c) \to \widetilde{C}_q(U, \mathcal{E}'_c)$ *defined by (5.2.9).*

Proof. Suppose $g \in \widetilde{C}_{q-1}(U, \mathcal{E}'_c)$ and $f \in \widetilde{C}^q(U, \mathcal{E}^\cdot)$. Then

$$
\begin{aligned}
\langle g, hf\rangle &= \frac{1}{q!}\sum_{\mathcal{I}\in\mathcal{N}^q}\sum_{\nu}\langle g_{\mathcal{I}}, \varphi_\nu f_{\nu\mathcal{I}}\rangle \\
&= \frac{1}{(q+1)!}\sum_{\mathcal{J}\in\mathcal{N}^{q+1}}\sum_{k=0}^{q}(-1)^k\langle \varphi_{\jmath_k} g_{\jmath_0\ldots\hat{\jmath}_k\ldots\jmath_q}, f_{\mathcal{J}}\rangle \\
&= \langle h'g, f\rangle,
\end{aligned}
$$

where the middle equality follows by letting $\mathcal{J} = (\imath_0, \ldots, \imath_{k-1}, \nu, \imath_{k+1}, \ldots, \imath_q)$. This finishes the proof.

□

Lemma 5.2.19 *$h'\partial + \partial h' = id$ on $\widetilde{C}_q(U, \mathcal{E}'_\iota)$.*

Proof. This follows from Lemmas 5.2.2, 5.2.17, and 5.2.18.

□

Corollary 5.2.20 *The complex $\widetilde{C}_{\cdot}(U, \mathcal{E}'_\iota)$ is exact. In particular, $H_0(U, \mathcal{E}'_\iota) \cong \mathcal{E}'_\iota(X)$ and $H_q(U, \mathcal{E}'_\iota) = 0$ for $q > 0$.*

Of course, Corollary 5.2.20 can be obtained also from Corollary 5.2.3 by using Lemmas 5.1.1 and 5.1.5.

Thus we have the bicomplex $C_{\cdot}(U, \mathcal{E}'_{\cdot})$ transposed to the bicomplex $C^{\cdot}(U, \mathcal{E}^{\cdot})$, and the corresponding "bordered" complex with exact rows:

$$
\begin{array}{ccccccccc}
 & & 0 & & 0 & & 0 & & \\
 & & \uparrow & & \uparrow & & \uparrow & & \\
0 & \leftarrow & \mathcal{E}'_0(X)/P'\mathcal{E}'_1(X) & \overset{\partial}{\leftarrow} & C^0(U, \mathcal{E}'_0/P'\mathcal{E}'_1) & \overset{\partial}{\leftarrow} & C^1(U, \mathcal{E}'_0/P'\mathcal{E}'_1) & \overset{\partial}{\leftarrow} & \cdots \\
 & & \uparrow \varphi & & \uparrow \varphi & & \uparrow \varphi & & \\
0 & \leftarrow & \mathcal{E}'_0(X) & \underset{h'}{\overset{\partial}{\leftrightarrows}} & C^0(U, \mathcal{E}'_0) & \underset{h'}{\overset{\partial}{\leftrightarrows}} & C^1(U, \mathcal{E}'_0) & \underset{h'}{\overset{\partial}{\leftrightarrows}} & \cdots \\
 & & \uparrow P' & \nearrow M' & \uparrow P' & \nearrow M' & \uparrow P' & \nearrow M' & \\
0 & \leftarrow & \mathcal{E}'_1(X) & \underset{h'}{\overset{\partial}{\leftrightarrows}} & C_0(U, \mathcal{E}'_1) & \underset{h'}{\overset{\partial}{\leftrightarrows}} & C_1(U, \mathcal{E}'_1) & \underset{h'}{\overset{\partial}{\leftrightarrows}} & \cdots \\
 & & \uparrow P' & \nearrow M' & \uparrow P' & \nearrow M' & \uparrow P' & \nearrow M' & \\
0 & \leftarrow & \mathcal{E}'_2(X) & \underset{h'}{\overset{\partial}{\leftrightarrows}} & C_0(U, \mathcal{E}'_2) & \underset{h'}{\overset{\partial}{\leftrightarrows}} & C_1(U, \mathcal{E}'_2) & \underset{h'}{\overset{\partial}{\leftrightarrows}} & \cdots \\
 & & \uparrow P' & \nearrow M' & \uparrow P' & \nearrow M' & \uparrow P' & \nearrow M' & \\
 & & \cdots & & \cdots & & \cdots & &
\end{array}
\tag{5.2.10}
$$

Define the mapping $M' : \widetilde{C}_{q-1}(U, \mathcal{E}'_{\iota+1}) \to \widetilde{C}_q(U, \mathcal{E}'_\iota)$ by $M'g = (h'P' - P'h')g$ for $g \in \widetilde{C}_{q-1}(U, \mathcal{E}'_{\iota+1})$, or explicitly

$$(M'g)_{\mathcal{I}} = \sum_{k=0}^{q} (-1)^k \left(\varphi_{\iota_k}(P' g_{\iota_0 \ldots \widehat{\iota_k} \ldots \iota_q}) - P'(\varphi_{\iota_k} g_{\iota_0 \ldots \widehat{\iota_k} \ldots \iota_q}) \right) \quad (\mathcal{I} \in \mathcal{N}^{q+1}). \tag{5.2.11}$$

Lemma 5.2.21 *The mapping $M : \widetilde{C}^q(U, \mathcal{E}^{\iota}) \to \widetilde{C}^{q-1}(U, \mathcal{E}^{\iota+1})$ is a continuous linear mapping with transposed mapping $M' : \widetilde{C}_{q-1}(U, \mathcal{E}'_{\iota+1}) \to \widetilde{C}_q(U, \mathcal{E}'_\iota)$ given by (5.2.11).*

Proof. This follows from Lemma 5.2.18 and Proposition 1.1.25.

□

In the next lemma we describe the cosheaf dual to the sheaf S^{ι}.

Lemma 5.2.22 *Assume that $\mathcal{O}$ is an open subset of X and the cohomology space $H^{i+1}(\mathcal{E}^{\cdot}(\mathcal{O}))$ is separated. Then*

$$S^i(\mathcal{O})' \overset{top}{\cong} \mathcal{E}'_i(\mathcal{O})/P'\mathcal{E}'_{i+1}(\mathcal{O}).$$

Proof. Applying Lemma 5.1.4 to the complex $0 \to \mathcal{E}^i(\mathcal{O}) \xrightarrow{P} \mathcal{E}^{i+1}(\mathcal{O})$, we get the topological isomorphism $S^i(\mathcal{O})' \cong \mathcal{E}'_i(\mathcal{O})/\overline{P'\mathcal{E}'_{i+1}(\mathcal{O})}$. Since $P\mathcal{E}^i(\mathcal{O})$ is closed in $\mathcal{E}^{i+1}(\mathcal{O})$, we may use Lemma 5.1.5 to conclude that $P'\mathcal{E}'_{i+1}(\mathcal{O})$ is closed in $\mathcal{E}'_i(\mathcal{O})$. Hence the desired statement follows.

□

Denote by $\mathcal{E}'_i/P'\mathcal{E}'_{i+1}$ the cosheaf $\sigma \to \mathcal{E}'_i(\sigma)/P'\mathcal{E}'_{i+1}(\sigma)$ where $\sigma \subset X$ is open. Every chain $g \in \widetilde{C}_q(U, \mathcal{E}'_i)$ determines a unique chain from $\widetilde{C}_q(U, \mathcal{E}'_i/P'\mathcal{E}'_{i+1})$ which will be denoted by $\overline{g}$. Note that if $g_j \in \widetilde{C}_q(U, \mathcal{E}'_i)$ $(j = 1, 2)$ determine the same element in $\widetilde{C}_q(U, \mathcal{E}'_i/P'\mathcal{E}'_{i+1})$, i.e. $\overline{g_1} = \overline{g_2}$, then $\langle g_1, f\rangle = \langle g_2, f\rangle$ for all $f \in \widetilde{C}^q(U, S^i)$. Hence the pairing (5.2.8) induces a pairing $\langle \overline{g}, f\rangle$ between $\widetilde{C}_q(U, \mathcal{E}'_i/P'\mathcal{E}'_{i+1})$ and $\widetilde{C}^q(U, S^i)$. Since $\widetilde{C}^q(U, S^i)$ is a closed subspace of $\widetilde{C}^q(U, \mathcal{E}^i)$, it is a Fréchet space. It follows from Lemma 5.2.22 that if $H^{i+1}(\mathcal{E}^{\cdot}(U_{\mathcal{I}}))$ is separated for each $\mathcal{I} \in \mathcal{N}^{q+1}$, then $\widetilde{C}_q(U, \mathcal{E}'_i/P'\mathcal{E}'_{i+1})$ is isomorphic to the dual of $C^q(U, \mathcal{E}^i)$ under the pairing (5.2.8).

Lemma 5.2.23 *The operators M' and P', and M' and ∂ anticommute. Hence M' induces a mapping HM' from $H_{q-1}(U, \mathcal{E}'_{i+1}/P'\mathcal{E}'_{i+2})$ into $H_q(U, \mathcal{E}'_i/P'\mathcal{E}'_{i+1})$ for all $q > 1$.*

Proof. This easily follows from Lemmas 5.2.4, 5.2.17, and 5.2.21, and Proposition 1.1.25.

□

Lemma 5.2.24 *The composition $M' \circ h' : \mathcal{E}'_{i+1}(X) \to C_1(U, \mathcal{E}'_i)$ induces the mapping $H(hM)'$ from $H_{i+1}(\mathcal{E}'(X))$ into $H_1(U, \mathcal{E}'_i/P'\mathcal{E}'_{i+1})$.*

Proof. Given any class $[g] \in H_{i+1}(\mathcal{E}'(X))$, choose some representative $g \in Z_{i+1}(\mathcal{E}'_{\cdot}(X))$ of $[g]$. Let $H(hM)'[g] = [\overline{M'h'g}]$ where $[\overline{M'h'g}]$ is the equivalence class in $H_1(U, \mathcal{E}'_i/P'\mathcal{E}'_{i+1})$ of the chain $M'h'g \in C_1(U, \mathcal{E}'_i)$.

To prove this is well-defined it is sufficient to check that:

1) $\overline{M'h'g} \in Z_1(C(U, \mathcal{E}'_i/P'\mathcal{E}'_{i+1}))$; and

2) the class $[\overline{M'h'g}]$ does not depend on the choice of a representative $g \in [g]$.

Let us prove 1). By Lemmas 5.2.23 and 5.2.19,

$$\begin{aligned}\partial(M'h'g) &= -M'(\partial h'g)\\ &= -M'g\\ &= P'(h'g),\end{aligned}$$

where $h'g \in C_0(U, \mathcal{E}'_{i+1})$. Therefore $\overline{\partial M'h'g} = 0$ in $C_1(U, \mathcal{E}'_i/P'\mathcal{E}'_{i+1})$, as required.

Let us prove 2). Assume that $g = P'v$ where $v \in \mathcal{E}'_{i+2}(X)$. Then we obtain by Lemmas 5.2.19 and 5.2.23

$$\begin{aligned} M'h'g &= M'h'P'v \\ &= M'h'P'(\partial h'v) \\ &= M'(1-\partial h')P'h'v \\ &= -P'(M'h'v) + \partial(M'h'P'h'v) \\ &= P'v_1 + \partial v_2, \end{aligned}$$

where $v_1 \in C_1(U, \mathcal{E}'_{i+1})$ and $v_2 \in C_2(U, \mathcal{E}'_i)$. Hence the class $[\overline{M'h'g}]$ is zero in $H_1(U, \mathcal{E}'_i/P'\mathcal{E}'_{i+1})$, as required.

The lemma is proved.

□

We will refer to the homomorphism HM' (or $H(hM)'$ if $q = 1$) as the adjoint Weil homomorphism (clearly, it is not the transpose to Weil's homomorphism in the general case):

$$\begin{aligned} H_q(U, \mathcal{E}'_i/P'\mathcal{E}'_{i+1}) &\overset{HM'}{\longleftarrow} H_{q-1}(U, \mathcal{E}'_{i+1}/P'\mathcal{E}'_{i+2}) \overset{HM'}{\longleftarrow} \cdots \\ &\overset{HM'}{\longleftarrow} H_1(U, \mathcal{E}'_{i+q-1}/P'\mathcal{E}'_{i+q}) \overset{H(hM)'}{\longleftarrow} H_{q+i}(\mathcal{E}'(X)). \end{aligned} \tag{5.2.12}$$

The adjoint Weil homomorphism can be described (equivalently) as follows.

Definition 5.2.25 *The class $[\overline{g^q}] \in H_q(U, \mathcal{E}'_i/P'\mathcal{E}'_{i+1})$ is the image of the class $[\overline{g^{q-1}}] \in H_{q-1}(U, \mathcal{E}'_{i+1}/P'\mathcal{E}'_{i+2})$ under the adjoint Weil homomorphism if there exists a chain $v^q \in C_q(U, \mathcal{E}'_i)$ such that: 1) $P'g^{q-1} = \partial v^q$ and 2) $[\overline{v^q}] = [\overline{g^q}]$ in $H_q(U, \mathcal{E}'_i/P'\mathcal{E}'_{i+1})$. If $q = 1$, then $[\overline{g^{q-1}}] \in H_{q-1}(U, \mathcal{E}'_{i+1}/P'\mathcal{E}'_{i+2})$ and 1) are replaced by $[g] \in H_{i+1}(\mathcal{E}'(X))$ and 3) $\partial v^1 = P'\partial^{-1}g$ respectively.*

The equivalence of both definitions follows from Corollary 5.2.20, according to which $\partial v^q = 0$ implies $v^q \in \partial C_{q+1}(U, \mathcal{E}'_i)$.

5.2.6 *Properties of the adjoint Weil homomorphism*

The surjectivity and injectivity of the adjoint Weil homomorphism do not follow from Lemmas 5.2.7 and 5.2.8 immediately because HM' is not the transpose to HM in the general case. We give independent proofs.

Lemma 5.2.26 *If $H^{i+2}(\mathcal{E}(U_\mathcal{I}))$ is separated and $H^{i+1}(\mathcal{E}(U_\mathcal{I})) = 0$ for all $\mathcal{I} \in \mathcal{N}^{q-1}$, then the adjoint Weil homomorphism,*

$$\begin{aligned} HM' &: H_{q-1}(U, \mathcal{E}'_{i+1}/P'\mathcal{E}'_{i+2}) \to H_q(U, \mathcal{E}'_i/P'\mathcal{E}'_{i+1}) \quad \text{for } q > 1, \\ H(hM)' &: H_{i+1}(\mathcal{E}'(X)) \to H_1(U, \mathcal{E}'_i/P'\mathcal{E}'_{i+1}) \quad \text{for } q = 1, \end{aligned}$$

is surjective.

Proof. First suppose $q > 1$ and $[\overline{g}] \in H_q(U, \mathcal{E}'_\iota/P'\mathcal{E}'_{\iota+1})$. If $g \in C_q(U, \mathcal{E}'_\iota)$ is a representative of the class $[\overline{g}]$, then $\partial g = P'v$ for some chain $v \in C_{q-1}(U, \mathcal{E}'_{\iota+1})$. Since $P'(\partial v) = \partial(P'v) = \partial(\partial g) = 0$ and the conditions of the lemma together with Poincaré's duality (5.1.2) imply $H_{\iota+1}(\mathcal{E}'(U_\mathcal{I})) \cong H^{\iota+1}(\mathcal{E}\,(U_\mathcal{I}))' = 0$ for all $\mathcal{I} \in \mathcal{N}^{q-1}$, we get $\overline{v} \in Z_{q-1}(C.(U, \mathcal{E}'_{\iota+1}/P'\mathcal{E}'_{\iota+2}))$. By Lemma 5.2.19

$$\begin{aligned} M'v &= h'P'v - P'h'v \\ &= h'\partial g - P'h'v \\ &= g - \partial(h'g) - P'(h'v), \end{aligned}$$

and hence $[\overline{v}] \in H_{q-1}(U, \mathcal{E}'_{\iota+1}/P'\mathcal{E}'_{\iota+2})$ and $HM'[\overline{v}] = [\overline{g}]$.

Let $q = 1$ and $[\overline{g}] \in H_1(U, \mathcal{E}'_\iota/P'\mathcal{E}'_{\iota+1})$. If $g \in C_1(U, \mathcal{E}'_\iota)$ is a representative of the class $[\overline{g}]$, then $\partial g = P'v$ for some chain $v \in C_0(U, \mathcal{E}'_{\iota+1})$. Define $v_1 = \partial v \in \mathcal{E}'_{\iota+1}(X)$. Then $P'v_1 = \partial(P'v) = \partial(\partial g) = 0$. By Lemmas 5.2.19 and 5.2.23

$$\begin{aligned} M'h'v_1 &= M'(h'\partial)v \\ &= M'(v - \partial h'v) \\ &= h'(P'v) - P'h'v + \partial M'h'v \\ &= (h'\partial)g - P'h'v + \partial M'h'v \\ &= g - P'h'v + \partial(M'h'v - h'g), \end{aligned}$$

therefore $[v_1] \in H_{\iota+1}(\mathcal{E}'_\cdot(X))$ and $H(hM)'[v_1] = [\overline{g}]$.

This completes the proof.

□

Lemma 5.2.27 *If $H^{\iota+2}(\mathcal{E}\,(U_\mathcal{I}))$ is separated and $H^{\iota+1}(\mathcal{E}\,(U_\mathcal{I})) = 0$ for all $\mathcal{I} \in \mathcal{N}^q$, then the adjoint Weil homomorphism*

$$\begin{aligned} HM' &: H_{q-1}(U, \mathcal{E}'_{\iota+1}/P'\mathcal{E}'_{\iota+2}) \to H_q(U, \mathcal{E}'_\iota/P'\mathcal{E}'_{\iota+1}) \qquad \textit{for } q > 1, \\ H(hM)' &: H_{\iota+1}(\mathcal{E}'_\cdot(X)) \to H_1(U, \mathcal{E}'_\iota/P'\mathcal{E}'_{\iota+1}) \qquad \textit{for } q = 1, \end{aligned}$$

is injective.

Proof. First suppose $q > 1$ and $[\overline{g}] \in H_{q-1}(U, \mathcal{E}'_{\iota+1}/P'\mathcal{E}'_{\iota+2})$ and $HM'[\overline{g}] = 0$ in $H_q(U, \mathcal{E}'_\iota/P'\mathcal{E}'_{\iota+1})$. If $g \in C_{q-1}(U, \mathcal{E}'_{\iota+1})$ is a representative of the class $[\overline{g}]$, then $\partial g = P'v$ for some chain $v \in C_{q-2}(U, \mathcal{E}'_{\iota+2})$, and $h'P'g = P'g_1 + \partial v_1$ for some chains $g_1 \in C_q(U, \mathcal{E}'_{\iota+1})$ and $v_1 \in C_{q+1}(U, \mathcal{E}'_\iota)$. Let $g_2 = g - \partial g_1 \in C_{q-1}(U, \mathcal{E}'_{\iota+1})$. Then by Lemma 5.2.19 we have

$$\begin{aligned} P'g_2 &= P'g - \partial(P'g_1) \\ &= P'g - \partial(h'P'g - \partial v_1) \\ &= P'g - \partial h'(P'g) \\ &= h'\partial(P'g) \\ &= h'P'(P'v) = 0. \end{aligned}$$

Since the conditions of the lemma and Poincaré's duality (5.1.2) imply

$$H_{\imath+1}(\mathcal{E}'_{\cdot}(U_{\mathcal{I}})) \cong H^{\imath+1}(\mathcal{E}^{\cdot}(U_{\mathcal{I}}))' = 0 \qquad \text{for all } \mathcal{I} \in \mathcal{N}^q,$$

there exists a chain $v_2 \in C_{q-1}(U, \mathcal{E}'_{\imath+2})$ such that $g_2 = P'v_2$. Hence $[\overline{g}] = 0$.

Let now $q = 1$, $[g] \in H_{\imath+1}(\mathcal{E}'_{\cdot}(X))$ and $H(hM)'[g] = 0$ in $H_1(U, \mathcal{E}'_{\imath}/P'\mathcal{E}'_{\imath+1})$. If $g \in \mathcal{E}'_{\imath+1}(X)$ is a representative of $[g]$, then $P'g = 0$ and $h'P'h'g = P'g_1 + \partial v_1$ for some chains $g_1 \in C_1(U, \mathcal{E}'_{\imath+1})$ and $v_1 \in C_2(U, \mathcal{E}'_{\imath})$. Let $g_2 = h'g - \partial g_1 \in C_0(U, \mathcal{E}'_{\imath+1})$. By Lemma 5.2.19 we obtain

$$\begin{aligned} P'g_2 &= P'h'g - \partial(P'g_1) \\ &= P'h'g - \partial(h'P'h'g - \partial v_1) \\ &= P'h'g - (1 - h'\partial)P'h'g \\ &= h'P'(\partial h'g) \\ &= h'P'g = 0. \end{aligned}$$

Using the conditions of the lemma and Poincaré's duality (5.1.2), as above, we deduce that $g_2 = P'v_2$ for some chain $v_2 \in C_0(U, \mathcal{E}'_{\imath+2})$. Define $v = \partial v_2 \in \mathcal{E}'_{\imath+2}(X)$, then $P'v = \partial g_2 = \partial(h'g - \partial g_1) = g$ and hence $[g] = 0$. The proof is complete.

□

Suppose that $V = \{V_\mu\}_{\mu \in \mathfrak{M}}$ is a refinement of the covering U, so that there exists a mapping $T : \mathfrak{M} \to \mathcal{N}$ with $V_\mu \subset U_{T(\mu)}$. Denote by $T^{\sharp\prime} : C_q(V, \mathcal{E}'_{\imath}) \to C^q(U, \mathcal{E}'_{\imath})$ the transposed mapping for the mapping $T^\sharp : C_q(U, \mathcal{E}^{\imath}) \to C^q(V, \mathcal{E}^{\imath})$. This transpose commutes with ∂ and P' and hence it induces a mapping $HT^{\sharp\prime} : H_q(V, \mathcal{E}'_{\imath}/P'\mathcal{E}'_{\imath+1}) \to H_q(U, \mathcal{E}'_{\imath}/P'\mathcal{E}'_{\imath+1})$. This mapping $HT^{\sharp\prime}$ can be shown to be transitive and independent of the particular mapping T.

Lemma 5.2.28 *The adjoint Weil homomorphism is natural in the sense, that if V is a refinement of the covering U, then there is the commutative diagram*

$$\begin{array}{ccccc}
H_q(U, \mathcal{E}'_{\imath}/P'\mathcal{E}'_{\imath+1}) & \xleftarrow{HM'} & H_{q-1}(U, \mathcal{E}'_{\imath+1}/P'\mathcal{E}'_{\imath+2}) & \xleftarrow{HM'} & \cdots \\
\uparrow HT^{\sharp\prime} & & \uparrow HT^{\sharp\prime} & & \\
H_q(V, \mathcal{E}'_{\imath}/P'\mathcal{E}'_{\imath+1}) & \xleftarrow{HM'} & H_{q-1}(V, \mathcal{E}'_{\imath+1}/P'\mathcal{E}'_{\imath+2}) & \xleftarrow{HM'} & \cdots \\
\xleftarrow{HM'} & H_1(U, \mathcal{E}'_{\imath+q-1}/P'\mathcal{E}'_{\imath+q}) & \xleftarrow{H(hM)'} & H_{q+\imath}(\mathcal{E}'_{\cdot}(X)) & \\
& \uparrow HT^{\sharp\prime} & & \uparrow 1 & \\
\xleftarrow{HM'} & H_1(V, \mathcal{E}'_{\imath+q-1}/P'\mathcal{E}'_{\imath+q}) & \xleftarrow{H(hM)'} & H_{q+\imath}(\mathcal{E}'_{\cdot}(X)). &
\end{array} \qquad (5.2.13)$$

Proof. One repeats the proof of Lemma 5.2.9 word for word.

□

Let $W'_q : \mathcal{E}'_{q+\imath}(X) \to C_q(U, \mathcal{E}'_{\imath})$ be the mapping transposed to the mapping $W_q : C^q(U, \mathcal{E}_{\imath}) \to \mathcal{E}^{q+1}(X)$. Obviously, $W'_q = (-1)^{\frac{q(q+1)}{2}} M'_q \circ \ldots \circ M'_1 \circ h'$.

Lemma 5.2.29 *For each section $g \in \mathcal{E}'_{q+i+1}(X)$, we have*

$$W'_q(P'g) = \partial W'_{q+1}(g) + (-1)^q P' W'_q(g). \tag{5.2.14}$$

Proof. Note that (5.2.14) is simply a transposed version of equality (5.2.6). □

We are now able to formulate the statements dual to Theorems 5.2.11 and 5.2.12.

Theorem 5.2.30 *$W' : \mathcal{E}'_\cdot(X) \to C_\cdot(U, \mathcal{E}'_0/P'\mathcal{E}'_1)$ is a chain mapping of complexes (of degree zero) which induces the homomorphism $HW' : H_\cdot(\mathcal{E}'_\cdot(X)) \to H_\cdot(U, \mathcal{E}'_0/P'\mathcal{E}'_1)$ independent of the choice of a partition of unity $\{\varphi_\nu\}_{\nu\in\mathcal{N}}$ and natural in the sense that, for each refinement V of the covering U, the following diagram is commutative:*

$$\begin{array}{ccc} H_\cdot(U, \mathcal{E}'_0/P'\mathcal{E}'_1) & \stackrel{HW'}{\longleftarrow} & H_\cdot(\mathcal{E}'_\cdot(X)) \\ \uparrow {\scriptstyle HT^{\mathfrak{l}\prime}} & & \uparrow {\scriptstyle 1} \\ H_\cdot(V, \mathcal{E}'_0/P'\mathcal{E}'_1) & \stackrel{HW'}{\longleftarrow} & H_\cdot(\mathcal{E}'_\cdot(X)). \end{array}$$

Proof. This follows from Lemmas 5.2.28 and 5.2.29. □

Theorem 5.2.31 *1) If the spaces $H^{i+1}(\mathcal{E}^\cdot(U_\mathcal{I}))$ are separated and $H^i(\mathcal{E}^\cdot(U_\mathcal{I})) = 0$ $(\mathcal{I} \in \mathcal{N}^{q-i})$ for every $i = 1, \ldots, q$, then $HW' : H_q(\mathcal{E}'_\cdot(X)) \to H_q(U, \mathcal{E}'_0/P'\mathcal{E}'_1)$ is an epimorphism.*

2) If the spaces $H^{i+1}(\mathcal{E}^\cdot(U_\mathcal{I}))$ are separated and $H^i(\mathcal{E}^\cdot(U_\mathcal{I})) = 0$ $(\mathcal{I} \in \mathcal{N}^{q-i+1})$ for every $i = 1, \ldots, q$, then $HW' : H_q(\mathcal{E}'_\cdot(X)) \to H_q(U, \mathcal{E}'_0/P'\mathcal{E}'_1)$ is a monomorphism.

Proof. Since

$$HW' = (-1)^{\frac{q(q+1)}{2}} HM'_q \circ \ldots \circ HM'_2 \circ H(hM_1)',$$

the theorem follows immediately from Lemmas 5.2.26 and 5.2.27. □

5.2.7 *A localized version of the adjoint Weil homomorphism*

Let us briefly discuss the localized version of the sequence (5.2.12).

Exercise 5.2.32 *Elaborate the principles of the homology theory of a topological space with values in a cosheaf.*

Hint. Dualize the main facts in Godement [72, Ch,II] for the cohomology theory of a topological space with values in a sheaf.

□

Thanks to Lemma 5.2.28, we are allowed to pass in (5.2.12) to the projective limit over refined coverings [3]. Then the adjoint Weil homomorphism HM' is raised to a homomorphism HM'_{loc} of homology groups of the manifold X:

$$H_q(X, \mathcal{E}'_i/P'\mathcal{E}'_{i+1}) \overset{HM'_{loc}}{\longleftarrow} H_{q-1}(X, \mathcal{E}'_{i+1}/P'\mathcal{E}'_{i+2}) \overset{HM'_{loc}}{\longleftarrow} \cdots \overset{HM'_{loc}}{\longleftarrow} H_1(X, \mathcal{E}'_{i+q-1}/P'\mathcal{E}'_{i+q}) \overset{H(hM)'_{loc}}{\longleftarrow} H_{q+i}(\mathcal{E}'(X)). \tag{5.2.15}$$

The following statements are analogous to Theorems 5.2.13 and 5.2.14.

Theorem 5.2.33 *Assume that there exists an open covering $\{U\}$ of the manifold X such that each open covering of X has a refinement consisting of elements of $\{U\}$, and $H^{q+1}(\mathcal{E}^{\cdot}(U))$ is separated and $H^1(\mathcal{E}^{\cdot}(U)) = \ldots = H^q(\mathcal{E}^{\cdot}(U)) = 0$ for each set $U \in \{U\}$. Then $HW'_{loc} : H_q(\mathcal{E}'_{\cdot}(X)) \to H_q(X, \mathcal{E}'_0/P'\mathcal{E}'_1)$ is an isomorphism.*

Proof. From Lemmas 5.2.26 and 5.2.27 by passing to the projective limit over refined coverings, we obtain that, if $H^{i+2}(\mathcal{E}^{\cdot}(U))$ is separated, and $H^{i+1}(\mathcal{E}^{\cdot}(U)) = 0$ for each set $U \in \{U\}$, then the mappings $HM'_{loc} : H_{q-1}(X, \mathcal{E}'_{i+1}/P'\mathcal{E}'_{i+2}) \to H_q(X, \mathcal{E}'_i/P'\mathcal{E}'_{i+1})$ for $q > 1$, and $H(hM)'_{loc} : H_{i+1}(\mathcal{E}'_{\cdot}(X)) \to H_1(X, \mathcal{E}'_i/P'\mathcal{E}'_{i+1})$ in the sequence (5.2.15) are isomorphisms. Hence the theorem follows.

□

Theorem 5.2.34 *If in addition to the conditions of Theorem 5.2.33 the spaces $H^{i+1}(\mathcal{E}^{\cdot}(U_{\mathcal{I}}))$ are separated, and $H^i(\mathcal{E}^{\cdot}(U_{\mathcal{I}})) = 0$ $(\mathcal{I} \in \mathcal{N}^{q-i} \cup \mathcal{N}^{q-i+1})$ for every $i = 1, \ldots, q$, then $HW' \circ (HW')^{-1}_{loc} : H_q(X, \mathcal{E}'_0/P'\mathcal{E}'_1) \to H_q(U, \mathcal{E}'_0/P'\mathcal{E}'_1)$ is an isomorphism.*

Proof. This immediately follows from Theorems 5.2.31 and 5.2.33.

□

5.2.8 *Pairing at sequences of Weil representatives*

The following concepts will prove useful in formulating the main result of this section.

They have been prompted by the investigations of Gleason [1] adapted to the Dolbeault complex by Harvey [1].

Definition 5.2.35 *A sequence $\{f^i\}_{i=0,\ldots,q}$ with $f^i \in Z^i(C(U, S^{q-i}))$ such that $[f^{i-1}]$ is the image of $[f^i] \in H^i(U, S^{q-i})$ under Weil's homomorphism for $i = 1, \ldots, q$ is called a sequence of Weil representatives of the class $[f^q] \in H^q(U, S^0)$. If $i = 1$, then "$[f^{i-1}]$ is the image of $[f^i]$" is replaced by "$[\delta^{-1} f^0]$ is the image of $[f^1]$."*

---

[3]See the notion of a projective (or inverse) limit, for example, in the book of Palamodov [175, Pt.2, §1]

Definition 5.2.36 *A sequence* $\{\overline{g^\imath}\}_{\imath=0,\dots,q}$ *with* $\overline{g^\imath} \in Z^\imath(C(U, \mathcal{E}'_{q-\imath}/P'\mathcal{E}'_{q-\imath+1}))$ *such that* $[\overline{g^\imath}] \in H_\imath(U, \mathcal{E}'_{q-\imath}/P'\mathcal{E}'_{q-\imath+1})$ *is the image of* $[\overline{g^{\imath-1}}]$ *under the adjoint Weil homomorphism for* $i = 1, \dots, q$, *is called an adjoint sequence of Weil representatives of the class* $[\partial g^0] \in H_q(\mathcal{E}'_\cdot(X))$.

Consider the pairing $\langle \overline{g}, f\rangle$ between $C_q(U, \mathcal{E}'_\imath/P'\mathcal{E}'_{\imath+1})$ and $C^q(U, S^\imath)$ restricted to $Z_q(C_\cdot(U, \mathcal{E}'_\imath/P'\mathcal{E}'_{\imath+1}))\otimes Z^q(C_\cdot(U, S^\imath))$. If $\overline{g} \in \partial C_{q+1}(U, \mathcal{E}'_\imath/P'\mathcal{E}'_{\imath+1})$ or $f \in \delta C^{q-1}(U, S^\imath)$, then this bilinear form vanishes. Indeed, by Lemma 5.2.17 we have $\langle \partial\overline{v}, f\rangle = \langle \overline{v}, \delta f\rangle = 0$ and similarly $\langle \overline{g}, \delta u\rangle = \langle \partial\overline{g}, u\rangle = 0$. Hence the bilinear form (5.2.8) induces a pairing $\langle [\overline{g}], [f]\rangle$ between $H_q(U, \mathcal{E}'_\imath/P'\mathcal{E}'_{\imath+1})$ and $H^q(U, S^\imath)$.

Now the main theorem of this section follows easily.

Theorem 5.2.37 *Let* $\{f^\imath\}_{\imath=0,\dots,q}$ *be a sequence of Weil representatives of a cohomology class* $[f^q] \in H^q(U, S^0)$ *and* $\{\overline{g^\imath}\}_{\imath=0,\dots,q}$ *be an adjoint sequence of Weil representatives of a homology class* $[\partial g^0] \in H_q(\mathcal{E}'_\cdot(X))$. *Then* $\langle \overline{g^\imath}, f^\imath\rangle$ *is the same for all* $i = 0, \dots, q$, *in particular*

$$\langle g^0, f^0\rangle_X = {\sum}'_{\mathcal{I}\in\mathcal{N}^{q+1}} \langle g^q_{\mathcal{I}}, f^q_{\mathcal{I}}\rangle_{U_{\mathcal{I}}}. \tag{5.2.16}$$

For the de Rham complex this theorem is due to Gleason [71] and for the Dolbeault complex it is due to Harvey [98].

Proof. As mentioned above the bilinear form (5.2.8) induces a pairing

$$H_q(U, \mathcal{E}'_\imath/P'\mathcal{E}'_{\imath+1}) \otimes H^q(U, S^\imath) \to \mathbb{C}.$$

Therefore, to prove the theorem it suffices to replace $\{f^\imath\}_{\imath=0,\dots,q}$ by any sequence of Weil representatives of the class $[f^q]$ and replace $\{\overline{g^\imath}\}_{\imath=0,\dots,q}$ by any adjoint sequence of Weil representatives of the class $[\partial g^0]$. Now replace $\{f^\imath\}_{\imath=0,\dots,q}$ by $\{(M_{\imath+1}\circ\dots\circ M_q)f^q\}_{\imath=0,\dots,q}$ and $\{\overline{g^\imath}\}_{\imath=0,\dots,q}$ by $\{(M'_\imath\circ\dots\circ M'_1)g^0\}_{\imath=0,\dots,q}$. Supposing $i \le q-1$, we get by Lemma 5.2.21

$$\begin{aligned}\langle \overline{g^\imath}, f^\imath\rangle &= \langle (M'_\imath\circ\dots\circ M'_1)g^0, M_{\imath+1}(M_{\imath+2}\circ\dots\circ M_q)f^q\rangle\\ &= \langle (M'_{\imath+1}\circ M'_\imath\circ\dots\circ M'_1)g^0, (M_{\imath+2}\circ\dots\circ M_q)f^q\rangle\\ &= \langle \overline{g^{\imath+1}}, f^{\imath+1}\rangle,\end{aligned}$$

as required. The proof is complete.

□

In conclusion we will show how Theorem 5.2.37 can be used to extend the Poincaré duality (5.1.2).

Corollary 5.2.38 *Let the spaces* $H^{\imath+1}(\mathcal{E}^\cdot(U_{\mathcal{I}}))$ *be separated and* $H^\imath(\mathcal{E}^\cdot(U_{\mathcal{I}})) = 0$ $(\mathcal{I} \in \mathcal{N}^{q-\imath} \cup \mathcal{N}^{q-\imath+1})$ *for every* $i = 1, \dots, q$. *Then the property that* $H^\imath(U, S^{q-\imath})$ *be separated for some* $i = 1, \dots, q$ *implies they are separated for all* $i = 0, \dots, q$, *and the fact that* $H_\imath(U, \mathcal{E}'_{q-\imath}/P'\mathcal{E}'_{q-\imath+1})$ *be separated for some* $i = 0, \dots, q$ *implies they are separated for all* $i = 0, \dots, q$, *and* $H^\imath(U, S^{q-\imath})' \overset{top.}{\cong} H_\imath(U, \mathcal{E}'_{q-\imath}/P'\mathcal{E}'_{q-\imath+1})$. *If* $i = 0$, *then* $H^\imath(U, S^{q-\imath})$ *is replaced by* $H^q(\mathcal{E}^\cdot(X))$ *and* $H_\imath(U, \mathcal{E}'_{q-\imath}/P'\mathcal{E}'_{q-\imath+1})$ *is replaced by* $H_q(\mathcal{E}'_\cdot(X))$.

For the Dolbeault complex this result was noted by Harvey [98].

Proof. It is clear that the property of being separated for $H^\imath(U, S^{q-\imath})$ or $H_\imath(U, \mathcal{E}'_{q-\imath}/P'\mathcal{E}'_{q-\imath+1})$ is equivalent to the non-degeneracy in the corresponding variable of the bilinear form

$$\langle [\overline{g}], [f] \rangle : \; H_\imath(U, \mathcal{E}'_{q-\imath}/P'\mathcal{E}'_{q-\imath+1}) \otimes H^\imath(U, S^{q-\imath}) \to \mathbb{C}.$$

Under the conditions of the corollary, both Weil's homomorphism and the adjoint Weil homomorphism are isomorphisms. Since Theorem 5.2.37 implies

$$\langle [\overline{g}], [f] \rangle = \langle (HM')^{\imath_0 - \imath}[\overline{g}], (HM)^{\imath - \imath_0}[f] \rangle,$$

the desired statement follows easily.

□

5.2.9 *A remark on hypoelliptic complexes*

All the results above are valid with obvious modifications when the sheaves $\mathcal{D}'^\imath$ of germs of distributional sections of $E^\imath$ are considered in place of $\mathcal{E}^\imath$ and the cosheaves $\sigma \to \mathcal{D}_\imath(\sigma) = \mathcal{D}(E^{\imath\prime}|_\sigma)$ dual to $\mathcal{D}'^\imath$ are considered in place of $\mathcal{E}'_\imath$. In particular, if the differential sheaves $\mathcal{E}^\cdot$ and $\mathcal{D}'$ are exact at positive degrees and the differential operator P_0 is hypoelliptic, then by Theorem 5.2.13 $H^\cdot(\mathcal{E}(E^\cdot)) \cong H^\cdot(\mathcal{D}'(E^\cdot))$ (cf. Palamodov [175, Ch,VII, §9, Corollary 2]). This result is of a different nature than Corollary 2.1.15 because on the one hand there are elliptic complexes such that the sheaf $\mathcal{E}^\cdot$ has non-trivial cohomologies at positive degrees and on the other hand a complex $E^\cdot$ satisfying the conditions above need not be elliptic nor even homotopically equivalent to an elliptic complex. Sometimes the equality $H^\cdot(\mathcal{E}(E^\cdot)) \cong H^\cdot(\mathcal{D}'(E^\cdot))$, where the isomorphism is arranged by the natural inclusion, is considered to be the definition of a hypoelliptic complex (see Andreotti, Hill, Lojasiewicz and MacKichan [11, §13]).

5.3 Integral Formulas Connected by the Weil Homomorphism

5.3.1 *U-resolution of a singular cycle*

Let $E^\cdot = \Lambda^\cdot T^*_{\mathbb{C}}(X)$ be the de Rham complex on a differentiable manifold X of dimension n. Sections of $\mathcal{E}'_\imath(X) = (\mathcal{E}(\Lambda^\imath T^*_{\mathbb{C}}(X)))'$ are currents of dimension $\imath$ with compact supports on X.

According to the results of de Rham [50, Theorems 14 and 16], each homology class of $H_\imath(\mathcal{E}'_\cdot(X))$ has a representative given by a (singular) cycle of dimension $\imath$. In this case the adjoint sequence of Weil representatives may be interpreted in a more perspicacious way suggested by Gleason [71].

Suppose that $U = \{U_\nu\}_{\nu \in \mathcal{N}}$ is some open covering of the manifold X. Following Gleason [71], a U-chain of degree q and dimension $\imath$ on X is a mapping c which assigns to every multi-index $\mathcal{I} \in \mathcal{N}^{q+1}$ a singular (of the class C^∞) chain $c_\mathcal{I}$ of dimension i with compact support in $U_\mathcal{I}$ so that $c_\mathcal{I}$ is an alternating function of $\mathcal{I}$, and so that $c_\mathcal{I} = 0$ except for a finite number of $\mathcal{I} \in \mathcal{N}^{q+1}$.

Obviously, every U-chain c of degree q and dimension i determines some element of $C_q(U, \mathcal{E}'_i)$ and the boundary mapping ∂ maps c to some U-chain of degree $(q-1)$ and dimension i.

The following definition is fundamental in the theory of Gleason [71].

Definition 5.3.1 *A U-resolution of a singular cycle c of dimension q on X is a sequence $\{c^i\}_{i=0,\ldots,q}$ such that:*

1) c^i is a U-chain of degree i and dimension $(q-i)$;

2) $c = \sum_{\nu\in\mathcal{N}} c^0_\nu$; and

3) boundary $(c^i_\mathcal{I}) = \sum_{\nu\in\mathcal{N}} c^{i+1}_{\nu\mathcal{I}}$ for all $\mathcal{I} \in \mathcal{N}^{i+1}$ $(i = 0, \ldots, q-1)$.

One can see that condition 2) means that $c = \partial c^0$, and condition 3) means that boundary $(c^i) = \partial c^{i+1}$. Since the boundary operator on chains coincides with the operator transposed by Stokes' Theorem to the exterior derivative on forms, we conclude that the U-resolution of a cycle c is a particular adjoint sequence of Weil representatives of the homology class $[c] \in H_q(\mathcal{E}'(X))$.

Therefore, for the de Rham complex, Theorem 5.2.37 can be reformulated in the following way.

Theorem 5.3.2 *Suppose that $\{f^i\}_{i=0,\ldots,q}$ is a sequence of Weil representatives of a class $[f^q] \in H^q(U, \mathbb{C})$ and $\{c^i\}_{i=0,\ldots,q}$ is a U-resolution of a q-dimensional singular cycle c on X. Then*

$$\int_c f^0 = \sum\nolimits'_{\mathcal{I}\in\mathcal{N}^{i+1}} \int_{c^i_\mathcal{I}} f^i_\mathcal{I} \quad \text{for all} \quad i = 0, \ldots, q.$$

Proof. This follows immediately from Theorem 5.2.37.

□

This result of Gleason [71] is so much the better since he proved the following theorem.

Theorem 5.3.3 *Let U be an open covering of a differentiable manifold X. Then every class of singular homologies of X may be represented by a C^∞ cycle which has a U-resolution.*

Proof. See Gleason [71].

□

Gleason [71] used these theorems for proving the Cauchy-Weil formula for holomorphic functions of several variables.

5.3.2 *Homologies generated by cycles and conservation laws*

Assume now that $E^{\cdot}$ is an arbitrary differential complex on the manifold X.

Lemma 5.3.4 *For every conservation law $G \in \mathrm{do}_g(E^q \to \Lambda^{n-1})$ for solutions of the system $P_q f = 0$, there exists a sequence $\{G_i\}_{i=0,\ldots,q}$ of differential operators $G \in \mathrm{do}_{g+p_{q-1}+\ldots+p_i-(q-i)}(E^i \to \Lambda^{n-(q-i)-1})$ such that:*
1) $G_q = G$, and
2) $dG_{i-1} = G_i \circ P_{i-1}$ for all $i = 1, \ldots, q$.

Proof. Use Theorem 2.4.11.

□

As follows from part 2) of the lemma, G_i is a conservation law for solutions of $P_i f = 0 \quad (i = 0, \ldots, q-1)$.

Exercise 5.3.5 *Prove that, if $\{G_i'\}$ and $\{G_i''\}$ are two sequences with properties 1) and 2), then there exists a sequence $\{D_i\}_{i=0,\ldots,q}$ of differential operators $D_i \in \mathrm{do}_{g+p_{q-1}+\ldots+p_i-(q-i)-1}(E^i \to \Lambda^{n-(q-i)-2})$ such that $G_i'' - G_i' = D_{i+1} \circ P_i + d \circ D_i$ $(i = 0, \ldots, q)$.*

Hint. Use Theorem 2.4.1 and Remark 2.4.2.

□

In particular, for $i = 0, \ldots, q$, the differences $(G_i'' - G_i')$ are trivial conservation laws for solutions of $P_i f = 0$.

Let c be a singular (of the class C^∞) cycle of dimension $(n-1)$ on X. Replacing c by a homologous cycle if necessary, one can assume that c has a U-resolution $\{c^i\}_{i=0,\ldots,n-1}$.

Fix some conservation law $G \in \mathrm{do}_g(E^q \to \Lambda^{n-1})$ of degree $(n-1)$ for solutions of the system $P_q f = 0$. We choose a sequence $\{G_i\}_{i=0,\ldots,q}$ for G in accordance with Lemma 5.3.4.

Remark 5.3.6 *It is easy to see that the functional $\langle g, f \rangle = \int_c G(f) \quad (f \in \mathcal{E}^q(X))$ determines a section $g \in Z_q(\mathcal{E}^{\prime}_{\cdot}(X))$.*

For $i = 0, \ldots, q$, we consider the i-chains $g^i \in C_i(U, \mathcal{E}'_{q-i})$ given by

$$\left\langle g^i_{\mathcal{I}}, f \right\rangle = \int_{c^i_{\mathcal{I}}} G_{q-i}(f) \quad \text{for all} \quad f \in \mathcal{E}^{q-i}(U_{\mathcal{I}}) \quad (\mathcal{I} \in \mathcal{N}^{i+1}). \tag{5.3.1}$$

Theorem 5.3.7 *The sequence $\{g^i\}_{i=0,\ldots,q}$ defined by (5.3.1) is an adjoint sequence of Weil representatives of the class $[g] \in H_q(\mathcal{E}^{\prime}_{\cdot}(X))$ of Remark 5.3.6.*

Proof. First of all, we have $g = \partial g^0$ because for each $f \in \mathcal{E}^q(X)$

$$\left\langle \partial g^0 f \right\rangle = \sum_{\nu \in \mathcal{N}} \int_{c^0_\nu} G_q(f) = \int_c G(f) = \langle g, f \rangle.$$

To complete the proof, it suffices to show by Definition 5.2.25 that $\partial g^i = P' g^{i-1}$ for $i = 1, \ldots, q$. In fact, if $\mathcal{I} \in \mathcal{N}^i$, then for all $f \in \mathcal{E}^{q-i}(U_{\mathcal{I}})$ we obtain by Stokes' Theorem

$$\begin{aligned} \left\langle (\partial g^i)_{\mathcal{I}}, f \right\rangle &= \left\langle \sum_{\nu \in \mathcal{N}} g^i_{\nu \mathcal{I}}, f \right\rangle \\ &= \sum_{\nu \in \mathcal{N}} \int_{c^i_{\nu \mathcal{I}}} G_{q-i}(f) \\ &= \int_{\partial c^{i-1}_{\mathcal{I}}} G_{q-i}(f) \\ &= \int_{c^{i-1}_{\mathcal{I}}} dG_{q-i}(f) \\ &= \int_{c^{i-1}_{\mathcal{I}}} G_{q-(i-1)}(P_{q-i} f) \\ &= \left\langle g^{i-1}_{\mathcal{I}}, Pf \right\rangle \\ &= \left\langle (P' g^{i-1})_{\mathcal{I}}, f \right\rangle, \end{aligned}$$

as desired.

□

It follows from the proof that Theorem 5.3.7 is also valid under weaker restrictions concerning the smoothness of the cycle c and its U-resolution $\{c^i\}_{i=0,\ldots,n-1}$. The smoothness of the class C^1 is sufficient. The main point is that Stokes' formula would be applicable. Later on we shall not stipulate this specially.

From Theorems 5.3.7 and 5.2.37, we get the following generalization of Theorem 5.3.2.

Corollary 5.3.8 *If $\{f^i\}_{i=0,\ldots,q}$ is a sequence of Weil representatives of a cohomology class $[f^q] \in H^q(U, S^0)$, then*

$$\int_c G(f^0) = \sum\nolimits'_{\mathcal{I} \in \mathcal{N}^{i+1}} \int_{c^i_{\mathcal{I}}} G_{q-i}(f^i_{\mathcal{I}}) \quad \textit{for all} \quad i = 0, \ldots, q. \tag{5.3.2}$$

5.3.3 *Abstract integral representations for solutions of overdetermined systems*

We now consider applications of the results of Section 5.2 to overdetermined systems. We assume that the differential operator P_0 has a left fundamental solution $\Phi_1 \in \mathcal{D}'(E^0 \boxtimes E^{1\prime})$ continuable and differentiable outside the diagonal of $X \times X$.

Example 5.3.9 If the Laplacian $P_0^* P_0$ is an elliptic differential operator with real analytic coefficients, then P_0 has a pseudo-differential left fundamental solution. In fact, according to the theorem of Malgrange [156, p.341], $P_0^* P_0$ has a pseudo-differential bilateral fundamental solution G_0. Then $\Phi_1 = G_0 P_0^*$ is a left fundamental solution of P_0, as desired.

□

Let K be a compact subset of X, and $U = \{U_\nu\}_{\nu \in \mathcal{N}}$ be an open covering of the set $Y = X \setminus K$. We shall apply the results of Section 5.2 to the (cochain) differential complex $F^\cdot$ on Y where $F^i = E^{q-i+1'}|_Y$ for $i = 0, \dots, q$ and $F^i = 0$ for other indices i, and the differential d_F is given by P'. In particular, HW_q is the homomorphism from $H^q(U, \mathcal{E}_{P'_q})$ to $H_1(\mathcal{E}_\cdot(Y))$ constructed in accordance with Theorem 5.2.11 for the complex $F^\cdot$.

Remark 5.3.10 *As follows from 2) of Theorem 5.2.12, if* $H_i(\mathcal{E}\,(U_\mathcal{I})) = 0$ $(\mathcal{I} \in \mathcal{N}^i)$ *for every* $i = 1, \dots, q$, *then* $HW : \; H^q(U, \mathcal{E}_{P'_q}) \to H_1(\mathcal{E}\,(Y))$ *is an epimorphism.*

Suppose that f is a solution of the system $P_0 f = 0$ in a neighborhood U' of K and c is an $(n-1)$-dimensional singular (of class C^1) cycle in $U' \setminus K$ covering K with multiplicity 1. This means that c is homologous in $U' \setminus K$ to the boundary of an open set containing K.

Define the section $[c]^P f$ of $\mathcal{E}'_1(Y)$ by the formula (2.4.21), i.e.,

$$\left\langle g, [c]^P f \right\rangle = \int_c G_{P_0}(g, f) \quad \text{for} \quad g \in \mathcal{E}_1(Y).$$

It follows from Theorem 3.1.26 that $P([c]^P f) = 0$ so $[c]^P f$ determines a unique cohomology class $[[c]^P f]$ of $H^1(\mathcal{E}'^\cdot(Y))$.

On the other hand, for a fixed $x \in K$, the kernel $\Phi_1(x, \cdot)$ satisfies $P'_0 \Phi_1(x, \cdot)$ on Y. Therefore it determines a unique homology class $[\Phi_1(x, .)]$ of $H_1(\mathcal{E}(E^0_x \otimes E^{\cdot\prime}|_Y))$. Assuming the hypothesis in Remark 5.3.10, we denote by $(HW)^{-1}[\Phi_1(x, \cdot)]$ some inverse image in $H^q(U, E^0_x \otimes \mathcal{E}_{P'_q})$ of the class $[\Phi_1(x, \cdot)]$ by the mapping HW.

Proposition 5.3.11 *Under the hypothesis of Remark 5.3.10, we have*

$$\begin{aligned} f(x) &= -\left\langle (HW)^{-1}[\Phi_1(x, \cdot)], (HW')\left[[c]^P f\right]\right\rangle & (5.3.3) \\ &= -\left\langle W\left((HW)^{-1}[\Phi_1(x, \cdot)]\right), [c]^P f\right\rangle \quad (x \in K). & (5.3.4) \end{aligned}$$

Proof. This follows from Green's formula (2.5.9) and Theorem 5.2.37. □

These are abstract integral representations for solutions of the overdetermined system $P_0 f = 0$.

The homology class $(HW')\left[[c]^P f\right]$ of $H_q(U, \mathcal{E}'^{q+1}/P\mathcal{E}'^q)$ can be realized as integration over an $(n-q-1)$-dimensional singular cycle in $U' \setminus K$ (a generalization of the Cauchy-Weil formula).

On the other hand, the kernel $W\left((HW)^{-1}[\Phi_1(x, \cdot)]\right)$ contains fairly arbitrary functions $\{\varphi_\nu\}_{\nu \in \mathcal{N}}$ (a generalization of the Cauchy-Fantappiè formula).

In the following subsections we shall give concrete expression to this program.

5.3.4 *Generalization of the Cauchy-Weil formula*

Let $K = \{x^0\}$ be a fixed point of X, and $U = \{U_\nu\}_{\nu \in \mathcal{N}}$ be an open covering of $Y = X \setminus K$.

If $x : U' \to \mathbb{R}^n$ is a coordinate map in a neighborhood of x^0 and $\varepsilon > 0$ is small enough, then the inverse image of the cube $\{|x_j - x_j^0| \le \varepsilon \ (j = 1, \ldots, n)\}$ determines an n-dimensional singular chain on X. The boundary b of this chain is an $(n-1)$-dimensional singular cycle in Y.

Suppose that c is an arbitrary $(n-1)$-dimensional singular cycle in Y which is homologous to zero on X. Then there exists an integer μ such that c is homologous to $\mu \cdot b$ in Y.

Remark 5.3.12 *The integer μ is called the multiplicity of c with respect to the point x^0.*

Choose some U-resolution $\{c^i\}_{i=0,\ldots,n-1}$ of the cycle c in Y whose existence is guaranteed by Theorem 5.3.3.

Below, we denote by G_i $(i \in \mathbb{Z}_+)$ the bidifferential operators of Corollary 2.4.13.

Theorem 5.3.13 *If $H_i(\mathcal{E}.(U_I)) = 0$ $(I \in \mathcal{N}^i)$ for every $i = 1, \ldots, q$, then for each solution $f \in \mathcal{E}_{P_0}(X)$ the following formula holds:*

$$\mu \cdot f(x^0) = -(-1)^{\frac{q(q+1)}{2}} \sum\nolimits'_{I \in \mathcal{N}^{q+1}} \int_{c_I^q} G_q\left(((HW)^{-1}[\Phi_1(x^0, \cdot)])_I, f\right). \tag{5.3.5}$$

Proof. Define the chains $f^i \in C_i(U, \mathcal{E}'^{i+1})$ $(i = 0, \ldots, q)$ by letting

$$\left\langle g, f_I^i \right\rangle_{U_I} = \int_{c_I^i} G_i(g, f) \quad \text{for all} \quad g \in \mathcal{E}_{i+1}(U_I) \quad (I \in \mathcal{N}^{i+1}).$$

It follows from Theorem 5.3.7 that the sequence $\{\overline{f^i}\}_{i=0,\ldots,q}$ is an adjoint sequence of Weil representatives of the class $\left[[c]^P f\right] \in H^1(\mathcal{E}'(Y))$. Therefore, we get $(HW')\left[[c]^P f\right] = (-1)^{\frac{q(q+1)}{2}} [\overline{f^q}]$, where $[\overline{f^q}]$ is the homology class of f^q in $H_q(U, \mathcal{E}'^{q+1}/P\mathcal{E}'^q)$.

To complete the proof, it suffices to invoke Proposition 5.3.11.

□

For holomorphic functions of several variables this theorem was obtained by Gleason [71].

Corollary 5.3.14 *Under the hypothesis of Theorem 5.3.13, if $f \in \mathcal{E}_{P_0}(X)$ vanishes up to order $(p_0 + \ldots + p_q - q - 1)$ on c^q (and $\mu \ne 0$), then $f(x^0) = 0$.*

Proof. In fact, via formula (5.3.5) the value $f(x^0)$ is completely determined by the $(p_0 + \ldots + p_q - q - 1)$-jets of f on the $(n - q - 1)$-dimensional singular chains c_I^q $(I \in \mathcal{N}^{q+1})$.

□

5.3.5 C^∞ *polyhedra in general position with respect to a covering*

Let now K be an arbitrary compact subset of X, and $U = \{U_\nu\}_{\nu \in \mathcal{N}}$ be an open covering of the set $Y = X \setminus K$.

A C^∞ polyhedron in general position with respect to U, as defined by Harvey [98], is a convenient interpretation of the U-resolution of Gleason [71].

Suppose that Π is a relatively compact open subset of X. Let, for every strictly increasing multi-index $\mathcal{I} \in \mathcal{N}^{i+1}$ $(i = 0, \ldots, n-1)$, a (not necessarily closed) C^∞ manifold $(\partial\Pi)_\mathcal{I} \subset U_\mathcal{I}$ of dimension $(n-i-1)$ on $\partial\Pi$ be given so that $(\partial\Pi)_\mathcal{I}$ do not intersect in pairs, $(\partial\Pi)_\mathcal{I} = \emptyset$ except for a finite number of $\mathcal{I}$, and $\partial\Pi = \cup_\mathcal{I}(\partial\Pi)_\mathcal{I}$. We extend the definition of $(\partial\Pi)_\mathcal{I}$ to all the multi-indices $\mathcal{I}$ by skew-symmetry. If moreover the following conditions are satisfied: 1) $\{\Pi, \cup_{\nu\in\mathcal{N}}(\partial\Pi)_\nu\}$ is an oriented (not necessarily closed) manifold with boundary, and 2) $\{(\partial\Pi)_\mathcal{I}, \cup_{\nu\in\mathcal{N}}(\partial\Pi)_{\nu\mathcal{I}}\}$ is an oriented (not necessarily closed) manifold with boundary, - then Π is said to be a C^∞ polyhedron in general position with respect to the covering U.

Remark 5.3.15 *If K and U are not too "bad" then for each neighborhood U' of K there exists a C^∞ polyhedron Π in general position with respect to U such that $K \subset \Pi \subset\subset U'$ (see Harvey [98] for details).*

The main result of this subsection consists in the following.

Theorem 5.3.16 *Let $H_i(\mathcal{E}.(U_\mathcal{I})) = 0$ $(\mathcal{I} \in \mathcal{N}^i)$ for every $i = 1, \ldots, q$, and $\Pi \subset X$ be a C^∞ polyhedron in general position with respect to U containing K. Then for each solution f of $P_0 f = 0$ in a neighborhood of $\overline{\Pi}$ we have*

$$f(x) = -(-1)^{\frac{q(q+1)}{2}} \sum\nolimits'_{\mathcal{I}\in\mathcal{N}^{q+1}} \int_{(\partial\Pi)_\mathcal{I}} G_q\left(\left((HW)^{-1}[\Phi_1(x,\cdot)]\right)_\mathcal{I}, f\right) \quad (x \in K). \quad (5.3.6)$$

Proof. We shall present a proof which is slightly incorrect in form, but nevertheless quite true in content.

Setting $c^i_\mathcal{I} = (\partial\Pi)_\mathcal{I}$ for $\mathcal{I} \in \mathcal{N}^{i+1}$, we define a U-chain c^i of degree i and dimension $(n-i-1)$ in Y. Since Π is a C^∞ polyhedron in general position with respect to U, the sequence $\{c^i\}_{i=0,..,n-1}$ is just a U-resolution of the cycle $\partial\Pi$ in Y.

To finish the proof, we may repeat the arguments of the proof of Theorem 5.3.13.

□

In particular, the values of f in K are completely determined by the $(p_0 + \ldots + p_q - q - 1)$-jet of f on the $(n-q-1)$-dimensional skeleton of the polyhedron Π. This is an effect of the overdeterminacy of the system $P_0 f = 0$. Under the hypothesis of Theorem 5.3.16, the overdeterminacy degree of the differential operator P_0 may be defined to be $\geq q$. Thus we see again that the longer the "right" compatibility complex for a differential operation P_0, the more overdetermined is the system $P_0 f = 0$.

Example 5.3.17 Let E be a Hilbert complex on an open set $X \subset \mathbb{R}^n$ associated with a $\mathcal{P}$-module M. We set q equal to the overdeterminacy degree of P_0, i.e., $\sigma(P_0) = n-d-1$ where d is the dimension of the algebraic variety $\mathfrak{M}(M)$. Then the condition of Theorem 5.3.16 is fulfilled for each convex (i.e., consisting of convex sets) open covering U of $X \setminus K$. In fact, $\text{Ext}^i(M, \mathcal{P}) = 0$ for $i \le n-d-1$ because of Lemma 5.3.18 (see below). And using Theorem 1.2.21 we conclude that $H_i(\mathcal{E}(U')) = 0$ $(i \le n-d-1)$ for every convex open set $U' \subset X$, as desired.

□

5.3.6 *Overdetermined homogeneous systems of differential equations with constant coefficients*

Let $P \in \text{do}_p(X \times \mathbb{C}^k \to X \times \mathbb{C}^l)$ be a differential operator with constant coefficients in a domain $X \subset \mathbb{R}^n$. Thus, it is given by an $(l \times k)$-matrix $P(D)$ of scalar partial differential operators with constant coefficients in X.

For the $\mathcal{P}$-module $M = \mathcal{P}^k / P(z)^t \mathcal{P}^l$, we denote by $r(0(M))$ the radical of zero submodule of M. That is, $r(0(M))$ is the ideal in the ring $\mathcal{P}$ consisting of all polynomials r with the following property: there exists an integer $i = i(r)$ such that $r^i \cdot p \in P(z)^t \mathcal{P}^l$ for all $p \in \mathcal{P}^k$.

The algebraic variety $\mathfrak{M}(M)$ in $\mathbb{C}^n$, formed by the common zeros of all polynomials of $r(0(M))$, is called the algebraic variety associated to M. It is known that $\mathfrak{M}(M) = \{z \in \mathbb{C}^n : \text{rank}_{\mathbb{C}} P(z) < k\}$ (see Palamodov [175, p.156]).

Lemma 5.3.18 *For every integer q $(0 \le q \le n)$, the union $\cup_{i \le q} \mathfrak{M}(\text{Ext}^i(M, \mathcal{P}))$ coincides with the union of all irreducible components of $\mathfrak{M}(M)$ of dimensions $\ge (n-q)$.*

Proof. See Palamodov [175, p.421].

□

Obviously, if a polynomial p vanishes on $\mathfrak{M}(M)$, then $p \in r(0(M))$. Also, $\mathfrak{M}(M) = \emptyset$ if and only if $M = 0$.

Definition 5.3.19 *The integer $\mathfrak{o}(P) = n - \dim \mathfrak{M}(M) - 1$ is said to be the overdeterminacy degree of the differential operator P.*

Let L be a plane of codimension $q \le \mathfrak{o}(P)$ in $\mathbb{R}^n$ which is given by an equation $T_1 x = c$ where $c \in \mathbb{R}^q$, and T_1 is a $(q \times n)$-matrix of real numbers of rank q. Choose a $((n-q) \times n)$-matrix of real numbers T_2 such that the matrix $T = \begin{pmatrix} T_1 \\ T_2 \end{pmatrix}$ is invertible. After the change of variables $y = Tx$, the plane L will be given by the equation $y' = c$ where $y' = (y_1, \ldots, y_q)$, $y''(y_{q+1}, \ldots, y_n)$ and $y = (y', y'')$. The differential operator $P(D_x)$ becomes the differential operator $\widetilde{P}(D_y) = P(T^t D_y)$ and the algebraic variety associated to the $\mathcal{P}$-module $\widetilde{M} = \mathcal{P}^k / \widetilde{P}(\zeta)^t \mathcal{P}^l$ is $\mathfrak{M}(\widetilde{M}) = (T^t)^{-1} \mathfrak{M}(M)$. Since $\dim \mathfrak{M}(\widetilde{M}) = \dim \mathfrak{M}(M) \le n-q-1$, it follows from the definition of the dimension of an algebraic variety that the polynomials $\zeta_{q+1}, \ldots, \zeta_n$ are algebraically dependent in the ring $\mathcal{P}/r(0(\widetilde{M}))$. In short, there exists a polynomial $r \ne 0$ of $(n-q)$ variables such that $r(\zeta'') \in r(0(\widetilde{M}))$.

Definition 5.3.20 *The plane L is said to be non-singular with respect to the differential operator P if there exists an elliptic polynomial $\tilde{r}(\zeta'') \in r(0(\widetilde{M}))$.*

Actually Definition 5.3.20 does not depend on the choice of the change of variables $y = Tx$.

Lemma 5.3.21 *The notion of non-singularity with respect to the differential operator P is well-defined, i.e., it does not depend on the particular choice of the matrix T.*

Proof. This is an easy exercise in linear algebra.

□

The following result provides a definition of non-singularity independent of the change of variables.

Proposition 5.3.22 *Let T_2 be a $((n-q) \times n)$-matrix of real numbers of rank $(n-q)$ satisfying $T_2 \cdot T_1^t = 0$. Then the plane L is non-singular with respect to P if and only if there exists an elliptic polynomial r of $(n-q)$ variables such that $r(T_2 z) = 0$ on $\mathfrak{M}(M)$.*

Proof. Consider the $(n \times n)$-matrix $T = \begin{pmatrix} T_1 \\ T_2 \end{pmatrix}$ where T_2 is given by the condition of the proposition.

Since

$$T\ T^t = \begin{pmatrix} T_1 \\ T_2 \end{pmatrix} \begin{pmatrix} T_1^t\ T_2^t \end{pmatrix} = \begin{pmatrix} T_1 T_1^t & 0 \\ 0 & T_2 T_2^t \end{pmatrix}, \tag{5.3.7}$$

we get $(\det T)^2 = \det(T_1 T_1^t) \cdot \det(T_2 T_2^t)$. Hence it follows that T is invertible because $\mathrm{rank}_{\mathbb{C}}(T_1 T_1^t) = q$ and $\mathrm{rank}_{\mathbb{C}}(T_2 T_2^t) = n - q$.

According to Lemma 5.3.21, we are allowed to use this matrix T in Definition 5.3.20. So the plane L is non-singular with respect to P if and only if there exists an elliptic polynomial $\tilde{r}(\zeta'')$ such that $\tilde{r}\left(((T^t)^{-1} z)''\right) = 0$ on $\mathfrak{M}(M)$.

In view of (5.3.7)

$$(T^t)^{-1} = \begin{pmatrix} (T_1 T_1^t)^{-1} & 0 \\ 0 & (T_2 T_2^t)^{-1} \end{pmatrix} \begin{pmatrix} T_1 \\ T_2 \end{pmatrix} = \begin{pmatrix} (T_1 T_1^t)^{-1}\ T_1 \\ (T_2 T_2^t)^{-1}\ T_2 \end{pmatrix}$$

so $((T^t)^{-1} z)'' = (T_2 T_2^t)^{-1} T_2 z$.

To complete the proof, it suffices to notice that the polynomials $\tilde{r}(\zeta'')$ and $r(\zeta'') = \tilde{r}\left((T_2 T_2^t)^{-1} \zeta''\right)$ are either elliptic or not elliptic simultaneously.

□

Clearly, the condition $T_2 \cdot T_1^t = 0$ is equivalent to the condition that $\{T_2 x = 0\}$ be the equation of the normal subspace to L in $\mathbb{R}^n$.

Definition 5.3.23 *The differential operator P is said to be elliptic in the wide sense if the space $\mathbb{R}^n$ is non-singular with respect to P.*

Of course, if P has injective symbol, then P is elliptic in the wide sense. Moreover, the following is true.

Lemma 5.3.24 *If the differential operator* det (P^*P) *is elliptic, then* P *is elliptic in the wide sense.*

Proof. It suffices to indicate an elliptic polynomial r such that $r \cdot p \in P(z)^t \mathcal{P}^l$ for every column $p \in \mathcal{P}^k$.

We assert that the polynomial $r(z) = \det\,(P^*(z)P(z))$ satisfies these conditions. In fact, r is elliptic. Moreover, for every $p \in \mathcal{P}^k$, we get

$$\begin{aligned} r \cdot p &= (r \cdot I_k)^t \cdot p \\ &= (\text{adj}\ (P^*(z)P(z)) \cdot (P^*(z)P(z)))^t \cdot p \\ &= P(z)^t\ (\text{adj}\ (P^*(z)P(z)) \cdot P^*(z))^t \cdot p \end{aligned}$$

so $r \cdot p \in P(z)^t \mathcal{P}^l$.

This finishes the proof.

□

As to the relation of Definition 5.3.23 with other definitions of ellipticity, we refer to the survey of Palamodov [174, §1].

Example 5.3.25 Consider the differential operator $P \in \text{do}_2(\mathbb{R}^4 \times \mathbb{C}^1 \to \mathbb{R}^4 \times \mathbb{C}^2)$ given by the matrix $P(D) = \begin{pmatrix} D_1^2 + D_3^2 + D_4^2 \\ D_2^2 - D_3^2 + D_4^2 \end{pmatrix}$. In this case, $\dim \mathfrak{M}(M) = 2$ and $\mathfrak{o}(P) = 1$. Using Lemma 5.3.24 we see that the differential operator P is elliptic in the wide sense. Prove that the hyperplane $\{x_1 = c\}$ is singular with respect to P. In order to do this, one must show that there exists no elliptic polynomial $r(z_2, z_3, z_4)$ vanishing on $\mathfrak{M}(M)$. In fact, let a polynomial r be independent of z_1 and equal to zero on $\mathfrak{M}(M)$. Then $r(z_2, z_3, z_4)$ vanishes on the set $\{z_2^2 - z_3^2 + z_4^2 = 0\}$. According to the Hilbert Theorem, there is an integer i and a polynomial $p(z_2, z_3, z_4)$ such that $r^i = (z_2^2 - z_3^2 + z_4^2) \cdot p$. Comparing the highest degree terms in the left-hand side and in the right-hand side of the equality, we conclude that the polynomial r is not elliptic, as desired. Finally we notice that the variety $\mathfrak{M}(M)$ is normally situated with respect to the coordinate system $(z_1, \ldots, z_4)$.

□

In practice the polynomial $r(\zeta'') \in r(0(\widetilde{M}))$ is obtained by excluding the variables ζ' from the system of algebraic equations $\text{rank}_{\mathbb{C}} P(T^t\zeta) < k$ in a certain way (see Van der Waerden [242, Ch.II]).

Suppose now that Π is a relatively compact ordinary polyhedron in X. Let $(\partial\Pi)_\nu$ $(\nu \in \mathcal{N})$ be some numeration of the $(n-1)$-dimensional edges (in other words, the sides) of Π with the orientations induced from $\partial\Pi$. For a multi-index $\mathcal{I} \in \mathcal{N}^{i+1}$ $(0 \le i \le n-1)$, we set $(\partial\Pi)_\mathcal{I} = (\partial\Pi)_{\iota_1} \cap \ldots \cap (\partial\Pi)_{\iota_{i+1}}$ so that $(\partial\Pi)_\mathcal{I}$ is a $(n-i-1)$ dimensional edge of Π. We choose orientations of $(\partial\Pi)_\mathcal{I}$ in such a way that the boundary of $(\partial\Pi)_\mathcal{I}$ is just $\cup_{\nu\in\mathcal{N}}(\partial\Pi)_{\nu\mathcal{I}}$, for all $\mathcal{I}$. It is also convenient to define $(\partial\Pi)_\mathcal{I} = \Pi$ for $\mathcal{I} \in \mathcal{N}^0$ $(= \emptyset?)$.

Theorem 5.3.26 *Let the edges $(\partial\Pi)_{\mathcal{I}}$ $(\mathcal{I} \in \mathcal{N}^{i})$ be non-singular with respect to P for every $i = 0, \dots q$, where $q \le \mathfrak{o}(P)$ is fixed. Then there is a bidifferential operator $G_q \in do_{g_q}((X \times \mathbb{C}^l, X \times \mathbb{C}^{l*}) \to \Lambda^{n-q-1})$ and a kernel-valued q-cochain $K_{\mathcal{I}} \in C_{loc}(\Pi \times (\partial\Pi)_{\mathcal{I}})$ $(\mathcal{I} \in \mathcal{N}^{q+1})$ such that, for each solution $f \in [C^{g_q}(\overline{\Pi})]^k$ of $Pf = 0$ in Π, we have*

$$f(x) = -(-1)^{\frac{q(q+1)}{2}} \sum\nolimits'_{\mathcal{I} \in \mathcal{N}^{q+1}} \int_{(\partial\Pi)_{\mathcal{I}}} G_q\left(K_{\mathcal{I}}(x,\cdot), f\right) \quad (x \in \Pi). \qquad (5.3.8)$$

Proof. We shall prove the theorem by induction on q.

Let $q = 0$. Since the space $\mathbb{R}^n$ is non-singular with respect to P, there exists an elliptic polynomial $r(z) \in r(0(M))$. By the definition of the ideal $r(0(M))$, we can find a natural number $i = i(r)$ such that $r^i \cdot I_k \in P(z)^t \mathcal{P}^{(l \times k)}$, i.e., $r^i \cdot I_k = \tilde{P}P$ for some $(k \times l)$-matrix of polynomials $\tilde{P}$. Consider the scalar differential operator $\Delta = (r(D))^i$. It is elliptic of order $g_0 + 1 = i \cdot \deg r$ because of the ellipticity of r, and each section f satisfying $Pf = 0$ satisfies also $\Delta f = 0$. Denote by $g(x)$ a fundamental solution of convolution type for the differential operator Δ. If $f \in [C^{g_0}(\overline{\Pi})]^k$ satisfies $Pf = 0$ in Π, then according to Green's formula (2.5.9) we obtain

$$f(x) = -\int_{\partial\Pi} G_\Delta\left(g(x - \cdot), f\right) = -\sum_{\nu \in \mathcal{N}} \int_{(\partial\Pi)_\nu} G_\Delta\left(g(x - \cdot), f\right) \quad (x \in \Pi).$$

This is just a representation of the type (5.3.8) for $q = 0$.

Assume now that Theorem 5.3.26 is valid with $(q-1)$ instead of q. Then there exists a bidifferential operator $G_{q-1} \in \mathrm{do}_{g_{q-1}}((X \times \mathbb{C}^l, X \times \mathbb{C}^{l*}) \to \Lambda^{n-q})$ and a kernel-valued $(q-1)$-cochain $K_{\mathcal{I}} \in C^\infty_{loc}(\Pi \times (\partial\Pi)_{\mathcal{I}})$ $(\mathcal{I} \in \mathcal{N}^q)$ such that, for each solution $f \in [C^{g_{q-1}}(\overline{\Pi})]^k$ of $Pf = 0$ in Π, we have

$$f(x) = -(-1)^{\frac{(q-1)q}{2}} \sum\nolimits'_{\mathcal{I} \in \mathcal{N}^q} \int_{(\partial\Pi)_{\mathcal{I}}} G_{q-1}\left(K_{\mathcal{I}}(x,\cdot), f\right) \quad (x \in \Pi).$$

Let $\mathcal{I} \in \mathcal{N}^q$ be fixed and the edge $(\partial\Pi)_{\mathcal{I}}$ belong to a plane L of codimension q in $\mathbb{R}^n$ given by the equation $T_1 x = c$ where $c \in \mathbb{R}^q$, and T_1 is a $(q \times n)$-matrix of real numbers of the rank q. Choose a $((n-q) \times n)$-matrix of real numbers T_2 of the rank $(n-q)$ such that $T_2 T_1^t = 0$. Perform the change of variables $y = Tx$ where $T = \begin{pmatrix} T_1 \\ T_2 \end{pmatrix}$. After this change, the edge $(\partial\Pi)_{\mathcal{I}}$ will lie in the plane $L = \{y' = c\}$ of variables $y'' \in \mathbb{R}^{n-q}$.

By hypothesis the plane L is non-singular with respect to P, i.e., there exists an elliptic polynomial $r(\zeta'') \in r(0(\widetilde{M}))$. Hence we can find a natural number $i = i(r)$ such that $r^i \cdot I_k \in P(T^t\zeta)^t \mathcal{P}^{(l \times k)}$, i.e., $r^i \cdot I_k = \tilde{P} \cdot P(T^t\zeta)$ for some $(k \times l)$-matrix of polynomials $\tilde{P}$. Consider the differential operator $\Delta = (r(D_{y''}))^i$. It is elliptic of order $i \cdot \deg r$ with respect to the variables y'', and each section f satisfying $P(T^t \cdot D_y)f = 0$ satisfies also $\Delta(f(c, y'')) = 0$.

Set $g_q = g_{q-1} + (\text{order } \Delta - 1)$, and consider a section $f \in [C^{g_q}(\overline{\Pi})]^k$ satisfying $Pf = 0$ in Π.

For every multi-index α $(|\alpha| \le g_{q-1})$, the derivative $D^\alpha f$ satisfies $P(D^\alpha f) = 0$ in Π because the differential operator P has constant coefficients. So $P(T^t \cdot$

$D_y)(D^\alpha f(T^{-1}y)) = 0$ in Π. As mentioned, this implies that the restriction of $D^\alpha f(T^{-1}y)$ to $(\partial\Pi)_\mathcal{I}$ satisfies the equation $\Delta(\cdot) = 0$. Therefore, if $g(y'')$ is a fundamental solution of convolution type for the differential operator $\Delta(D_{y''})$, then according to Green's formula (2.5.9) we obtain for y'' in the interior of $(\partial\Pi)_\mathcal{I}$:

$$D^\alpha f(T^{-1}y)|_{y'=c} = -\int_{boundary\ ((\partial\Pi)_\mathcal{I})} G_\Delta\left(g(y''-w''), D^\alpha f(T^{-1}w)|_{w'=c}\right). \quad (5.3.9)$$

Under the change of variables $y = Tx$, we have $D_{y''} = (T_2T_2^t)^{-1}T_2 \cdot D_x$. Denote by $\tilde{\Delta}$ the differential operator $\Delta((T_2T_2^t)^{-1}T_2 \cdot D_x)$, so that $\tilde{\Delta}$ is a differential operator with constant coefficients in X. Returning to the variables $x = T^{-1}y$ in formula (5.3.9), we get for x in the interior of $(\partial\Pi)_\mathcal{I}$:

$$D^\alpha f(x) = -\sum_{\nu\in\mathcal{N}} \int_{(\partial\Pi)_{\nu\mathcal{I}}} G_{\tilde{\Delta}}\left(g(T_2(x-\cdot)), D^\alpha f\right). \quad (5.3.10)$$

Since $T^t(T^t)^{-1} = T_1^t(T_1T_1^t)^{-1}T_1 + T_2^t(T_2T_2^t)^{-1}T_2 = I_n$, we see that

$$\left|(x-y) - T_1^t(T_1T_1^t)^{-1}T_1(x-y)\right| \le \left\|T_2^t(T_2T_2^t)^{-1}\right\| \; |T_2(x-y)|$$

where $\|\cdot\|$ means the norm of a linear mapping.

In particular,

$$|T_2(x-y)| \ge \frac{1}{\|T_2^t(T_2T_2^t)^{-1}\|}\,|x-y| \qquad \text{for } x, y \in L,$$

so that the derivatives up to order (order $\Delta - 1$) of $g(T_2(x-y))$ are locally integrable with respect to $x \in L$, for each fixed $y \in L$. Hence it follows that we may by the Fubini Theorem rearrange the integrals $\int_{(\partial\Pi)_\mathcal{I}}$ and $\int_{(\partial\Pi)_{\nu\mathcal{I}}}$ when substituting the derivatives $D^\alpha f$ $(|\alpha| \le g_{q-1})$ of (5.3.10) into the equality (5.3.8) with $(q-1)$ instead of q. Then we get precisely formula (5.3.8) with the kernels $K_{\nu\mathcal{I}}$ being linear combinations of the integrals $\int_{(\partial\Pi)_\mathcal{I}} D^{\alpha'}K_\mathcal{I}(x,\cdot)\cdot D^{\alpha''}g(T_2(\cdot - y))$, where $|\alpha''| \le$ order $\Delta - 1$. Hence the desired properties of $K_\mathcal{I}$ $(\mathcal{I} \in \mathcal{N}^{q+1})$ follow.

□

One can see from the proof that the exponent g_q may be found by using elimination theory.

Corollary 5.3.27 *Under the condition of Theorem 5.3.26, if the j-th component of the solution f vanishes up to order g_q on the $(n-q-1)$-dimensional skeleton of Π for some $1 \le j \le k$, then this component vanishes everywhere in Π.*

Proof. This follows from formula (5.3.8) immediately.

□

According to Theorem 2 of Palamodov [175, p.457], if Π is a bounded ordinary polyhedron in $\mathbb{R}^n$ and f is a weak solution of the system $Pf = 0$ in a neighborhood of $\overline{\Pi}$ which vanishes in a neighborhood of the $(n - \mathfrak{o}(P) - 1)$-dimensional skeleton of Π, then $f \equiv 0$ in Π. Thus, Corollary 5.3.27 sharpens this general result.

5.3.7 A Cauchy-Fantappiè formula for Koszul complexes

We assume that the variables $x = (x_1, \ldots, x_n)$ of $\mathbb{R}^n$ are partitioned into N disjoint groups $x^{(1)} \in \mathbb{R}^{n_1}, \ldots, x^{(N)} \in \mathbb{R}^{n_N}$ such that $n_1 + \ldots + n_N = n$. Let $P^{(i)} = P^{(i)}(D_{x^{(i)}})$ $(1 \le i \le N)$ be scalar differential operators with constant coefficients in $\mathbb{R}^n$ and each $P^{(i)}$ be elliptic of order $p^{(i)}$ with respect to the group $x^{(i)}$ of variables.

Denote by $E^{\cdot} = \{E^q, P_q\}$ the Koszul complex in $\mathbb{R}^n$ related to the differential operators $\{P^{(i)}\}$ by the formulas $E^q = \mathbb{R}^n \times \mathbb{C}^{k_q}$ where $k_q = \binom{N}{q}$, and (1.2.22). In particular, the differential operator $P = P_0$ is given by the column $\begin{pmatrix} P^{(1)} \\ \ldots \\ P^{(N)} \end{pmatrix}$ (see Subsection 1.2.7 for details). It is easy to see that the differential operator P is elliptic in the wide sense, and $\mathfrak{o}(P) = N - 1$.

We start with a version of Theorem 5.3.26 for solutions of the system $Pf = 0$. For $1 \le i \le N$, we denote by $g^{(i)}(x^{(i)})$ a fundamental solution of convolution type of the differential operator $P^{(i)}(D_{x^{(i)}})$. Each distribution $g^{(i)}$ is actually a real analytic function outside the plane $\{x^{(i)} = 0\}$ in $\mathbb{R}^n$, and $g^{(i)}$ has a singularity of finite order on the plane.

Suppose that Π_i $(1 \le i \le N)$ are bounded domains in $\mathbb{R}^{n_i}$ with piecewise smooth boundaries, and $\Pi = \Pi_1 \times \ldots \times \Pi_N$.

Lemma 5.3.28 *For each solution $f \in \mathcal{E}_P(\Pi)$ whose derivatives $D^{\alpha^{(1)}}_{x^{(1)}} \ldots D^{\alpha^{(N)}}_{x^{(N)}} f$ $(|\alpha^{(1)}| \le p^{(1)} - 1, \ldots, |\alpha^{(N)}| \le p^{(N)} - 1)$ are continuous on $\Pi \cup (\partial\Pi_1 \times \ldots \times \partial\Pi_N)$, the following formula holds:*

$$f(x) = (-1)^N \int_{\partial\Pi_1 \times \ldots \times \partial\Pi_N} G_{P^{(1)}}\left(g^{(1)}(x^{(1)} - y^{(1)}), \ldots, G_{P^{(N)}}(g^{(N)}(x^{(N)} - y^{(N)}), f(y)) \ldots\right) \qquad (x \in \Pi). \tag{5.3.11}$$

Proof. Formula (5.3.11) is obtained by successively applying Green's formula (2.5.9) to each of the domains $\{\Pi_i\}$. Because of the continuity of the appropriate derivatives of f on the skeleton $\partial\Pi_1 \times \ldots \times \partial\Pi_N$, the repeated integral may be replaced by the multiple one, as desired.

□

For holomorphic functions of several variables, (5.3.11) is the well-known Cauchy formula (see Gunning and Rossi [96, p.14]).

Exercise 5.3.29 *Prove that, for each choice of open sets $U_i \subset \mathbb{R}^{n_i}$ $(i = 1, \ldots, N)$, the complex $\mathcal{E}(U_1 \times \ldots \times U_N)$ is exact at dimensions $< N$.*

Hint Use the Künneth formula (1.4.3).

□

Let K_i $(i = 1, \ldots, N)$ be compact subsets of Π_i in $\mathbb{R}^{n_i}$, and $K = K_1 \times \ldots \times K_N$. Consider the set of indices $\mathcal{N} = \{1, \ldots, N\}$. For $\nu \in \mathcal{N}$, if we define $U_\nu = \{x \in \mathbb{R}^n : x^{(\nu)} \notin K_\nu\}$, or

$$U_\nu = \mathbb{R}^{n_1} \times \ldots \times \mathbb{R}^{n_{\nu-1}} \times (\mathbb{R}^{n_\nu} \setminus K_\nu) \times \mathbb{R}^{n_{\nu+1}} \times \ldots \times \mathbb{R}^{n_N},$$

then $U = \{U_\nu\}_{\nu \in \mathcal{N}}$ is an open covering of $Y = \mathbb{R}^n \setminus K$.

Because of (5.2.4), we have a sequence of Weil homomorphisms corresponding to the covering U:

$$H^{N-1}(U, \mathcal{E}_{P'_{N-1}}) \xrightarrow{HM} H^{N-2}(U, \mathcal{E}_{P'_{N-2}}) \xrightarrow{HM} \dots \xrightarrow{HM} H^1(U, \mathcal{E}_{P'_1}) \xrightarrow{H(hM)} H_1(\mathcal{E}(Y)). \tag{5.3.12}$$

Remark 5.3.30 *It follows from Theorem 5.2.12 and Exercise 5.3.29 that each mapping in the sequence (5.3.12) is an isomorphism.*

Set $g(x) = g^{(1)}(x^{(1)}) \cdot \ldots \cdot g^{(N)}(x^{(N)})$. Since $g^{(i)}$ is a fundamental solution of the differential operator $P^{(i)}$, the function $g(x - \cdot)$ is in $\mathcal{E}_{P'_{N-1}}(U_1 \cap \ldots \cap U_N)$ for each fixed point $x \in K$. Moreover we have $C^N(U, \mathcal{E}_{P'_{N-1}}) = 0$ because U consists of N elements. Therefore $Z^{N-1}(U, \mathcal{E}_{P'_{N-1}}) = C^{N-1}(U, \mathcal{E}_{P'_{N-1}})$, and $g(x - \cdot)$ determines a unique cocycle in $Z^{N-1}(U, \mathcal{E}_{P'_{N-1}})$ which will also be denoted by $g(x - \cdot)$.

Theorem 5.3.31 *For any solution $f \in \mathcal{E}_P(\Pi)$ whose derivatives $D^{\alpha^{(1)}}_{x^{(1)}} \ldots D^{\alpha^{(N)}}_{x^{(N)}} f$ ($|\alpha^{(1)}| \le p^{(1)} - 1, \ldots, |\alpha^{(N)}| \le p^{(N)} - 1$) are continuous on $\overline{\Pi}$, the following formula holds:*

$$f(x) = -(-1)^{\frac{(N-1)N}{2}} \int_{\partial\Pi} G_P(W_{N-1}\, g(x - \cdot), f) \quad (x \in K). \tag{5.3.13}$$

Proof. Rewrite the integrand in the right-hand side of (5.3.11) in the following way

$$\begin{aligned}
&(-1)^N\, G_{P^{(1)}}\Big(g^{(1)}(x^{(1)} - y^{(1)}), \ldots, G_{P^{(N)}}(g^{(N)}(x^{(N)} - y^{(N)}), f(y)) \ldots\Big) \\
&= (\sqrt{-1})^N \sum\nolimits'_{\substack{|\beta^{(1)}+\gamma^{(1)}+1_{j_1}| \le p^{(1)} \\ |\beta^{(N)}+\gamma^{(N)}+1_{j_N}| \le p^{(N)}}} (-1)^{|\beta^{(1)}|+\ \ +|\beta^{(N)}|} \\
&\quad \times \Big(D^{\beta^{(1)}}_{y^{(1)}} \ldots D^{\beta^{(N)}}_{y^{(N)}} g(x - y)\Big)\Big(P^{(1)}_{\beta^{(1)}+\gamma^{(1)}+1_{j_1}} \cdots P^{(N)}_{\beta^{(N)}+\gamma^{(N)}+1_{j_N}}\Big)\Big(D^{\gamma^{(1)}}_{y^{(1)}} \ldots D^{\gamma^{(N)}}_{y^{(N)}} f\Big) \\
&\qquad \times \Big(\star dy^{(1)}_{j_1} \wedge \ldots \wedge \star dy^{(N)}_{j_N}\Big) \\
&= -G_{N-1}(g(x - y), f(y)),
\end{aligned}$$

where G_{N-1} is a bidifferential operator of type $(E^{N\prime}, E^0) \to \Lambda^{n-N}$ and of order $(p^{(1)} + \ldots + p^{(N)} - N)$ in $\mathbb{R}^n$.

Since the covering U consists just of N elements, we may define a unique $(N-1)$-cochain f^{N-1} for U with values in $\mathcal{E}'^N$ by setting

$$\langle g, f^{N-1}_{(1,\ \ ,N)} \rangle = \int_{\partial\Pi_1 \times\ \ \times \partial\Pi_N} G_{N-1}(g, f) \quad \text{for all} \quad g \in \mathcal{E}_N(U_{(1,\ \ ,N)}).$$

Then the proof consists of verifying the fact that the homotopy class of f^{N-1} in $H_{N-1}(U, \mathcal{E}'^N / P\mathcal{E}'^{N-1})$ is the image under the adjoint Weil homomorphism of the cohomology class of $[\partial\Pi]^P f$ in $H^1(\mathcal{E}'(Y))$.

For $\nu \in \mathcal{N}$, we define

$$(\partial\Pi)_\nu = \overline{\Pi_1} \times \ldots \times \overline{\Pi_{\nu-1}} \times \partial\Pi_\nu \times \overline{\Pi_{\nu+1}} \times \ldots \times \overline{\Pi_N},$$

and for a strictly increasing multi-index $\mathcal{I} \in \mathcal{N}^{i+1}$ $(0 \le i \le N-1)$, we set $(\partial\Pi)_\mathcal{I} = (\partial\Pi)_{i_1} \cap \ldots \cap (\partial\Pi)_{i_{i+1}}$. Finally, we extend the definition of $(\partial\Pi)_\mathcal{I}$ to all multi-indices $\mathcal{I}$ by skew-symmetry. The orientations of $(\partial\Pi)_\mathcal{I}$ are chosen in such a way that the boundary of $(\partial\Pi)_\mathcal{I}$ coincides with $\cup_{\nu\in\mathcal{N}}(\partial\Pi)_{\nu\mathcal{I}}$ for all $\mathcal{I}$.

Since $(\partial\Pi)_\mathcal{I} \subset U_\mathcal{I}$ for all $\mathcal{I}$, we have endowed Π with the structure similar to the one of a C^∞ polyhedron in general position with respect to the covering U.

Direct calculations show that there exist bidifferential operators G_i of type $(E^{i+1\prime}, E^0) \to \Lambda^{n-i-1}$ and $\widetilde{G}_i$ of type $(E^{i+2\prime}, E^1) \to \Lambda^{n-i-1}$ $(i = 0, \ldots, N-2)$ such that $dG_i(g, f) = G_{i-1}(P'g, f) + \widetilde{G}_{i-1}(g, Pf)$ for all $g \in \mathcal{E}_{i+1}(X)$ and $f \in \mathcal{E}^0(X)$ $(i = 1, \ldots, N-1)$, and G_0 is moreover a Green operator for the differential operator P_0 (cf. Corollary 2.4.13).

Define now the chains $f^i \in C_i(U, \mathcal{E}^{n+1})$ $(i = 0, \ldots, N-2)$ by letting

$$\left\langle g, f^i_\mathcal{I}\right\rangle_{U_\mathcal{I}} = \int_{(\partial\Pi)_\mathcal{I}} G_i(g, f) \quad \text{for all} \quad g \in \mathcal{E}_{i+1}(U_\mathcal{I}) \quad (\mathcal{I} \in \mathcal{N}^{i+1}).$$

It follows from Theorem 5.3.7 that the sequence $\{\overline{f^i}\}_{i=0,\ldots,N-1}$ is an adjoint sequence of Weil representatives of the class $\left[[\partial\Pi]^P f\right] \in H^1(\mathcal{E}'^{\cdot}(Y))$, as desired.

To complete the proof, it suffices to invoke Lemma 5.3.28 and Theorem 5.2.37. □

We shall now make use of the construction of the Weil homomorphism. Let $\{\varphi_\nu\}_{\nu\in\mathcal{N}}$ be a partition of unity on Y subordinated to the covering U. According to (5.2.3), we get

$$\begin{aligned}(-1)^{\frac{(N-1)N}{2}} & W_{N-1}\, g(x - \cdot) \\ = & \sum_{\nu_0,\ldots,\nu_{N-1}=1}^{N} \varphi_{\nu_{N-1}}(P_{N-2}(\varphi_{\nu_{N-2}}(\ldots P_0(\varphi_{\nu_0} g_{(\nu_0,\ldots,\nu_{N-1})}(x - \cdot))\ldots)))).\end{aligned} \tag{5.3.14}$$

As follows from Definition 5.2.6, the condition $\operatorname{supp} \varphi_\nu \subset U_\nu$ $(\nu \in \mathcal{N})$ can be omitted provided $M^i g(x - \cdot)$ is in $C^{N-i-1}(U, \mathcal{E}_{N-i})$ for every $i = 0, \ldots, N-1$. Thus, the sequence of Weil representatives of the cohomology class $[g(x - \cdot)] \in H^{N-1}(U, \mathcal{E}_{P'_{N-1}})$ can be chosen in such a way that the kernel in formula (5.3.13) will contain N rather arbitrary functions $\{\varphi_1, \ldots, \varphi_N\}$ connected by the single condition $\sum_{\nu=1}^N \varphi_\nu = 1$ on Y.

Remark 5.3.32 *Supposing $p^{(1)} = \ldots = p^{(N)} = 1$, we choose smooth functions $\psi_\nu(x, \cdot)$ $(\nu = 1, \ldots, N)$ such that $\psi_\nu(x, \cdot) = \sum_{\nu=1}^N \frac{\psi_\nu(x,\cdot)}{g^{(\nu)}(x^{(\nu)} -)}$ does not vanish on Y. Then one can take $\varphi_\nu(x, \cdot) = \frac{1}{\psi(x,\cdot)} \cdot \frac{\psi_\nu(x,\cdot)}{g^{(\nu)}(x^{(\nu)} - \cdot)}$ $(\nu = 1, \ldots N)$, for fixed $x \in K$.*

As Harvey [98] proved, for holomorphic functions of several variables this path leads to the well-known formula of Leray [148] called the Cauchy-Fantappiè formula by Leray himself.

As to the applications of Theorem 5.3.31 for the de Rham complex, we refer to the author's papers [221] and [229], and to Section 6.1 below.

Finally, we remark that one can replace the cycle $\partial\Pi$ in (5.3.13) by any $(n-1)$-dimensional cycle in Y, homologous to $\partial\Pi$ in the domain of the solution f, because $W_{N-1}g(x-\cdot)$ satisfies the transposed equation $P'(\cdot)=0$ in Y.

5.3.8 *Čech parametrices*

Let E be an elliptic complex on a differentiable manifold X, countable at infinity. As Toledo and Tong [239] observed, Čech cohomology gives a very natural way of constructing a global parametrix for E from local ones without introducing an auxiliary Laplacian. Non-uniqueness of local parametrices gives rise to a sequence of kernel-valued Čech cochains called a Čech parametrix. A global parametrix may be obtained quite easy from here.

Let us recall that, in order to define a parametrix for E, one constructs the complex $(E \boxtimes E')$ on $X\times X$ whose grading is from $-N$ to N, and whose differential $d_{E\boxtimes E'}$ is denoted simply by $\widehat{P}$ (see Subsection 1.5.6). Then global parametrices of E are solutions $K\in\mathcal{D}'((E\boxtimes E')^{-1})$ of the equation $\widehat{P}K=\operatorname{tr}\Delta^*$ modulo smooth kernels on $X\times X$.

In is not true in general that the complex $(E \boxtimes E')$ is elliptic. However it inherits the main property of E.

Lemma 5.3.33 *The complex* $(E \boxtimes E')$ *has a properly supported pseudo-differential parametrix.*

Proof. In fact, let Π be a properly supported pseudo-differential parametrix of the complex E whose existence is guaranteed by Theorem 2.1.12. Then the tensor product $\Pi\otimes\Pi'$, defined in an appropriate way, provides a properly supported pseudo-differential parametrix of the complex $(E \boxtimes E')$, as desired.

□

We shall apply the construction of Weil's homomorphism to the complex $(E \boxtimes E')$. Denote by $\mathcal{E}^r$ (respectively, $\mathcal{D}'^r$) the sheaf of germs of differentiable sections (respectively, of distributional sections) of the bundle $(E\boxtimes E')^r$ over $X\times X$. Lemma 5.3.33 implies, because of Theorem 2.1.9, that the cohomology of the complexes $\mathcal{E}(X\times X)$ and $\mathcal{D}'(X\times X)$ are actually the same, i.e.,

$$H(\mathcal{E}(X\times X)) \overset{top}{\cong} H(\mathcal{D}'(X\times X)). \tag{5.3.15}$$

Let $U=\{U_\nu\}_{\nu\in\mathcal{N}}$ be an open covering of $X\times X$, and for every $\nu\in\mathcal{N}$ let a local parametrix $K^0_\nu\in\mathcal{D}'^{-1}(U_\nu)$ of the complex E be given, i.e., $\widehat{P}K^0_\nu=\operatorname{tr}\Delta^*$ modulo smooth kernels on U_ν. We would like to construct a global parametrix of E by "pasting together" the local parametrices in the following way.

One can consider K^0 as a O-cochain for U with values in $\mathcal{D}'^{-1}$. Choose a partition of unity $\{\varphi_\nu\}_{\nu\in\mathcal{N}}$ on $X\times X$ subordinated to the covering U, and set $\Pi^0=hK^0=\sum_{\nu\in\mathcal{N}}\varphi_\nu K^0_\nu$. Then $\Pi^0\in\mathcal{D}'^{-1}(X\times X)$, and

$$\begin{aligned}\widehat{P}\Pi^0 &= h(\widehat{P}K^0)+(\widehat{P}h-h\widehat{P})K^0\\ &= \operatorname{tr}\Delta^*+MK^0 \pmod{\mathcal{E}^0(X\times X)}.\end{aligned}$$

The kernel K^0 is not a parametrix of E because we can not guarantee that the remainder $MK^0 \in \mathcal{D}'^0(X \times X)$ is smooth. However the remainder has a very special form since we get $MK^0 = (h\delta)MK^0 = -(hM)\delta K^0 = W_1(\delta K^0)$ according to Lemmas 5.2.2 and 5.2.4.

We obtain the following approximation of the parametrix by setting $\Pi^1 = \Pi^0 + W_1(K^1)$ where the cochain $K^1 \in C^1(U, \mathcal{D}'^{-2})$ is a solution of the equation $\widehat{P}K^1 = \delta K^0 \pmod{C^1(U, \mathcal{E}^{-1})}$. Indeed, because of Lemma 5.2.10

$$\begin{aligned} \widehat{P}\Pi^1 &= \widehat{P}\Pi^0 + \widehat{P}W_1(K^1) \\ &= \operatorname{tr}\Delta^* + W_1(\delta K^0) + W_2(\delta K^1) - W_1(\widehat{P}K^1) \quad \pmod{\mathcal{E}^0(X \times X)} \\ &= \operatorname{tr}\Delta^* + W_2(\delta K^1) \quad \pmod{\mathcal{E}^0(X \times X)}. \end{aligned}$$

Let us consider the question of solvability of the equation

$$\widehat{P}K^1 = \delta K^0 \quad \pmod{C^1(U, \mathcal{E}^{-1})}.$$

We assert that the equation is solvable provided $H^{-1}(\mathcal{E}\,(U_{\mathcal{I}})) = 0$ for all $\mathcal{I} \in \mathcal{N}^2$. In fact, $\widehat{P}(\delta K^0) = \delta(\widehat{P}K^0)$ is of class $C^1(U, \mathcal{E}^0)$. Because of (5.3.15), there exists a cochain $S \in C^1(U, \mathcal{E}^{-1})$ such that $\widehat{P}S = \widehat{P}(\delta K^0)$. Using now the hypothesis and the isomorphism (5.3.15), we find a cochain $K^1 \in C^1(U, \mathcal{D}'^{-2})$ satisfying $\widehat{P}K^1 = \delta K^0 - S$, as desired.

It is now clear how to proceed. Step by step, for $\imath = 2, \dots, N-1$, we solve the following equations with respect to $K^\imath \in C^\imath(U, \mathcal{D}'^{(-\imath-1)})$:

$$\widehat{P}K^\imath = \delta K^{\imath-1} \pmod{C^\imath(U, \mathcal{E}^{-\imath})}. \tag{5.3.16}$$

The equation (5.3.16) is solvable provided $H^{-\imath}(\mathcal{E}\,(U_{\mathcal{I}})) = 0$ for all multi-indices $\mathcal{I} \in \mathcal{N}^{\imath+1}$. In fact, assuming the previous equation to be solvable, we see that $\widehat{P}(\delta K^{\imath-1}) = \delta(\widehat{P}K^{\imath-1})$ is of class $C^\imath(U, \mathcal{E}^{-(\imath-1)})$. Because of (5.3.15), there is a cochain $S \in C^\imath(U, \mathcal{E}^{-\imath})$ such that $\widehat{P}S = \widehat{P}(\delta K^{\imath-1})$. Finally, using the hypothesis and the isomorphism (5.3.15), we find a cochain $K^\imath \in C^\imath(U, \mathcal{D}'^{(-\imath-1)})$ satisfying $\widehat{P}K^\imath = \delta K^{\imath-1} - S$, as desired.

Definition 5.3.34 *A sequence $\{K^\imath\}_{\imath=0,\dots,N-1}$ of cochains $K^\imath \in C^\imath(U, \mathcal{D}'^{(-\imath-1)})$ satisfying (5.3.16) for $\imath = 0, \dots, N-1$ (if $\imath = 0$, then $\delta K^{\imath-1}$ is replaced by tr Δ^*), is called a Čech parametrix of E.*

The main result of this section is the following.

Theorem 5.3.35 *If $\{K^\imath\}_{\imath=0,\dots,N-1}$ is a Čech parametrix of E, then*

$$\Pi = \sum_{\imath=0}^{N-1} (-1)^{\frac{(\imath-1)\imath}{2}} W_\imath(K^\imath)$$

is a parametrix of the same complex.

Proof. First of all, we have $\Pi \in \mathcal{D}'^{-1}(X \times X)$. Using Lemma 5.2.10, we get

$$\begin{aligned}
\hat{P}\Pi &= \sum_{\imath=0}^{N-1} (-1)^{\frac{(\imath-1)\imath}{2}} \hat{P} W_\imath(K^\imath) \\
&= \sum_{\imath=0}^{N-1} (-1)^{\frac{(\imath-1)\imath}{2}} \left(W_{\imath+1}(\delta K^\imath) + (-1)^\imath W_\imath(\hat{P} K^\imath) \right) \\
&= W_0(\mathrm{tr}\ \Delta^*) + \sum_{\imath=0}^{N-1} (-1)^{\frac{\imath(\imath+1)}{2}} W_\imath(\hat{P} K^\imath - \delta K^{\imath-1}) - (-1)^{\frac{N(N+1)}{2}} W_N(\delta K^{N-1}).
\end{aligned}$$

Hence, in view of (5.3.16)

$$\hat{P}\Pi = \mathrm{tr}\ \Delta^* - (-1)^{\frac{N(N+1)}{2}} W_N(\delta K^{N-1}) \pmod{\mathcal{E}^0(X \times X)}.$$

Consider the chain δK^{N-1} of $C^N(U, \mathcal{D}'^{-N})$. Since $\hat{P}(\delta K^{N-1}) = \delta(\hat{P} K^{N-1})$ is in $C^N(U, \mathcal{E}^{-(N-1)})$ because of (5.3.16) and the differential operator $\widehat{P_{-N}}$ is hypoelliptic, we conclude that δK^{N-1} lies actually in $C^N(U, \mathcal{E}^{-N})$. Therefore, $W_N(\delta K^{N-1})$ belongs to $\mathcal{E}^0(X \times X)$. Thus Π is a parametrix of $E^\cdot$.

The proof is complete.

□

Notice that, if $H^{-\imath}(\mathcal{E}\,(U_{\mathcal{I}})) = 0 \quad (\mathcal{I} \in \mathcal{N}^{\imath+1})$ for every $\imath = 0, \dots, N-1$, then all the equations $\tilde{P} K^\imath = \delta K^{\imath-1} \quad (i = 0, \dots, N-1)$ can be solved exactly and we obtain a parametrix Π of $E^\cdot$ with $\hat{P}\Pi = \mathrm{tr}\ \Delta^* - (-1)^{\frac{N(N+1)}{2}} W_N(\delta K^{N-1})$.

Remark 5.3.36 *This follows also from Theorem 5.2.12, according to which* $HW: H^N(U, \mathcal{E}_{\widehat{P_{-N}}}) \to H^0(\mathcal{D}'(X \times X))$ *is an epimorphism.*

5.4 Grothendieck's Theorem on Cohomology Classes Regular at Infinity

5.4.1 *History of the question*

It is well-known that each holomorphic function outside a compact subset of the complex plane may be uniquely represented as the sum of an entire function and a function vanishing at infinity. This elementary, but very useful observation was extended by Grothendieck [89] to solutions of arbitrary elliptic systems. Just as one might expect, the analogy spreads much further. It turns out that one obtains complete results in the context of differential complexes (see author's [233] paper).

5.4.2 *Sections regular at infinity*

Assume that $E^\cdot$ is a differential complex on X which has a biregular fundamental solution at degree q which we denote by Φ and fix. Let K be a compact subset of X.

Definition 5.4.1 *A section $f \in Z^q(\mathcal{D}'(E^\cdot|_{X\setminus K}))$ is called regular at infinity if for each section $g \in Z_q(\mathcal{D}(E^{\cdot\prime}))$ and for each function $\varphi \in \mathcal{D}(X)$ being equal to* 1 *in a neighborhood of $K \cup supp\ g$,*

$$\left\langle \Phi'_{q+1} g, P_q(\varphi f)\right\rangle_{X\setminus K} = 0. \tag{5.4.1}$$

We note that Definition 5.4.1 depends essentially on the choice of a fundamental solution Φ.

If $f \in Z^q(\mathcal{E}(E^\cdot|_{X\setminus K}))$ then formula (5.4.1) is valid when $g \in Z_q(\mathcal{E}'(E^{\cdot\prime}))$ or φ is replaced by the characteristic function of some precompact neighborhood U of the set $K \cup \operatorname{supp} g$. Then the integral in the left-hand side of (5.4.1) is equal to

$$\left\{-\int_{\partial U} G_{P_q}(\Phi'_{q+1} g, f)\right\}.$$

Notice further, since $g = P'_q \Phi'_{q+1} g$, that $\Phi'_{q+1} g$ is a solution of the equation $P'_q v = 0$ on the open set $X \setminus \operatorname{supp} g$. Moreover, the integral in the left-hand side of (5.4.1) may be computed also on $X \setminus (K \cup \operatorname{supp} g)$ because $P_q(\varphi f) = 0$ on $(X \setminus K) \cap \operatorname{supp} g$. Hence the left-hand side of (5.4.1) does not depend on the choice of φ and we can take φ equal to 1 on a compact set as large as possible. Thus, if the section $f \in Z^q(\mathcal{D}'(E^\cdot|_{X\setminus K}))$ is regular at infinity, then its restriction to each open subset of $X \setminus K$ with a compact complement is still regular at infinity. So the question is really a property of regularity at infinity.

It is obvious that if X is compact, then each section $f \in Z^q(\mathcal{D}'(E^\cdot|_{X\setminus K}))$ is regular at infinity because we can take as φ in (5.4.1) the function equal to 1 everywhere on X.

Finally, if we consider the left-hand side of (5.4.1) as the integral over $X \setminus (K \cup \operatorname{supp} g)$, then it obviously equals $-\langle P'_q(\varphi \Phi'_{q+1} g), f\rangle_{X\setminus(K\cup supp\ g)}$. Therefore, formula (5.4.1) is equivalent to

$$\left\langle P'_q(\varphi \Phi'_{q+1} g), f\right\rangle_{X\setminus(K\cup supp\ g)} = 0. \tag{5.4.2}$$

Lemma 5.4.2 *Suppose that the space $H^q(\mathcal{D}'(E^\cdot))$ is separated and a section $f \in \mathcal{D}'(E^q)$ satisfies $P_q f = 0$ on $X \setminus K$. Then f is regular at infinity if and only if*

$$f = \Phi_{q+1} P_q f \quad in \quad H^q(\mathcal{D}'(E^\cdot)). \tag{5.4.3}$$

Notice that the the right-hand side of (5.4.3) is well-defined because $P_q f \in \mathcal{E}'(E^{q+1})$.

Proof. Since the space $H^q(\mathcal{D}'(E^\cdot))$ is separated, by the Poincaré duality we have $H^q(\mathcal{D}'(E^\cdot)) \overset{tr}{\cong} H_q(\mathcal{D}(E^{\cdot\prime}))$ (see Remark 5.1.9). Hence equality (5.4.3) is equivalent to the condition that for all $g \in Z_q(\mathcal{D}(E^{\cdot\prime}))$

$$\langle g, f\rangle = \langle g, \Phi_{q+1} P_q f\rangle. \tag{5.4.4}$$

Let $\varphi \in \mathcal{D}(X)$ equal to 1 in a neighborhood of $K \cup \operatorname{supp} g$. Then the left-hand side of (5.4.3) is written, by homotopy formula (2.3.2) at degree q, as

$$\begin{aligned} \langle g, \varphi f\rangle &= \langle g, \Phi_{q+1} P_q(\varphi f) + P_{q-1}\Phi_q(\varphi f)\rangle \\ &= \langle \Phi'_{q+1} g, P_q(\varphi f)\rangle. \end{aligned}$$

Since the right-hand side of (5.4.4) equals $\langle \Phi'_{q+1}g, P_q f\rangle$, (5.4.4) is equivalent to $\langle \Phi'_{q+1}g, P_q(\varphi f) - P_q f\rangle = 0$. However the difference $(P_q(\varphi f) - P_q f$ vanishes in some neighborhood of K, so that (5.4.4) is reduced precisely to (5.4.1).

□

Corollary 5.4.3 *If the space $H^q(\mathcal{D}'(E^\cdot))$ is separated, then each section $f \in Z^q(\mathcal{D}'(E^\cdot))$ which is regular at infinity has zero image in $H^q(\mathcal{D}'(E^\cdot))$.*

Corollary 5.4.4 *Assume that the space $H^q(\mathcal{D}'(E^\cdot))$ is separated. If a section $f \in Z^q(\mathcal{D}'(E^\cdot|_{X\setminus K}))$ is regular at infinity, then for each function $\varphi \in \mathcal{E}(X)$, equal to 0 in a neighborhood of K and equal to 1 in a neighborhood of infinity, we have*

$$\varphi f = \Phi_{q+1}P_q(\varphi f) \quad in \quad H^q(\mathcal{D}'(E^\cdot)). \tag{5.4.5}$$

Notice, if f is a differentiable section, then we can take as φ the characteristic function of any set $U \subset X \setminus K$ with a compact complement such that $\overline{U} \subset X \setminus K$. Then, formula (5.4.5) can be rewritten as

$$\chi_U f = -\Phi_{q+1}([\partial U]^P f) \quad \text{in} \quad H^q(\mathcal{D}'(E^\cdot)). \tag{5.4.6}$$

Lemma 5.4.5 *Assume that the space $H^q(\mathcal{D}'(E^\cdot))$ is separated and Φ is also a fundamental solution of the complex $E^\cdot$ at degree $q+1$. Then a section $f \in Z^q(\mathcal{D}'(E^\cdot|_{X\setminus K}))$ is regular at infinity iff in the complement of a suitable compact set K' we have for some $H \in Z^{q+1}(\mathcal{E}'(E^\cdot))$:*

$$f = \Phi_{q+1}H \quad in \quad H^q(\mathcal{D}'(E^\cdot|_{X\setminus K'})). \tag{5.4.7}$$

Proof. *Necessity* follows from formula (5.4.5) when $H = P_q(\varphi f)$. Let us prove the *sufficiency*.

Since $H = P_q\Phi_{q+1}H$, by setting $f' = \Phi_{q+1}H$ we get $P_q f' = 0$ in $X \setminus \operatorname{supp} H$ and $\Phi_{q+1}P_q f' = \Phi_{q+1}H = f'$. So Lemma 5.4.2 implies that the section f' is regular at infinity.

By (5.4.7), to prove the lemma, it remains to show that, if in the complement of a suitable compact set $K' \supset K$ we have $f = P_{q-1}u$ where $u \in \mathcal{D}'(E^{q-1}|_{X\setminus K'})$, then the section f is regular at infinity.

That is, for each section $g \in Z_q(\mathcal{D}'(E^{\cdot\prime}))$ and for each function $\varphi \in \mathcal{D}(X)$ equal to 1 in a neighborhood of $K \cup \operatorname{supp} g$, equality (5.4.1) holds.

Let K'' be a compact subset of X which contains a neighborhood of $K' \cup \operatorname{supp} g$, and let $\varphi_1 \in \mathcal{D}(X)$ equal to 1 in a neighborhood of K''. Choose a section $u_1 \in \mathcal{D}'(E^{q-1})$ which coincides with u on $X \setminus K''$ and equals zero in a neighborhood of $K' \cup \operatorname{supp} g$. Then we have

$$\begin{aligned}
\langle \Phi'_{q+1}g, P_q(\varphi f)\rangle_{X\setminus K} &= \langle \Phi'_{q+1}g, P_q(\varphi_1 f)\rangle_{X\setminus K} \\
&= \langle \Phi'_{q+1}g, P_q(\varphi_1 f)\rangle_{X\setminus K''} \\
&= \langle \Phi'_{q+1}g, P_q(\varphi_1 P_{q-1}u_1)\rangle \\
&= \langle P'_q\Phi'_{q+1}g, \varphi_1 P_{q-1}u_1\rangle \\
&= \langle g, P_{q-1}u_1\rangle \\
&= 0,
\end{aligned}$$

as required.

□

5.4.3 *Cohomology classes regular at infinity*

We have seen from Lemma 5.4.5 that it would be more exact to speak of the regularity at infinity of cohomology classes from $H^q(\mathcal{D}'(E\,|_{X\setminus K}))$.

Namely, a cohomology class $[f] \in H^q(\mathcal{D}'(E\,|_{X\setminus K}))$ is called regular at infinity if some representative $f \in Z^q(\mathcal{D}'(E\,|_{X\setminus K}))$ of this class is regular at infinity.

Then by Lemma 5.4.5 the same is true for each other representative of the class $[f]$, so this notion is well-defined.

Denote by $H^q_{reg}(\mathcal{D}'(E\,|_{X\setminus K}))$ the corresponding subspace of $H^q(\mathcal{D}'(E\,|_{X\setminus K}))$ with the induced topology. The following theorem appeared first in the author's paper [233].

Theorem 5.4.6 *If the space $H^q(\mathcal{D}'(E))$ is separated and E has a biregular fundamental solution Φ at degrees q and $q+1$, then*

$$H^q(\mathcal{D}'(E\,|_{X\setminus K})) = H^q(\mathcal{D}'(E)) \overset{top}{\oplus} H^q_{reg}(\mathcal{D}'(E\,|_{X\setminus K})). \tag{5.4.8}$$

We emphasize that the space $H^q(\mathcal{D}'(E))$ is considered here as a subspace of $H^q(\mathcal{D}'(E\,|_{X\setminus K}))$.

It follows from this theorem in particular that in the natural cases the space $H^q_{reg}(\mathcal{D}'(E\,|_{X\setminus K}))$ on the whole does not depend on which fundamental solution Φ has been used to define regularity at infinity in 5.4.1. The choice of Φ determines uniquely the topological complement of $H^q(\mathcal{D}'(E))$ in $H^q(\mathcal{D}'(E\,|_{X\setminus K}))$.

Proof. Fix some precompact neighborhood U of the compact set K. Let $[f] \in H^q(\mathcal{D}'(E\,|_{X\setminus K}))$ where $f \in Z^q(\mathcal{D}'(E\,|_{X\setminus K}))$ is a representative of $[f]$. By changing f in the layer $U \setminus K$, we find a section $\tilde{f} \in \mathcal{D}'(E^q)$ which coincides with f in the complement of U.

Consider for each precompact open set $\mathcal{O} \subset X$ the section

$$f_{\mathcal{O}} = \Phi_{q+1} P_q((\varphi - 1)f) + P_{q-1}\Phi_q(\varphi\tilde{f}) \quad \text{on} \quad \mathcal{O}, \tag{5.4.9}$$

where the function $\varphi \in \mathcal{D}(X)$ equals to 1 in a neighborhood of $\overline{\mathcal{O} \cup U}$. First we remark that $f_{\mathcal{O}}$ is well-defined because $P_q((\varphi-1)f) \in \mathcal{E}'(E^{q+1})$.

Let now $\mathcal{O}_1$ and $\mathcal{O}_2$ be two precompact open subsets of X. We denote by φ_1 and φ_2 the corresponding functions φ, that is, $\varphi_j \in \mathcal{D}(X)$ is equal to 1 in a neighborhood of $\overline{\mathcal{O}_j \cup U}$ $(j = 1, 2)$. Then according to the homotopy formula (2.3.2), at degree q we obtain

$$\begin{aligned} f_{\mathcal{O}_2} - f_{\mathcal{O}_1} &= \Phi_{q+1} P_q((\varphi_2 - \varphi_1)f) + P_{q-1}\Phi_q((\varphi_2 - \varphi_1)f) \\ &= (\varphi_2 - \varphi_1)f \\ &= 0 \quad \text{on} \quad \mathcal{O}_1 \cap \mathcal{O}_2. \end{aligned}$$

Hence formula (5.4.9) determines as a matter of fact the global section $f_e \in \mathcal{D}'(E^q)$ which does not depend on the choice of the function φ.

We claim that $P_q f_e = 0$. Indeed it is sufficient to show that this equality holds on each precompact set $\mathcal{O} \subset X$, and then it holds in general from formula (5.4.9) and the fact that Φ is a fundamental solution of the complex $E^{\cdot}$ at degree $(q+1)$. We thus have

$$P_q f_e = P_q f_{\mathcal{O}} = P_q \Phi_{q+1}(P_q((\varphi - 1)f)) = P_q((\varphi - 1)f) = 0 \quad \text{on} \quad \mathcal{O}$$

as required.

We now show that the cohomology class $[f_e]$ of the section f_e in $H^q(\mathcal{D}'(E^{\cdot}))$ does not depend on how f has been extended from $X \setminus U$ to a section $\tilde{f} \in \mathcal{D}'(E^q)$. Let $\tilde{f}$ and $\tilde{\tilde{f}}$ be two such extensions. Then, we have $(\tilde{\tilde{f}} - \tilde{f}) \in \mathcal{E}'(E^q)$ and $\operatorname{supp}(\tilde{\tilde{f}} - \tilde{f}) \subset \overline{U}$. Therefore, if $\widetilde{f_{\mathcal{O}}}$ and $\tilde{\tilde{f}}_{\mathcal{O}}$ are the corresponding sections of E^q over $\mathcal{O}$ from formula (5.4.9), then we get

$$\tilde{\tilde{f}}_{\mathcal{O}} - \tilde{f}_{\mathcal{O}} = P_{q-1}\Phi_q(\varphi(\tilde{\tilde{f}} - \tilde{f})) = P_{q-1}\Phi_q(\tilde{\tilde{f}} - \tilde{f}) \quad \text{on} \quad \mathcal{O},$$

and $\Phi_q(\tilde{\tilde{f}} - \tilde{f}) \in \mathcal{D}'(E^{q-1})$. Hence $[\widetilde{f_e}] = [\widetilde{\widetilde{f_e}}]$.

Let us now show now that $[f_e]$ does not depend on the choice of a representative $f \in [f]$. It is sufficient to prove that, if $f = P_{q-1}u$ where $u \in \mathcal{D}'(E^{q-1}|_{X\setminus K})$, then $[f_e] = 0$ in $H^q(\mathcal{D}'(E^{\cdot}))$. Indeed, by changing u in the layer $U \setminus K$ we find a section $\tilde{u} \in \mathcal{D}'(E^{q-1})$ which coincides with u on $X \setminus U$. Then one can take $\tilde{f} = P_{q-1}\tilde{u}$ as an extension of f, and we get

$$\begin{aligned} f_{\mathcal{O}} &= \Phi_{q+1}P_q((\varphi - 1)P_{q-1}\tilde{u}) + P_{q-1}\Phi_q(\varphi P_{q-1}\tilde{u}) \\ &= \Phi_{q+1}P_q(\varphi P_{q-1}\tilde{u}) + P_{q-1}\Phi_q(\varphi P_{q-1}\tilde{u}) \\ &= \varphi P_{q-1}\tilde{u} \\ &= P_{q-1}\tilde{u}. \end{aligned}$$

We used formula (2.3.2) at degree q. Hence $[f_e] = 0$ again.

Consequently, if we set $\pi([f]) = [f_e]$, then we obtain a well-defined mapping $\pi : H^q(\mathcal{D}'(E^{\cdot}|_{X\setminus K})) \to H^q(\mathcal{D}'(E^{\cdot}))$.

Further we will prove that the section $(f - f_e) \in Z^q(\mathcal{D}'(E^{\cdot}|_{X\setminus K}))$ is regular at infinity. Then each class $[f] \in H^q(\mathcal{D}'(E^{\cdot}|_{X\setminus K}))$ will be the sum of the class $[f_e] \in H^q(\mathcal{D}'(E^{\cdot}))$ and the class $[f - f_e] \in H^q_{reg}(\mathcal{D}'(E^{\cdot}|_{X\setminus K}))$, that is $H^q(\mathcal{D}'(E^{\cdot}|_{X\setminus K}))$ will be the sum of its subspaces $H^q(\mathcal{D}'(E^{\cdot}))$ and $H^q_{reg}(\mathcal{D}'(E^{\cdot}|_{X\setminus K}))$.

One must show that, for each section $g \in Z_q(\mathcal{D}(E^{\cdot\prime}))$ and each function $\varphi \in \mathcal{D}(X)$ equal to 1 in a neighborhood of $K \cup \operatorname{supp} g$, the following equality holds:

$$\langle \Phi'_{q+1}g, P_q(\varphi f)\rangle\Big|_{X\setminus K} = \langle \Phi'_{q+1}g, P_q(\varphi f_e)\rangle\Big|_{X\setminus K}. \tag{5.4.10}$$

Now φf_e is a section of $\mathcal{E}'(E^q)$ which is defined on all of the manifold X and $P_q(\varphi f_e) = 0$ in a neighborhood of K. Thus the integral in the right-hand side of (5.4.10) can be taken over all of X. Then, by the transposition rule we get

$$\begin{aligned} \langle \Phi'_{q+1}g, P_q(\varphi f_e)\rangle_{X\setminus K} &= \langle P'_q\Phi'_{q+1}g, \varphi f_e\rangle \\ &= \langle g, \varphi f_e\rangle \\ &= \langle g, f_e\rangle. \end{aligned}$$

But it is precisely the left-hand side of (5.4.10) because

$$\begin{aligned} \langle g, f_e \rangle &= \langle g, \Phi_{q+1} P_q((\varphi - 1)f) + P_{q-1}\Phi_q(\varphi_1 \tilde{f})\rangle \\ &= \langle \Phi'_{q+1} g, P_q((\varphi_1 - 1)f)\rangle \\ &= \langle \Phi'_{q+1} g, P_q(\varphi_1 f)\rangle_{X\setminus K} \\ &= \langle \Phi'_{q+1} g, P_q(\varphi f)\rangle_{X\setminus K}, \end{aligned}$$

where $\varphi_1 \in \mathcal{D}(X)$ equals 1 in a neighborhood of $\overline{U} \cup \operatorname{supp} g$. This proves (5.4.10).

On the other hand, according to Corollary 5.4.3 we have

$$H^q(\mathcal{D}'(E^\cdot)) \cap H^q_{reg}(\mathcal{D}'(E^\cdot|_{X\setminus K})) = \{0\}.$$

So $H^q(\mathcal{D}'(E^\cdot|_{X\setminus K}))$ is the algebraic direct sum of $H^q(\mathcal{D}'(E^\cdot))$ and $H^q_{reg}(\mathcal{D}'(E^\cdot|_{X\setminus K}))$.

Moreover, $H^q(\mathcal{D}'(E^\cdot|_{X\setminus K}))$ is the topological direct sum of these two subspaces. Indeed, this means that the projection $\pi : H^q(\mathcal{D}'(E^\cdot|_{X\setminus K})) \to H^q(\mathcal{D}'(E^\cdot))$ is continuous which can be shown immediately by formula (5.4.9).

The proof is complete.

□

Theorem 5.4.6 contains as a very special case the result of Grothendieck [89] for elliptic differential operators on X.

Remark 5.4.7 *The results obtained above are valid with obvious modifications for cohomology classes of the complex $\mathcal{E}(E^\cdot)$ also. The proofs repeat the ones above word for word. However, if the fundamental solution Φ is differentiable off the diagonal, then the results are the same because for each open set $\mathcal{O} \subset X$ we have $H^\cdot(\mathcal{D}'(E^\cdot|_\mathcal{O})) \overset{top.}{\cong} H^\cdot(\mathcal{E}(E^\cdot|_\mathcal{O}))$ by Theorem 2.1.9 and Corollary 2.1.11.*

5.4.4 *Example*

Let $P \in \mathrm{do}_p(\mathbb{R}^n \times \mathbb{C}^k \to \mathbb{R}^n \times \mathbb{C}^k)$ be a homogeneous elliptic differential operator of order $p < n$ with constant coefficients in $\mathbb{R}^n$.

Then P has only one fundamental solution of convolution type

$$\Phi(x) \in [\mathcal{D}'(\mathbb{R}^n)]^{(k\times k)}$$

which is a matrix-valued function (positively) homogeneous of degree $(p-n)$, and real analytic outside the origin. For example, it can be taken in the form $\Phi(x) = (\operatorname{adj} P(D))g(x)$ where $g(x)$ is a fundamental solution of the (scalar) elliptic differential operator $(\det P(D))$ constructed by formula (6) of Shilov [200, §22].

Given any compact $K \subset \mathbb{R}^n$, it is easy to show by Lemma 5.4.5 that a section $f \in [\mathcal{D}'(\mathbb{R}^n \setminus K)]^k$, satisfying the equation $Pf = 0$ in $\mathbb{R}^n \setminus K$, is regular at infinity with respect to Φ if and only if $f(\infty) = 0$. Another way is to use Theorem 5.4.6, Corollary 5.4.4 and the Liouville type theorem (see Exercise 5 of Shilov [200, §16]).

Then decomposition (5.4.8) means that every section $f \in [\mathcal{D}'(\mathbb{R}^n \setminus K)]^k$, satisfying $Pf = 0$ in $\mathbb{R}^n \setminus K$, can be represented uniquely in the form $f = f_e + f_r$, where the section $f_e \in [\mathcal{E}(\mathbb{R}^n)]^k$ satisfies $Pf_e = 0$ in $\mathbb{R}^n$ and the section $f_r \in [\mathcal{E}(\mathbb{R}^n \setminus K)]^k$ vanishes at infinity. Moreover, f_e and f_r depend on f continuously.

Exercise 5.4.8 *Find explicit formulas for f_e and f_r.*

Hint. Use (5.4.9)

□

It is clear from this example that Corollary 5.4.3 and Theorem 5.4.6 are straightforward generalizations of the Liouville type theorems (see Exercise 5 of Shilov [200, §16]; Weck [251] and other).

5.5 Grothendieck Duality for Elliptic Complexes

5.5.1 *Survey of results*

In the theory of functions of a complex variable, the Köthe-Grothendieck-Sebastiao e Silva theorem on the general form of a continuous linear functional for the space of holomorphic functions is well-known. It says that, if Y is a bounded domain in $\mathbb{C}^1$, then $\mathcal{O}^{(0)}(Y)' \overset{top.}{\cong} \mathcal{O}^{(0)}_{reg}(\mathbb{C}^1 \setminus Y)$, where $\mathcal{O}^{(0)}_{reg}(\mathbb{C}^1 \setminus Y)$ is the space of functions holomorphic in a neighborhood of $\mathbb{C}^1 \setminus Y$ and vanishing at infinity. For further results in complex analysis, see the survey of Havin [102].

The following important step was taken by Grothendieck [89]. Let P be a (scalar) elliptic differential operator on a manifold X and Y be a relatively compact open subset of X. Grothendieck [89] showed that, if P has a fundamental solution and satisfies the so-called "approximation condition" on Y, then there is a topological isomorphism $\mathcal{E}_P(\mathcal{O})' \cong \vec{\mathcal{E}}_{P'}(X \setminus \mathcal{O})/\mathcal{E}_{P'}(X)$ where $\vec{\mathcal{E}}_{P'}(X \setminus \mathcal{O})$ was the space of solutions of the transposed equation in a neighborhood of $X \setminus \mathcal{O}$ endowed with the inductive limit topology. According to Theorem 5.4.6, the quotient space in the right-hand side is just the space of solutions of $P'g = 0$ in a neighborhood of $X \setminus \mathcal{O}$ which are regular at infinity.

In particular, the conditions for P are fulfilled provided all the objects X and P are real analytic. Under this assumption, Schapira [192] found an independent proof of Grothendieck's duality.

Kiselman [122] applied Grothendieck's duality to the problem of supports of linear functionals on the space $\mathcal{E}_P(\mathbb{R}^n)$ where P is a scalar differential operator with constant coefficients in $\mathbb{R}^n$.

Palamodov [175, §14] generalized the Grothendieck duality theorem to the case of "rectangular" differential operators P with constant coefficients in $\mathbb{R}^n$ such that the $\mathcal{P}$-module $\operatorname{Ext}^1(\operatorname{coker} P(z)^t, \mathcal{P})$ is elliptic. Common examples of such differential operators are operators with surjective symbols. Palamodov [175] proved that, if K is a convex compact set in $\mathbb{R}^n$ and X is a neighborhood of K, then $\left(\mathcal{E}_P(X \setminus K)/\widehat{\mathcal{E}_P(X)}\right)' \overset{top.}{\cong} \vec{\mathcal{D}}_{P'}(K)$ where $\widehat{\mathcal{E}_P(X)}$ is the subspace of $\mathcal{E}_P(X \setminus K)$ consisting of limits of sequences $\{f_\nu\} \subset \mathcal{E}_P(X)$ which stabilize outside any neighborhood of K.

There are also known explicit forms of Grothendieck's duality for particular differential operators P, for instance, in the hypercomplex function theory (see Delanghe and Brackx [51]).

The author [222, 233] found the meaning of Grothendieck's duality in the context of arbitrary elliptic complexes.

Notice that Grothendieck's duality is closely related to Alexander-Pontryagin duality (5.1.5), in particular, it completes (5.1.5) in the extreme case $q = 0$.

5.5.2 *The space* $H^q(\mathcal{E}(E\,|_Y))$

Let E be an elliptic complex on a differentiable manifold X countable at infinity.

We assume that E has a fundamental solution at degree $(q+1)$. Denote this fundamental solution by Φ. According to Proposition 2.3.9, one can consider that $\Phi_i \in \mathrm{pdo}_{-p_{i-1}}(E^i \to E^{i-1})$ $(i \in \mathbb{Z})$.

Let Y be a relatively compact open subset of X.

Denote by $Z^q(\widehat{\mathcal{E}(E)})$ the subspace of $Z^q(\mathcal{E}(E\,|_Y))$ consisting of limits of sequences $\{f_\nu\} \subset Z^q(\mathcal{E}(E))$ of the form $f_\nu = \Phi(H_\nu) \pmod{P\mathcal{E}(E^{q-1}|_{X\setminus K^{(\nu)}})}$ in the complement of an appropriate compact set $K^{(\nu)} \subset X$, where $H_\nu \in Z^{q+1}(\mathcal{E}'(E))$, which stabilize on every compact set $K \subset Y$. In order to remove a touch of mystery from the condition on $\{f_\nu\}$, we notice that by Lemma 5.4.5 it is equivalent to regularity at infinity of $\{f_\nu\}$ at least under natural assumptions.

As usual, we endow the quotient $Z^q(\mathcal{E}(E\,|_Y))/Z^q(\widehat{\mathcal{E}(E)})$ with the quotient topology, in general not separated.

Proposition 5.5.1 *If Φ is a fundamental solution of E at degrees q and $(q+1)$, then there is the embedding $B^q(\mathcal{E}(E\,|_Y)) \hookrightarrow Z^q(\widehat{\mathcal{E}(E)})$ with dense range.*

Proof. Suppose that a section $f \in Z^q(\mathcal{E}(E))$ has the form $f = \Phi(H) + Pu$ in the complement of an appropriate compact set $K \subset X$, where $H \in Z^{q+1}(\mathcal{E}'(E))$ and $u \in \mathcal{E}(E^{q-1})$.

Set $r = f - \Phi(H) - Pu$. Then $r \in \mathcal{E}'(E^q)$ is supported in K. Applying the operator P to both sides of the equality, we get $Pr = -H$. Hence it follows by the homotopy formula that $f = P(u + \Phi(r))$ where $(u + \Phi(r)) \in \mathcal{D}'(E^{q-1})$. Because of $f \in \mathcal{E}(E^q)$, we conclude from the first isomorphism of Corollary 2.1.15 that $f \in B^q(\mathcal{E}(E))$.

Therefore, under the hypothesis of the proposition, $Z^q(\widehat{\mathcal{E}(E)})$ is the subspace of $Z^q(\mathcal{E}(E\,|_Y))$ consisting of limits of sequences $\{f_\nu\} \subset B^q(\mathcal{E}(E))$ that stabilize on every compact set $K \subset Y$.

Take now a section f of $B^q(\mathcal{E}(E\,|_Y))$, i.e., $f = Pu$ for some section $u \in \mathcal{E}(E^{q-1}|_Y)$. According to Proposition 1.1.1, there exists a sequence $\{U_\nu\}$ of relatively compact open sets in Y such that $\overline{U_\nu} \subset U_{\nu+1}$ for all ν, and $\cup_\nu U_\nu = Y$. Fix some functions $\varphi_\nu \in \mathcal{D}(Y)$ equal to 1 in a neighborhood of $\overline{U_\nu}$, and set $f_\nu = P(\varphi_\nu u)$. Then the sequence $\{f_\nu\} \subset B^q(\mathcal{D}(E))$ stabilizes on every compact set $K \subset Y$, and it approximates f in the topology of $\mathcal{E}(E^q|_Y)$.

It becomes clear from here that we have the embedding

$$B^q(\mathcal{E}(E\,|_Y)) \hookrightarrow Z^q(\widehat{\mathcal{E}(E)})$$

with dense range, as desired.

□

Thus, if Φ is a fundamental solution of the complex $E^\cdot$ at degrees q and $(q+1)$, then the dual spaces to the quotients $H^q(\mathcal{E}(E^\cdot|_Y))$ and $Z^q(\mathcal{E}(E^\cdot|_Y))/\widehat{Z^q(\mathcal{E}(E^\cdot))}$ are identical. Moreover, the following is true.

Corollary 5.5.2 *Under the hypothesis of Proposition 5.5.1, if moreover the space $H^q(\mathcal{E}(E^\cdot|_Y))$ is separated, then we have*

$$H^q(\mathcal{E}(E^\cdot|_Y)) = \frac{Z^q(\mathcal{E}(E^\cdot|_Y))}{\widehat{Z^q(\mathcal{E}(E^\cdot))}}.$$

Proof. In fact, if $H^q(\mathcal{E}(E^\cdot|_Y))$ is separated, then according to Proposition 1.1.12 the subspace $B^q(\mathcal{E}(E^\cdot|_Y))$ is closed in $\mathcal{E}(E^q|_Y)$. Therefore Proposition 5.5.1 implies that $B^q(\mathcal{E}(E^\cdot|_Y)) = \widehat{Z^q(\mathcal{E}(E^\cdot))}$, which completes the proof.

□

5.5.3 *The space* $H^{reg}_{q+1}(\vec{\mathcal{E}}(E^{\cdot\prime}|_{X\setminus Y}))$

Let us recall that the (chain) complex of topological vector spaces $\vec{\mathcal{E}}(E^{\cdot\prime}|_{X\setminus Y})$ was defined in Subsection 5.1.3.

Since Φ is a fundamental solution at degree $(q+1)$ of the complex $E^\cdot$, the transposed operator Φ' provides the fundamental solution at dimension $(q+1)$ of the complex $E^{\cdot\prime}$. In this situation we may apply the results of Section 5.4 to the complex $E^{\cdot\prime}$.

What we need is the subspace of $H_{q+1}(\vec{\mathcal{E}}(E^{\cdot\prime}|_{X\setminus Y}))$ consisting of homology classes regular at infinity. We denote this subspace by $H^{reg}_{q+1}(\vec{\mathcal{E}}(E^{\cdot\prime}|_{X\setminus Y}))$.

However, we shall use a slight modification of Definition 5.4.1 of regularity at infinity. Namely, a class $[g] \in H_{q+1}(\vec{\mathcal{E}}(E^{\cdot\prime}|_{X\setminus Y}))$ is considered to be regular at infinity if some representative g of the class is representable by the formula $g = -\Phi'([\partial U]^{P'}g)$ (mod $B_{q+1}(\mathcal{E}(E^{\cdot\prime}|_U))$) where U is a neighborhood of $X \setminus Y$ with piecewise smooth boundary.

It follows from Lemma 5.4.2 that both definitions are equivalent when the space $H_{q+1}(\mathcal{E}(E^{\cdot\prime}))$ is separated. But in general the definition, accepted in this subsection, is more restrictive.

Proposition 5.5.3 *If the space $H_{q+1}(\mathcal{E}(E^{\cdot\prime}))$ is separated and Φ is a fundamental solution of $E^\cdot$ at degrees q and $(q+1)$, then*

$$H^{reg}_{q+1}(\vec{\mathcal{E}}(E^{\cdot\prime}|_{X\setminus Y})) \overset{top.}{\cong} \frac{H_{q+1}(\vec{\mathcal{E}}(E^{\cdot\prime}|_{X\setminus Y}))}{H_{q+1}(\mathcal{E}(E^{\cdot\prime}))}.$$

Proof. Apply Theorem 5.4.6 to the complex $E^{\cdot\prime}$, and make use of the hypoellipticity of the complex.

□

5.5.4 *The approximation condition*

To formulate the main result of this subsection we also need to consider, in the context of elliptic complexes, the "approximation condition" of Grothendieck [89].

Definition 5.5.4 *Given a complex $E^\cdot$, the approximation condition is said to be fulfilled on Y at degree q if for each open set $Y_1 \subset\subset Y$ there exists an open set $Y_2 \subset\subset Y$ containing Y_1 and such that the space $H^q(\mathcal{E}(E^\cdot|_Y))$ is dense in $H^q(\mathcal{E}(E^\cdot|_{Y_2}))$ in the topology of $H^q(\mathcal{E}(E^\cdot|_{Y_1}))$.*

It follows that the approximation condition means the validity of Runge type theorems for solutions of the overdetermined system $P_q f = 0$ on Y. This topic is of high interest to study.

Proposition 5.5.5 *If the differential operator P_{N-1} has real analytic coefficients, then the approximation condition for $E^\cdot$ is fulfilled at degree $(N-1)$ on every open set $Y \subset X$.*

Proof. Let Y_1 be a relatively compact open subset of Y. There exists a relatively compact open subset Y_2 of Y containing Y_1 and such that $Y \setminus Y_2$ has no compact connected components. In fact, one can take Y_2 equal to the union of Y_1 and of all compact connected components of $Y \setminus Y_1$. According to Runge's theorem for solutions of systems with surjective symbols (see Tarkhanov [228]), the space $Z^{N-1}(\mathcal{E}(E^\cdot|_Y))$ is dense in $Z^{N-1}(\mathcal{E}(E^\cdot|_{Y_2}))$. Hence the statement of the proposition follows.

□

5.5.5 *Grothendieck duality*

Theorem 5.5.6 *Suppose that $Y \subset\subset X$ is the union of an increasing sequence of open sets which are $(q+1)$-convex with respect to $E^\cdot$, and the approximation condition for $E^\cdot$ is fulfilled on Y at degree q. Then*

$$\left(\frac{Z^q(\mathcal{E}(E^\cdot|_Y))}{\widehat{Z^q(\mathcal{E}(E^\cdot))}}\right)' \overset{top.}{\cong} H^{reg}_{q+1}(\vec{\mathcal{E}}(E^{\cdot\prime}|_{X\setminus Y})). \tag{5.5.1}$$

Isomorphism (5.5.1) is said to be the Grothendieck duality for the complex $E^\cdot$. The mechanism of the duality consists in the following. Define the bilinear form $[\partial Y]^P$ on $Z_{q+1}(\vec{\mathcal{E}}(E^{\cdot\prime}|_{X\setminus Y})) \otimes Z^q(\mathcal{E}(E^\cdot|_Y))$ by the equality $[\partial Y]^P(g \otimes f) = \lim_{\nu\to\infty} \int_{\partial Y_\nu} G_{P_q}(g, f)$, where $\{Y_\nu\}$ is an exhaustion of Y by an increasing sequence of open sets with piecewise smooth boundaries. This limit of course exists, since from the definition of the Green operator it follows that the sequence $\{\int_{\partial Y_\nu} G_{P_q}(g, f)\}$ stabilizes starting with some ν. The bilinear form is lowered to the pairing of the quotient spaces $H^{reg}_{q+1}(\mathcal{E}(E^{\cdot\prime}|_{X\setminus Y})) \otimes \frac{Z^q(\mathcal{E}(E^\cdot|_Y))}{\widehat{Z^q(\mathcal{E}(E^\cdot))}} \to \mathbb{C}$ that is non-degenerate in the first argument under the conditions of the theorem. Thus we obtain the embedding of $H^{reg}_{q+1}(\mathcal{E}(E^{\cdot\prime}|_{X\setminus Y}))$ into the space of continuous linear functionals on $\left(\frac{Z^q(\mathcal{E}(E^\cdot|_Y))}{\widehat{Z^q(\mathcal{E}(E^\cdot))}}\right)$.

The mapping inverse to this embedding is given by associating to a functional $g \in \left(\frac{Z^q(\mathcal{E}(E\,|_Y))}{Z^q(\mathcal{E}(E\,))}\right)'$ its "Fantappiè indicatrix", i.e., the image in $H^{reg}_{q+1}(\vec{\mathcal{E}}(E^{\cdot\prime}|_{X\setminus Y}))$ of the potential $(-\Phi'(\tilde{g}))$ where $\tilde{g} \in \mathcal{E}'(E^{q'}|_Y)$ is some extension of the functional g.

5.5.6 *A preparatory lemma*

To prove Theorem 5.5.6 we need special decompositions of the kernel $\Phi_{q+1}(x,y)$. They are similar to those given by Lemma 4.5.3.

Lemma 5.5.7 *Let Y be an open subset of X $(q+1)$-convex with respect to the complex $E\,|_Y$. Then there exist operators $K^{(-1)}_{q+1} \in \mathcal{L}(\mathcal{E}'(E^{q+1}|_{X\setminus\overline{Y}}) \to \mathcal{E}(E^q|_Y))$ and $K^{(-2)}_{q+2} \in \mathcal{L}(\mathcal{E}'(E^{q+2}|_{X\setminus\overline{Y}}) \to \mathcal{E}(E^q|_Y))$ such that: 1) $P_q K^{(-1)}_{q+1} = 0$, and 2) $\Phi_{q+1} = K^{(-1)}_{q+1} - K^{(-2)}_{q+2} P_{q+1}$ on $\mathcal{E}'(E^{q+1}|_{X\setminus\overline{Y}})$.*

Proof. Using the connection between kernels and operators given by Theorem 1.5.1, we with to construct kernels $K^{(-1)}_{q+1} \in \mathcal{E}(E^q \boxtimes E^{q+1\prime}|_{Y\times(X\setminus\overline{Y})})$ and $K^{(-2)}_{q+2} \in \mathcal{E}(E^q \boxtimes E^{q+2\prime}|_{Y\times(X\setminus\overline{Y})})$ such that:

1) $P_q(x)K^{(-1)}_{q+1} = 0$, and

2) $\Phi_{q+1} = K^{(-1)}_{q+1} - P'_{q+1}(y)K^{(-2)}_{q+2}$ in the product $Y \times (X \setminus \overline{Y})$.

For this purpose we copy equation (2.1.3) for those components of the fundamental solution Φ which are of interest for us:

$$P'_i(y)\Phi_{i+1}(x,y) + P_{i-1}(x)\Phi_i(x,y) = \operatorname{tr}\Delta^* - S_i(x,y) \quad (i = q+1,\dots,N). \quad (5.5.2)$$

Here the kernels $S_i \in \mathcal{E}(E^i \boxtimes E^{i\prime})$ satisfy the system (2.3.7), and $S_{q+1} = 0$.

First we would like to find a correction $M \in \mathcal{E}((E \boxtimes E\,')^{-1}|_{Y\times X})$ supported at the degrees $\geq (q+2)$ (i.e., $M_i = 0$ for $i < q+2$), such that the kernel $\Phi^{(q+1)} = \Phi + M$ satisfies system (5.5.2) on $Y \times X$ with zero errors S_i. For this it is required that the components of the kernel M satisfy the following system of differential equations considered in the product $Y \times X$:

$$P'_i(y)M_{i+1}(x,y) + P_{i-1}(x)M_i(x,y) = S_i(x,y) \quad (i = q+1,\dots,N). \quad (5.5.3)$$

System (5.5.3) is solved by the chain rule. Namely, using the condition $H^N(\mathcal{E}(E\,|_Y)) = 0$ and Theorem 1.4.9 we may find $M_N \in \mathcal{E}(E^{N-1} \boxtimes E^{N\prime}|_{Y\times X})$ from the last equation of (5.5.3) (i.e., for $i = N$). This gives automatically the compatibility condition for the next to last equation of (5.5.3) (i.e., for $i = N-1$) relatively to M_{N-1}. Using the condition $H^{N-1}(\mathcal{E}(E^\cdot|_Y)) = 0$ and Theorem 1.4.9 we find then $M_{N-1} \in \mathcal{E}(E^{N-2} \boxtimes E^{N-1\prime}|_{Y\times X})$ from the next to last equation of (5.5.3). This gives the compatibility condition for the previous to next to last equation of (5.5.3) relatively to M_{N-2}, etc. Such considerations have already been used in the proof of Theorem 2.3.12.

But in the last step we meet the new difficulty that we have to find a solution $M_{q+2} \in \mathcal{E}(E^{q+1} \boxtimes E^{q+2\prime}|_{Y\times X})$ of the equation $P_{q+1}(x)M_{q+2}(x,y) = S_{q+2}(x,y) -$

$P'_{q+2}(y)M_{q+3}(x,y)$ which also satisfies $P'_{q+1}(y)M_{q+2}(x,y)=0$ on $Y\times X$. In order to do this, Theorem 1.4.9 is used in full force.

Consider the complex $L^\cdot$ on $Y\times X$ endowed with the differential $P(x)$, where $L^\cdot$ is the subspace of $\mathcal{E}(E^\cdot\boxtimes E^{q+2\prime}|_{Y\times X})$ consisting of all sections f satisfying $P'_{q+1}(y)f=0$. Since $H^{q+2}(\mathcal{E}(E^\cdot|_Y))=0$, we get by Theorem 1.4.9 that $H^{q+2}(L^\cdot)=0$. Therefore, for the kernel $(S_{q+2}-P'_{q+2}(y)M_{q+3})\in Z^{q+2}(L^\cdot)$ there exists a kernel M_{q+2} of L^{q+1} such that $P_{q+1}(x)M_{q+2}=S_{q+2}-P'_{q+2}(y)M_{q+3}$ on $Y\times X$, which is what we wanted to prove.

Thus the components of the kernel $\Phi^{(q+1)}=\Phi+M$ satisfy the following system of differential equations on $Y\times X$:

$$P'_i(y)\Phi^{(q+1)}_{i+1}(x,y)+P_{i-1}(x)\Phi^{(q+1)}_i(x,y)=\mathrm{tr}\;\Delta^* \quad (i=q+1,\dots,N). \tag{5.5.4}$$

This system is solved by analogy with the system (5.5.3).

Consider the complex $\mathcal{E}(E^\cdot\boxtimes E^{N\prime}|_{Y\times(X\setminus\overline{Y})})$ endowed with the differential $P(x)$. Since $H^{N-1}(\mathcal{E}(E^\cdot|_Y))=0$, Theorem 1.4.9 implies that $H^{N-1}(\mathcal{E}(E^\cdot\boxtimes E^{N\prime}|_{Y\times(X\setminus\overline{Y})}))=0$. It follows from the last equation of (5.5.4) (i.e., for $i=N$) that $\Phi^{(q+1)}_N\in Z^{N-1}(\mathcal{E}(E^\cdot\boxtimes E^{N\prime}|_{Y\times(X\setminus\overline{Y})}))$ So for the kernel $\Phi^{(q+1)}_N$ we can find a kernel $K^{(-2)}_N\in\mathcal{E}(E^{N-2}\boxtimes E^{N\prime}|_{Y\times(X\setminus\overline{Y})})$ such that $P_{N-2}(x)K^{(-2)}_N=\Phi^{(q+1)}_N$ on the set $Y\times(X\setminus\overline{Y})$.

Define now the kernel $K^{(-1)}_{N-1}\in\mathcal{E}(E^{N-2}\boxtimes E^{N-1\prime}|_{Y\times(X\setminus\overline{Y})})$ from the equality $\Phi^{(q+1)}_{N-1}=K^{(-1)}_{N-1}-P'_{N-1}(y)K^{(-2)}_N$ on $Y\times(X\setminus\overline{Y})$. Then, using the next to last equality of (5.5.4) (i.e., for $i=N-1$), we obtain

$$P_{N-2}(x)K^{(-1)}_{N-1}=P_{N-2}(x)\Phi^{(q+1)}_{N-1}+P'_{N-1}(y)(P_{N-2}(x)K^{(-2)}_N)=0 \quad\text{on}\quad Y\times(X\setminus\overline{Y}).$$

Consider the complex $\mathcal{E}(E^\cdot\boxtimes E^{N-1\prime}|_{Y\times(X\setminus\overline{Y})})$ endowed with the differential $P(x)$. Since $H^{N-2}(\mathcal{E}(E^\cdot|_Y))=0$, Theorem 1.4.9 yields $H^{N-2}(\mathcal{E}(E^\cdot\boxtimes E^{N-1\prime}|_{Y\times(X\setminus\overline{Y})}))=0$. Therefore, for the kernel $K^{(-1)}_{N-1}\in Z^{N-2}(\mathcal{E}(E^\cdot\boxtimes E^{N-1\prime}|_{Y\times(X\setminus\overline{Y})}))$ we can find a kernel $K^{(-2)}_{N-1}\in\mathcal{E}(E^{N-3}\boxtimes E^{N-1\prime}|_{Y\times(X\setminus\overline{Y})})$ such that $P_{N-3}(x)K^{(-2)}_{N-1}=K^{(-1)}_{N-1}$ in the product $Y\times(X\setminus\overline{Y})$.

Define the kernel $K^{(-1)}_{N-2}\in\mathcal{E}(E^{N-3}\boxtimes E^{N-2\prime}|_{Y\times(X\setminus\overline{Y})})$ from the equality $\Phi^{(q+1)}_{N-2}=K^{(-1)}_{N-2}-P'_{N-2}(y)K^{(-2)}_{N-1}$ on $Y\times(X\setminus\overline{Y})$. Then, using the previous to next to last equality of (5.5.4), we get $P_{N-3}(x)K^{(-1)}_{N-2}=0$ on $Y\times(X\setminus\overline{Y})$.

Continuing this process as far as the $(q+1)$-convexity of Y allows, we construct kernels $K^{(-2)}_i\in\mathcal{E}(E^{i-2}\boxtimes E^{i\prime}|_{Y\times(X\setminus\overline{Y})})$ $(i=q+2,\dots,N)$ such that $\Phi^{(q+1)}_i=P_{i-2}(x)K^{(-2)}_i-P'_i(y)K^{(-2)}_{i+1}$ $(i\ge q+2)$ on the set $Y\times(X\setminus\overline{Y})$.

Define the kernel $K^{(-1)}_{q+1}\in\mathcal{E}(E^q\boxtimes E^{q+1\prime}|_{Y\times(X\setminus\overline{Y})})$ from the equality $\Phi^{(q+1)}_{q+1}=K^{(-1)}_{q+1}-P'_{q+1}(y)K^{(-2)}_{q+2}$ on $Y\times(X\setminus\overline{Y})$. Then, using the first equality of (5.5.4), we obtain on the set $Y\times(X\setminus\overline{Y})$:

$$\begin{aligned}P_q(x)K^{(-1)}_{q+1} &= P_q(x)\Phi^{(q+1)}_{q+1}+P'_{q+1}(y)\left(P_q(x)K^{(-2)}_{q+2}\right)\\ &= P_q(x)\Phi^{(q+1)}_{q+1}+P'_{q+1}(y)\left(\Phi^{(q+1)}_{q+2}+P'_{q+2}(y)K^{(-2)}_{q+3}\right)\\ &= 0.\end{aligned}$$

To complete the proof of the lemma it suffices to recall that $\Phi_{q+1}^{(q+1)} = \Phi_{q+1}$ on $Y \times X$.

□

5.5.7 *Proof of the Grothendieck duality theorem*

It follows from Proposition 2.4.6 that the bilinear form $[\partial Y]^P$ given initially on $Z_{q+1}(\vec{\mathcal{E}}(E^{\cdot\prime}|_{X\setminus Y})) \otimes Z^q(\mathcal{E}(E^\cdot|_Y))$ induces the pairing

$$H^{reg}_{q+1}(\vec{\mathcal{E}}(E^{\cdot\prime}|_{X\setminus Y})) \otimes \left(\frac{Z^q(\mathcal{E}(E^\cdot|_Y))}{\widehat{Z^q(\mathcal{E}(E^\cdot))}}\right) \to \mathbb{C}.$$

We also note that this pairing is continuous in the second argument.

We now show that the pairing $[\partial Y]^P$ is non-degenerate in the first argument on $H^{reg}_{q+1}(\vec{\mathcal{E}}(E^{\cdot\prime}|_{X\setminus Y})) \otimes \left(\frac{Z^q(\mathcal{E}(E^\cdot|_Y))}{\widehat{Z^q(\mathcal{E}(E^\cdot))}}\right)$.

Suppose that U_g is a neighborhood of $X\setminus Y$ and $g \in Z_{q+1}(\mathcal{E}(E^{\cdot\prime}|_{U_g}))$ is a solution of $P'_q g = 0$ in U_g regular at infinity such that

$$[\partial Y]^P(g \otimes f) = 0 \quad \text{for all} \quad f \in Z^q(\mathcal{E}(E^\cdot|_Y)). \tag{5.5.5}$$

Choose a smaller neighborhood U of $X \setminus Y$ with piecewise smooth boundary, i.e., such that $\overline{U} \subset U_g$.

By hypothesis, there exists an increasing sequence of open sets $(q+1)$-convex with respect to $E^\cdot$ that approximate Y from the inside. Let $\{Y_\nu\}$ be this sequence, i.e., $\overline{Y_\nu} \subset Y_{\nu+1}$ and $\cup_\nu Y_\nu = Y$. Then we can take the number ν so large that $\partial Y_\nu \subset U$.

Since the solution g is regular at infinity, we have by the Green formula

$$g = -\Phi'([\partial U]^{P'} g) \pmod{B_{q+1}(\mathcal{E}(E^{\cdot\prime}|_U))}. \tag{5.5.6}$$

To transform the integral we apply Lemma 5.5.7 on the set Y_ν. Denoting by $K_\nu^{(-1)}$ and $K_\nu^{(-2)}$ the operators guaranteed by this lemma, we obtain

$$\Phi'([\partial U]^{P'} g) = K_\nu^{(-1)\prime}([\partial U]^{P'} g) - P' K_\nu^{(-2)\prime}([\partial U]^{P'} g) \quad \text{in} \quad X \setminus \overline{Y_\nu}. \tag{5.5.7}$$

Combining (5.5.6) and (5.5.7) and using Proposition 2.4.5, we arrive at

$$g(y) = [\partial U]^P (g \otimes K_\nu^{(-1)}(\cdot, y)) \pmod{B_{q+1}(\mathcal{E}(E^{\cdot\prime}|_{X\setminus\overline{Y_\nu}}))}. \tag{5.5.8}$$

According to Lemma 5.5.7, the kernel $K_\nu^{(-1)}$ satisfies $P_q K_\nu^{(-1)}(\cdot, y) = 0$ in Y_ν for each fixed point $y \in X \setminus \mathcal{O}_\nu$. Hence for sufficiently large ν the right-hand side of (5.5.8) vanishes by (5.5.5) and Definition 5.5.4. So the quotient class of g in $H^{reg}_{q+1}(\vec{\mathcal{E}}(E^{\cdot\prime}|_{X\setminus Y}))$ is zero as desired. Thus, we have embedded $H^{reg}_{q+1}(\vec{\mathcal{E}}(E^{\cdot\prime}|_{X\setminus Y}))$ into the space of continuous liner functionals on $\left(\frac{Z^q(\mathcal{E}(E^\cdot|_Y))}{\widehat{Z^q(\mathcal{E}(E^\cdot))}}\right)$.

We now verify that this embedding is an epimorphism. By the well-known property of Fréchet-Schwartz spaces, $\left(\frac{Z^q(\mathcal{E}(E^\cdot|_Y))}{Z^q(\widehat{\mathcal{E}(E^\cdot}))}\right)'$ is isomorphic as a topological vector space to the annihilator of the subspace $Z^q(\widehat{\mathcal{E}(E^\cdot}))$ in $Z^q(\mathcal{E}(E^\cdot|_Y))$ that is denoted by $\left(Z^q(\widehat{\mathcal{E}(E^\cdot}))\right)^\perp$ (see Grothendieck [91]). Let $g \in \left(Z^q(\widehat{\mathcal{E}(E^\cdot}))\right)^\perp$, i.e., $g \in (Z^q(\mathcal{E}(E^\cdot|_Y)))'$ and g vanishes on $Z^q(\widehat{\mathcal{E}(E^\cdot}))$. By the Hahn-Banach Theorem, g can be extended to a continuous linear functional $\tilde{g}$ on the space $\mathcal{E}(E^q|_Y)$. That is, there exists a section $\tilde{g} \in \mathcal{E}'(E^{q\prime}|_Y)$ equal to g on $Z^q(\widehat{\mathcal{E}(E^\cdot}))$. We notice at once that $\tilde{g}$ satisfies $P'\tilde{g} = 0$ on X independently of the extension because $B^q(\mathcal{E}(E^\cdot))$ belongs to $Z^q(\widehat{\mathcal{E}(E^\cdot}))$.

Let $U \subset\subset Y$ be a neighborhood of the support of $\tilde{g}$ with piecewise smooth boundary. If $f \in Z^q(\mathcal{E}(E^\cdot|_Y))$, then by (2.4.22)

$$-\Phi([\partial U]^P f) + P\Phi(\chi_U f) = \chi_U f - S(\chi_U f). \tag{5.5.9}$$

Since Φ is a fundamental solution for the complex $E^\cdot$ at degree $(q+1)$, we have $S_{q+1} = 0$ and consequently $P_q S_q = S_{q+1} P_q = 0$. Therefore the section $S(\chi_U f)$ is in $Z^q(\mathcal{E}(E^\cdot))$. Moreover, it is seen from formula (5.5.9) that in the complement of the compact set $K = \overline{U}$ we have $S(\chi_U f) = \Phi([\partial U]^P f)$ modulo $B^q(\mathcal{E}(E^\cdot|_{X\setminus K}))$ and $[\partial U]^P f \in Z^{q+1}(\mathcal{E}'(E^\cdot))$. Thus the section $S(\chi_U f)$ of (5.5.9) lies in $Z^q(\widehat{\mathcal{E}(E^\cdot}))$.

Further, it is easy to see that the section $P\Phi(\chi_U f)$ coincides in some neighborhood of the support of $\tilde{g}$ with an element of $B^q(\mathcal{E}(E^\cdot)) \subset Z^q(\widehat{\mathcal{E}(E^\cdot}))$ because $\Phi(\chi_U f) \in \mathcal{E}(E^{q-1}|_U)$.

Applying now the functional $\tilde{g}$ to both sides of (5.5.9), we have

$$g(f) = \tilde{g}(-\Phi([\partial U]^P f)) = [\partial U]^P((-\Phi'(\tilde{g})) \otimes f). \tag{5.5.10}$$

Because of the pseudo-locality property of pseudo-differential operators (see Proposition 1.5.12), the potential $(-\Phi'(\tilde{g})) \in \mathcal{D}'(E^{q+1\prime})$ is differentiable in the complement of the support of $\tilde{g}$. In particular, it is differentiable in some neighborhood of $X \setminus Y$.

We now show that $P'(-\Phi'(\tilde{g})) = 0$ in the neighborhood. That is,

$$\langle -\Phi'(\tilde{g}), Pu\rangle_X = 0$$

for all sections $u \in \mathcal{D}(E^q)$ supported in the complement of the support of $\tilde{g}$. Let u be such a section. Using the homotopy formula (2.1.4), we obtain

$$\begin{aligned}\langle -\Phi'(\tilde{g}), Pu\rangle_X &= \langle -\tilde{g}, \Phi Pu\rangle_X \\ &= \langle -\tilde{g}, u - Su - P\Phi u\rangle_X \\ &= \langle -\tilde{g}, Su + P\Phi u\rangle_X.\end{aligned}$$

Notice that the section $(Su + P\Phi u)$ belongs to $Z^q(\widehat{\mathcal{E}(E^\cdot}))$. Indeed, we have on the one hand obviously that $(Su + P\Phi u) \in Z^q(\mathcal{E}(E^\cdot))$, and on the other hand

$(Su + P\Phi u) = \Phi(-Pu)$ in the complement of the support of u where $(-Pu) \in Z^{q+1}(\mathcal{D}(E^{\cdot}))$. Therefore $\langle -\Phi'(\tilde{g}), Pu\rangle_X = \langle g, Su + P\Phi u\rangle_X = 0$, as desired.

Thus the potential $(-\Phi'(\tilde{g}))$ is a differentiable solution of the system $P_q'(\cdot) = 0$ in a neighborhood of $X \setminus Y$ and by definition it is regular at infinity. Returning to formula (5.5.10), we see that $g([f]) = [\partial Y]^P([-\Phi'(\tilde{g})], [f])$ for all $[f] \in \left(\frac{Z^q(\mathcal{E}(E^{\cdot}|_Y))}{Z^q(\widehat{\mathcal{E}(E^{\cdot}}))}\right)$ where $[-\Phi'(\tilde{g})]$ is the homology class of the section$(-\Phi'(g))$ in $H^{reg}_{q+1}(\vec{\mathcal{E}}(E^{\cdot\prime}|_{X\setminus Y}))$.

In spite of the non-uniqueness of the extension $\tilde{g}$, the class $[-\Phi'(\tilde{g})]$, is determined uniquely because it is equal to the homology class of the section $(-K_\nu^{(-1)\prime}(g))$ in $H^{reg}_{q+1}(\vec{\mathcal{E}}(E^{\cdot\prime}|_{X\setminus Y}))$, where the number ν is taken so large that the support of $\tilde{g}$ is contained in Y_ν.

Thus, the mapping of $H^{reg}_{q+1}(\mathcal{E}(E^{\cdot\prime}|_{X\setminus Y}))$ to the dual space of $\left(\frac{Z^q(\mathcal{E}(E^{\cdot}|_Y))}{Z^q(\widehat{\mathcal{E}(E^{\cdot}}))}\right)$, given by $[g] \to [\partial Y]^P([g] \otimes \cdot)$, is an isomorphism of vector spaces (without topologies). Moreover the mapping inverse to this one is given by associating to a functional $g \in \left(\frac{Z^q(\mathcal{E}(E^{\cdot}|_Y))}{Z^q(\widehat{\mathcal{E}(E^{\cdot}}))}\right)'$ its "Fantappiè indicatrix" $[-\Phi'(\tilde{g})]$, where $\tilde{g} \in \mathcal{E}'(E^{q\prime}|_Y)$ is some extension of the functional g.

To complete the proof it remains only to observe that each of these mappings is continuous if $\left(\frac{Z^q(\mathcal{E}(E^{\cdot}|_Y))}{Z^q(\widehat{\mathcal{E}(E^{\cdot}}))}\right)'$ is provided with the strong dual topology.

□

5.5.8 *Some consequences*

Corollary 5.5.8 *Let X have no compact connected components, and P be a differential operator with surjective symbol and real analytic coefficients on X. Then for each open set $Y \subset\subset X$ we have*

$$\left(\frac{\mathcal{E}_P(Y)}{\widehat{\mathcal{E}_P(X)}}\right)' \overset{top.}{\cong} \frac{\mathcal{E}_{P'}(X \setminus Y)}{\mathcal{E}_{P'}(X)}. \tag{5.5.11}$$

Proof. Denote by $E^{\cdot}$ the transposed complex for a compatibility complex for the differential operator P'. We have $P = P_{N-1}$ where N is the length of $E^{\cdot}$. It follows from Corollaries 2.1.6 and 2.1.14 that the complex $E^{\cdot}$ has a (properly supported) pseudo-differential parametrix on X. Further, according to the well-known theorem of Malgrange [156, p.341] the mapping $\Delta : \mathcal{E}(E^N) \to \mathcal{E}(E^N)$ is surjective where $\Delta = PP^*$. Therefore the mapping $P : \mathcal{E}(E^{N-1}) \to \mathcal{E}(E^N)$ is also surjective, i.e., $H^N(\mathcal{E}(E^{\cdot})) = 0$. We see that the conditions of Theorem 2.3.12 are fulfilled for $q = N$. According to this theorem, $E^{\cdot}$ has a pseudo-differential fundamental solution Φ at degree N. Moreover, every open subset of X is N-convex with respect to the complex $E^{\cdot}$. Also, because of Proposition 5.5.5 the approximation condition for $E^{\cdot}$ is fulfilled at degree $(N-1)$ on every open set $Y \subset X$. We conclude that all the conditions of Theorem 5.5.6 are satisfied for $q = N - 1$. Combining now Theorem 5.5.6 and Proposition 5.5.3 we obtain the isomorphism (5.5.11), as desired.

□

In particular, if P is an elliptic differential operator with real analytic coefficients on X, then $\widehat{\mathcal{E}_P(X)} = \{0\}$ because of Corollary 5.4.3. In this case the isomorphism (5.5.11) gives just the duality theorem of Grothendieck [89].

There are qualitative new consequences of Theorem 5.5.6 connected with overdetermined systems.

Corollary 5.5.9 *If X is a Stein manifold of dimension n and $Y \subset\subset X$ is a holomorphically convex open set, then*

$$\mathcal{O}^{(p)}(Y)' \overset{top.}{\cong} \frac{H^{n-1}(X \setminus Y, \mathcal{O}^{(n-p)})}{H^{n-1}(X, \mathcal{O}^{(n-p)})}. \tag{5.5.12}$$

Proof. Let $E^{\cdot} = \Lambda^{p,\cdot}T^*_{\mathbb{C}}(X)$ be the Dolbeault complex with coefficients in $\Lambda^{p,0}T^*_{\mathbb{C}}(X)$. According to Theorem B of Cartan (see Gunning and Rossi [96, p.310]), the manifold X is 1-convex with respect to $E^{\cdot}$.

Moreover, by the pairing $\Lambda^{n-p,n-q}T^*_{\mathbb{C}}(X) \otimes \Lambda^{p,q}T^*_{\mathbb{C}}(X) \to \Lambda^{n,n}T^*_{\mathbb{C}}(X)$ the transposed complex $E^{\cdot\prime}$ can be identified with the complex $\Lambda^{n-p,\cdot}T^*_{\mathbb{C}}(X)$ (see Example 1.5.23). Therefore

$$H_0(\mathcal{E}(E^{\cdot\prime})) \cong H^n(\mathcal{E}(\Lambda^{n-p,\cdot}T^*_{\mathbb{C}}(X))) = 0$$

because of the same Theorem B of Cartan. We see that all the conditions of Corollary 2.3.14 are satisfied and so the complex $E^{\cdot}$ has a pseudo-differential fundamental solutions Φ on X. Further, if $Y \subset\subset X$ is a holomorphically convex open set, then Y can be exhausted by an increasing sequence $\{Y_\nu\}$ of special analytic polyhedra (ibid, p.276). Each polyhedron Y_ν is a Stein manifold and hence it is 1-convex with respect to $E^{\cdot}$. Finally, it follows from the Oka-Weil Theorem (ibid, p.267) that the approximation condition for $E^{\cdot}$ is fulfilled on Y at degree 0. We conclude that all the conditions of Theorem 5.5.6 are satisfied for $q = 0$. Combining now Theorem 5.5.6 and Propositions 5.5.3 and 1.2.49, we obtain the isomorphism (5.5.12), as desired. □

For $n = 1$, the isomorphism (5.5.12) expresses just the Köthe-Grothendieck-Sebastiao e Silva duality mentioned in Subsection 5.5.1. As to the case $n > 1$, we have $H^{n-1}(X, \mathcal{O}^{(n-p)}) = 0$ because of Theorem B of Cartan. Therefore, one has $\mathcal{O}^{(p)}(Y)' \overset{top.}{\cong} H^{n-1}(X \setminus Y, \mathcal{O}^{(n-p)})$. This is apparently part of the folk-lore of complex analysis.

Chapter 6

The Atiyah-Bott-Lefschetz Theorem on Fixed Points for Elliptic Complexes

6.1 The Argument Principle for Elliptic Complexes

6.1.1 *A brief history*

The logarithmic residue formula in complex analysis obtained independently by Yuzhakov and Roos (see [6, §3]) has found applications to implicit functions, in algebraic geometry, in combinatorial analysis, in number theory, etc. For a recent account of the theory, we refer to the survey of Aizenberg, Tsikh and Yuzhakov [3]. A more general formula in the language of currents was earlier proved by King [121]. In the complex plane, this is just the Poincaré-Lelong formula. Another attractive particular case of King's formula was later found by Luppaciolu [153]. An argument principle turned out to be valid also for solutions of elliptic systems, and even in the context of elliptic complexes (see author's papers [221, 232]). However, such formulas had earlier been used by Toledo [238].

6.1.2 *Chains defined by smooth mappings*

Let X be a differentiable manifold of dimension n countable at infinity. In the sequel, $y = m(x)$ is a mapping of X to an open set $Y \subset \mathbb{R}^N$ ($N \le n$) containing 0. As to smoothness, m is supposed to be of class C^s_{loc} where $s = \max(n - N + 1, 2)$.

Set $dm = dm_1 \wedge \ldots \wedge dm_N$, where m_j ($1 \le j \le N$) is the j-th component of m. In general, dm might vanish at some points of the set $m^{-1}(0)$ which is defined to consist of all $x \in X$ such that $m(x) = 0$. However $m^{-1}(0)$ is here assumed to be a submanifold of X of dimension $(n - N)$ (and of class C^1_{loc}. If $n = N$, this means just that $m^{-1}(0)$ is discrete.

Remark 6.1.1 *According to Sard's Theorem, almost all points $y \in Y$ are regular values of m (see Pontryagin [183, p.34]). This means $dm \neq 0$ at each point of $m^{-1}(y)$, and so $m^{-1}(y)$ is an $(n-N)$-dimensional submanifold of X of class C^s_{loc} (possibly, $m^{-1}(y) = \emptyset$).*

Given a regular value $y \in Y$ of the mapping m, we endow the manifold $m^{-1}(y)$ with the orientation defined by the differential form

$$i^\sharp\left(\frac{dm}{|dm|}\rfloor dx\right)$$

of maximal degree on $m^{-1}(y)$, where i stands for the inclusion map $m^{-1}(y) \hookrightarrow X$.

Lemma 6.1.2 *Let $\mathcal{D}$ be a subdomain of X with piecewise smooth boundary such that $dm \neq 0$ on $m^{-1}(0) \cap \partial\mathcal{D}$, and $m^{-1}(0) \cap \partial\mathcal{D}$ has zero Lebesgue measure on $m^{-1}(0)$. Then, for every differential form $f \in C^1_{comp}(\Lambda^{n-N}T^*_{\mathbb{C}}(\overline{\mathcal{D}}))$, the limit*

$$\lim_{y_\nu \to 0} \int_{m^{-1}(y_\nu)\cap\mathcal{D}} f$$

exists and does not depend on the sequence $\{y_\nu\}$ of regular values of m tending to 0.

Proof. We limit ourselves to a sketch of the proof.

Let $m^{-1}(0) \cup c_j$ be the decomposition of $m^{-1}(0)$ into the union of connected components, so that $c_i \cap c_j = \emptyset$ $(i \neq j)$. Every component c_j is a submanifold of X of dimension $(n-N)$ and of class C^1_{loc}. We endow each of the components with some orientation.

Moreover we provide the manifolds c_j with multiplicities μ_j in the following way. Let x be a fixed point of c_j and $N_x(c_j)$ be a smooth submanifold of X of dimension N normal to c_j at the point x. The orientation of $N_x(c_j)$ is determined by the condition that at the point x it coheres with the orientation of c_j.

The point x is an isolated zero of the mapping $m : N_x(c_j) \to Y$. We define $\mu_j(x)$ to be equal to the local degree of the mapping $m : N_x(c_j) \to Y$ at the point x (see Pontryagin [183, §9]).

Using properties of the local degree of a mapping, one can easily prove that the function $\mu_j(x)$ is locally constant on c_j. Hence $\mu_j(x)$ is constant on c_j because c_j is connected.

The multiplicity μ_j certainly depends on the orientation of c_j. If the orientation of c_j is replaced by the opposite one, then μ_j is multiplied by (-1). Hence it follows that for any differential form $f \in C^1_{comp}(\Lambda^{n-N}T^*_{\mathbb{C}}(\overline{\mathcal{D}}))$ the integral $\mu_j \cdot \int_{c_j\cap\mathcal{D}} f$ does not depend on the particular choice of the orientation of c_j, and it is uniquely determined by m and f.

To finish the proof it suffices to observe that under the conditions of the lemma the following equality holds:

$$\lim_{y_\nu \to 0} \int_{m^{-1}(y_\nu)\cap\mathcal{D}} f = \sum_j \mu_j \cdot \int_{c_j\cap\mathcal{D}} f. \tag{6.1.1}$$

□

Given a differential form $f \in C^1_{comp}(\Lambda^{n-N}T^*_{\mathbb{C}}(\overline{\mathcal{D}}))$, the integral $\int_{m^{-1}(0)\cap\mathcal{D}} f$ is defined to be the limit $\lim_{y_\nu \to 0}(\int_{m^{-1}(y_\nu)\cap\mathcal{D}} f)$ where $\{y_\nu\} \subset Y$ is a sequence of regular values of m converging to 0. Another expression is given by (6.1.1). In particular, if 0 is a regular value of the mapping m, then the integral $\int_{m^{-1}(0)\cap\mathcal{D}} f$ is equal to the integral of f over the manifold with boundary $m^{-1}(0) \cap \overline{\mathcal{D}}$ which is defined in the usual way.

6.1.3 *The argument principle*

Let $P \in \mathrm{do}_1(E \to F)$ be a first order differential operator with injective symbol on Y. Here $E = Y \times \mathbb{C}^k$ and $F = Y \times \mathbb{C}^l$ are (trivial) vector bundles over Y, and P is given by an $(l \times k)$-matrix $P(y, D) = \sum_{j=1}^N P^{(j)}(y) \cdot D_j + P_0(y)$ of scalar differential operators on Y.

We assume that P has a left fundamental solution $\Phi \in \mathrm{pdo}_{-1}(F \to E)$ on Y. This is the case, for instance, if the coefficients of P are real analytic on Y (see Example 2.3.15).

As above, $m^\sharp$ denotes the "pull-back" operator on differential forms induced by the mapping m.

How should one "lower" the equation $Pf = 0$ from Y to X using the mapping m? Of course, an answer should depend on the purpose. We suggest the following one.

Given $q \in \mathbb{Z}$, define the differential operator $P^{(m)}$ of type $m^*E \otimes \Lambda^q T^*_{\mathbb{C}}(X) \to m^*F \otimes \Lambda^{q+N} T^*_{\mathbb{C}}(X)$ and order 1 on X by the equality

$$P^{(m)} f = \frac{(-1)^{N-1}}{\sqrt{-1}} m^\sharp \left(\sum_{j=1}^N P^{(j)} \star dy_j \right) \wedge df + m^\sharp (P_0 dy) \wedge f \quad \text{for} \quad f \in [\mathcal{E}(\Lambda^q T^*_{\mathbb{C}}(X))]^k . \tag{6.1.2}$$

Exercise 6.1.3 *Prove that, given a section $f \in \mathcal{E}(E)$, the equality*

$$P^{(m)}(m^* f \cdot \varphi) = m^*(Pf) \cdot dm \wedge \varphi$$

holds for every closed differential form $\varphi \in \mathcal{E}(\Lambda^q T^*_{\mathbb{C}}(X))$.

Hint. Use Proposition 1.2.38.

□

It follows that the construction (6.1.2) gives rise to complexes of differential operators on X.

Exercise 6.1.4 *Prove that, if $Q \in do_1(F \to G)$ is a differential operator on Y such that $Q \circ P = 0$, then $Q^{(m)} \circ P^{(m)} = 0$.*

Hint. In the only non-obvious case $N = 1$ use the equality

$$Q^{(1)} \cdot D_1 P^{(1)} + Q^{(1)} \cdot D_1 P_0 + Q^{(1)} \cdot P_0 = 0$$

which follows from $Q \circ P = 0$.

□

We may now formulate an argument principle for solutions of the system $Pf = 0$.

Theorem 6.1.5 *Let $\mathcal{D} \subset X$ be a domain with piecewise smooth boundary such that $dm \neq 0$ on $m^{-1}(0) \cap \partial\mathcal{D}$ and $m^{-1}(0) \cap \partial\mathcal{D}$ has zero Lebesgue measure on $m^{-1}(0)$. Then for every ($\mathbb{C}^k$-valued) differential form $f \in \left[C^1_{comp}(\Lambda^{n-N}T^*_{\mathbb{C}}(\overline{\mathcal{D}}))\right]^k$ satisfying $P^{(m)}f = 0$ in $\mathcal{D}$ the following formula holds:*

$$-\int_{\partial\mathcal{D}} m^\sharp\left(G_P(\Phi(0,\cdot), I_k)\right) \wedge f = \int_{m^{-1}(0)\cap\mathcal{D}} f. \tag{6.1.3}$$

The theorem becomes quite transparent when applied to sections of the form $f(m(\cdot))\varphi$ where $f \in \mathcal{E}_P(Y)$, and $\varphi \in C^1_{comp}(\Lambda^{n-N}T^*_{\mathbb{C}}(\overline{\mathcal{D}}))$ is a closed differential form. We get

$$-\int_{\partial\mathcal{D}} m^\sharp(G_P(\Phi(0,\cdot), f)) \wedge \varphi = f(0) \int_{m^{-1}(0)\cap\mathcal{D}} \varphi$$

because of Exercise 6.1.3. In such a form the statement seems to be extendible to arbitrary differential operators P (not necessarily of order 1).

Exercise 6.1.6 *Extend the argument principle to an arbitrary differential operator $P \in do_p(E \to F)$ with injective symbol and left fundamental solution on Y.*

Hint. If P has a left fundamental solution on Y, then according to Proposition 2.3.9 P has also one in $\text{pdo}_{-p}(F \to E)$. Use now the remark after Theorem 6.1.5.

□

6.1.4 *Proof of Theorem 6.1.5*

It suffices to prove Theorem 6.1.5 only for domains $\mathcal{D}$ relatively compact in X.

We divide the proof into two parts: 1) 0 is a regular value of m, and 2) the general case.

1) Assume that $dm \neq 0$ everywhere on the set $m^{-1}(0)$.

Since $\|\Phi(0,y)\| \leq c\,\frac{1}{|y|^{N-1}}$ for y small enough and the set $m^{-1}(0) \cap \partial\mathcal{D}$ has zero $(n-N)$-dimensional Lebesgue measure, the differential form $m^\sharp(G_P(\Phi(0,\cdot), I_k)) \wedge f$ is absolutely integrable over $\partial\mathcal{D}$. Therefore the integral on the left-hand side of (6.1.3) is well-defined.

We now show that the ($\mathbb{C}^k$-valued) differential form $m^\sharp(G_P(\Phi(0,\cdot), I_k)) \wedge f$ is closed in $\mathcal{D} \setminus m^{-1}(0)$.

Using Proposition 1.2.38 and the definition of the Green operator we obtain

$$\begin{aligned} &d\left(m^\sharp(G_P(\Phi(0,\cdot), I_k)) \wedge f\right) \\ &\quad = m^\sharp((\Phi(0,\cdot) \cdot PI_k - P'\Phi(0,\cdot))dy) \wedge f + (-1)^{N-1} m^\sharp(G_P(\Phi(0,\cdot), I_k)) \wedge df. \end{aligned}$$

Since for a first order differential operator P there exists a unique Green operator G_P given by formula (2.4.13), we see that

$$d\left(m^{\sharp}(G_P(\Phi(0,\cdot),I_k))\wedge f\right) = -m^{\sharp}(\delta(y)dy)\wedge f + \Phi(0,m(\cdot))\cdot P^{(m)}f. \qquad (6.1.4)$$

By hypothesis we have $P^{(m)}f = 0$ in $\mathcal{D}$. It follows that the differential form $m^{\sharp}(G_P(\Phi(0,\cdot),I_k))\wedge f$ is closed in $\mathcal{D}\setminus m^{-1}(0)$, as desired.

If now U is a neighborhood with piecewise smooth boundary of the set $m^{-1}(0)$ in X, then by Stokes' Theorem we get

$$\int_{\partial\mathcal{D}\setminus U} m^{\sharp}(G_P(\Phi(0,\cdot),I_k))\wedge f = \int_{\partial U\cap\mathcal{D}} m^{\sharp}(G_P(\Phi(0,\cdot),I_k))\wedge f. \qquad (6.1.5)$$

We construct such a neighborhood U by using the function $\varepsilon = |m(x)|$ on X. Since $0\in\mathbb{R}^N$ is a regular value of the mapping m, the image $m(X)$ contains some neighborhood of 0 in $\mathbb{R}^N$. Therefore the image $|m|(X)$ contains some neighborhood of 0 in $\overline{\mathbb{R}^+}$. Moreover, almost all point of the neighborhood are regular values of $|m|$ because of Sard's Theorem (see Remark 6.1.1).

We take $U = \{x\in X:\ |m(x)|<\varepsilon\}$ where $\varepsilon>0$ is a sufficiently small regular value of $|m|$. It follows from the Implicit Function Theorem that the boundary of U given by $\partial U = \{x\in X:\ |m(x)|=\varepsilon\}$ is a hypersurface of class C^s_{loc} in X.

Let us start with the calculation of the integral in the right-hand side of (6.1.5). Since $dm\neq 0$ on the set $m^{-1}(0)$, for each point $p\in m^{-1}(0)$ there exists a neighborhood $U(p)$ in X and a smooth mapping $z = z^{(p)}(x)$ of $U(p)\to\mathbb{R}^{n-N}$ such that the mapping $x^{(p)}:\ U(p)\to\mathbb{R}^n$ given by $x^{(p)}(x) = (m(x), z^{(p)}(x))$ provides a coordinate system in $U(p)$. From compactness considerations, we may choose a finite number of points $\{p_\nu\}$ in $m^{-1}(0)$ and a regular value $\varepsilon>0$ of $|m|$ so small that $U\cap\mathcal{D}$ is a relatively compact subset of $\cup_\nu U(p_\nu)$. Set $U_\nu = U(p_\nu)$ and let $x^{(\nu)}(x) = (m(x), z^{(\nu)}(x))$ be the corresponding coordinate systems in U_ν.

Therefore, if $\{\varphi_\nu\}$ is a partition of unity on $\cup_\nu U_\nu$ subordinate to the covering $\{U_\nu\}$, then we get

$$\begin{aligned}
&\int_{\partial U\cap\mathcal{D}} m^{\sharp}(G_P(\Phi(0,\cdot),I_k))\wedge f\\
&\quad = \sum_\nu \int_{\partial U\cap\mathcal{D}} m^{\sharp}(G_P(\Phi(0,\cdot),I_k))\wedge(\varphi_\nu f)\\
&\quad = \sum_\nu \int_{x^{(\nu)}(\partial U\cap\mathcal{D}\cap U_\nu)} G_P(\Phi(0,\cdot),I_k)\wedge({x^{(\nu)}}^{-1})^{\sharp}(\varphi_\nu f).
\end{aligned} \qquad (6.1.6)$$

Let H be the hypersurface in $\mathbb{R}^n$ given by $\{(y,z)\in\mathbb{R}^n:\ |y|=\varepsilon\}$. Then $x^{(\nu)}(\partial U\cap\mathcal{D}\cap U_\nu) = H\cap x^{(\nu)}(\mathcal{D}\cap U_\nu)$. We now notice that H is a cylindrical hypersurface and it can be realized as the product $\partial B(0,\varepsilon)\times\mathbb{R}^{n-N}_z$, where $B(0,\varepsilon)$ is the ball with center 0 and radius ε in the space $\mathbb{R}^N_y$.

Hence it follows that if supp $\varphi_\nu\cap\partial\mathcal{D}=\emptyset$, we have

$$\int_{x^{(\nu)}(\partial U\cap\mathcal{D}\cap U_\nu)} G_P(\Phi(0,\cdot),I_k)\wedge({x^{(\nu)}}^{-1})^{\sharp}(\varphi_\nu f)$$

$$= \int_H G_P(\Phi(0,\cdot), I_k) \wedge (x^{(\nu)^{-1}})^\sharp(\varphi_\nu f)$$

$$= \int_{\mathbb{R}_z^{n-N}} \left\{ \int_{\partial B(0,\varepsilon)_y} G_P(\Phi(0,\cdot), I_k) \wedge (x^{(\nu)^{-1}})^\sharp(\varphi_\nu f) \right\}.$$

In order to transform the integral in the braces, we use Green's formula (2.5.9) and the condition $P^{(m)} f = 0$ in $\mathcal{D}$:

$$\begin{aligned}
&\int_{x^{(\nu)}(\partial U \cap \mathcal{D} \cap U_\nu)} G_P(\Phi(0,\cdot), I_k) \wedge (x^{(\nu)^{-1}})^\sharp(\varphi_\nu f) \\
&= \int_{\mathbb{R}_z^{n-N}} \Big\{ -(x^{(\nu)^{-1}})^\sharp(\varphi_\nu f)\Big|_{y=0} \\
&\qquad + \int_{B(0,\varepsilon)_y} (-1)^{N-1} G_P(\Phi(0,\cdot), I_k) \wedge (x^{(\nu)^{-1}})^\sharp(d\varphi_\nu \wedge f) \Big\} \\
&= -\int_{m^{-1}(0)\cap\mathcal{D}\cap U_\nu} (\varphi_\nu f) + \int_{U\cap\mathcal{D}\cap U_\nu} (-1)^{N-1} m^\sharp(G_P(\Phi(0,\cdot), I_k)) \wedge (d\varphi_\nu \wedge f).
\end{aligned} \tag{6.1.7}$$

Here the orientation of the manifold $m^{-1}(0)$ is defined by means of the differential form $i^\sharp\left(\frac{dm}{|dm|} \rfloor dx\right)$ of maximal degree on $m^{-1}(0)$, where i is the inclusion $m^{-1}(0) \hookrightarrow X$.

Suppose now that $\operatorname{supp} \varphi_\nu \cap \partial\mathcal{D} \neq \emptyset$. We write the integral over the set $H \cap x^{(\nu)}(\mathcal{D}\cap U_\nu)$ as an iterated integral where the exterior integral is taken over $\mathbb{R}_z^{n-N}$ and the inner integral is taken over the sections $\{y \in \partial B(0,\varepsilon) : (y,z) \in x^{(\nu)}(\mathcal{D}\cap U_\nu)\}$ for each fixed $z \in \mathbb{R}^{n-N}$.

Further, we complete every such section to the boundary of the open set $\sigma_z^{(\nu)} = \{y \in B(0,\varepsilon) : (y,z) \in x^\nu(\mathcal{D}\cap U_\nu)\}$ in $\mathbb{R}_y^N$ by adding $\{y \in B(0,\varepsilon) : (y,z) \in x^{(\nu)}(\partial(\mathcal{D}\cap U_\nu))\}$ to the section. The cycle $\partial\sigma_z^{(\nu)}$ is obviously piecewise smooth for almost all $z \in \mathbb{R}^{n-N}$.

Thus, we obtain

$$\begin{aligned}
&\int_{x^{(\nu)}(\partial U \cap \mathcal{D} \cap U_\nu)} G_P(\Phi(0,\cdot), I_k) \wedge (x^{(\nu)^{-1}})^\sharp(\varphi_\nu f) \\
&= \int_{\mathbb{R}_z^{n-N}} \left\{ \int_{\partial\sigma_z^{(\nu)}} G_P(\Phi(0,\cdot), I_k) \wedge (x^{(\nu)^{-1}})^\sharp(\varphi_\nu f) \right\} \\
&\quad - \int_{\mathbb{R}_z^{n-N}} \left\{ \int_{\substack{y\in B(0,\varepsilon):\\ (y,z)\in x^{(\nu)}(\partial\mathcal{D}\cap U_\nu)}} G_P(\Phi(0,\cdot), I_k) \wedge (x^{(\nu)^{-1}})^\sharp(\varphi_\nu f) \right\}.
\end{aligned} \tag{6.1.8}$$

Since the set $m^{-1}(0) \cap \partial\mathcal{D}$ has zero Lebesgue measure on $m^{-1}(0)$, the Lebesgue measure of the set $\{z \in \mathbb{R}^{n-N} : 0 \in \partial\sigma_z^{(\nu)}\}$ is zero also. Therefore, by calculating the integral over $\mathbb{R}^{n-N}$ we may disregard the points z such that $0 \in \mathbb{R}^N$ is on the boundary of $\sigma_z^{(\nu)}$. Concerning other points $z \in \mathbb{R}^{n-N}$, we can apply Green's formula (2.5.9) in order to transform the integral over $\partial\sigma_z^{(\nu)}$ in the right-hand side of (6.1.8) just as in (6.1.7).

By the same reasoning, the second integral in the right-hand side of (6.1.8) is well-defined and it is equal to the integral over $U \cap \partial\mathcal{D} \cap U_\nu$.

Hence

$$\int_{x^{(\nu)}(\partial U \cap \mathcal{D} \cap U_\nu)} G_P(\Phi(0,\cdot), I_k) \wedge (x^{(\nu)^{-1}})^\sharp(\varphi_\nu f)$$
$$= -\int_{m^{-1}(0)\cap\mathcal{D}\cap U_\nu} (\varphi_\nu f) + \int_{U\cap\mathcal{D}\cap U_\nu} (-1)^{N-1} m^\sharp(G_P(\Phi(0,\cdot), I_k)) \wedge (d\varphi_\nu \wedge f)$$
$$- \int_{U\cap\partial\mathcal{D}\cap U_\nu} m^\sharp(G_P(\Phi(0,\cdot), I_k)) \wedge (\varphi_\nu f). \tag{6.1.9}$$

We now substitute the expressions (6.1.7) and (6.1.9) into (6.1.6) and we take into account that $\sum_\nu \varphi_\nu = 1$ in a neighborhood of $\overline{U \cap \mathcal{D}}$, so $\sum_\nu d\varphi_\nu = 0$ there. Thus, after summation over ν, we get

$$\int_{\partial U \cap \mathcal{D}} m^\sharp(G_P(\Phi(0,\cdot), I_k)) \wedge f$$
$$= -\int_{m^{-1}(0)\cap\mathcal{D}} f - \int_{U\cap\partial\mathcal{D}} m^\sharp(G_P(\Phi(0,\cdot), I_k)) \wedge f. \tag{6.1.10}$$

Combining (6.1.5) and (6.1.10), we derive formula (6.1.3) in the case when 0 is a regular value of the mapping m. This is what we wanted to prove.

2) Now let m be an arbitrary mapping of X to Y satisfying the hypotheses of the theorem.

Since $dm \neq 0$ on $m^{-1}(0) \cap \partial\mathcal{D}$, the mapping $y = m(x)$ may be completed to a coordinate system in a neighborhood of each point $x \in m^{-1}(0) \cap \partial\mathcal{D}$ on X. So the same reasoning as in 1) shows that the integral in the left-hand side of (6.1.3) is well-defined.

Let U be a neighborhood of the point 0 in $\mathbb{R}^N$. Define by $U^{(1)}$ the set of critical values of m in U and by $U^{(2)}$ the set of those regular values $y \in U$ of m such that $m^{-1}(y) \cap \partial\mathcal{D}$ has non-zero Lebesgue measure on $m^{-1}(y)$. As mentioned, $U^{(1)}$ has zero Lebesgue measure in $\mathbb{R}^N$ (see Remark 6.1.1). One can also prove that the set $U^{(2)}$ has zero Lebesgue measure in $\mathbb{R}^N$. Therefore, the set $U^{(3)} = U \setminus (U^{(1)} \cup U^{(2)})$ is dense in U.

Choose a sequence $\{y_\nu\}$ of points of $U^{(3)}$ which converges to 0. For every ν, we set $m^{(\nu)}(x) = m(x) - y_\nu$. Then each mapping $y = m^{(\nu)}(x)$ satisfies the conditions of the theorem and 0 is a regular value of $m^{(\nu)}$. By 1), formula (6.1.3) holds for $m^{(\nu)}$, i.e.,

$$-\int_{\partial\mathcal{D}} (m - y_\nu)^\sharp(G_P(\Phi(0,\cdot), I_k)) \wedge f = \int_{m^{-1}(y_\nu)\cap\mathcal{D}} f. \tag{6.1.11}$$

Since $dm \neq 0$ on $m^{-1}(0) \cap \partial\mathcal{D}$ and $m^{-1}(0) \cap \partial\mathcal{D}$ has zero Lebesgue measure on $m^{-1}(0)$, it is easy to see that

$$\lim_{y_\nu \to 0} \int_{\partial\mathcal{D}} (m - y_\nu)^\sharp(G_P(\Phi(0,\cdot), I_k)) \wedge f = \int_{\partial\mathcal{D}} m^\sharp(G_P(\Phi(0,\cdot), I_k)) \wedge f.$$

Therefore, passing in (6.1.11) to the limit as $\nu \to \infty$ we get formula (6.1.3) in the general case, as desired.

□

Let us consider some examples and applications.

6.1.5 *The logarithmic residue formula for holomorphic functions*

Let X be a complex manifold of (complex) dimension n. Consider a holomorphic mapping $\zeta = m(z)$ of X to $\mathbb{C}^N$ $(N \le n)$. We assume that m is generic, i.e., $m^{-1}(0)$ is an analytic subvariety of X of dimension $(n-N)$.

We consider as P the Cauchy-Riemann operator in $\mathbb{C}^N$. I.e.,

$$P(D) = \begin{pmatrix} \partial/\partial\overline{\zeta}_1 \\ \dots \\ \partial/\partial\overline{\zeta}_N \end{pmatrix}$$

where the derivatives with respect to the complex variables $\zeta_j = y_j + \sqrt{-1}y_{N+j}$ $(j = 1, \dots, N)$ are given by $\frac{\partial}{\partial\zeta_j} = \frac{1}{2}\left(\frac{\partial}{\partial y_j} - \sqrt{-1}\frac{\partial}{\partial y_{N+j}}\right)$.

The left fundamental solution of convolution type $\Phi_1^{(0)}$ for P was constructed in Subsection 2.5.3, so that

$$-m^\sharp(G_P(\Phi_1^{(0)}(0,\cdot),1)) = \frac{(N-1)!}{(2\pi\sqrt{-1})^N}\sum_{j=1}^{N}(-1)^{j-1}\frac{\overline{m_j}}{|m|^{2N}}\, dm \wedge d\overline{m}[j].$$

This is just $((-1)$ times) the pull-back of the Martinelli-Bochner kernel in $\mathbb{C}^N$ under the mapping m i.e., $-m^\sharp(\Phi_1^{(0)}(0,\cdot))$.

Given a differential form $f \in C^1_{loc}(\Lambda^{n-N,q}T^*_{\mathbb{C}}(X))$, the condition $P^{(m)}f = 0$ is equivalent to $dm \wedge d\overline{m}[j] \wedge \overline{\partial} f = 0$ $(j = 1, \dots, N)$. In particular, the condition is fulfilled if f is $\overline{\partial}$-closed.

Thus, we get the following result proved earlier by Luppaciolu [153] under stronger conditions.

Corollary 6.1.7 *Let $\mathcal{D} \subset X$ be a domain with piecewise smooth boundary such that $dm \wedge d\overline{m} \ne 0$ on $m^{-1}(0) \cap \partial\mathcal{D}$ and $m^{-1}(0) \cap \partial\mathcal{D}$ has zero Lebesgue measure on $m^{-1}(0)$. Then for each $\overline{\partial}$-closed differential form $f \in C^1_{comp}(\Lambda^{n-N,n-N}T^*_{\mathbb{C}}(\overline{\mathcal{D}}))$, the following formula holds:*

$$-\int_{\partial\mathcal{D}} m^\sharp(\Phi_1^{(0)}(0,\cdot)) \wedge f = \int_{m^{-1}(0)\cap\mathcal{D}} f. \tag{6.1.12}$$

Proof. This follows immediately from Theorem 6.1.5.

□

If $N = n$, formula (6.1.12) is just the logarithmic residue formula for holomorphic functions due to Yuzhakov and Roos (see Aizenberg and Yuzhakov [6, §3])

Formula (6.1.12) also implies the well-known formula of Wirtinger [256] for the $(n-N)$-dimensional Lebesgue volume of $m^{-1}(0) \cap \mathcal{D}$ (see Luppaciolu [153]).

6.1.6 A generalization of the Kronecker formula

Let X be a differentiable manifold of dimension n. Consider a mapping $y = m(x)$ of X to $\mathbb{R}^n$ of class C^1_{loc}.

We assume that: 1) $m^{-1}(0)$ is discrete and 2) the n-dimensional Lebesgue measure of the set $\{x \in X : dm(x) = 0\}$ is equal to zero.

For a point $p \in m^{-1}(0)$, we denote by $\mu(p)$ the local degree of the mapping m at the point p (see Pontryagin [183, §9]).

If $\mathcal{D}$ is a relatively compact subdomain of X with piecewise smooth boundary such that $m^{-1}(0) \cap \partial\mathcal{D} = \emptyset$, then $\mu(m(\partial\mathcal{D}))$ means the multiplicity of the cycle $m(\partial\mathcal{D}) \subset \mathbb{R}^n \setminus \{0\}$ with respect to the point 0 (see Remark 5.3.12). The number $\mu(m(\partial\mathcal{D}))$ is also called the Kronecker index or the rotation of the vector field m on the boundary of $\mathcal{D}$ (see Krasnosel'skii and Zabreiko [138, §3]).

Remark 6.1.8 *The following formula is well-known as Kronecker's formula:*

$$\mu(m(\partial\mathcal{D})) = \sum_{p \in m^{-1}(0)\cap\mathcal{D}} \mu(p) \tag{6.1.13}$$

(ibid, §4).

In the author's paper [229], a series of formulas was found representing the number $\mu(m(\partial\mathcal{D}))$ by means of integrals over cycles of dimensions from 0 to $(n-1)$, lying on $\partial\mathcal{D}$.

Given $0 \le q \le n-1$, we define

$$K^{(q)}(y) = \frac{(-1)}{2^q \sigma_{n-q}} \sum_{\substack{1 \le j_0 \le n-q \le j_1 < \ldots < j_q \le n \\ j_0 < j_1 < \ldots < j_q}} (-1)^{j_0 - 1} \times \frac{y_{j_0} y_{j_1} \cdots y_{j_q}}{(y_1^2 + \ldots + y_{n-q-1}^2 + y_{n-q}^2 \cdot \ldots \cdot y_n^2)^{\frac{n-q}{2}}} dy[j_0, j_1, \ldots, j_q]. \tag{6.1.14}$$

Exercise 6.1.9 *Prove that this differential form $K^{(q)}$ is closed outside the planes $\{y_1 = \ldots = y_{n-q-1} = y_j = 0\}$ $(n-q \le j \le n)$ in $\mathbb{R}^n$.*

Hint. Use direct calculations.

□

Choose

$$\varepsilon < \min_{x \in \partial\mathcal{D}} \max \left(\left(\sum_{j=1}^{n-q} m_j(x)^2 \right)^{1/2} ; \ |m_j(x)| \ (n-q+1 \le j \le n) \right).$$

Then the image of the cycle $c^q = \{x \in \partial\mathcal{D} : \ |m_{n-q+1}(x)| = \ldots = |m_n(x)| = \varepsilon\}$ under the mapping m lies in the domain of the form $K^{(q)}$. We endow the cycle c^q with the orientation coherent with the orientation of the surface $\{x \in \overline{\mathcal{D}} :$

$|m_{n-q+1}(x)| = \ldots = |m_n(x)| = \varepsilon\}$, which is determined by the order of the parameters $(m_1, \ldots, m_{n-q})$. The transversality of the intersection with $\partial\mathcal{D}$ is certainly assumed here; otherwise we slightly deform $\mathcal{D}$.

Theorem 6.1.10 *For $0 \le q \le n-1$, the following formula holds:*

$$-\int_{c^q} m^\sharp(K^{(q)}) = \mu(m(\partial\mathcal{D})). \tag{6.1.15}$$

Proof. Formula (6.1.15) is well-known in the case $q = 0$ (see Example 6.1.11). For $q \neq 0$, it follows from Theorem 5.3.2 and this particulal case. For an independent proof, see Tarkhanov [229].

□

Complex analogs of (6.1.15) may be found in the paper of Yuzhakov and Kuprikov [257].

Example 6.1.11 Assuming $q = 0$, we have $K^{(0)} = \Phi_1(0, \cdot)$ where Φ_1 is the left fundamental solution of the gradient operator in $\mathbb{R}^n$ that was constructed in Subsection 2.5.2. Using the equality $d\frac{m}{|m|}[j] = \sum_{j_0=1}^{n}(-1)^{j_0-j}\frac{m_j m_{j_0}}{|m|^{n+1}}\, dm[j_0]$, we derive the well-known Poincaré formula from (6.1.15):

$$\mu(m(\partial\mathcal{D})) = \int_{\partial\mathcal{D}} \frac{1}{\sigma_n}\sum_{j=1}^{n}(-1)^{j-1}\frac{m_j}{|m|}\, d\frac{m}{|m|}[j]. \tag{6.1.16}$$

□

Example 6.1.12 For $q = n-1$, formula (6.1.15) becomes

$$\begin{aligned}\mu(m(\partial\mathcal{D})) &= \sum_{x\in c^{n-1}} \operatorname{sign}\left(\det\frac{\partial(|m_1|, \ldots, |m_n|)}{\partial(x_1, \ldots, x_n)}\right)\cdot\frac{1}{2^n}\operatorname{sign}(m_1\cdot\ldots\cdot m_n)\\ &= \sum_{x\in c^{n-1}}\frac{1}{2^n}\operatorname{sign}\left(\det\frac{\partial m}{\partial x}\right),\end{aligned} \tag{6.1.17}$$

and the sum in the right-hand side is finite. In particular, if $n = 2$, formula (6.1.17) is equivalent to formula (3.4) of Krasnosel'skii and Zabreiko [138, p.18]. Formula (6.1.13) has proved to be useful in systems of algebraic equations (A.M. Kytmanov, 1991).

□

Theorem 6.1.5 implies the following generalization of Kronecker's formula.

Corollary 6.1.13 *Let $y = m(x)$ be a mapping of X to $\mathbb{R}^N$ $(N \le n)$ of class C^{n-N+1}_{loc} and $\mathcal{D}$ be a subdomain of X with piecewise smooth boundary such that $dm \neq 0$ on $m^{-1}(0)\cap\partial\mathcal{D}$ and $m^{-1}(0)\cap\partial\mathcal{D}$ has zero Lebesgue measure on $m^{-1}(0)$. Then for every closed differential form $f \in C^1_{comp}(\Lambda^{n-N}T^*_{\mathbb{C}}(\overline{\mathcal{D}}))$ we have*

$$\int_{\partial\mathcal{D}}\left(\frac{1}{\sigma_N}\sum_{j=1}^{N}(-1)^{j-1}\frac{m_j}{|m|}\, d\frac{m}{|m|}[j]\right)\wedge f = \int_{m^{-1}(0)\cap\mathcal{D}} f. \tag{6.1.18}$$

Proof. It suffices to apply Theorem 6.1.5 in the case when P is the gradient operator in $\mathbb{R}^N$ and $\Phi = \Phi_1$ is the left fundamental solution of P constructed in Subsection 2.5.2.

□

6.1.7 *Mappings preserving the dimension*

The case $N = n$ is attractive because of its simplicity. We are considering a mapping $y = m(x)$ of X to an open set $Y \subset \mathbb{R}^n$ containing 0, of class C^2_{loc}, such that $m^{-1}(0)$ is discrete.

Corollary 6.1.14 *If $\mathcal{D} \subset\subset X$ is a domain with piecewise smooth boundary such that $m^{-1}(0) \cap \partial\mathcal{D} = \emptyset$, then for each section $f \in [C^1(\overline{\mathcal{D}})]^k$ satisfying $P^{(m)} f = 0$ in $\mathcal{D}$ we have*

$$-\int_{\partial\mathcal{D}} m^{\sharp}(G_P(\Phi(0,\cdot), I_k))f = \sum_{p \in m^{-1}(0)\cap\mathcal{D}} \mu(p) \cdot f(p). \tag{6.1.19}$$

Proof. Since the differential form $m^{\sharp}(G_P(\Phi(0,\cdot), I_k))f$ is closed, we may assume that $m^{-1}(0) \cap \mathcal{D}$ consists of a single point p.

Let $y \in \mathbb{R}^n$ be a regular value of the mapping m, such that $|y| < \min_{x \in \partial\mathcal{D}} |m(x)|$. Then $m(x) - t \cdot y \neq 0$ on $\partial\mathcal{D}$ for all $0 \leq t \leq 1$, so the mappings m and $(m - y)$ are homotopic as mappings from $\partial\mathcal{D}$ to $\mathbb{R}^n \setminus \{0\}$.

According to a property of the Kronecker index (see Krasnosel'skii and Zabreiko [138, p.16]), this implies the equality $\mu(m(\partial\mathcal{D})) = \mu((m - y)(\partial\mathcal{D}))$. We now use Kronecker's formula (6.1.13) to see that

$$\mu(p) = \sum_{x \in m^{-1}(y)\cap\mathcal{D}} \operatorname{sign} \left(\det \frac{\partial m}{\partial x}\right).$$

Therefore, because of the continuity of f we obtain

$$\begin{aligned}
\int_{m^{-1}(0)\cap\mathcal{D}} f &= \lim_{y\to 0} \int_{m^{-1}(y)\cap\mathcal{D}} f \\
&= \lim_{y\to 0} \left(\sum_{x \in m^{-1}(y)\cap\mathcal{D}} \operatorname{sign} \left(\det \frac{\partial m}{\partial x}\right) \cdot f(x)\right) \\
&= \mu(p) \cdot f(p).
\end{aligned}$$

To finish the proof, it suffices to apply formula (6.1.3).

□

Notice that the condition $P^{(m)} f = 0$ in $\mathcal{D}$ is easily verified to be equivalent to the condition that $Pf(m^{-1}(y)) = 0$ at each regular value $y \in m(\mathcal{D})$ of m.

6.1.8 *An analog of the Cauchy-Fantappiè formula for closed differential forms*

Given n-dimensional vectors $v_1, \ldots, v_N$ with entries in an algebra, we denote by $D_{n_1,\ldots,n_N}(v_1, \ldots, v_N)$ the determinant of order n whose first n_1 columns are v_1, the second n_2 columns are v_2 etc., the last n_N columns are v_N, where $n_1 + \ldots + n_N = n$.

Remark 6.1.15 *The determinant is calculated by columns, i.e.,* $\det(v_{ij}) = \sum_I (-1)^{\varepsilon_I} v_{i_1 1} \ldots v_{i_n n}$ *where* ε_I *is the parity of the rearrangement* $I = (i_1, \ldots, i_n)$ *of the numbers* $(1, \ldots, n)$.

For a smooth mapping $w(x,y,t) = \begin{pmatrix} w_1(x,y,t) \\ \ldots \\ w_n(x,y,t) \end{pmatrix}$ of an open set $U \subset \mathbb{R}^n_x \times \mathbb{R}^n_y \times \mathbb{R}^1_t$ to $\mathbb{R}^n$, we consider the double differential forms $K_q(w(x,y,t))$ $(1 \leq q \leq n)$ of degree $(q-1)$ in x and of degree $(n-q)$ in y and t given by

$$K_q(w(x,y,t)) = \frac{(-1)^q}{\sigma_n (q-1)!(n-q)!} \times D_{1,q-1,n-q}\left(\begin{pmatrix} \frac{w_1(x,y,t)}{|w(x,y,t)|} \\ \ldots \\ \frac{w_n(x,y,t)}{|w(x,y,t)|} \end{pmatrix}, \begin{pmatrix} d_x \frac{w_1(x,y,t)}{|w(x,y,t)|} \\ \ldots \\ d_x \frac{w_n(x,y,t)}{|w(x,y,t)|} \end{pmatrix}, \begin{pmatrix} (d_y + d_t) \frac{w_1(x,y,t)}{|w(x,y,t)|} \\ \ldots \\ (d_y + d_t) \frac{w_n(x,y,t)}{|w(x,y,t)|} \end{pmatrix} \right). \tag{6.1.20}$$

Moreover, we set $K_0 = K_{n+1} = 0$.

Remark 6.1.16 *The kernels of the Norguet representation (2.5.5) in* $\mathbb{R}^n$ *are just the double forms (6.1.20) with a special choice of the mapping* $w(x,y,t)$. *Namely, we have* $K_q(y-x) = \Phi_q(x,y)$ *for all* q.

Notice also that $K_q(w(x,y,t))$ is the component of degree $(q-1)$ in x of the "pull-back" $w^\sharp(\Phi_1(0,\cdot))$ times $(-1)^{(n-1)(q-1)}$ where we first have to take full differentials in x, y and t, then to place all the differentials in x after the differentials in y and t, and finally to declare the form to be a double one.

Definition 6.1.17 *Continuous mappings* $m_i : T \to S$ $(i = 0, 1)$ *of topological spaces are said to be homotopic if there exists a continuous mapping* $m_t : T \times [0,1] \to S$ *with coincides with* m_0 *for* $t = 0$ *and with* m_1 *for* $t = 1$.

When it is possible to speak of differentiability and m_i $(i = 0, 1)$ are homotopic and differentiable, then one can choose the homotopy m_t to be differentiable too, roughly speaking, of the same class as m_i (see Theorem 8 of Pontryagin [183, p.64]).

From the topological point of view, given a relatively compact domain $\mathcal{D} \subset X$ with smooth boundary, if smooth mappings m_i $(i = 0, 1)$ of $\partial\mathcal{D}$ to $\mathbb{R}^n \setminus \{0\}$ are homotopic, then they have the same rotations on $\partial\mathcal{D}$ (see Krasnosel'skii and Zabreiko [138, p.16]).

From the analytical point of view, it follows from the Poincaré formula (6.1.16) that $\int_{\partial\mathcal{D}} (m_1^\sharp \Phi_1(0,\cdot) - m_0^\sharp \Phi_1(0,\cdot)) = 0$. Hence the difference $m_1^\sharp \Phi_1(0,\cdot) - m_0^\sharp \Phi_1(0,\cdot)$ is exact on $\partial\mathcal{D}$ by the de Rham Theorem.

We would like to extend this obvious observation to the kernels $K_q(w(x,y,t))$.

Lemma 6.1.18 *Let $w = w(x,y,t)$ be a mapping of an open set $U \subset \mathbb{R}^n_x \times \mathbb{R}^n_y \times \mathbb{R}^1_t$ to $\mathbb{R}^n$ of class C^2_{loc}. Then the following equality holds on the set $U \setminus w^{-1}(0)$:*

$$d(x)K_q(w(x,y,t)) + (-1)^{q+1}(d(y)+d(t))K_{q+1}(w(x,y,t)) = 0. \tag{6.1.21}$$

Proof. As mentioned, $K_q(w(x,y,t))$ is the component of degree $(q-1)$ in x of the "pull-back" $w^\sharp(\Phi_1(0,\cdot))$ times $(-1)^{(n-1)(q-1)}$. Using now Proposition 1.2.38 and the equality $d\Phi_1(0,\cdot) = -\delta(\cdot)d(\cdot)$ in $\mathbb{R}^n$, we obtain for $(x,y,t) \in U \setminus w^{-1}(0)$

$$\begin{aligned}
&d(x)\ K_q(w(x,y,t)) + (-1)^{q+1}\ (d(y)+d(t))\ K_{q+1}(w(x,y,t)) \\
&\quad = (-1)^{n-q}\ d(x)\left((-1)^{(n-1)(q-1)}\ w^\sharp(\Phi_1(0,\cdot))\right) \\
&\qquad + (-1)^{q+1}\ (d(y)+d(t))\left((-1)^{(n-1)q}\ w^\sharp(\Phi_1(0,\cdot))\right) \\
&\quad = (-1)^{nq+1}\ d\ w^\sharp(\Phi_1(0,\cdot)) \\
&\quad = (-1)^{nq}\ w^\sharp(\delta(\cdot)d(\cdot)) \\
&\quad = 0,
\end{aligned}$$

as desired. □

For another proof of Lemma 6.1.18, see the proof of Lemma 1.2 in the book of Aizenberg and Dautov [5].

Lemma 6.1.19 *If w_i $(i = 0,1)$ are homotopic mappings of a set $\sigma \subset \mathbb{R}^n_x \times \mathbb{R}^n_y$ to $\mathbb{R}^n \setminus \{0\}$ of class C^2_{loc} and w_t $(t \in [0,1])$ is a C^2_{loc} homotopy between w_0 and w_1, then for every $0 \le q \le n-1$ we have on σ:*

$$\begin{aligned}
&K_{q+1}\,(w_1(x,y)) - K_{q+1}\,(w_0(x,y)) \\
&\quad = d(x)\left((-1)^q \int_0^1 (dt\rfloor K_q(w_t(x,y)))\,dt\right) \\
&\qquad - (-1)^q\ d(y)\left((-1)^{q+1} \int_0^1 (dt\rfloor K_{q+1}(w_t(x,y)))\,dt\right).
\end{aligned} \tag{6.1.22}$$

Proof. It suffices to integrate equality (6.1.21) over $t \in [0,1]$ and to take into account that

$$d(y)\int_0^1 (dt\rfloor K_{q+1}(w_t(x,y)))\,dt = -\int_0^1 (dt\rfloor d(y)K_{q+1}(w_t(x,y)))\,dt.$$

□

Lemma 6.1.19 for $w_1(x,y) = y - x$ provides decompositions of the kernels $\Phi_{q+1}(x,y)$ like those given by Lemmas 4.5.3 and 5.5.7.

Theorem 6.1.20 *Let $\mathcal{D}$ be a domain in $\mathbb{R}^n$ with piecewise smooth boundary and $w = w(x,y)$ be a mapping of $\mathcal{D}_x \times \partial\mathcal{D}_y$ to $\mathbb{R}^n \setminus \{0\}$ of class C^2_{loc}. If w is homotopic to the mapping $(y-x)$ on $\mathcal{D}_x \times \partial\mathcal{D}_y$ and w_t $(t \in [0,1])$ is a C^2_{loc} homotopy between w and $(y-x)$, then for every form $f \in C^1_{comp}(\Lambda^q T^*_{\mathbb{C}}(\overline{\mathcal{D}}))$ we have*

$$f(x) = -\int_{\partial\mathcal{D}} f \wedge K_{q+1}(w(x,\cdot)) + hdf + dhf \quad (x \in \mathcal{D}), \tag{6.1.23}$$

where

$$(h_q f)(x) = -\int_{\partial\mathcal{D}\times[0,1]} f \wedge ((-1)^q K_q(w_t(x,\cdot))) + \int_{\mathcal{D}} f \wedge \Phi_q(x,\cdot).$$

Proof. Setting $w_1 = y - x$ and $w_0 = w$, we decompose the kernel $\Phi_{q+1}(x,y)$ on the set $\mathcal{D}_x \times \partial\mathcal{D}_y$ in accordance with equality (6.1.22). Substituting this decomposition into the integral over the boundary in (2.5.5) and using Stokes' Theorem, we derive formula (6.1.23).

□

Thus, any smooth deformation of the mapping $(y - x)$ on $\mathcal{D}_x \times \partial\mathcal{D}_y$ leads to an integral representation of differential forms in $\mathcal{D}$.

Exercise 6.1.21 *Prove that, if* $w = w(x,y)$ *is a continuous mapping of* $\mathcal{D}_x \times \partial\mathcal{D}_y$ *to* $\mathbb{R}^n \setminus \{0\}$ *such that* $\langle w, y - x\rangle_{(x,y)} > -|w||y - x|$ *on the set, then* w *is homotopic to* $(y - x)$.

Hint. Consider the continuous mapping $w_t = t(y-x) + (1-t)w$ of $\mathcal{D}_x \times \partial\mathcal{D}_y \times [0,1]_t$ to $\mathbb{R}^n$.

In order to prove that $w_t \neq 0$, estimate the value $|w_t|^2 = \langle w_t, w_t\rangle_{(x,y)}$ from below.

□

Finally, we consider a sample application of Theorem 6.1.20.

Example 6.1.22 Let $\mathcal{D}$ be a convex subdomain of $\mathbb{R}^n$ with twice differentiable boundary. Choose a real-valued function ρ of class C^2_{loc} in a neighborhood of $\overline{\mathcal{D}}$, such that $\mathcal{D} = \{x : \rho(x) < 0\}$ and $\nabla p(y) \neq 0$ for $y \in \partial\mathcal{D}$ where ∇ is the gradient operator in $\mathbb{R}^n$. Since the tangential hyperplane to $\partial\mathcal{D}$ at a point $y \in \partial\mathcal{D}$ does not intersect $\mathcal{D}$, we have $\langle\nabla p(y), x - y\rangle_{(x,y)} < 0$ on the set $\mathcal{D}_x \times \partial\mathcal{D}_y$. Using now Exercise 6.1.21 we conclude that the mapping $w = \nabla\rho(y)$ of $\mathcal{D}_x \times \partial\mathcal{D}_y$ to $\mathbb{R}^n \setminus \{0\}$ is homotopic to $(y - x)$ on the set. Moreover the homotopy constructed above is $w_t(x,y) = t(y-x) + (1-t)w(x,y)$ which is smooth. Applying now Theorem 6.1.20 we derive the following series of formulas valid in $\mathcal{D}$:

$$f(x) = -\int_{\partial\mathcal{D}} f \wedge K_{q+1}(\nabla\rho) + hdf + dhf \quad \text{for all} \quad f \in C^1_{comp}(\Lambda^q T^*_{\mathbb{C}}(\overline{\mathcal{D}})). \quad (6.1.24)$$

Since $\nabla\rho$ does not depend on x, the kernels $K_{q+1}(\nabla\rho)$ vanish provided $q > 0$. Therefore the equalities (6.1.24) all together mean that the operator h is a fundamental solution of the de Rham complex $\Lambda^\cdot T^*_{\mathbb{C}}(\overline{\mathcal{D}})$ at positive degrees. We should emphasize that $\Lambda^\cdot T^*_{\mathbb{C}}(\overline{\mathcal{D}})$ is a differential complex on the *bordered* manifold. Another fundamental solution is given by (1.2.19) but it has much worse local properties.

□

Thus Theorem 6.1.20 shows a way to find new fundamental solutions of the de Rham complex on a manifold with boundary.

We mention also the author's paper [221] where Theorem 6.1.20 was applied to obtain a successful integral representation in $\mathcal{D}$ for differential forms of finite order of growth near the boundary $\partial\mathcal{D}$.

Exercise 6.1.23 *Let $w = w(y)$ be a continuous mapping of $\partial\mathcal{D}$ to $\mathbb{R}^n \setminus \{0\}$ such that, for each point $y \in \partial\mathcal{D}$, the ray $y + tw(y)$ $(t \geq 0)$ does not intersect $\mathcal{D}$. Prove that the mapping $w : \mathcal{D}_x \times \partial\mathcal{D}_y \to \mathbb{R}^n \setminus \{0\}$ is homotopic to $(y - x)$.*

Hint. Check that

$$\langle w, y - x\rangle_{(x,y)} > -|w||y - x| \qquad \text{for all } (x, y) \in \mathcal{D} \times \partial\mathcal{D}.$$

□

In particular, if $\mathcal{D}$ is starlike with respect to a point $x^0 \in \mathcal{D}$, we can take $w(y) = y - x^0$.

6.2 An Integral Formula for the Lefschetz Number

6.2.1 *A brief survey of the Lefschetz theory for elliptic complexes*

The original *Fixed Point Theorem* of Lefschetz [147] states that if $m : X \to X$ is a continuous mapping of a compact manifold X into itself and all the fixed points of m are isolated, then the Lefschetz number $L(m^\sharp)$ of the "pull-back" operator $m^\sharp$ is given by $L(m^\sharp) = \sum_{p \in Fix\, m} \mu(p)$. Here $\mu(p)$ is the local degree of the mapping $1 - m$ at p, and the sum runs over the fixed points p of m.

If m is a smooth mapping and the fixed point p is *simple* in the sense that $\det\left(1 - \frac{\partial m}{\partial x}(p)\right) \neq 0$, then the local degree of $1 - m$ at p has the infinitesimal description $\mu(p) = \operatorname{sign} \det\left(1 - \frac{\partial m}{\partial x}(p)\right)$.

Atiyah and Bott [19, 20] showed that the Lefschetz theory also makes sense in the context of elliptic complexes. They proved that if m induces a cochain mapping $m^\sharp$ of an elliptic complex $\mathcal{E}(E^\cdot)$ over X and the fixed points of m are simple, then the Lefschetz number of $m^\sharp$ is given by $L(m^\sharp) = \sum_{p \in Fix\, m} \mu(p)$, where $\mu(p)$ are infinitesimal invariants of m at p.

Note that in general $\mu(p)$ is a complex number, and not an integer. The classical Lefschetz formula, where $\mu(p) = \pm 1$, is highly special in this direction.

Kotake [137] found an analytic representation of the Lefschetz number $L(m^\sharp)$ in terms of fundamental solutions of the heat operators $\left\{\frac{\partial}{\partial t} - \Delta_\imath\right\}$, where $\{\Delta_\imath\}$ are the Laplacians of the complex $E^\cdot$ given by (2.1.8). Hence he derived a very elegant proof of the theorem of Atiyah and Bott [19].

Brenner and Shubin [32] extended the theorem of Atiyah and Bott [19] to elliptic complexes on compact manifolds with boundaries.

In his thesis, Toledo [238] presented a general approach to fixed point theory for elliptic complexes. This method is based on a classical formula of de Rham [50, §33] which expresses intersection numbers on Riemannian manifolds in terms of the Green kernel. It leads to an integral representation for the Lefschetz number from which the Atiyah-Bott theorem can be derived by some delicate but quite elementary analysis.

Moreover, if the complex is sheaf exact, then also a cohomological expression for the index of an isolated fixed point can be derived. For simple fixed points this reduces of course to the infinitesimal description.

It is worth pointing out that the approach of Toledo [238] is an exposition in the analytic context of the original intersection-theoretic proof of Lefschetz [147] for the classical theorem.

Remark 6.2.1 *From this point of view, Lefschetz theory centers around the study of the Thom class, i.e., a generator for $H^n((X \times X, (X \times X) \setminus \Delta), \mathbb{R}) \cong \mathbb{R}$, where n is the dimension of X (cf. also Nestke [171]).*

For further results in this direction, we refer to the survey of Gilkey [70].

6.2.2 The Lefschetz number of an endomorphism of an elliptic complex

Let E be an elliptic complex on a compact differentiable manifold X. Every cochain mapping $e : \mathcal{E}(E) \to \mathcal{E}(E)$ of degree 0 is said to be an *endomorphism* of the complex $\mathcal{E}(E)$.

Each endomorphism e of $\mathcal{E}(E)$ preserves the spaces of cocycles $Z^i(\mathcal{E}(E))$ and coboundaries $B^i(\mathcal{E}(E))$. Therefore, after passing to quotient spaces it induces the homomorphisms $(He)_i : H^i(\mathcal{E}(E)) \to H^i(\mathcal{E}(E))$, where $i \in \mathbb{Z}$.

Since the vector spaces $H^i(\mathcal{E}(E))$ are finite-dimensional by Corollary 2.2.4, the traces $\operatorname{tr}(He)_i$ of the linear mappings $(He)_i$ are well-defined.

Definition 6.2.2 *The Lefschetz number of an endomorphism e of $\mathcal{E}(E)$ is defined as $L(e) = \sum_{i=0}^{N} (-1)^i \operatorname{tr}(He)_i$.*

The following example is classical.

Example 6.2.3 Let $e = I$ be the identity endomorphism of $\mathcal{E}(E)$. Then we have $\operatorname{tr}(He)_i = \dim H^i(\mathcal{E}(E))$ for $i \in \mathbb{Z}$. Hence the Lefschetz number $L(e)$ is just the *Euler characteristic* of $\mathcal{E}(E)$, i.e., $\chi(\mathcal{E}(E)) = \sum_{i=0}^{N} (-1)^i \dim H^i(\mathcal{E}(E))$.

□

The generalized Lefschetz problem consists in evaluating the Lefschetz number $L(e)$ in terms of topological invariants of the manifold X, the complex $\mathcal{E}(E)$, and the endomorphism e (see Gilkey [70]).

Recall that an operator $K : \mathcal{D}(E) \to \mathcal{D}'(E)$ is said to be smoothing if K has its kernel in $\mathcal{E}(E \boxtimes E')$. Denoting the kernel by $K(x, y)$ we have $\Delta^* K = K(x, x)$ in $\mathcal{E}(E \otimes E')$, and $\operatorname{tr} \Delta^* K$ in $\mathcal{E}(\Omega)$.

Definition 6.2.4 *Given a smoothing operator $K : \mathcal{D}(E) \to \mathcal{D}'(E)$ on X, the trace of K is the number $\operatorname{tr} K = \int_X \operatorname{tr} \Delta^* K$.*

If K has finite-dimensional range, then the definition agrees with the purely algebraic definition of the trace.

Exercise 6.2.5 *Given a smoothing operator $K : \mathcal{D}(E) \to \mathcal{D}'(E)$ on X of finite-dimensional range, let $\|K_{ij}\|$ $(K_{ij} \in \mathbb{C})$ be a matrix representation of the corresponding linear mapping $\frac{\mathcal{D}(E)}{\ker K} \to \operatorname{im} K$. Prove that* $\operatorname{tr} K = \sum_i K_{ii}$.

Hint. See Atiyah and Bott [19, §7].

□

Since the operators with finite-dimensional range are dense in $\mathcal{L}_b(\mathcal{E}'(E) \to \mathcal{E}(E))$ and the above defined $\operatorname{tr} K$ is a continuous function of K in this space, Definition 6.2.4 is the unique extension of tr to smoothing operators.

Exercise 6.2.6 *Prove that if K is a smoothing operator of type $F \to E$ and D is a differential operator of type $E \to F$ on X, then* $\operatorname{tr} K \circ D = \operatorname{tr} D \circ K$.

Hint. Prove first this equality for operators K of finite-dimensional range using Exercise 6.2.5.

□

The following generalization of Euler's identity was proved by Atiyah and Bott [19].

Lemma 6.2.7 *If $e = \{e_i\}$ is a smoothing endomorphism of the complex $\mathcal{E}(E^\cdot)$, then*

$$L(e) = \sum_{i=0}^{N} (-1)^i \operatorname{tr} e_i. \tag{6.2.1}$$

Proof. Formula (6.2.1) is easily verified for endomorphisms of $\mathcal{E}(E^\cdot)$ with finite-dimensional range. Using Theorem 2.2.2 one can show that smoothing endomorphisms of $\mathcal{E}(E^\cdot)$ with finite-dimensional range are dense in the space of all smoothing endomorphisms endowed with the topology of $\mathcal{L}_b(\mathcal{E}'(E^\cdot) \to \mathcal{E}(E^\cdot))^0$. Then formula (6.2.1) for arbitrary smoothing endomorphisms follows by passing to the limit. For more details, we refer to Atiyah and Bott [19, §7].

□

6.2.3 *Main integral formula for the Lefschetz number*

Fix some (properly supported) pseudo-differential parametrix $K = \{K_i\}$ for the complex $E^\cdot$, i.e., $\widehat{P}K = \operatorname{tr} \Delta^* - S$ for some kernel $S \in \mathcal{E}((E^\cdot \boxtimes E^{\cdot\prime})^0)$. As mentioned after Definition 2.1.4, the corresponding operator S is a smoothing endomorphism of the complex $\mathcal{E}(E^\cdot)$.

Lemma 6.2.8 *Given any endomorphism $e = \{e_i\}$ of $\mathcal{E}(E^\cdot)$, we have*

$$L(e) = \int_X \operatorname{tr} \Delta^* \left(\sum_{i=0}^{N} (-1)^i e_i(x) \, S_i(x,y) \right). \tag{6.2.2}$$

Proof. By the homotopy formula (2.1.4),

$$KPf + PKf = f - Sf \qquad \text{for all } f \in \mathcal{E}(E^i).$$

Applying the endomorphism e to both sides of this equality, we get

$$(eK)\,Pf + P\,(eK)f = ef - (eS)f \qquad \text{for all } f \in \mathcal{E}(E^i). \tag{6.2.3}$$

Since $(eK) \in \mathcal{L}(\mathcal{E}(E^\cdot) \to \mathcal{E}(E^\cdot))^{(-1)}$, (6.2.3) means that the endomorphisms e and (eS) of the complex $\mathcal{E}(E^\cdot)$ are homotopic. Hence it follows, by Proposition 1.1.13, that $He = H(eS)$ and so $L(e) = L(eS)$.

On the other hand, (eS) is a smoothing endomorphism of the complex $\mathcal{E}(E^\cdot)$ because $e_i S_i$ is extendible to a continuous mapping $\mathcal{E}'(E^i) \to \mathcal{E}(E^i)$, for every i. Moreover, the kernel of (eS) is just $e(x)S(x,y)$.

To finish the proof it suffices to apply Lemma 6.2.7.

□

6.2.4 *Geometric endomorphisms*

In the sequel we consider only geometric endomorphisms of the complex $\mathcal{E}(E^\cdot)$ in the sense of Atiyah and Bott [19].

Let $m : X \to X$ be a differentiable mapping of X. Then we have the induced "pull-back" bundles m^*E^i ($i \in \mathbb{Z}$) over X whose fibers over $x \in X$ are $E^i_{m(x)}$, as well as the linear mappings $m^* : \mathcal{E}(E^i) \to \mathcal{E}(m^*E^i)$ defined by $(m^*f)(x) = f(m(x))$.

Assume that there are also given (smooth) bundle homomorphisms $h_i : m^*E^i \to E^i$ ($i \in \mathbb{Z}$). It will cause no confusion if we use the same letters h_i to designate the induced mappings $\mathcal{E}(m^*E^i) \to \mathcal{E}(E^i)$. Then the composition $m^\sharp = h \circ m^*$ defines a mapping $\mathcal{E}(E^\cdot) \to \mathcal{E}(E^\cdot)$ of degree 0. More precisely, we have $(m^\sharp f)(x) = h_i(x)\, f(m(x))$ for $f \in \mathcal{E}(E^i)$.

Definition 6.2.9 *The sequence $m^\sharp = \{m^\sharp_i\}$ is said to be a geometric endomorphism of $\mathcal{E}(E^\cdot)$ determined by the mapping m and the bundle homomorphisms $\{h_i\}$ if $P_i \circ m^\sharp_i = m^\sharp_{i+1} \circ P_i$ for all $i \in \mathbb{Z}$.*

One says also that $m^\sharp$ is a *lift* of m to $\mathcal{E}(E^\cdot)$ by means of the homomorphisms $\{h_i\}$.

Example 6.2.10 For the de Rham complex $E^\cdot = \Lambda^\cdot T^*_{\mathbb{C}}(X)$ on X, all smooth mappings $m : X \to X$ have a natural lift to $\mathcal{E}(E^\cdot)$. Indeed, the tangent mapping $\frac{\partial m}{\partial x}$ maps $T_x(X)$ to $T_{m(x)}(X)$ for each $x \in X$. Thus the transposed mapping $\left(\frac{\partial m}{\partial x}\right)'$ is a bundle homomorphism $m^*T^*_{\mathbb{C}}(X) \to T^*_{\mathbb{C}}(X)$ and the i-th exterior power of $\left(\frac{\partial m}{\partial x}\right)'$ provides the (smooth) bundle homomorphism $m^*\Lambda^i T^*_{\mathbb{C}}(X) \to \Lambda^i T^*_{\mathbb{C}}(X)$ We set $h_i = \bigwedge^i \left(\frac{\partial m}{\partial x}\right)'$. Then the mapping $m^\sharp = h \circ m^*$ is the usual pull-back of differential forms. According to Proposition 1.2.38, $m^\sharp$ is an endomorphism of $\mathcal{E}(E^\cdot)$. And $L(m^\sharp)$ is the usual Lefschetz number.

□

The definitions for a general complex $E^\cdot$ are of course designed to generalize this situation.

Example 6.2.11 For the Dolbeault complex $E^\cdot = \Lambda^{p,\cdot}T^*_{\mathbb{C}}(X)$ with coefficients in the bundle $\Lambda^{p,0}T^*_{\mathbb{C}}(X)$ over a complex manifold X, only holomorphic mappings $m: X \to X$ have a natural lift to $\mathcal{E}(E^\cdot)$. In fact, the natural lift $m^\sharp$ of m to the de Rham complex $\mathcal{E}(\Lambda^\cdot T^*_{\mathbb{C}}(X))$ preserves the decompositions $\mathcal{E}(\Lambda^i T^*_{\mathbb{C}}(X)) = \oplus_{p+q=i}\mathcal{E}(\Lambda^{p,q}T^*_{\mathbb{C}}(X))$. Therefore, for every p, it induces an endomorphism of the complex $\mathcal{E}(\Lambda^{p,\cdot}T^*_{\mathbb{C}}(X))$. In the spirit of the Kronecker formula, this endomorphism can be denoted by $m^{\sharp(p)}$.

□

In the general case, when the bundles E^i are not connected with the geometry of X, there is no natural construction for the h_i, and their existence must be postulated. Thus the existence of a lift $m^\sharp$ will in general put strong restrictions on the mappings m for which the Lefschetz theory can be formulated.

6.2.5 *A modification of the main formula for geometric endomorphisms*

Assume that $m^\sharp$ is a geometric endomorphism of the complex $\mathcal{E}(E^\cdot)$ determined by the mapping m and bundle homomorphisms $h_i : m^*E^i \to E^i$ $(i \in \mathbb{Z})$.

Denote by $\operatorname{Fix} m = \{x \in X : x = m(x)\}$ the set of all fixed points of m. This can be geometrically interpreted as the intersection of the graph $\Gamma_m = \{(x, m(x)) \in X \times X\}$ with the diagonal Δ.

If m lifts to $\mathcal{E}(E^\cdot)$, then the mapping $m \times 1 : X \times X \to X \times X$ given by $(m \times 1)(x,y) = (m(x), y)$ has a natural lift to the complex $\mathcal{E}((E^\cdot \boxtimes E'\,)^\cdot)$. Namely, we can set $((m \times 1)^\sharp K)(x,y) = \oplus h_{i+r}(x)\, K_i(m(x),y)$ for a kernel $K = \oplus K_i$ in $\mathcal{E}((E^\cdot \boxtimes E'\,)^r)$. It is obvious that $\hat{P}\,(m \times 1)^\sharp = (m \times 1)^\sharp\, \hat{P}$.

Lemma 6.2.12 *For any pseudo-differential parametrix $K = \{K_i\}$ of the complex $E^\cdot$, we have*

$$L(m^\sharp) = -\int_X \operatorname{tr} \Delta^* \sum_{i=1}^N (-1)^i \left(P_{i-1}(x)(m \times 1)^\sharp K_i - P'_{i-1}(y)(m \times 1)^\sharp K_i\right). \quad (6.2.4)$$

Proof. Since $\hat{P}K = -S$ outside the diagonal of $X \times X$, we have $(m \times 1)^\sharp S = -\hat{P}(m \times 1)^\sharp K$ outside the graph Γ_m.

Hence, it follows that

$$\sum_{i=0}^N (-1)^i\,(m \times 1)^\sharp S_i = -\sum_{i=1}^N (-1)^i \left(P_{i-1}(x)(m \times 1)^\sharp K_i - P'_{i-1}(y)(m \times 1)^\sharp K_i\right). \quad (6.2.5)$$

Observe that even though each summand

$$P_{i-1}(x)(m \times 1)^{\sharp} K_i - P'_{i-1}(y)(m \times 1)^{\sharp} K_i$$

is defined only for $y \neq m(x)$, the alternating sum on the right-hand side of (6.2.5) extends smoothly to all of $X \times X$, namely, by the left-hand side of the equality.

To finish the proof it now suffices to apply Lemma 6.2.8.

□

6.2.6 *Localization of the Lefschetz number*

We start with a simple observation.

Example 6.2.13 Let $E^{\cdot} = \Lambda^{\cdot} T^*_{\mathbb{C}}(X)$ be the de Rham complex on X. Then the complex $(E^{\cdot} \boxtimes E^{\cdot\prime})^{\cdot}$ can be identified with the de Rham complex on $X \times X$ by comparing a double form $K = \{K_i\}$ in $\mathcal{E}((E \boxtimes E')^r)$ with the differential form $\sum(-1)^i K_i$ in $\mathcal{E}(\Lambda^{n+r} T^*_{\mathbb{C}}(X \times X))$ (see Example 1.5.21). The operation $\operatorname{tr} \Delta^*$ is the usual restriction of an n-form on $X \times X$ to an n-form on X via the diagonal mapping $\Delta : X \to X \times X$, i.e., $\operatorname{tr} \Delta^* = \Delta^{\sharp}$. Hence it follows that outside the set $\operatorname{Fix} m$ we have the equality

$$\begin{aligned}
&\operatorname{tr} \Delta^* \sum_{i=1}^{N} (-1)^i \left(P_{i-1}(x)(m \times 1)^{\sharp} K_i - P'_{i-1}(y)(m \times 1)^{\sharp} K_i\right) \\
&= \Delta^{\sharp} \sum_{i=1}^{N} (-1)^i \left(d(x)(m \times 1)^{\sharp} K_i + (-1)^{i-1} d(y)(m \times 1)^{\sharp} K_i\right) \\
&= d\left(\sum_{i=1}^{N} (-1)^i \Delta^{\sharp} (m \times 1)^{\sharp} K_i\right).
\end{aligned}$$

□

In other words, for the de Rham complex the integrand in Lemma 6.2.12 is exact on $X \setminus \operatorname{Fix} m$. The same argument holds for the Dolbeault complex with coefficients in the bundle $\Lambda^{p,0} T^*_{\mathbb{C}}(X)$.

For a general complex $E^{\cdot}$ there is no natural $(n-1)$-form defined on $X \setminus \operatorname{Fix} m$ whose exterior derivative is the integrand of (6.2.4) restricted to $X \setminus \operatorname{Fix} m$. But forms with this property can always be constructed as follows.

Let $G_{P'_{i-1}}(\cdot, \cdot)$ be a Green operator for the differential operator P'_{i-1}, where $i = 1, \ldots, N$. According to Theorem 2.4.7, there exists a unique differential operator $G_{P'_{i-1}}(\cdot) \in \mathrm{do}_{p_{i-1}-1}(E^{i-1} \boxtimes E^{i\prime} \to \pi_1^* \Lambda^{n-1} T^*_{\mathbb{C}}(X))$ such that

$$\Delta^* G_{P'_{i-1}}(f \otimes g) = G_{P'_{i-1}}(f, g) \qquad \text{for all } f \in \mathcal{E}(E^{i-1}),\ g \in \mathcal{E}(E^{i\prime});$$

here $\pi_1 : X \times X \to X$ is the projection onto the first factor. The role of these operators is clear from the following technical lemma.

Lemma 6.2.14 *Assume that U is an open subset of $X \times X$ and $K \in \mathcal{E}((E \boxtimes E^{\cdot\prime})^{-1}|_U)$. Then over $\Delta^{-1}(U)$*

$$\operatorname{tr}\Delta^* \sum_{i=1}^{N}(-1)^i \left(P_{i-1}(x)K_i - P'_{i-1}(y)K_i\right) = d\left(-\sum_{i=1}^{N}(-1)^i \Delta^* G_{P'_{i-1}}(K_i)\right). \quad (6.2.6)$$

Proof. One need consider only the case of homogeneous kernels, i.e., $K \in \mathcal{E}(E^{i-1} \boxtimes E^{i\prime}|_U)$.

Let $x \in \Delta^{-1}(U)$ and let U' be a neighborhood of x in X such that $U' \times U' \subset U$. If $K(x,y) = f(x) \otimes g(y)$ in $U' \times U'$ with $f \in \mathcal{E}(E^{i-1}|_{U'})$ and $g \in \mathcal{E}(E^{i\prime}|_{U'})$, then by the definition of the Green operator we get in U':

$$\begin{aligned}\operatorname{tr}\Delta^*\left(P_{i-1}(x)K - P'_{i-1}(y)K\right) &= \operatorname{tr}\Delta^*\left(Pf(x) \otimes g(y) - f(x) \otimes P'g(y)\right)\\ &= \langle Pf, g\rangle_x - \langle f, P'g\rangle_x\\ &= -dG_{P'_{i-1}}(f,g)\\ &= d\left(-\Delta^* G_{P'_{i-1}}(K)\right).\end{aligned}$$

This is precisely equality (6.2.6).

Since $\mathcal{E}(E^{i-1}|_{U'}) \otimes \mathcal{E}(E^{i\prime}|_{U'})$ is dense in $\mathcal{E}(E^{i-1} \boxtimes E^{i\prime}|_{U'\times U'})$, the equality (6.2.6) is still valid on U' for all kernels $K \in \mathcal{E}(E^{i-1} \boxtimes E^{i\prime}|_{U'\times U'})$ as desired.

□

We are now ready to formulate the main result of this subsection.

For K a smooth kernel of type (-1) on an open set $U \subset X \times X$, we consider the following differential form of degree $(n-1)$ on $\Gamma_m^{-1}(U) = \{x \in X : (x, m(x)) \in U\}$:

$$\varphi(K) = -\sum_{i=1}^{N}(-1)^i \Delta^* G_{P'_{i-1}}((m \times 1)^\sharp K_i). \quad (6.2.7)$$

Theorem 6.2.15 *Given any pseudo-differential parametrix K of the complex $E^\cdot$, the exterior derivative $d\varphi(K)$ of the form $\varphi(K) \in \mathcal{E}(\Lambda^{n-1}T^*_{\mathbb{C}}(X \setminus \operatorname{Fix} m))$ extends smoothly to a global form $\widetilde{d\varphi} \in \mathcal{E}(\Omega)$ such that*

$$L(m^\sharp) = -\int_X \widetilde{d\varphi}.$$

Proof. Using Lemma 6.2.14 and (6.2.5), we obtain over $X \setminus \operatorname{Fix} m$

$$d\varphi(K) = -\operatorname{tr}\Delta^*\left(\sum_{i=0}^{N}(-1)^i (m \times 1)^\sharp S_i\right).$$

To complete the proof, it suffices to apply Lemma 6.2.8.

□

6.2.7 *Local index of an isolated component of the set of fixed points*

Theorem 6.2.15 gives at once the most elementary result of Lefschetz theory.

Corollary 6.2.16 *If m has no fixed points, then $L(m^\sharp) = 0$.*

Proof. Indeed,

$$L(m^\sharp) = -\int_X d\varphi(K) = 0$$

because $\varphi(K)$ is smooth on X.

□

If the set of fixed points is not empty, the theorem shows that $L(m^\sharp)$ depends only on the behaviour of the mapping m and the homomorphisms $\{h_\iota\}$ near the set $\mathrm{Fix}\, m$.

Corollary 6.2.17 *If $\{U_\nu\}$ is a sequence of neighborhoods of* $\mathrm{Fix}\, m$ *with piecewise smooth boundary such that $\overline{U_{\nu+1}} \subset U_\nu$ and $\cap_\nu U_\nu = \mathrm{Fix}\, m$, then*

$$L(m^\sharp) = \lim_{\nu\to\infty} \int_{\partial U_\nu} \varphi(K) - \int_{\mathrm{Fix}\, m} \widetilde{d\varphi}. \tag{6.2.8}$$

Proof. Using Theorem 6.2.15 and Stokes' Theorem, we get

$$\begin{aligned} L(m^\sharp) &= -\int_{X\setminus U_\nu} \widetilde{d\varphi} - \int_{U_\nu} \widetilde{d\varphi} \\ &= \int_{\partial U_\nu} \varphi(K) - \int_{U_\nu} \widetilde{d\varphi} \\ &= \lim_{\nu\to\infty} \int_{\partial U_\nu} \varphi(K) - \int_{\mathrm{Fix}\, m} \widetilde{d\varphi}, \end{aligned}$$

as desired.

□

In this way one can define a local index for each isolated component of $\mathrm{Fix}\, m$. Namely, given any isolated component $\mathfrak{c}$ of the set of fixed points, we let

$$\mu(\mathfrak{c}) = \lim_{\nu\to\infty} \int_{\partial U_\nu(\mathfrak{c})} \varphi(K),$$

where $\{U_\nu(\mathfrak{c})\}$ is a sequence of neighborhoods of $\mathfrak{c}$ with piecewise smooth boundary such that $\overline{U_{\nu+1}} \subset U_\nu$ and $\cap_\nu U_\nu = \mathfrak{c}$.

Exercise 6.2.18 *Prove that, for each isolated component $\mathfrak{c}$ of* $\mathrm{Fix}\, m$, *the local index $\mu(\mathfrak{c})$ is well-defined, i.e., the limit exists and does not depend on the particular choice of the sequence $\{U_\nu(\mathfrak{c})\}$.*

Hint. Choose the number ν large enough so that $U_\nu(\mathfrak{c}) \cap \operatorname{Fix} m = \mathfrak{c}$, and choose $\mu \geq \nu$. Then by Stokes' formula

$$\begin{aligned}\left| \int_{\partial U_\nu(\mathfrak{c})} \varphi(K) - \int_{\partial U_\mu(\mathfrak{c})} \varphi(K) \right| &= \left| \int_{U_\nu(\mathfrak{c}) \setminus U_\mu(\mathfrak{c})} \widetilde{d\varphi} \right| \\ &\leq \max \left| \frac{\widetilde{d\varphi}}{dv} \right| \cdot \operatorname{vol}(U_\nu(\mathfrak{c}) \setminus \mathfrak{c}),\end{aligned}$$

whence it follows that $\left\{ \int_{\partial U_\nu(\mathfrak{c})} \varphi(K) \right\}$ is a Cauchy sequence.

On the other hand, if both $\{U'_\nu(\mathfrak{c})\}$ and $\{U''_\nu(\mathfrak{c})\}$ satisfy the condition above, then

$$\left| \int_{\partial U'_\nu(\mathfrak{c})} \varphi(K) - \int_{\partial U''_\mu(\mathfrak{c})} \varphi(K) \right| \leq \max \left| \frac{\widetilde{d\varphi}}{dv} \right| \cdot \operatorname{vol}(U'_\nu(\mathfrak{c}) \setminus \mathfrak{c})$$

for μ large enough. Therefore, the sequences $\left\{ \int_{\partial U'_\nu(\mathfrak{c})} \varphi(K) \right\}$ and $\left\{ \int_{\partial U''_\nu(\mathfrak{c})} \varphi(K) \right\}$ have the same limit as desired.

□

Further properties of the *local index* will be adduced in Section 6.4.

Corollary 6.2.19 *If all the components of the set of fixed points are isolated, then*

$$L(m^\sharp) = \sum_{\{\mathfrak{c}\} \in \operatorname{Fix} m} \mu(\mathfrak{c}) - \int_{\operatorname{Fix} m} \widetilde{d\varphi}. \tag{6.2.9}$$

Proof. This follows from Corollary 6.2.17.

□

To derive computable formulas for the local index, one needs a good deal of information on the singularities of $\varphi(K)$ on $\operatorname{Fix} m$. The simplest case in which this information is available is when $\operatorname{Fix} m$ consists of isolated points.

This is discussed in details in the next section. For the present we present the following starting point.

Corollary 6.2.20 *Suppose that* $\mathfrak{p}$ *be an isolated fixed point of* m *and* $P(x, D) = \sum_{|\alpha| \leq p} P_\alpha(x)\, D^\alpha$ *be a representation of the differential in a neighborhood of* $\mathfrak{p}$. *Then we have*

$$\begin{aligned}\mu(\mathfrak{p}) = \lim_{\varepsilon \to 0} \sum_{i=1}^{N} (-1)^i \int_{\partial B(\mathfrak{p}, \varepsilon)} \frac{1}{\sqrt{-1}} \\ \times \sum\nolimits'_{|\beta+\gamma+1_j| \leq p_{i-1}} (-1)^{|\beta|} \operatorname{tr} \Delta^* \, D_x^\gamma D_y^\beta \left(P_{\beta+\gamma+1_j}(y)\, h_{i-1}(x)\, K_i(m(x), y) \right) \star dx_j.\end{aligned} \tag{6.2.10}$$

Proof. According to Proposition 2.4.5 and formula (2.4.12), we can write the Green operator for P'_{i-1} in a neighborhood of the point $\mathfrak{p}$ in the form

$$G^{(\mathfrak{p})}_{P'_{i-1}}(f, g) = -\frac{1}{\sqrt{-1}} \sum\nolimits'_{|\beta+\gamma+1_j| \leq p_{i-1}} (-1)^{|\beta|} D^\beta (g\, P_{\beta+\gamma+1_j})\, D^\gamma f \star dx_j.$$

Using then formula (2.4.17), we get

$$G^{(\mathfrak{p})}_{P'_{\imath-1}}(f(x)\otimes g(y))$$
$$= -\frac{1}{\sqrt{-1}}\sum\nolimits'_{|\beta+\gamma+1_\jmath|\leq p_{\imath-1}}(-1)^{|\beta|}\mathrm{tr}\, D_x^\gamma D_y^\beta \left(P_{\beta+\gamma+1_\jmath}(y)\, f(x)\otimes g(y)\right)\star dx_\jmath.$$

As follows from the proof of Theorem 2.4.1, there exists a global Green operator $G_{P'_{\imath-1}}(\cdot,\cdot)$ for $P'_{\imath-1}$ which agrees with $G^{(\mathfrak{p})}_{P'_{\imath-1}}(\cdot,\cdot)$ on a neighborhood of $\mathfrak{p}$. The same is still true for $G^{(\mathfrak{p})}_{P'_{\imath-1}}(\cdot,\cdot)$.

To finish the proof it now suffices to use equality (6.2.7).

□

We emphasize that the individual integrals in formula (6.2.10) do not converge as $\varepsilon\to 0$. It is only the alternating sum which converges.

6.3 The Atiyah-Bott Formula for Simple Fixed Points

6.3.1 *Preliminaries*

As before, let $m^\sharp = h\circ m^*$ be a geometric endomorphism of $\mathcal{E}(E^\cdot)$ defined by means of a mapping $m:\ X\to X$ and bundle homomorphisms $h_\imath:\ m^*E^\imath\to E^\imath$ $(i\in\mathbb{Z})$.

Suppose that p is a simple fixed point of m in the sense that

$$\det\left(1-\frac{\partial m}{\partial x}(p)\right)\neq 0.$$

Geometrically this means that the graph of m and the diagonal intersect transversally at the point $(p,p)\in X\times X$.

Then the local index $\mu(p)$ can be computed explicitly by going to the limit in the expression of Corollary 6.2.20. The result is precisely the formula of Atiyah and Bott [19, Theorem A].

Remark 6.3.1 *Since this formula involves the values of the $\{h_\imath\}$ at p, the Dirac measure at p (evaluation at p) must be relevant to the formula.*

We work in a coordinate patch $U(p)$ centered at p and with fixed trivializations of the $E^\imath$ over $U(p)$. We write the kernels $K_\imath$, $S_\imath$, etc. simply as matrices of distributions on $U(p)\times U(p)$ with respect to these trivializations. Though, strictly speaking, we should write $K_\imath\otimes dy$, $S_\imath\otimes dy$, etc. where dy is the volume element given by the local coordinates.

Remark 6.3.2 *Since p is an isolated point of the set* $\mathrm{Fix}\, m$, *we can choose a smaller neighborhood $U'(p)$ of p such that $m(U'(p))\subset U(p)$ and $U'(p)\cap\mathrm{Fix}\, m=\{p\}$.*

If needed, the neighborhood $U'(p)$ could be reduced so that the mapping $y = x - m(x)$ be a diffeomorphism in $U'(p)$ because p is a simple fixed point of m.

In Section 6.4 we show that the local index $\mu(p)$ does not depend on the particular choice of the pseudo-differential parametrix K^{loc} of the complex $E^{\cdot}|_{U(p)}$ which is used in the definition of $\mu(p)$. At present we fix such a parametrix with the regularity properties as in Theorem 2.1.12, i.e.,

$$K_i^{loc} \in \mathrm{pdo}_{-p_{i-1}}(E^i|_{U(p)} \to E^{i-1}|_{U(p)}), \qquad i \in \mathbb{Z}.$$

Lemma 6.3.3 *For $i \in \mathbb{Z}$, each derivative $D_x^\alpha D_y^\beta (m \times 1)^* K_i^{loc}$ with $|\alpha| + |\beta| < p_{i-1}$ is a locally integrable kernel on $U'(p) \times U(p)$, and even $D_x^\alpha D_y^\beta (m \times 1)^* K_i^{loc}$ is in $L^q_{loc}(E^{i-1} \boxtimes E^{i'}|_{U'(p) \times U(p)})$ with any $q < \frac{n}{n-(p_{i-1}-|\alpha|-|\beta|)}$.*

Proof. This follows from Theorem 2.1.12 and the *Fubini Theorem.*

□

For simple fixed points, the following statement is sharper.

Lemma 6.3.4 *Given any $i \in \mathbb{Z}$, each derivative $D_x^\alpha D_y^\beta (m \times 1)^* K_i^{loc}$ has a restriction to the diagonal of $U'(p) \times U'(p)$ belonging to $L^q_{loc}(E^{i-1} \otimes E^{i'}|_{U'(p)})$ with any exponent $q < \frac{n}{n-(p_{i-1}-|\alpha|-|\beta|)}$, provided that $|\alpha| + |\beta| < p_{i-1}$.*

Proof. It suffices to consider only the case when $n - (p_{i-1} - |\alpha| - |\beta|) > 0$, i.e., $|\alpha| + |\beta| > p_{i-1} - n$.

Let σ be a compact subset of $U'(p)$. By Theorem 2.1.12, there exists a constant $c' > 0$ such that

$$\left|(D_x^\alpha D_y^\beta (m \times 1)^* K_i^{loc})(x, x)\right| \le \frac{c'}{|x - m(x)|^{n-(p_{i-1}-|\alpha|-|\beta|)}} \qquad \text{for all } x \in \sigma.$$

On the other hand, since p is a simple fixed point of m and $U'(p)$ does not contain fixed points of m different from p, there is a constant $c'' > 0$ such that $|x - m(x)| \ge c'' |x|$ for all $x \in \sigma$.

Combining these estimates we obtain a constant c such that

$$\left|(D_x^\alpha D_y^\beta (m \times 1)^* K_i^{loc})(x, x)\right| \le \frac{c}{|x|^{n-(p_{i-1}-|\alpha|-|\beta|)}} \qquad \text{for all } x \in \sigma.$$

Hence the lemma follows.

□

The important consequence of Lemma 6.3.4 is that the form $\varphi(K^{loc})$ given by (6.2.7) is well-defined, not only outside the point p in $U'(p)$ but also on the whole neighborhood $U'(p)$, provided it is considered as a current.

Corollary 6.3.5 *Given a parametrix $K_i^{loc} \in \mathrm{pdo}_{-p_{i-1}}(E^i|_{U(p)} \to E^{i-1}|_{U(p)})$, $i \in \mathbb{Z}$, of the complex $E^{\cdot}|_{U(p)}$, the form $\varphi(K^{loc})$ is well-defined as an element of $L^q_{loc}(\Lambda^{n-1} T^*_{\mathbb{C}}(U'(p)))$ $(q < \frac{n}{n-1})$ smooth in $U'(p) \setminus \{p\}$.*

Proof. This follows directly from Lemma 6.3.4 because

$$\varphi(K^{loc}) = -\sum_{i=1}^{N}(-1)^i \,\Delta^* G_{P'_{i-1}}(h_{i-1}(x)\,(m\times 1)^* K_i^{loc}) \qquad \text{in } U'(p).$$

□

Thus, we may try to calculate the differential of the current $\varphi(K)$ in $U'(p)$. This is done in the following lemma.

Recall that $\widehat{P}K^{loc} = \operatorname{tr}\Delta^* - S^{loc}$ in $U(p)\times U(p)$, where S^{loc} is a smooth kernel of type 0 on this set.

Lemma 6.3.6 *Let* K^{loc} *be a parametrix of the complex* $E\,|_{U(p)}$, *with* $K_i^{loc} \in \operatorname{pdo}_{-p_{i-1}}(E^i|_{U(p)} \to E^{i-1}|_{U(p)})$, $i\in\mathbb{Z}$. *Then over* $U'(p)$

$$\begin{aligned} &d\varphi(K^{loc})\\ &= \left(\sum_{i=0}^{N}(-1)^i \operatorname{tr} h_i(x)\right)\delta(x-m(x))\,dx - \left(\sum_{i=0}^{N}(-1)^i \operatorname{tr}\Delta^*(m\times 1)^\sharp S_i^{loc}\right)dx. \end{aligned} \tag{6.3.1}$$

Proof. By formula (6.2.7),

$$d\varphi(K^{loc}) = -\sum_{i=1}^{N}(-1)^i d\,\Delta^* G_{P'_{i-1}}((m\times 1)^\sharp K_i^{loc}).$$

If a kernel $K \in \mathcal{D}'(E^{i-1}\boxtimes E^{i'}|_{U'(p)\times U'(p)})$ has the form $K(x,y) = f(x)\otimes g(y)$, where $f\in \mathcal{E}(E^{i-1}|_{U'(p)})$ and $g\in\mathcal{E}(E^{i'}|_{U'(p)})$, then according to Theorem 2.4.7 we obtain

$$\begin{aligned} d\,\Delta^* G_{P'_{i-1}}(K) &= d\,G_{P'_{i-1}}(f,g)\\ &= \left(\langle f, P'_{i-1}g\rangle_x - \langle P_{i-1}f, g\rangle_x\right)dx\\ &= \operatorname{tr}\Delta^*\left(P'_{i-1}(y)K - P_{i-1}(x)K\right)dx. \end{aligned} \tag{6.3.2}$$

Since the subspace $\mathcal{E}(E^{i-1}|_{U'(p)})\otimes\mathcal{E}(E^{i'}|_{U'(p)})$ is dense in the space $\mathcal{D}'(E^{i-1}\boxtimes E^{i'}|_{U'(p)\times U'(p)})$, the equality (6.3.2) holds also for all the kernels K in $\mathcal{D}'(E^{i-1}\boxtimes E^{i'}|_{U'(p)\times U'(p)})$ whose singularities are transversal to the diagonal of $U'(p)\times U'(p)$ (see Hörmander [113]). This is just the case for the kernel $(m\times 1)^\sharp K_i^{loc}$ whose singular support is $\Gamma_m\cap(U'(p)\times U'(p))$.

Therefore,

$$\begin{aligned} d\varphi(K^{loc}) &= -\sum_{i=1}^{N}(-1)^i\operatorname{tr}\Delta^*\left(P'_{i-1}(y)(m\times 1)^\sharp K_i^{loc} - P_{i-1}(x)(m\times 1)^\sharp K_i^{loc}\right)dx\\ &= \sum_{i=0}^{N}(-1)^i\operatorname{tr}\Delta^*\left(P'_{i-1}(x)(m\times 1)^\sharp K_i^{loc} + P'_i(y)(m\times 1)^\sharp K_{i+1}^{loc}\right)dx. \end{aligned}$$

We now use the equality

$$\widehat{P}(m \times 1)^{\sharp} K^{loc} = (m \times 1)^{\sharp} \operatorname{tr} \Delta^* - (m \times 1)^{\sharp} S^{loc}$$

on $U'(p) \times U'(p)$. Since the homogeneous component of the kernel $(m \times 1)^{\sharp} \operatorname{tr} \Delta^*$ at degree $\imath$ is

$$\begin{aligned} (m \times 1)^{\sharp} \operatorname{tr} \Delta^* \Big|_{\mathcal{D}(E^{\imath\prime} \boxtimes E^{\imath}|_{U'(p) \times U'(p)})} &= (m \times 1)^{\sharp} \left(\delta(y-x)\, I_{k_\imath}\right) \\ &= h_\imath(x)\, \delta(y - m(x)), \end{aligned} \tag{6.3.3}$$

we obtain

$$\begin{aligned} & d\varphi(K^{loc}) \\ &= \sum_{\imath=0}^{N} (-1)^{\imath} \operatorname{tr} \Delta^* \left(h_\imath(x) \delta(y - m(x)) - (m \times 1)^{\sharp} S_\imath^{loc} \right) dx \\ &= \left(\sum_{\imath=0}^{N} (-1)^{\imath} \operatorname{tr} h_\imath(x) \right) \delta(x - m(x))\, dx - \left(\sum_{\imath=0}^{N} (-1)^{\imath} \operatorname{tr} \Delta^* (m \times 1)^{\sharp} S_\imath^{loc} \right) dx. \end{aligned}$$

This completes the proof.

□

6.3.2 *The Atiyah-Bott formula*

The main result of Atiyah and Bott [19, Theorem A] follows quite easily from Lemma 6.3.6.

Theorem 6.3.7 *If* $\mathfrak{p}$ *is a simple fixed point of* m*, then*

$$\mu(\mathfrak{p}) = \frac{\sum_{\imath=0}^{N} (-1)^{\imath} \, tr\, h_\imath(\mathfrak{p})}{\left| \det \left(1 - \frac{\partial m}{\partial x}(\mathfrak{p}) \right) \right|}. \tag{6.3.4}$$

Notice that, for a point $x \in X$, the value $h_\imath(x)$ is a linear mapping of the fiber $(m^* E^\imath)_x = E^\imath_{m(x)}$ to the fiber $E^\imath_x$. Therefore, the trace of $h_\imath(x)$ is well-defined at each fixed point $x = \mathfrak{p}$ of m.

Proof. We start with the formula

$$\mu(\mathfrak{p}) = \lim_{\nu \to \infty} \int_{\partial U_\nu} \varphi(K^{loc}),$$

where $\{U_\nu\}$ is a system of neighborhoods of p with piecewise smooth boundary such that $\overline{U_{\nu+1}} \subset U_\nu$ and $\cap_\nu U_\nu = \{\mathfrak{p}\}$.

Choose the number ν so large that $\overline{U_\nu} \subset U'(\mathfrak{p})$ and $y = x - m(x)$ is a diffeomorphism in a neighborhood of $\overline{U_\nu}$.

Let $R^{(\varepsilon)}$ ($\varepsilon > 0$) be the regularization in the space of currents on $U'(p)$ given by Theorem 12 in de Rham [50]. This means that $R^{(\varepsilon)}$ is a smoothing endomorphism of the complex $\mathcal{D}'(\Lambda\, T^*_{\mathbb{C}}(U'(\mathfrak{p})))$ whose kernel is supported in the ε-neighborhood of the diagonal of $U'(\mathfrak{p}) \times U'(\mathfrak{p})$, and such that, for every current f on $U'(p)$, we have $R^{(\varepsilon)}f \to f$ ($\varepsilon \to +0$) in the topology corresponding to the smoothness of f.

Since $\varphi(K^{loc})$ is smooth in $U'(\mathfrak{p})$ except for the point $\mathfrak{p}$, the regularization $R^{(\varepsilon)}(\varphi(K^{loc}))$ converges to $\varphi(K^{loc})$ in the topology of $\mathcal{E}(\Lambda^{n-1}T^*_{\mathbb{C}}(U'(\mathfrak{p}) \setminus \mathfrak{p}))$. Therefore,

$$\begin{aligned}\int_{\partial U_\nu} \varphi(K^{loc}) &= -\left\langle \varphi(K^{loc}), d\chi_{U_\nu} \right\rangle \\ &= \lim_{\varepsilon \to 0}(-1)\left\langle R^{(\varepsilon)}(\varphi(K^{loc})), d\chi_{U_\nu} \right\rangle.\end{aligned}$$

The characteristic function χ_{U_ν} of the set U_ν is considered here as a current of degree 0. Since this current has compact support and the regularization $R^{(\varepsilon)}(\varphi(K^{loc}))$ is smooth on $U'(\mathfrak{p})$, we may apply the transposition rule on the right-hand side. Then we obtain

$$\begin{aligned}\int_{\partial U_\nu} \varphi(K^{loc}) &= \lim_{\varepsilon \to 0}\left\langle d\, R^{(\varepsilon)}(\varphi(K^{loc})), \chi_{U_\nu} \right\rangle \\ &= \lim_{\varepsilon \to 0}\left\langle R^{(\varepsilon)}(d\,\varphi(K^{loc})), \chi_{U_\nu} \right\rangle.\end{aligned}$$

Lemma 6.3.6 now implies that

$$\begin{aligned}&\int_{\partial U_\nu} \varphi(K^{loc}) \\ &= \lim_{\varepsilon \to 0}\left\langle R^{(\varepsilon)}\left(\left(\sum_{i=0}^{N}(-1)^i \operatorname{tr} h_i(x)\right)\delta(x - m(x))\,dx\right), \chi_{U_\nu}\right\rangle \\ &- \lim_{\varepsilon \to 0}\left\langle R^{(\varepsilon)}\left(\left(\sum_{i=0}^{N}(-1)^i \operatorname{tr} \Delta^*(m \times 1)^\sharp S_i^{loc}\right)dx\right), \chi_{U_\nu}\right\rangle.\end{aligned} \tag{6.3.5}$$

The regularization

$$R^{(\varepsilon)}\left(\left(\sum_{i=0}^{N}(-1)^i \operatorname{tr} h_i(x)\right)\delta(x - m(x))\,dx\right)$$

is supported in the ball $B(\mathfrak{p}, \varepsilon)$ and it converges to

$$\left(\sum_{i=0}^{N}(-1)^i \operatorname{tr} h_i(x)\right)\delta(x - m(x))\,dx$$

in the topology of $\mathcal{E}'(\Lambda^n T^*_{\mathbb{C}}(U'(\mathfrak{p})))$. Therefore the first limit on the right-hand side of (6.3.5) is equal to

$$\begin{aligned}&\left\langle \left(\sum_{i=0}^{N}(-1)^i \operatorname{tr} h_i(x)\right)\delta(x - m(x))\,dx, 1\right\rangle \\ &= \left\langle \delta(y)dy, \left(\sum_{i=0}^{N}(-1)^i \operatorname{tr} h_i((1-m)^{-1}y)\right)\left|\frac{\partial(1-m)^{-1}}{\partial y}\right|\right\rangle \\ &= \frac{\sum_{i=0}^{N}(-1)^i \operatorname{tr} h_i(\mathfrak{p})}{\left|\det\left(1 - \frac{\partial m}{\partial x}(\mathfrak{p})\right)\right|}.\end{aligned}$$

As to the second limit on the right-hand side of (6.3.5), the regularization there converges in the topology of $\mathcal{E}(\Lambda^n T^*_{\mathbb{C}}(U'(\mathfrak{p})))$. Hence it follows that

$$\int_{\partial U_\nu} \varphi(K^{loc}) = \frac{\sum_{i=0}^{N}(-1)^i \operatorname{tr} h_i(\mathfrak{p})}{\left|\det\left(1 - \frac{\partial m}{\partial x}(\mathfrak{p})\right)\right|} - \int_{U_\nu}\left(\sum_{i=0}^{N}(-1)^i \operatorname{tr} \Delta^*(m \times 1)^\sharp S_i^{loc}\right) dx. \qquad (6.3.6)$$

In order to derive formula (6.3.4) from (6.3.6), it suffices to go to the limit as $\nu \to \infty$ on both sides of the equality.

□

It is worth pointing out that the formula for $\mu(p)$ does not explicitly involve the differential operators P_i. Thus the formula for the Lefschetz number is much simpler than the *index formula* (cf. Atiyah [18]). Of course, the P_i are implicitly involved by the condition $m^\sharp P = P m^\sharp$.

6.4 Isolated Components of the Set of Fixed Points

6.4.1 *General remarks*

The results of this section follow easily by the techniques developped in Chapter 2.

Let $m^\sharp$ be a geometric endomorphism of the complex $\mathcal{E}(E\,)$ defined by means of a mapping $m:\ X \to X$ and bundle homomorphisms $h_i:\ m^*E^i \to E^i$ ($i \in \mathbb{Z}$).

Assume that p is an isolated fixed point of m, nor necessarily simple. In this case there is no hope of carrying out an explicit computation of $\mu(p)$ as in Theorem 6.3.7.

First of all, the change of variable $y = x - m(x)$ is no longer available. This gives rise to some trouble in the interpretation of the current $\delta(x - m(x))dx$.

But, what is more significant, in the absence of the estimate $|x - m(x)| \geq c\,|x|$ the information on the singularities of K^{loc} near the diagonal is totally useless in interpreting the form $\varphi(K^{loc})$ as a current on a whole neighborhood of p.

For this reason we want to avoid taking the limit in the definition of $\mu(p)$. This can be done if the integrand $\varphi(K^{loc})$ in that expression can be replaced by a closed differential form.

The topological argument of Toledo [238] suggests that this is possible provided we can replace K^{loc} by a cohomology class of $H^{-1}(\mathcal{E}((E \boxtimes E\,')\,|_{U(p)\times U(p)\setminus\Delta}))$ - the local *Thom class*.

6.4.2 *Uniqueness of the local index*

Assume that $\mathfrak{c}$ is an isolated component of the set of fixed points of m.

Let $U(\mathfrak{c})$ be a fixed neighborhood of $\mathfrak{c}$ in X. We choose a smaller neighborhood $U'(\mathfrak{c})$ of $\mathfrak{c}$ such that $m(U'(\mathfrak{c})) \subset U(\mathfrak{c})$ and $U'(\mathfrak{c}) \cap \operatorname{Fix} m = \mathfrak{c}$.

Given a pseudo-differential parametrix K^{loc} of the complex $E^{\cdot}|_{U(\mathfrak{c})}$, the differential form $\varphi(K^{loc})$ is well-defined in $U'(\mathfrak{c}) \setminus \mathfrak{c}$ by equality (6.2.7).

We define the local index of $\mathfrak{c}$ to be the limit

$$\mu(\mathfrak{c}) = \lim_{\nu \to \infty} \int_{\partial U_\nu(\mathfrak{c})} \varphi(K^{loc}),$$

where $\{U_\nu(\mathfrak{c})\}$ is a sequence of neighborhoods of $\mathfrak{c}$ with piecewise smooth boundary such that $\overline{U_{\nu+1}} \subset U_\nu$ and $\cap_\nu U_\nu = \mathfrak{c}$.

It follows from Proposition 2.4.4 that $\mu(\mathfrak{c})$ is independent of the particular choice of the Green operators $G_{P'_{i-1}}$ which occur in the definition of the form $\varphi(K^{loc})$. However the question arises of how much $\mu(\mathfrak{c})$ depends on the parametrix K^{loc}.

We note that if $E^{\cdot}$ consists of just one operator P (necessarily elliptic), i.e., $N = 1$, then the uniqueness of the local index is trivial. In this case, the parametrix K^{loc} is determined up to smoothing operators. But smooth kernels do not affect the local index.

If the complex consists of more than one operator, then the equation $\widehat{P}K^{loc} = \operatorname{tr}\Delta^* - S^{loc}$ on $U(\mathfrak{c}) \times U(\mathfrak{c})$ no longer determines the symbol of K^{loc} and K^{loc} is determined only up to operators K satisfying $\widehat{P}K = 0$ modulo smooth kernels on $U(\mathfrak{c}) \times U(\mathfrak{c})$. Nevertheless Theorem 6.4.2 below states the uniqueness of the local index in the class of local pseudo-differential parametrices of $E^{\cdot}$.

We start with a technical lemma which has often been used.

Lemma 6.4.1 *Let K be a smooth kernel of type (-1) on $U(\mathfrak{c}) \times U(\mathfrak{c}) \setminus \Delta$ satisfying $\widehat{P}K = 0$. Then the differential form $\varphi(K)$ is closed on $U'(\mathfrak{c}) \setminus \mathfrak{c}$.*

Proof. We have

$$d\varphi(K) = -\operatorname{tr}\Delta^* \left(\sum_{i=0}^{N} (-1)^i (m \times 1)^\sharp (\widehat{P}K)_i \right) \qquad \text{in } U'(\mathfrak{c}) \setminus \mathfrak{c},$$

where $(\widehat{P}K)_i$ is the homogeneous component in $\mathcal{E}(E^i \boxtimes E^{i'}|_{U(\mathfrak{c}) \times U(\mathfrak{c}) \setminus \Delta})$ of the kernel $\widehat{P}K$.

Hence the lemma follows.

□

Theorem 6.4.2 *The index $\mu(\mathfrak{c}) = \lim_{\nu \to \infty} \int_{\partial U_\nu(\mathfrak{c})} \varphi(K^{loc})$ of an isolated component $\mathfrak{c}$ of measure zero of the set Fixm does not depend on the particular choice of the pseudo-differential parametrix K^{loc} of the complex $E^{\cdot}|_{U(\mathfrak{c})}$.*

Proof. Let Π be a properly supported parametrix of the complex $E^{\cdot}|_{U(\mathfrak{c})}$ with regularity properties as in Theorem 2.1.12, so that $\widehat{P}\Pi = \operatorname{tr}\Delta^* - R$ with some smooth kernel R on $U(\mathfrak{c}) \times U(\mathfrak{c})$.

From Corollary 2.1.2 we obtain as in the proof of Proposition 2.3.2 that any parametrix K of the complex $E^{\cdot}|_{U(\mathfrak{c})}$ has the form

$$K = (\Pi - S\Pi + \Pi R - \Pi S R + RKR) - \widehat{P}(\Pi\Pi - \Pi S \Pi + RK\Pi - \Pi K).$$

Therefore, if K^j $(j = 1,2)$ are two pseudo-differential parametrices of the complex $E^{\cdot}|_{U(\mathfrak{c})}$, i.e., $\widehat{P}K^j = \operatorname{tr}\Delta^* - S^j$ for some smooth kernels S^j $(j = 1,2)$ on $U(\mathfrak{c}) \times U(\mathfrak{c})$, then we have $K^2 - K^1 = K^{(-1)} + \widehat{P}K^{(-2)}$, where

$$\begin{aligned} K^{(-1)} &= (S^2 - S^1)\Pi - \Pi(S^2 - S^1)R + R(K^2 - K^1)R, \\ K^{(-2)} &= \Pi(S^2 - S^1)\Pi - R(K^2 - K^1)\Pi + \Pi(K^2 - K^1). \end{aligned}$$

It is easy to check that the operator $K^{(-1)}$ extends to a continuous linear mapping of $\mathcal{E}'(E^{\cdot}|_{U(\mathfrak{c})})$ to $\mathcal{E}(E^{\cdot}|_{U(\mathfrak{c})})$ of degree (-1). Hence it follows by Theorem 1.5.1 that this operator is smoothing, i.e., $K^{(-1)} \in \mathcal{E}((E^{\cdot} \boxtimes E^{\cdot\prime})^{-1}|_{U(\mathfrak{c})\times U(\mathfrak{c})})$.

On the other hand, the operator $K^{(-2)}$ is pseudo-differential. Therefore, its kernel in $\mathcal{D}'((E^{\cdot} \boxtimes E^{\cdot\prime})^{-2}|_{U(\mathfrak{c})\times U(\mathfrak{c})})$ is infinitely differentiable outside the diagonal of $U(\mathfrak{c}) \times U(\mathfrak{c})$.

Thus

$$\varphi(K^2) - \varphi(K^1) = \varphi(K^{(-1)}) + \varphi(\widehat{P}K^{(-2)}). \tag{6.4.1}$$

Since the form $\varphi(K^{(-1)})$ is smooth in $U'(\mathfrak{c})$ and the measure of $U_\nu(\mathfrak{c})$ $(\nu = 1,2,\ldots)$ tends to zero, we get by Stokes' Theorem

$$\lim_{\nu\to\infty} \int_{\partial U_\nu(\mathfrak{c})} \varphi(K^{(-1)}) = 0. \tag{6.4.2}$$

Consider now the form $\varphi(\widehat{P}K^{(-2)})$, which is obviously smooth in $U'(\mathfrak{c}) \setminus \mathfrak{c}$. It follows from Lemma 6.4.1 that this form is closed in $U'(\mathfrak{c})\setminus\mathfrak{c}$, because $\widehat{P}(\widehat{P}K^{(-2)}) = 0$ on $U(p)\times U(p)\setminus\Delta$. Therefore, if the number ν_0 is large enough, we obtain by Stokes' Theorem

$$\lim_{\nu\to\infty} \int_{\partial U_\nu(\mathfrak{c})} \varphi(\widehat{P}K^{(-2)}) = \int_{\partial U_{\nu_0}(\mathfrak{c})} \varphi(\widehat{P}K^{(-2)}).$$

Choose now a kernel $\widetilde{K}^{(-2)}$ in $\mathcal{E}((E^{\cdot} \boxtimes E^{\cdot\prime})^{-2}|_{U(\mathfrak{c})\times U(\mathfrak{c})})$ which coincides with $K^{(-2)}$ in a neighborhood of the diagonal of $U(\mathfrak{c}) \times U(\mathfrak{c})$ so small that

$$\varphi(\widehat{P}\widetilde{K}^{(-2)}) = \varphi(\widehat{P}K^{(-2)}) \qquad \text{on } \partial U_{\nu_0}(\mathfrak{c}).$$

Then the differential form $\varphi(\widehat{P}\widetilde{K}^{(-2)})$ is smooth and closed everywhere in $U'(\mathfrak{c})$, so

$$\lim_{\nu\to\infty} \int_{\partial U_\nu(\mathfrak{c})} \varphi(\widehat{P}K^{(-2)}) = \int_{\partial U_{\nu_0}(\mathfrak{c})} \varphi(\widehat{P}\widetilde{K}^{(-2)}) = 0. \tag{6.4.3}$$

Combining equalities (6.4.1) – (6.4.3), we conclude that

$$\lim_{\nu\to\infty} \int_{\partial U_\nu(\mathfrak{c})} \left(\varphi(K^2) - \varphi(K^1)\right) = 0,$$

which finishes the proof.

□

The formulation of Corollary 6.2.19 now becomes satisfactory. The indices $\mu(\mathfrak{c})$ there were defined in terms of the global parametrix K of the complex $E^{\cdot}$, whose construction involves the whole manifold X. But Theorem 6.4.2 shows that $\mu(\mathfrak{c})$ has a purely local description, at least if $\mathfrak{c}$ is of measure zero.

6.4.3 *A cohomological formula for the local index*

Both for aesthetic and practical reasons, Theorem 6.4.2 gives a satisfactory description of the local index of an isolated component $\mathfrak{c}$ of measure zero of the set of fixed points. In order to find $\mu(\mathfrak{c})$, it suffices to know some local parametrix of the complex $E^{\cdot}$ in a neighborhood of $\mathfrak{c}$.

The description becomes especially fruitful if the complex $E^{\cdot}$ has a fundamental solution in a neighborhood $U(\mathfrak{c})$ of the set $\mathfrak{c}$. Under such a choice of the parametrix K^{loc}, the form $\varphi(K^{loc})$ turns out to be closed in $U'(\mathfrak{c}) \setminus \mathfrak{c}$. So we avoid taking the limit in the expression for $\mu(\mathfrak{c})$.

To do this we need a local cohomological condition on $E^{\cdot}$.

Lemma 6.4.3 *If* $H^0(\mathcal{E}((E^{\cdot} \boxtimes E^{\cdot\prime})^{\cdot}|_{U(\mathfrak{c})\times U(\mathfrak{c})})) = 0$, *then the complex* $E^{\cdot}|_{U(\mathfrak{c})}$ *has a pseudo-differential fundamental solution* Φ^{loc}.

Proof. In fact, let K^{loc} be a pseudo-differential parametrix of $E^{\cdot}|_{U(\mathfrak{c})}$, i.e., $\hat{P}K^{loc} = \operatorname{tr}\Delta^* - S^{loc}$ for some smooth kernel S^{loc} of type 0 on $U(\mathfrak{c}) \times U(\mathfrak{c})$. Then $\hat{P}S^{loc} = 0$ on the set, so there exists a kernel $M^{loc} \in \mathcal{E}((E^{\cdot} \boxtimes E^{\cdot\prime})^{-1}|_{U(\mathfrak{c})\times U(\mathfrak{c})})$ such that $\hat{P}M^{loc} = S^{loc}$. Setting now $\Phi^{loc} = K^{loc} + M^{loc}$, we obtain a pseudo-differential fundamental solution of $E^{\cdot}|_{U(\mathfrak{c})}$, as desired.

□

It follows from Theorem 1.4.9 that if the set $U(\mathfrak{c})$ is 1-convex with respect to $E^{\cdot}$ and $H_0(\mathcal{E}(E^{\cdot\prime}|_{U(\mathfrak{c})})) = 0$, then $H^0(\mathcal{E}((E^{\cdot} \boxtimes E^{\cdot\prime})^{\cdot}|_{U(\mathfrak{c})\times U(\mathfrak{c})})) = 0$.

Theorem 6.4.4 *Given a pseudo-differential fundamental solution* Φ^{loc} *of the complex* $E^{\cdot}|_{U(\mathfrak{c})}$, *the differential form* $\varphi(\Phi^{loc})$ *is closed in* $U'(\mathfrak{c}) \setminus \mathfrak{c}$, *and for* ν *large enough we have*

$$\mu(\mathfrak{c}) = \int_{\partial U_\nu(\mathfrak{c})} \varphi(\Phi^{loc}). \tag{6.4.4}$$

Proof. Since the kernel Φ^{loc} is smooth on $U(\mathfrak{c}) \times U(\mathfrak{c}) \setminus \Delta$ and $\hat{P}\Phi^{loc} = 0$ on this set, we see from Lemma 6.4.1 that the form $\varphi(\Phi^{loc})$ is closed in $U'(\mathfrak{c}) \setminus \mathfrak{c}$.

Choose the number ν_0 so large that $\overline{U_{\nu_0}(\mathfrak{c})} \subset U'(\mathfrak{c})$. If $\nu > \nu_0$, then the closure of $U_{\nu_0}(\mathfrak{c}) \setminus U_\nu(\mathfrak{c})$ lies in $U'(\mathfrak{c}) \setminus \mathfrak{c}$. So by Stokes' Theorem

$$\int_{\partial U_{\nu_0}(\mathfrak{c})} \varphi(\Phi^{loc}) - \int_{\partial U_\nu(\mathfrak{c})} \varphi(\Phi^{loc}) = \int_{\partial(U_{\nu_0}(\mathfrak{c})\setminus U_\nu(\mathfrak{c}))} \varphi(\Phi^{loc}) = 0.$$

Thus, the sequence

$$\left\{\int_{\partial U_\nu(\mathfrak{c})} \varphi(\Phi^{loc})\right\}$$

stabilizes starting with the number ν_0.

To complete the proof, it suffices to invoke Theorem 6.4.2.

□

Thus, the kernel Φ^{loc} may be considered as a representative of the *local Thom class* (see Remark 6.2.1).

Exercise 6.4.5 *Prove that the cohomology class in*

$$H^{-1}(\mathcal{E}((E^{\cdot} \boxtimes E^{\cdot\prime})^{\cdot}|_{U(\mathfrak{c})\times U(\mathfrak{c})\setminus\Delta})) \quad \textit{modulo} \quad H^{-1}(\mathcal{E}((E^{\cdot} \boxtimes E^{\cdot\prime})^{\cdot}|_{U(\mathfrak{c})\times U(\mathfrak{c})}))$$

is the same for all pseudo-differential fundamental solutions Φ^{loc} of the complex $E^{\cdot}|_{U(\mathfrak{c})}$.

Hint. Use Proposition 2.3.2.

□

For the local index $\mu(\mathfrak{c})$, all that matters is the cohomology class of the form $\varphi(\Phi^{loc})$ in

$$H^{n-1}(\mathcal{E}(\Lambda^{\cdot} T^*_{\mathbb{C}}(U'(\mathfrak{c}) \setminus \mathfrak{c}))) \quad \text{modulo} \quad H^{n-1}(\mathcal{E}(\Lambda^{\cdot} T^*_{\mathbb{C}}(U'(\mathfrak{c})))).$$

Exercise 6.4.6 *Prove that the cohomology class of the form $\varphi(\Phi^{loc})$ in*

$$H^{n-1}(\mathcal{E}(\Lambda^{\cdot} T^*_{\mathbb{C}}(U'(\mathfrak{c}) \setminus \mathfrak{c}))) \quad \textit{modulo} \quad H^{n-1}(\mathcal{E}(\Lambda^{\cdot} T^*_{\mathbb{C}}(U'(\mathfrak{c}))))$$

depends only on the cohomology class of the kernel Φ^{loc} in

$$H^{-1}(\mathcal{E}((E^{\cdot} \boxtimes E^{\cdot\prime})^{\cdot}|_{U(\mathfrak{c})\times U(\mathfrak{c})\setminus\Delta})) \quad \textit{modulo} \quad H^{-1}(\mathcal{E}((E^{\cdot} \boxtimes E^{\cdot\prime})^{\cdot}|_{U(\mathfrak{c})\times U(\mathfrak{c})})).$$

Proof. Use Theorem 2.4.7 and Proposition 2.4.6.

□

6.5 Some Examples for the Classical Complexes

In conclusion we show a few examples illustrating aur approach to the Atiyah-Bott-Lefschetz theory for elliptic complexes.

6.5.1 *The classical Lefschetz formula*

Let $E^{\cdot} = \Lambda^{\cdot} T^*_{\mathbb{C}}(X)$ be the de Rham complex on a compact differentiable manifold X of dimension n.

As explained in Example 6.2.10, each smooth mapping $m: X \to X$ has a natural lift to the complex $\mathcal{E}(\Lambda^{\cdot} T^*_{\mathbb{C}}(X))$. It is defined by $m^{\sharp} = h \circ m^*$ where h_i is the i-th exterior power of the mapping transposed to the tangential mapping, i.e., $h_i = \Lambda^i \left(\frac{\partial m}{\partial x}\right)'$ $(i \in \mathbb{Z})$.

For this case, $(Hm^{\sharp})_i$ are the usual homomorphisms of the cohomology spaces $H^i(X, \mathbb{C})$ of the manifold X induced by the mapping m. And $L(m^{\sharp})$ is just the classical Lefschetz number.

Corollary 6.5.1 *If $\mathfrak{p}$ is a simple fixed point of m, then*

$$\mu(\mathfrak{p}) = sign \ \det \left(1 - \frac{\partial m}{\partial x}(\mathfrak{p})\right).$$

Proof. By Theorem 6.3.7,

$$\mu(\mathfrak{p}) = \frac{\sum_{\imath=0}^{N}(-1)^{\imath} tr \bigwedge^{\imath}\left(\frac{\partial m}{\partial x}(\mathfrak{p})\right)'}{\left|\det\left(1-\frac{\partial m}{\partial x}(\mathfrak{p})\right)\right|}.$$

However, given any endomorphism E of an n-dimensional vector space, we have

$$\sum_{\imath=0}^{n}(-1)^{\imath}\mathrm{tr} \bigwedge^{\imath} E = \det(1-E).$$

Therefore,

$$\begin{aligned}\mu(\mathfrak{p}) &= \frac{\det\left(1-\frac{\partial m}{\partial x}(\mathfrak{p})\right)'}{\left|\det\left(1-\frac{\partial m}{\partial x}(\mathfrak{p})\right)\right|} \\ &= \text{sign } \det\left(1-\frac{\partial m}{\partial x}(\mathfrak{p})\right),\end{aligned}$$

as desired.

□

If an isolated fixed point $\mathfrak{p}$ is not simple, formula (6.3.4) is no longer applicable and we resort to the theory of Section 6.4.

Corollary 6.5.2 *For any isolated fixed point p of m, the following is true:*

$$\mu(\mathfrak{p}) = \left\{ \begin{array}{c} \textit{local degree of} \\ (1-m) \ \textit{at} \ \mathfrak{p} \end{array} \right\}.$$

Proof. By formula (6.4.4),

$$\mu(\mathfrak{p}) = \int_{\partial B(\mathfrak{p},\varepsilon)} \varphi(\Phi^{loc}),$$

where Φ^{loc} is a pseudo-differential fundamental solution of the de Rham complex in a neighborhood of $\mathfrak{p}$ and $\varepsilon > 0$ is small enough.

Fix a coordinate neighborhood $U(\mathfrak{p})$ centered at $\mathfrak{p}$ with coordinates $x = x(\cdot)$. We identify $U(\mathfrak{p})$ with its image $x(U(\mathfrak{p}))$ in $\mathbb{R}^n$. We may assume the point $\mathfrak{p}$ corresponds to the origin $x = 0$.

By the *invariance property* of the exterior derivative, the de Rham complex $\Lambda T^*_{\mathbb{C}}(X)$ restricted to $U(\mathfrak{p})$ is actually the de Rham complex $\Lambda T^*_{\mathbb{C}}(x(U(\mathfrak{p})))$ on the open subset of $\mathbb{R}^n$. So we may use the fundamental solution of convolution type $\Phi(x,y)$ of the complex $\Lambda T^*_{\mathbb{C}}(x(U(\mathfrak{p})))$ constructed in Subsection 2.5.2.

Let us recall that $\Phi_\imath(x,y)$ are double forms of degree $(\imath-1)$ in x and degree $(n-\imath)$ in y in $(\mathbb{R}^n \times \mathbb{R}^n) \setminus \Delta$, associated by certain rules with the convolution kernel $\Phi(x) = d^*(-g(x)dx)$ in $\mathbb{R}^n$. Namely, $\Phi_\imath(x,y)$ is the component of degree $(n-\imath)$ in y of $(-1)^n\Phi(x-y)$ times $(-1)^\imath$, where one must first rearrange all the differentials in x before the differentials in y and then declare the form to be double.

Using now the formula of Example 6.2.13, we obtain

$$\begin{aligned}
\mu(\mathfrak{p}) &= \int_{\partial B(\mathfrak{p},\varepsilon)} \varphi(\Phi) \\
&= \int_{\partial B(\mathfrak{p},\varepsilon)} \sum_{i=1}^{n} (-1)^i \, \Delta^\sharp (m \times 1)^\sharp \Phi_i \\
&= \int_{\partial B(\mathfrak{p},\varepsilon)} \Delta^\sharp (m \times 1)^\sharp \left((-1)^n \Phi(x-y)\right) \\
&= -\int_{\partial B(\mathfrak{p},\varepsilon)} (1-m)^\sharp \Phi \\
&= -\int_{\partial B(\mathfrak{p},\varepsilon)} (1-m)^\sharp \Phi_1(0,\cdot). \qquad (6.5.1)
\end{aligned}$$

To complete the proof it suffices to apply formula (6.1.19) which implies in particular that the right-hand side of (6.5.1) is equal to the local degree of $(1-m)$ at the point $\mathfrak{p}$.

□

Corollary 6.5.2 gives the well-known result of Lefschetz [147] in algebraic topology.

Instead of the de Rham complex $\Lambda\, T^*_{\mathbb{C}}(X)$, one can consider the tensor product $E \otimes \Lambda\, T^*_{\mathbb{C}}(X)$ with any locally constant $\mathbb{C}$-vector bundle E over X. The operator $1 \otimes d$ is well-defined on $\mathcal{E}(E \otimes \Lambda^i T^*_{\mathbb{C}}(X))$ $(i \in \mathbb{Z})$ and we again have an elliptic complex on X.

If $m : X \to X$ is a smooth mapping and $h : m^*E \to E$ is a smooth bundle homomorphism, then the tensor products $h \otimes \Lambda^i \left(\frac{\partial m}{\partial x}\right)'$ $(i \in \mathbb{Z})$ provide the lift of m to $\mathcal{E}(E \otimes \Lambda\, T^*_{\mathbb{C}}(X))$. This endomorphism of $\mathcal{E}(E \otimes \Lambda\, T^*_{\mathbb{C}}(X))$ is denoted by $m^{\sharp(E)}$.

Exercise 6.5.3 *Prove that if the set* Fix m *is discrete, then*

$$L(m^{\sharp(E)}) = \sum_{\mathfrak{p} \in Fix\, m} \mu(\mathfrak{p}),$$

where

$$\mu(\mathfrak{p}) = tr\, h(\mathfrak{p}) \cdot \left\{ \begin{array}{c} \textit{local degree of} \\ (1-m) \ \textit{at} \ \mathfrak{p} \end{array} \right\}.$$

Hint. Construct a local fundamental solution of convolution type for the complex $E \otimes \Lambda\, T^*_{\mathbb{C}}(X)$ using the results of Subsection 2.5.2. Follow then the proof of Corollary 6.5.2.

□

This formula is more or less well-known (see Atiyah and Bott [20]).

6.5.2 *The holomorphic Lefschetz formula*

The formula for the local index for the de Rham complex as a local degree is very special. It is a consequence of the fact that the local cohomology of the complex is one-dimensional.

For the Dolbeault complex $E = \Lambda^{p,\cdot} T^*_{\mathbb{C}}(X)$ with coefficients in $\Lambda^{p,0}T^*_{\mathbb{C}}(X)$ on a compact complex manifold X of dimension n, the situation is more interesting because the local cohomology is infinite-dimensional (the space of local holomorphic p-forms).

Let $m : X \to X$ be a holomorphic mapping. As mentioned in Example 6.2.11, m has a natural lift to the complex $\mathcal{E}(\Lambda^{p,\cdot} T^*_{\mathbb{C}}(X))$. We would like to make this more precise.

Denote by $T'(X)$ the complex subbundle of the complexified tangent bundle $T_{\mathbb{C}}(X) = \mathbb{C} \otimes_{\mathbb{R}} T(X)$ of X which is generated by the vectors $\left\{\frac{\partial}{\partial z_j}\right\}_{1\leq j\leq n}$ in local holomorphic coordinates $z = z(p)$ on X, and by $T''(X)$ the complex subbundle of $T_{\mathbb{C}}(X)$ generated by the vectors $\left\{\frac{\partial}{\partial \bar{z}_i}\right\}_{1\leq i\leq n}$. We obtain the decomposition $T_{\mathbb{C}}(X) = T'(X) \oplus T''(X)$ so that the bundles $\Lambda^{p,q}T^*_{\mathbb{C}}(X)$ $(q \in \mathbb{Z})$ can be identified with the tensor products $\bigwedge^p(T'(X))^* \otimes \bigwedge^q(T''(X))^*$.

Since the mapping m is holomorphic, the complexified tangent mapping $1 \otimes \frac{\partial m}{\partial x}$ (denoted simply by $\frac{\partial m}{\partial x}$) preserves the decomposition $T_{\mathbb{C}}(X) = T'(X) \oplus T''(X)$. For this reason, $\frac{\partial m}{\partial x} = \frac{\partial m}{\partial z} \oplus \overline{\frac{\partial m}{\partial z}}$, with homomorphisms

$$\frac{\partial m}{\partial z}(p) : \; T'_p(X) \to T'_{m(p)}(X),$$
$$\overline{\frac{\partial m}{\partial z}}(p) : \; T''_p(X) \to T''_{m(p)}(X).$$

We thus obtain

$$\bigwedge^i \left(\frac{\partial m}{\partial x}\right)' = \sum_{p+q=i} \bigwedge^p \left(\frac{\partial m}{\partial z}\right)' \otimes \bigwedge^q \left(\overline{\frac{\partial m}{\partial x}}\right)', \qquad i \in \mathbb{Z}.$$

The tensor products $\bigwedge^p \left(\frac{\partial m}{\partial z}\right)' \otimes \bigwedge^q \left(\overline{\frac{\partial m}{\partial z}}\right)'$, $q \in \mathbb{Z}$, now provide smooth bundle homomorphisms $m^*\Lambda^{p,q}T^*_{\mathbb{C}}(X) \to \Lambda^{p,q}T^*_{\mathbb{C}}(X)$ which lift m to the complex $\mathcal{E}(\Lambda^{p,\cdot} T^*_{\mathbb{C}}(X))$. We denote this natural lift by $m^{\sharp(p)}$.

The endomorphisms $m^{\sharp(p)}$ is easily seen to be the usual "pull-back" operator on $(p,\cdot)$-forms induced by the mapping m. The equality $\overline{\partial}^{(p)} m^{\sharp(p)} = m^{\sharp(p)} \overline{\partial}^{(p)}$ was stated in Proposition 1.2.46

Remark 6.5.4 *By the Dolbeault Theorem, the homomorphisms $(Hm^{\sharp(p)})_q$ may also be considered to act on the cohomology spaces $H^q(X, \mathcal{O}^{(p)})$ of X. The Lefschetz number $L(m^{\sharp(p)})$ is the alternating sum of their traces.*

Let us calculate the local index of an isolated fixed point $\mathfrak{p}$ of m with respect to the complex $\Lambda^{p,\cdot} T^*_{\mathbb{C}}(X)$. To avoid confusion, we denote the index by $\mu^{(p)}(\mathfrak{p})$.

Corollary 6.5.5 *If $\mathfrak{p}$ is a simple fixed point of m, then*

$$\mu^{(p)}(\mathfrak{p}) = \frac{\operatorname{tr} \bigwedge^p\left(\frac{\partial m}{\partial z}(\mathfrak{p})\right)'}{\det\left(1 - \frac{\partial m}{\partial z}(\mathfrak{p})\right)}.$$

Proof. By Theorem 6.3.7,

$$\begin{aligned}\mu^{(p)}(\mathfrak{p}) &= \frac{\sum_{q=0}^{n}(-1)^q \mathrm{tr}\, \bigwedge^p\left(\frac{\partial m}{\partial z}(\mathfrak{p})\right)' \otimes \bigwedge^q\left(\overline{\frac{\partial m}{\partial z}}(\mathfrak{p})\right)'}{\left|\det\left(1-\frac{\partial m}{\partial x}(\mathfrak{p})\right)\right|} \\ &= \mathrm{tr} \bigwedge^p \left(\frac{\partial m}{\partial z}(\mathfrak{p})\right)' \cdot \frac{\det\left(1-\overline{\frac{\partial m}{\partial z}}(\mathfrak{p})\right)}{\left|\det\left(1-\frac{\partial m}{\partial x}(\mathfrak{p})\right)\right|}.\end{aligned}$$

Since $\det\left(1-\overline{\frac{\partial m}{\partial z}}(\mathfrak{p})\right) = \overline{\det\left(1-\frac{\partial m}{\partial x}(\mathfrak{p})\right)}$, we get

$$\begin{aligned}\det\left(1-\frac{\partial m}{\partial x}(\mathfrak{p})\right) &= \det\left(1-\frac{\partial m}{\partial z}(\mathfrak{p})\right)\cdot \det\left(1-\overline{\frac{\partial m}{\partial z}}(\mathfrak{p})\right) \\ &= \left|\det\left(1-\frac{\partial m}{\partial z}(\mathfrak{p})\right)\right|^2 \\ &> 0,\end{aligned}$$

whence the corollary follows.

□

If an isolated fixed point $\mathfrak{p}$ is not simple, Theorem 6.3.7 is no longer applicable. However we can invoke the theory of Section 6.4.

To obtain the local index, we need some local fundamental solution to the complex $\Lambda^{p,\cdot}T^*_{\mathbb{C}}(X)$.

Fix a coordinate neighborhood $U(\mathfrak{p})$ centered at $\mathfrak{p}$, with holomorphic coordinates $z = z(\cdot)$. By the invariance of the $\overline{\partial}$-derivative under holomorphic changes of variables, the complex $\Lambda^{p,\cdot}T^*_{\mathbb{C}}(X)$ restricted to $U(\mathfrak{p})$ is just the Dolbeault complex $\Lambda^{p,\cdot}T^*_{\mathbb{C}}(z(U(\mathfrak{p})))$ on the open subset $z(U(\mathfrak{p}))$ of $\mathbb{C}^n$. Therefore, we may use the fundamental solution of convolution type $\Phi^{(p)}(z,\zeta)$ to the complex $\Lambda^{p,\cdot}T^*_{\mathbb{C}}(z(U(\mathfrak{p})))$ which has been constructed in Subsection 2.5.3.

Recall that $\Phi^{(p)}_q(z,\zeta)$ ($q \in \mathbb{Z}$) is the component of bidegree $(n-p, n-q)$ in ζ of $\Phi(z-\zeta)$ times $(-1)^{p+q}$, where $\Phi(z) = \overline{\partial}^*\left(-2g(z)\left(\frac{\sqrt{-1}}{2}\right)^n dz \wedge d\overline{z}\right)$. Moreover, $\Phi^{(p)}_q(z,\zeta)$ is considered to be a double form in z and ζ, where one has first to rearrange all the differentials in z and $\overline{z}$ before the differentials in ζ and $\overline{\zeta}$, and then to declare the form to be double.

Corollary 6.5.6 *If $\mathfrak{p}$ is an isolated fixed point of m, then for all sufficiently small $\varepsilon > 0$ we have*

$$\mu^{(p)}(\mathfrak{p}) = \int_{\partial B(\mathfrak{p},\varepsilon)} \Delta^\sharp(m\times 1)^\sharp \left(\sum_{q=1}^{n}(-1)^q\, \Phi^{(p)}_q\right). \qquad (6.5.2)$$

Proof. This follows immediately from Theorem 6.4.4.

□

If $n = 1$, the Dolbeault complex consists of just one operator. Then formulas (6.5.2) become

$$\mu^{(0)}(\mathfrak{p}) = \frac{1}{2\pi\sqrt{-1}}\int_{\partial B(\mathfrak{p},\varepsilon)} \frac{dz}{z-m(z)},$$

$$\mu^{(1)}(\mathfrak{p}) = \frac{1}{2\pi\sqrt{-1}} \int_{\partial B(\mathfrak{p},\varepsilon)} \frac{dm(z)}{z - m(z)}.$$

In higher dimensions there are analogous formulas in terms of the *Grothendieck residue* rather than (6.5.2) (see Toledo [238]). The properties of the Grothendieck residue imply that $\mu^{(p)}(\mathfrak{p})$ depends only on a finite (however large) number of derivatives of m at the point p.

Exercise 6.5.7 *Prove that if* $\mathfrak{p}$ *is an isolated fixed point of* m, *then*

$$\sum_{p=0}^{n} \mu^{(p)}(\mathfrak{p}) = \left\{ \begin{matrix} \text{local degree of} \\ (1-m) \ \text{ at } \ \mathfrak{p} \end{matrix} \right\}. \tag{6.5.3}$$

Hint. Use Corollary 6.5.6. Verify the equality

$$\sum_{p=0}^{n}(-1)^p \left(\Delta^\sharp (m \times 1)^\sharp \left(\sum_{q=1}^{n} (-1)^q \, \Phi_q^{(p)} \right) \right) = -(1-m)^\sharp \, \Phi_1^{(0)}.$$

Finally, apply the logarithmic residue formula (6.1.12).

□

For simple fixed points $\mathfrak{p}$, this result follows from Corollary 6.5.5 directly.

Thus, even though each $\mu^{(p)}(\mathfrak{p})$ is not expressible as a local degree, their alternating sum is. Therefore $\sum_{p=0}^{n}(-1)^p \mu^{(p)}(\mathfrak{p})$ is a topological, rather than holomorphic, invariant of m at $\mathfrak{p}$. This is not surprising because for the Lefschetz numbers it is easy to see that $L(m^\sharp) = \sum_{p=0}^{n}(-1)^p L(m^{\sharp(p)})$, where $m^\sharp$ is the natural lift of m to the de Rham complex over X. Equality (6.5.3) just says that this relation also holds for the local indices.

The complex $\Lambda^{p,\cdot}T^*_{\mathbb{C}}(X)$ is equal to the tensor product $E \otimes \Lambda^{0,\cdot}T^*_{\mathbb{C}}(X)$, where $E = \Lambda^{p,0}T^*_{\mathbb{C}}(X)$ is the bundle of exterior forms of bidegree $(p,0)$ over X. More generally, we may consider the tensor product of the Dolbeault complex $\Lambda^{0,\cdot}T^*_{\mathbb{C}}(X)$ with any holomorphic vector bundle E over X. The operator $1 \otimes \overline{\partial}^{(0)}$ is well-defined on $\mathcal{E}(E \otimes \Lambda^{0,q}T^*_{\mathbb{C}}(X))$ ($q \in \mathbb{Z}$) and we again get an elliptic complex over X.

If $m : X \to X$ is a holomorphic mapping and $h : m^*E \to E$ is a holomorphic bundle homomorphism, then the tensor products $h \otimes \Lambda^q \left(\frac{\partial m}{\partial z}\right)'$ ($q \in \mathbb{Z}$) provide a lift of m to $\mathcal{E}(E \otimes \Lambda^{0,\cdot}T^*_{\mathbb{C}}(X))$. This geometric endomorphism of $\mathcal{E}(E \otimes \Lambda^{0,\cdot}T^*_{\mathbb{C}}(X))$ is denoted by $m^{\sharp(E)}$.

Exercise 6.5.8 *Prove that if the set* Fix m *is discrete, then*

$$L(m^{\sharp(E)}) = \sum_{\mathfrak{p} \in \mathrm{Fix}\, m} \mu^{(E)}(\mathfrak{p}),$$

where

$$\mu^{(E)}(\mathfrak{p})$$
$$= \int_{\partial B(\mathfrak{p},\varepsilon)} \operatorname{tr} h(z) \left(\frac{(n-1)!}{(2\pi\sqrt{-1})^n} \sum_{j=1}^{n} (-1)^{j-1} \frac{\overline{z_j - m_j(z)}}{|z - m(z)|^{2n}} \, dz \wedge d\overline{(z - m(z))}\,[j] \right). \tag{6.5.4}$$

Hint. Construct a local fundamental solution for the complex $E \otimes \Lambda^{0,\cdot} T^*_{\mathbb{C}}(X)$ using the results of Subsection 2.5.3. Apply then Theorem 6.4 4.

□

Given any simple fixed point $\mathfrak{p}$ of m, the integral on the right-hand side of (6.5.4) is easily calculated by the logarithmic residue formula (6.1.12). This gives a well-known formula of Atiyah and Bott [20], namely,

$$\mu^{(E)}(\mathfrak{p}) = \frac{\operatorname{tr} h(\mathfrak{p})}{\det\left(1 - \frac{\partial m}{\partial z}(\mathfrak{p})\right)}.$$

6.5.3 *Other examples and applications*

In connection with other examples and applications of Lefschetz theory for elliptic complexes, we refer to the original paper of Atiyah and Bott [20].

Finally, we discuss coincidence theorems for elliptic complexes.

Let E be an elliptic complex over a compact manifold X and let F be an elliptic complex over a compact manifold Y.

Given a smooth mapping $y = m_1(x)$ of X to Y, we assume that m_1 may be lifted to a cochain mapping $m_1^\sharp$ of $\mathcal{E}(F)$ to $\mathcal{E}(E)$. Analogously, given a smooth mapping $y = m_2(x)$ of X to Y, we assume that m_2 may be lifted to a chain mapping $m_2^\sharp$ of $\mathcal{E}(F')$ to $\mathcal{E}(E')$.

For every $\imath \in \mathbb{Z}$, the composition $\left(Hm_2^\sharp\right)'_\imath \left(Hm_1^\sharp\right)_\imath$ is well-defined to be a homomorphism of the cohomology space $H^\imath(\mathcal{E}(F))$, where $\left(Hm_2^\sharp\right)'_\imath : H^\imath(\mathcal{E}(E)) \to H^\imath(\mathcal{E}(F))$ is the mapping transposed to $\left(Hm_2^\sharp\right)_\imath : H^\imath(\mathcal{E}(F')) \to H^\imath(\mathcal{E}(E'))$ by Poincaré duality (see Example 5.1.11).

The number

$$L(m_1^\sharp, m_2^\sharp) = \sum_\imath (-1)^\imath \operatorname{tr} \left(Hm_2^\sharp\right)'_\imath \left(Hm_1^\sharp\right)_\imath .$$

is said to be the *Lefschetz coincidence number.*

Exercise 6.5.9 *Characterize* $L(m_1^\sharp, m_2^\sharp)$ *in terms of the coincidence point set* $\{x \in X : m_1(x) = m_2(x)\}$.

Hint. See Toledo [238].

□

Taking $X = Y$, $E = F$ and $m_2 = 1$, we can reduce this to $L(m_1^\sharp)$.

Bibliography

[1] Asada A. Currents and residue exact sequences. *J. Fac. Sci. Shinshu Univ.*, 3(2): 85–151, 1968.

[2] S. Agmon, A. Douglis, and L. Nirenberg. Estimates near the boundary for solutions of elliptic partial differential equations satisfying general boundary conditions. Pt. 1. *Comm. Pure Appl. Math.*, 12: 623–727, 1959.

[3] Aizenberg, L. A., A. K. Tsikh, and A. P. Yuzhakov. Many-dimensional residues and their applications. In: *Modern problems of mathematics. Fundamental directions. Vol. 8*, VINITI, Moscow, 1985, 5–64 (Russian).

[4] L. A. Aizenberg. A remark on Morera's theorem. In: *Holomorphic functions of several complex variables*, Inst. Fiz. SO Akad. Nauk SSSR, Krasnoyarsk, 1972, 165–166 (Russian).

[5] L. A. Aizenberg and Sh. A. Dautov. *Differential forms orthogonal to holomorphic functions or forms, and their properties.* Nauka, Novosibirsk, 1975 (Russian). (English transl.: *Amer. Math. Soc.*, Providence, R.I., 1983.)

[6] L. A. Aizenberg and A. P. Yuzhakov. *Integral representations and residues in many-dimensional complex analysis.* Nauka, Novosibirsk, 1979 (Russian).

[7] W. Ambrose. Products of distributions with values in distributions. *Z. für die reine und angewandte Mathematik*, 315: 73–91, 1980.

[8] A. Andreotti. *Complexes of partial differential operators.* Yale Univ. Press, New Haven and London, 1975.

[9] A. Andreotti and C. D. Hill. E.E.Levi convexity and the Hans Lewy problem. Part 1: Reduction to vanishing theorems. *Ann. Scuola Norm. Super. Pisa*, 26(3): 325–363, 1972.

[10] A. Andreotti and C. D. Hill. E.E.Levi convexity and the Hans Lewy problem. Part 2: Vanishing theorems. *Ann. Scuola Norm. Super. Pisa*, 26(4): 747–806, 1972.

[11] A. Andreotti, C. D. Hill, S. Łojasiewicz, and B. MacKichan. Complexes of differential operators. The Mayer-Vietoris sequence. *Invent. Math*, 35(1): 43–86, 1976.

[12] A. Andreotti and A. Kas. Duality on complex spaces. *Ann. Scuola Norm. Super. Pisa*, 27(2): 187–263, 1973.

[13] A. Andreotti and M. Nacinovich. Analytic convexity and the principle of Phragmen-Lindelöf. *Quaderni Scuola Norm. Super. Pisa*, 1980, 181 pp.

[14] A. Andreotti and M. Nacinovich. Non-characteristic hypersurfaces for complexes of differential operators. *Ann. Mat. Pura e Appl.*, 125(IV): 13–83, 1980.

[15] A. Andreotti and M. Nacinovich. On analytic and C^∞ Poincaré lemma. In: *Math. Anal. Appl. Part A*, New York et al., 1981, 41–93.

[16] A. M. Aronov and A. M. Kytmanov. On holomorphy of functions represented by the Martinelli-Bochner integral. *Funkz. Analiz*, 9(3): 83–84, 1975.

[17] N. Aronszain. Theory of reproducing kernels. *Trans. Amer. Math. Soc.*, 68: 337–404, 1950.

[18] M. F. Atiyah. *K-theory.* Harvard Univ., Cambridge, Mass., 1965.

[19] M. F. Atiyah and R. Bott. A Lefschetz fixed point formula for elliptic complexes. 1. *Ann. Math.*, 86(2): 374–407, 1967.

[20] M. F. Atiyah and R. Bott. A Lefschetz fixed point formula for elliptic complexes. 2: Applications. *Ann. Math.*, 88(3): 451–491, 1968.

[21] F.V. Atkinson. Normal solvability of linear equations in normed spaces. *Mat. Sb.*, 28(1): 3–14, 1951.

[22] L. Bers, F. John, and M. Schechter. *Partial differential equations.* Interscience Publ., New York-London-Sydney, 1964.

[23] A. V. Bitsadze. *Boundary problems for elliptic equations of second order.* Nauka, Moscow, 1966.

[24] S. Bochner. Analytic and meromorphic continuation by means of Green's formula. *Ann. Math.*, 44(2): 652–673, 1943.

[25] R. Bott and L. W. Tu. *Differential forms in algebraic topology.* Springer-Verlag, New York-Heidelberg-Berlin, 1982.

[26] N. Bourbaki. Sur certains espaces vectoriels topologiques. *Ann. Inst. Fourier*, 2: 5–16, 1950.

[27] N. Bourbaki. *Topologie generale. Ch I-II.* Nauka, Moscow, 1968.

[28] N. Bourbaki. *Topologie generale. Ch IX-X.* Nauka, Moscow, 1975.

[29] N. Bourbaki. *Varietes differentielles et analytiques.* Mir, Moscow, 1975.

[30] N. Bourbaki. *Algebre. Ch X: Algebre homologique.* Nauka, Moscow, 1987.

[31] H. Bremermann. *Distributions, complex variables and Fourier transformations.* Mir, Moscow, 1968.

[32] A.V. Brenner and M.A. Shubin. The Atiyah-Bott-Lefschetz theorem for manifolds with boundary. *Funkz. Analiz*, 15(4): 67–68, 1981.

[33] L. Bungart. Boundary kernel functions for domains on complex manifolds. *Pacif. J. Math.*, 14(4): 1151–1164, 1964.

[34] L. Bungart. Holomorphic functions with values in locally convex spaces and applications to integral formulas. *Trans. Amer. Math. Soc.*, 111(2): 317–344, 1964.

[35] M. Cantor. Elliptic operators and the decomposition of tensor fields. *Bull. Amer. Math. Soc.*, 5(3): 235–262, 1981.

[36] H. Cartan. *Calcul differentiel. Formes differentielles.* Hermann, Paris, 1967.

[37] B. Cenkl. Vanishing theorem for an elliptic differential operator. *J. Differential Geom.*, 1(3): 381–418, 1967.

[38] E. M. Chirka. An analytic representation of CR-functions. *Mat. Sb.*, 98(4): 591–623, 1975 (Russian).

[39] E. M. Chirka. Regularization and $\overline{\partial}$-homotopy on a complex manifold. *Dokl. AN SSSR*, 244(2): 300–303, 1979 (Russian).

[40] J. F. Colombeau. A multiplication of distributions. *J. Math. Anal. and Appl.*, 94(1), 1983. 96–115.

[41] P. E. Conner. The Neumann problem for differential forms on Riemannian manifolds *Mem Amer Math Soc* , 20, 1953 1–58

[42] M. Craioveanu and M. Puta. A chain homotopy theorem for a class of first order linear differential operators In: *Proc. of the Conf. on Diff. Geom. and its Appl.*, Univ. Karlova, Praha, 1981, 9–16.

[43] N. S. Dairbekov and A. P. Kopylov. ξ-stability of mappings' classes and systems of linear partial differential equations. *Sibirsk. Mat. Zh.*, 26(2): 73–90, 1985.

[44] Sh. A. Dautov. On $\bar{\partial}$-closed forms of the type $(p, n-1)$ as analogs of holomorphic functions of one complex variable. In: *Holomorphic functions of several complex variables*, Inst. Fiz. SO Akad. Nauk SSSR, Krasnoyarsk, 1972, 21–36 (Russian).

[45] Sh. A. Dautov and A. M. Kytmanov. On boundary values of the Martinelli-Bochner type integral. In: *Some properties of holomorphic functions of several complex variables*, Inst. Fiz. SO Akad. Nauk SSSR, Krasnoyarsk, 1973, 49–54 (Russian).

[46] Sh. A. Dautov and S. V. Znamenskii. Morera's theorem in many-dimensional complex analysis. *Izv. VUZ. Mat.*, 5: 17–19, 1975.

[47] E. de Giorgi. Solutions analytiques des equations aux derivees partielles a coefficients constants. In: *Seminaire Goulauic-Schwartz (1971-72), Exposé 29*, 1972.

[48] E. de Giorgi and L. Cattabriga. Una dimostrazione diretta del l'esistenza di soluzioni analitiche nel piano reale. *Bull. U.M.I.*, 4(4): 1015–1027, 1971.

[49] E. de Giorgi and L. Cattabriga. Una formula di rappresentazione per funzioni analitiche in $\mathbb{R}^n$. *Bull. U.M.I.*, 4(4): 1010–1014, 1971.

[50] Georges de Rham. *Variétés différentiables. Formes, courants, formes harmoniques.* Actualites Sci. Indust, no. 1222. Hermann, Paris, 1955.

[51] R. Delanghe and F. Brackx. Duality in hypercomplex function theory. *J. Funct. Anal.*, 37(2): 164–181, 1980.

[52] A. A. Desin. *General questions of the theory of boundary problems.* Nauka, Moscow, 1980.

[53] K. Diederich. Some recent developments in the theory of the Bergman kernel function: a survey. In: *Proc. Sympos. Pure Math. Vol. 30*, Amer. Math. Soc., Providence, R.I., 1977, 127–137.

[54] J. Dieudonné and L. Schwartz. La dualité dans les espaces (F) et (DF). *Ann. Inst. Fourier*, 1: 61–101, 1949.

[55] P. Dolbeault. Sur la cohomologie des varietés analytiques complexes. *C. R. Acad. Sci. Paris*, 236(2): 175–177, 1953.

[56] A. Dold. *Lectures on algebraic topology.* Springer-Verlag, Berlin-Heildelberg-New York, 1972.

[57] P. I. Dudnikov and S. N. Samborskii. Noetherian boundary problems for overdetermined systems of partial differential equations. Technical Report Preprint N 81.47, Inst. of Math., Kiev, 1981. 34 pp.

[58] P. I. Dudnikov and S. N. Samborskii. Overdetermined systems of linear partial differential equations, boundary and initial boundary value problems for them. In: *Modern problems of mathematics. Fundamental directions. Vol. 65*, VINITI, Moscow, 1991, 6–94 (Russian).

[59] G. F. Duff and D. C. Spencer. Harmonic tensors on Riemannian manifolds with boundary. *Ann. Math.*, 56(1): 128–156, 1952.

[60] A. S. Dynin. Elliptic boundary problems for pseudo-differential complexes. *Funkz. Analiz*, 6(1): 75–76, 1972.

[61] E. M. Dyn'kin. Methods of the theory of singular integrals. (Hilbert transform and the Calderon-Zygmund theory.). In: *Modern problems of mathematics. Fundamental directions. Vol. 15*, VINITI, Moscow, 1986, 197–292 (Russian).

[62] R. Edwards. *Functional analysis. Theory and applications.* Holt, Rinehart and Winston, New York et al., 1965. 781 pp.

[63] Jn. V. Egorov. *Linear differential equations of the principal type.* Nauka, Moscow, 1984.

[64] L. Ehrenpreis. Solution of some problems of division. *Amer. J. Math.*, 76: 883–903, 1954.

[65] L. Ehrenpreis. Analytic functions and the Fourier transform of distributions. pt. i. *Ann. Math.*, 63(1): 129–159, 1956.

[66] L. Ehrenpreis. Analytic functions and the Fourier transform of distributions. pt. ii. *Trans. Amer. Math. Soc.*, 89: 450–483, 1958.

[67] J. Fredholm. Sur l'intégrale fondamentale d'une equation différentielle elliptique a coefficients constants. *Rend. Circ. Mat. Palermo*, 25: 346–351, 1908.

[68] A. Friedman. *Partial differential equations of parabolic type.* Prentice-Hall, Englewood Cliffs, N.J., 1964.

[69] A. Gasiev. On behavior of the Martinelli-Bochner type integral near by the surface of integration. In: *Questions of math. analysis and its applications*, Samarkand Univ., Samarkand, 1981, 111–119 (Russian).

[70] P. B. Gilkey. Lefschetz fixed point formulas and the heat equation. In: *Part. Diff. Eq. and Geom. (Proc. Park City Conf., 1977)*, New York-Basel, 1979, 91–147.

[71] A. M. Gleason. The Cauchy-Weil theorem. *J. Math. and Mech.*, 3(12): 429–444, 1963.

[72] R. Godement. *Topologie algébrique et théorie des faisceaux.* Hermann, Paris, 1958.

[73] H. Goldschmidt. Existence theorems for analytic linear partial differential equations. *Ann. Math.*, 86(2): 246–270, 1967.

[74] H. Goldschmidt. A prolongation theorem for systems of linear partial differential equations. 1: A conjecture of E. Cartan. *Ann. Sci. Ecole Norm. Super. (4 série)*, 1: 417–444, 1968.

[75] H. Goldschmidt. Prolongations of linear partial differential equations. 2: Inhomogeneous equations. *Ann. Sci. Ecole Norm. Super. (4 série)*, 1: 617–625, 1968.

[76] V. M. Gol'dshtein, V. I. Kuz'minov, and I. A. Shvedov. Differential forms on a Lipschitz manifold. *Sibirsk. Mat. Zh.*, 23(2): 16–30, 1982.

[77] V. M. Gol'dshtein, V. I. Kuz'minov, and I. A. Shvedov. Functional spaces of generalized differential forms. *Sibirsk. Mat. Zh.*, 24(6): 209–210, 1983.

[78] V. M. Gol'dshtein, V. I. Kuz'minov, and I. A. Shvedov. On a theorem of Yolf. *Sibirsk. Mat. Zh.*, 24(5): 31–42, 1983.

[79] V. M. Gol'dshtein, V. I. Kuz'minov, and I. A. Shvedov. On a property of the regularization operators of de Rham. *Sibirsk. Mat. Zh.*, 25(2): 104–111, 1984.

[80] V. M. Gol'dshtein, V. I. Kuz'minov, and I. A. Shvedov. Dual spaces for spaces of differential forms. *Sibirsk. Mat. Zh.*, 27(1): 45–56, 1986.

[81] V. M. Gol'dshtein, V. I. Kuz'minov, and I. A. Shvedov. On normal and compact solvability of the exterior differentiation operator under homogeneous boundary conditions. *Sibirsk. Mat. Zh.*, 28(4): 82–96, 1987.

[82] V. M. Gol'dshtein, V. I. Kuz'minov, and I. A. Shvedov. On normal and compact solvability of linear operators. *Sibirsk. Mat. Zh.*, 30(5): 49–59, 1989.

[83] V. M. Gol'dshtein, V. I. Kuz'minov, and I. A. Shvedov. L^p-cohomologies of curved cylinders. *Sibirsk. Mat. Zh.*, 31(6): 55–63, 1990.

[84] V. M. Gol'dshtein, V. I. Kuz'minov, and I. A. Shvedov. Reduced L^p-cohomologies of curved cylinders. *Sibirsk. Mat. Zh.*, 31(5)· 10–23, 1990.

[85] V. M. Gol'dshtein, V. I. Kuz'minov, and I A. Shvedov. On approximation of exact and closed differential forms by compactly supported ones. *Sibirsk. Mat. Zh.*, 33(2): 49–65, 1992.

[86] V. D. Golovin. *Cohomologies of analytic sheaves and duality theorems.* Nauka, Moscow, 1986.

[87] M. Golubitsky and V Guillemin. *Stable mappings and their singularities.* Springer-Verlag, New York-Heidelderg-Berlin, 1973.

[88] C. Grosu and F. H. Vasilescu. The Künneth formula for Hilbert complexes. *Integral Equations and Operator Theory*, 5(1): 1–17, 1982.

[89] A. Grothendieck. Sur les espaces de solutions d'une classe générale d'équations aux derivées partielles. *J. Anal. Math.*, 2: 243–280, 1952–1953.

[90] A. Grothendieck. Opérations algébriques sur les distributions à valeurs vectorielles, théorème de Künneth. In: *Produits tensoriels topologiques d'espaces vectoriels topologiques. Espaces vectoriels topologiques nucléares. Applications (Séminaire Schwartz de la Faculté des Sciences de Paris, 1953/1954)*, Secrét. Math., Paris, 1954, 144 pp.

[91] A. Grothendieck. Sur les espaces (F) et (DF). *Summa Bras. Math.*, 3: 57–123, 1954.

[92] A. Grothendieck. Produits tensoriels topologiques et espaces nucléaires. *Mem. Amer. Math. Soc.*, 6, 1955.

[93] V. K. Gugenheim and D. C. Spenser. Chain homotopy and the de Rham theory. *Proc. Amer. Math. Soc.*, 7(1): 144–152, 1956.

[94] V. Guillemin. Some algebraic results concerning the characteristics of overdetermined partial differential equations. *Amer. J. Math.*, 90: 270–284, 1968.

[95] V. Guillemin and I. M. Singer. Differential equations and G-structures. In: *Proc. U.S.-Japan Seminar Diff. Geom. (Kyoto, 1965)*, Nippon Hyoronsha Co., Ltd, Tokyo, 1966, 34–36.

[96] R. Gunning and H. Rossi. *Analytic functions of several complex variables.* Prentice-Hall, Englewood Cliffs, N.J., 1965.

[97] J. Hadamard. *Le problème de Cauchy et les équations aux dérivées partielles linéaires hyperboliques.* Ganthier-Villars, Paris, 1932.

[98] R. Harvey. Integral formulae connected by Dolbeault's isomorphism. *Rice University Studies*, 56: 77–97, 1970.

[99] R. Harvey and J. C. Polking. Removable singularities of solutions of linear partial differential equations. *Acta Math*, 125(1–2): 39–56, 1970.

[100] R. Harvey and J. C. Polking. Fundamental solution in complex analysis. Part 1: The Cauchy-Riemann operator. *Duke Math. J.*, 46(2): 253–300, 1979.

[101] R. Harvey and J. C. Polking. Fundamental solution in complex analysis. Part 2: The Cauchy-Riemann operator. *Duke Math J*, 46(2): 301–340, 1979.

[102] V. P. Havin. Spaces of analytic functions. In: *Mathematical analysis*, VINITI, Moscow, 1966, 76–164.

[103] G. M. Henkin. The H. Lewy equation and analysis on pseudoconvex manifold. *Uspekhi Mat. Nauk*, 32(3): 57–118, 1977.

[104] G. M. Henkin. The method of integral representation in complex analysis. In: *Modern problems of mathematics. Fundamental directions. Vol. 7*, VINITI, Moscow, 1985, 167–226 (Russian).

[105] G. Herglotz. Über die Integration linearer, partieller Differentialgleichungen mit konstanten Koeffizienten. *Ber. Verh. Sächs. Acad. Wiss.*, 78: 93–126 and 287–318, 1926.

[106] D. Hilbert. *Grundzüge einer allgemeinen Theorie der linearen Integralgleichungen.* Leipzig-Berlin, Teubner, 1912.

[107] W. V. D. Hodge. *The theory and application of harmonic integrals.* Cambridge Univ. Press, New York, 1941.

[108] L. Hörmander. The Frobenius-Nirenberg theorem. *Arkiv Math*, 5(5): 425–432, 1964.

[109] L. Hörmander. L^2 estimates and existence theorems for the $\overline{\partial}$ operator. *Acta Math.*, 113(1–2): 89–152, 1965.

[110] L. Hörmander. Pseudo-differential operators and hypoelliptic equations. *Amer. Math. Soc. Symp. on Singular Integrals*, pages 129–209, 1966.

[111] L. Hörmander. Pseudo-differential operators and non-elliptic boundary problems. *Ann. Math.*, 83(1): 129–209, 1966.

[112] L. Hörmander. Fourier integral operators. I. *Acta Math*, 127(1–2): 79–183, 1971.

[113] L. Hörmander. *The analysis of linear partial differential operators. Vol. 1: Distribution theory and Fourier analysis.* Springer-Verlag, Berlin et al., 1983.

[114] L. Hörmander. *The analysis of linear partial differential operators. Vol. 2: Differential operators with constant coefficients.* Springer-Verlag, Berlin et al., 1983.

[115] M. Itano. On the multiplicative products of distributions. *J. Sci. Hiroshima Univ. Sep. A-1*, 29(1): 51–74, 1965.

[116] V. K. Ivanov. Hyper-distributions and multiplication of Schwartz's distributions. *Dokl. AN SSSR*, 204(5): 1045–1048, 1972 (Russian).

[117] V. K. Ivanov. On multiplication of homogeneous generalized functions of several variables. *Dokl. AN SSSR*, 257(1): 29–33, 1981 (Russian).

[118] Ch. Iwasaki. Parametrix of $\Box_b$. In: *Non-linear Funct. Anal. and Appl. (Proc. Summer Res. Inst. Berkeley, 1983. Pt. 1)*, Providence, R.I., 1986, 521–524.

[119] F. John. On linear partial differential equations with analytic coefficients. *Comm. Pure Appl. Math.*, 2: 209–253, 1949.

[120] V. A. Kakichev. Character of continuity of boundary values of the Martinelli-Bochner integral. *Sci. Trans. of N. K. Krupsky MOPI*, 96(6): 145–150, 1960.

[121] J. R. King. The currents defined by analytic varieties. *Acta Math.*, 127(3–4): 185–220, 1971.

[122] C. O. Kiselman. Supports des fonctionnelles sur un espace de solution d'une équation aux dérivées partielles à coefficients constants. In: *Séminaire P. Lelong (Analyse)*, volume 8, 1967–1968, 118–126.

[123] K. Kodaira. Harmonic fields in Riemannian manifolds. *Ann. Math.*, 50(2): 587–665, 1949.

[124] K. Kodaira. On a differential-geometric method in the theory of analytic stacks. *Proc. Nat. Acad. Sci.*, 39: 1268–1273, 1953.

[125] J. J. Kohn. Harmonic integrals on strongly pseudoconvex manifolds. I. *Ann. Math.*, 78(1): 112–148, 1963.

[126] J. J. Kohn. Harmonic integrals on strongly pseudoconvex manifolds. II. *Ann. Math.*, 79(3): 450–472, 1964.

[127] J. J. Kohn. Differential complexes. In: *Trans. Intern. Congress of Math. (1966)*, Mir, Moscow, 1968, 57–64.

[128] J. J. Kohn. Subellipticity of the $\bar{\partial}$-Neumann problem on pseudoconvex domains: sufficient conditions. *Acta Math.*, 142(1–2): 70–122, 1979.

[129] J. J. Kohn. Boundary regularity of $\bar{\partial}$. *Ann. Math. Stud.*, 100: 243–260, 1981.

[130] J. J. Kohn. Several complex variables from the point of view of linear partial differential equations. In: *Proc. Beijing Symp. Diff. Geom. and Diff. Eq. (1980). Vol. 2*, Beijing-New York, 1982, 70–122.

[131] J. J. Kohn and L. Nirenberg. An algebra of pseudo-differential operators. *Comm. Pure Appl. Math.*, 18(1–2): 269–305, 1965.

[132] J. J. Kohn and L. Nirenberg. Non-coercive boundary value problems. *Comm. Pure Appl. Math.*, 18: 443–492, 1965.

[133] J. J. Kohn and H. Rossi. On the extension of holomorphic functions from the boundary of a complex manifold. *Ann. Math.*, 81(3): 451–472, 1965.

[134] J. J. Kohn and D. C. Spencer. Complex Neumann problem. *Ann. Math.*, 66(1): 89–140, 1957.

[135] H. Komatsu. Resolutions by hyperfunctions of sheaves of solutions of differential equations with constant coefficients. *Math. Ann.*, 176: 77–86, 1968.

[136] W. Koppelman. The Cauchy integral for differential forms. *Bull. Amer. Math. Soc.*, 73(4): 554–556, 1967.

[137] T. Kotake. The fixed point theorem of Atiyah-Bott via parabolic operators. *Comm. Pure Appl. Math.*, 22(6): 789–806, 1969.

[138] M. A. Krasnosel'sky and P. P. Zabreiko. *Geometrical methods of non-linear analysis.* Nauka, Moscow, 1975.

[139] P. Kree. Les noyaux des opérateurs pseudo-différentiels. In: *Publ. du Séminaire de Math. de l'Univ. de Bari.* L'Univ. de Bari, 1969.

[140] S. F. Krendelev. *Boundary value problems for elliptic systems in* $\mathbb{R}^n$. PhD thesis, Novosibirsk Univ., Novosibirsk, 1984 (Russian).

[141] M. Kuranishi. On the complexes on the boundary induced by elliptic complexes of differential operators. In: *Contribs. Anal.*, New York-London, 1974, 215–240.

[142] V. I. Kuz'minov and I. A. Shvedov. On compactly-supported approximation of closed differential forms on Riemannian manifolds of special type. *Sibirsk. Mat. Zh.*, 34(3): 102–117, 1993.

[143] V. I. Kuz'minov and I. A. Shvedov. On compactly-supported approximation of differential forms in weighted spaces of Sobolev type. *Sibirsk. Mat. Zh.*, 34(6): 91–112, 1993.

[144] A. M. Kytmanov. Representation and multiplication of distributions of several variables by means of harmonic functions. *Izv. VUZ. Mat.*, 1: 36–42, 1978.

[145] A. M. Kytmanov. *The Bochner-Martinelli integral, and its applications.* Nauka, Novosibirsk, 1992 240 pp. (Russian).

[146] Ch. Laurent-Thiebaut. Produits de courants et formule des résidus. *Bull. Sci. Math.*, 2(105): 113–158, 1981.

[147] S. Lefschetz. Intersections and transformations of complexes and manifolds. *Trans. Amer. Math. Soc.*, 28: 1–49, 1926.

[148] J. Leray. *Differential and integral calculus on a complex analytic manifold.* IL, Moscow, 1961.

[149] E. E. Levi. Sulle equazione lineari totalmente ellitique alle derivate parziali. *Rend. Circ. Mat. Palermo*, 24: 257–317, 1907.

[150] H. Lewy. An example of a smooth linear differential equations without solutions. *Ann. Math.*, 66(2): 155–158, 1957.

[151] I. Lieb and R. M. Range. On integral representations and a priori Lipschitz estimates for the canonical solution of the $\bar{\partial}$-equation. *Math. Ann.*, 265(2): 221–251, 1983.

[152] C. H. Look and T. D. Zhong. An extension of Privalov theorem. *Acta Math.*, 7(1): 144–165, 1957.

[153] G. Luppaciolu. On the argument principle in many-dimensional complex manifolds. *Rend. della Academia Nazionale dei Lincei*, 66(5): 323–330, 1979.

[154] B. MacKichan. A generalization to overdetermined systems of the notion of diagonal operators. 1: Elliptic operators. *Acta Math.*, 126(1–2): 83–119, 1971.

[155] B. MacKichan. A generalization to overdetermined systems of the notion of diagonal operators. 2: Hyperbolic operators. *Acta Math.*, 134(3–4): 240–274, 1975.

[156] B. Malgrange. Existence et approximations des équations aux dérivées partielles et des équations de convolution. *Ann. Inst. Fourier (Grenoble)*, 6: 271–355, 1955–1956.

[157] B. Malgrange. Systèmes différentielles à coefficients constants. *Séminaire Bourbaki*, 15(1): 246/01–246/11, 1962–1963.

[158] B. Malgrange. Sur les systèmes différentiels à coefficients constants. *Colloq. Internat. Centre Nat. Rech. Scient.*, (117): 113–122, 1963.

[159] A. Martineau. Distributions et valeurs au bord des fonctions holomorphes. In: *Proc. of the Intern. Summer Inst.*, Lisbon, 1964.

[160] W. Massey. *Homology and cohomology theory.* Marcel Dekker, New York-Basel, 1978.

[161] S. G. Mikhlin. *Many-dimensional singular integrals and integral equations.* Fizmatgiz, Moscow, 1962 234 pp. (Russian).

[162] Gr. C. Moisil and N. Theodoresco. Fonctions Holomorphes dans l'espace. *Mathematica*, 5: 142–153, 1931.

[163] C. B. Morrey. A variational method in the theory of harmonic integrals. II. *Amer. J. Math.*, 78(1): 137–170, 1956.

[164] C. B. Morrey and J. Eells. A variational method in the theory of harmonic integrals. I. *Amer. J. Math.*, 63(2): 91–128, 1956.

[165] N. I. Muskhelishvili. *Singular integral equations.* Nauka, Moscow, 1968 (Russian).

[166] M. Nacinovich. Complex analysis and complexes of differential operators. *Springer Lecture Notes in Math.*, 950: 105–195, 1982.

[167] M. Nacinovich. On weighted estimates for some systems of partial differential operators. *Rend. Sem. Univ. Radova*, 69: 221–232, 1983.

[168] M. Nacinovich. On boundary Hilbert differential complexes. *Ann. Pol. Math.*, 46: 213–235, 1985.

[169] M. Nacinovich. Cauchy problem for overdetermined systems. *Ann. Mat. Pura ed Appl.*, 156(IV): 265–321, 1990.

[170] I. Naruki. Localization principle for differential complexes and its applications. *Publ. RIMS Kyoto Univ.*, 8(1): 43–110, 1973.

[171] A. Nestke. Cohomology of elliptic complexes and the fixed point formula of Atiyah and Bott. *Math. Nachr.*, 104: 289–313, 1981.

[172] F. Norguet. Introduction aux fonctions de plusieurs variables complexes répresentations intégrales. *Springer Lecture Notes in Math.*, 409: 74–82, 1974.

[173] V. P. Palamodov. On systems of differential equations with constant coefficients. *Dokl. AN SSSR*, 148(3): 523–526, 1963.

[174] V. P. Palamodov. Systems of linear differential equations. In: *Mathematical analysis*, VINITI, Moscow, 1966, 5–37.

[175] V. P. Palamodov. *Linear differential operators with constant coefficients.* Nauka, Moscow, 1967 488 pp. (Russian). (English transl.: Springer-Verlag, 1970.)

[176] V. P. Palamodov. Differential operators in the class of convergent power series and the preparatory lemma of Weierstrass. *Funkz. Analiz*, 2(3): 58–69, 1968.

[177] J. Peetre. Rectifications à l'article "Une caractérisation abstraite des opérateurs différentiels". *Math. Scand*, 8: 116–120, 1960.

[178] I. G. Petrowsky. On the diffusion of waves and the lacunas for systems of hyperbolic equations. *Mat. Sb*, 17: 289–370, 1945.

[179] I. G. Petrowsky. *Selected works*, pages 435–448. Nauka, Moscow, 1986 (Russian).

[180] L. Piccinini. Non-surjectivity of the Cauchy-Riemann operator on the space of analytic functions on $\mathbb{R}^n$. Generalization to parabolic operators. *Bull. U.M.I.*, 87(4): 12–28, 1973.

[181] V. Pillat and B.-W. Schulze. Elliptische Randwert-Probleme für Komplexe von Pseudodifferential operatoren. *Math. Nachr.*, 94: 173–210, 1980.

[182] J. Pommaret. *Systems of partial differential equations and Lie pseudogroups.* Gordon and Breach Sci. Publ., New York-London-Paris, 1978.

[183] L. S. Pontryagin. *Smooth manifolds and their applications in the theory of homotopies.* Nauka, Moscow, 1976.

[184] B. B. Prenov and N. N. Tarkhanov. A remark on the jump of the Martinelli-Bochner integral. *Sibirsk. Mat. Zn.*, 29(6), 1988.

[185] B. B. Prenov and N. N. Tarkhanov. On the singular Martinelli-Bochner integral. *Sibirsk. Mat. Zn.*, 33(2): 202–205, 1992.

[186] D. C. Quillen. Formal properties of overdetermined systems of linear partial differential equations. Technical report, Harvard, 1964.

[187] R. M. Range. The $\bar{\partial}$-Neumann operator on the unit ball in $\mathbb{C}^n$. *Math. Ann.*, 266(4): 449–456, 1984.

[188] St. Rempel and B.-W. Schulze. *Index theory of elliptic boundary problems.* Akademie-Verlag, Berlin, 1982.

[189] F. Riesz and B. Sz.-Nagy. *Lecons d'analyse fonctionnelle.* Acad. des Sci. de Hongrie, Budapest, 1952.

[190] Ya. A. Roitberg. On values on the boundary of domain of generalized solutions of elliptic equations. *Mat. Sb.*, 86(2): 248–267, 1971.

[191] S. N. Samborskii. Boundary value problems for overdetermined systems of equations with partial derivatives. Technical Report Preprint N 81.48, Inst. of Math., Kiev, 1981. 46 pp.

[192] P. Schapira. *Théorie des hyperfonctions.* Springer-Verlag, Berlin-Heidelberg-New York, 1970.

[193] B.-W. Schulze and G. Wildenhain. *Methoden der Potentialtheorie für elliptische Differentialgleichungen beliebiger Ordnung.* Academie-Verlag, Berlin, 1977.

[194] L. Schwartz. Espaces de fonctions différentiables à valeurs vectorielles. *J. Anal. Math. Jerusalem*, 4: 88–148, 1954.

[195] L. Schwartz. Théorie des distributions à valeurs vectorielles. Chap. 1. *Ann. Inst. Fourier*, 7: 1–141, 1957.

[196] L. Schwartz. Théorie des distributions à valeurs vectorielles. Chap. 2. *Ann. Inst. Fourier*, 8: 1–209, 1958.

[197] R. Seeley. Topics in pseudo-differential operators. In Cremonese, editor, *Pseudo-differential operators*, pages 167–306. CIME, Roma, 1969.

[198] Jean-Pierre Serre. Un théorème de dualité. *Comment. Math. Helv.*, 29(1): 9–26, 1955.

[199] V. I. Shevchenko. On the Hilbert problem for a holomorphic vector in many-dimensional space. In: *Dif. and Int. Eq.*, Tbilisi Univ. Press, 1979, 279–291.

[200] G. E. Shilov. *Mathematical analysis. Second special course.* Moscow Univ. Press, 2nd ed. edition, 1984.

[201] M. A. Shubin. *Pseudodifferential operators and spectral theory.* Springer-Verlag, Berlin et al., 1987.

[202] I.V. Skrypnik. a-harmonic forms on a Riemannian space. In: *Complex methods in mathematical physics (Trans, of All-Union school)*, Donezk, 1984, 96–102.

[203] D. C. Spencer. Deformation of structures on manifolds defined by transitive, continuous pseudogroups. I. *Ann. Math.*, 76(2): 306–398, 1962.

[204] D. C. Spencer. Deformation of structures on manifolds defined by transitive, continuous pseudogroups. II. *Ann. Math.*, 76(3): 399–445, 1962.

[205] D. C. Spencer. Harmonic integrals and Neumann problems associated with linear partial differential equations. In: *Outlines of the Joint Soviet-American Symp. on Part. Diff. Eq.*, Novosibirsk, 1963, 253–260.

[206] D. C. Spencer. Existence of local coordinates for structures defined by elliptic pseudogroups. In: *Diff. Anal.*, Oxford Univ. Press, London, 1964, 135–162.

[207] D. C. Spencer. Deformation of structures on manifolds defined by transitive, continuous pseudogroups. III. *Ann. Math.*, 81(3): 389–450, 1965.

[208] D. C. Spencer. Overdetermined systems of linear partial differential equations. *Bull. Amer. Math. Soc.*, 75(2): 179–239, 1969.

[209] M. Spivak. *A comprehensive introduction to differential geometry. Vol. I.* Publish. or Perish, Inc., Waltham, Mass., 1970.

[210] W. J. Sweeney. A non-compact Dirichlet norm. *Proc. Nat. Acad. Sei. USA*, 58: 2193–2195, 1967.

[211] W. J. Sweeney. The D-Neumann problem. *Acta Math.*, 120(3–4): 223–277, 1968.

[212] W. J. Sweeney. A uniqueness theorem for the Neumann problem. *Ann. Math.*, 90(2): 353–360, 1969.

[213] W. J. Sweeney. Estimates for elliptic complexes of higher order. *J. Differential Equations*, 10: 112–140, 1971.

[214] W. J. Sweeney. Coerciveness in the Neumann problem. *J. Differential Geometry*, 6: 375–393, 1971/72.

[215] W. J. Sweeney. A characterization of local subellipticity for pseudodifferential complexes. *J. Differential Equations*, 13: 495–505, 1973.

[216] W. J. Sweeney. A condition for subellipticity in Spencer's Neumann problem. *J. Differential Equations*, 21(2): 316–362, 1976.

[217] W. J. Sweeney. Subellipticity for Poincaré complexes with non-degenerate Levi form. *J. Differential Equations*, 36(1): 99–116, 1980.

[218] N. N. Tarkhanov. A characteristic property of closed exterior differential forms. In: *On holomorphic functions of several complex variables*, Inst. Fiz. SO Akad. Nauk SSSR, Krasnoyarsk, 1976, 203–209 (Russian).

[219] N. N. Tarkhanov. A formula and estimates for solutions of the equation $du = f$ in a domain and on its boundary. *Izv. VUZ. Mat.*, 5: 58–66, 1980.

[220] N. N. Tarkhanov. On an integral representation for solutions of systems of first order partial differential equations and some its applications. In: *Some questions of many-dimensional complex analysis*, Inst. Fiz. SO Akad. Nauk SSSR, Krasnoyarsk, 1980, 147–160 (Russian).

[221] N. N. Tarkhanov. An analogue of the logarithmic residue formula for solutions of first order elliptic systems, and weight estimates for solutions of the d-problem. *Sibirsk. Mat. Zh.*, 23(3): 188–197, 1982 (Russian). (English transl.: *Siberian Math. J.* 23, 1982.)

[222] N. N. Tarkhanov. Grothendieck duality theorem for elliptic complexes. In: *Analytic Functions (Conf., Blazejewko)*, Univ. of Łodz, 1982, 51.

[223] N. N. Tarkhanov. Fundamental solutions of elliptic complexes and their applications. *Izv. VUZ. Mat.*, 6: 33–42, 1983 (Russian). (English transl.: *Soviet Math. (Iz. VUZ).*)

[224] N. N. Tarkhanov. Jump theorems and the exterior product of currents in $D'^q(M)$ by means of their harmonic representations. *Sibirsk. Mat. Zh.*, 24(2): 203–204, 1983 (Russian). (English transl.: *Siberian Math. J.* 24, 1983.)

[225] N. N. Tarkhanov. *The Koppelman-type integral representations and other methods of complex analysis by application to elliptic complexes of differential operators.* PhD thesis, Ekaterinburg Univ., Ekaterinburg, 1983 150 pp. (Russian).

[226] N. N. Tarkhanov. The Sokhotskii-Plemelj formulas for elliptic complexes, and their applications. Technical Report Preprint N 24, Inst. of Physics, Krasnoyarsk, 1983. 48 pp.

[227] N. N. Tarkhanov. On a problem of Kohn and Nirenberg. *Sibirsk. Mat. Zh.*, 25(5): 200–203, 1984 (Russian). (English transl.: *Siberian Math. J.* 25, 1984.)

[228] N. N. Tarkhanov. On approximating properties of solutions of systems transposed to elliptic systems. In: *Complex analysis and its applications to partial differential equations*, Martin Lüther Univ., Halle (Saale), 1984, 34.

[229] N. N. Tarkhanov. On evaluation of the Poincaré index. *Izv. VUZ. Mat.*, 9: 47–50, 1984 (Russian). (English transl.: *Soviet Math. (Iz. VUZ).*)

[230] N. N. Tarkhanov. On a kernel approach in solving the equation $Pu = f$ for an elliptic complex P. *Sibirsk. Mat. Zh.*, 25(4): 179–191, 1984 (Russian). (English transl.: *Siberian Math. J.* 25, 1984.)

[231] N. N. Tarkhanov. On Poincaré duality for elliptic complexes. In: *Many-dimensional complex analysis*, Inst. Fiz. SO Akad. Nauk SSSR, Krasnoyarsk, 1985, 173–185 (Russian). (English transl.: *Amer. Math. Soc. Transl* 146: 75–83, 1990.)

[232] N. N. Tarkhanov. On the argument principle for elliptic complexes. *Izv. VUZ. Mat.*, 12: 74–75, 1985 (Russian). (English transl.: *Soviet Math. (Iz. VUZ).*)

[233] N. N. Tarkhanov. On Alexander duality for elliptic complexes. *Mat. Sb.*, 130(1): 62–85, 1986 (Russian). (English transl.: *Math. USSR Sbornik* 58: 59–82, 1986.)

[234] N. N. Tarkhanov. On Laurent series for elliptic complexes. *Dokl. Akad. Nauk SSSR*, 291(1): 40–44, 1986 (Russian). (English transl.: *Soviet Math. Dokl.* 34: 437–440, 1987.)

[235] N. N. Tarkhanov. A remark on the Moisil-Theodoresko system. *Sibirsk. Mat. Zh.*, 28(3): 208–213, 1987 (Russian). (English transl.: *Siberian Math. J.* 28, 1987.)

[236] M. Taylor. *Pseudodifferential operators.* Princeton Univ. Press, Princeton, 1983. 451 pp.

[237] H. G. Tillmann. Darstellung der Schwartzschen Distributionen durch analytische Functionen. *Math. Z.*, 77(2): 106–124, 1961.

[238] D. Toledo. On the Atiyah-Bott formula for isolated fixed points. *J. Differential Geom.*, 8(3): 401–436, 1973.

[239] D. Toledo and Y. L. Tong. A parametrix for $\overline{\partial}$ and Riemann-Roch in Čech theory. *Topology*, 15(4): 273–301, 1976.

[240] F. Treves. *Topological vector spaces, distributions and kernels.* Acad. Press, New York - London, 1967.

[241] H. Triebel. *Interpolation theory, function spaces, differential operators.* VEB Wiss. Verlag, Berlin, 1978. 528 pp.

[242] B. L. Van der Waerden. *Moderne Algebra.* Leipzig, 1940.

[243] A. M. Vinogradov. Geometry of non-linear differential equations. In: *Problems of Geometry. Vol. 11*, VINITI, Moscow, 1980, 89–134 (Russian).

[244] A. M. Vinogradov, I. S. Krasil'shchik, and V. V. Lychagin. *Introduction to geometry of non-linear differential equations.* Nauka, Moscow, 1986.

[245] V. S. Vinogradov. An analog of the Cauchy type integral for analytic functions of several complex variables. *Dokl. AN SSSR*, 178(2): 282–285, 1968 (Russian).

[246] V. S. Vladimirov. The problem of linear conjugation of holomorphic functions of several complex variables. *Izv. AN SSSR. Ser. mat.*, 29(4): 807–834, 1965.

[247] V. S. Vladimirov. The Laplace transformation of temperate distributions. In: *Modern problems of mathematics. Vol. 1*, VINITI, Moscow, 1972, 61–84 (Russian).

[248] V. S. Vladimirov and V. V. Zharinov. Closed forms associated with linear differential operators. *Diff. Eq.*, 16(5): 845–867, 1980.

[249] D. Vogt. Distributionen auf dem $\mathbb{R}^n$ als Randverteilungen holomorpher Funktionen. *J. reine und angew. Math.*, 261: 131–145, 1973.

[250] R. Wawak. Some remarks about the product of distributions. *Bull. Pol. Acad. Sci. Math.*, 32(3–4): 179–183, 1884.

[251] N. Weck. Liouville theorems for linear elliptic systems. *Proc. Roy. Soc Edinburgh*, 94(3–4 A): 309–322, 1883.

[252] A. Weil. Sur les théorèmes de de Rham. *Comment. Math. Helv.*, 26: 119–145, 1852.

[253] B. M. Weinstock. Inhomogeneous Cauchy-Riemann systems with smooth dependence on parameters. *Duke Math. J.*, 40(2): 307–312, 1973.

[254] R. Wells. *Differential analysis on complex manifolds.* Prentice-Hall, Englewood Cliffs, N.J., 1973.

[255] G. Wildenhain. Darstellung von Lösungen linearer elliptische Differentialgleichungen. *Math. Res.*, 8, 1981, 92 pp.

[256] W. Wirtinger. Eine Determinantenidentität und ihre Anwendung auf analytische Gebielde in Euklidischer und Hermitischer Massbestimmung. *Monatsh. Math. und Phys.*, 44: 343–365, 1936.

[257] A. P. Yuzhakov and A. V. Kuprikov. On the logarithmic residue in $\mathbb{C}^n$. In: *Some properties of holomorphic functions of several complex variables*, Inst. Fiz. SO Akad. Nauk SSSR, Krasnoyarsk, 1973, 181–191 (Russian).

[258] A. I. Zaslavskii. *Multiplication of distributions and the method of asymptotic expansions, Thesis.* PhD thesis, Ekaterinburg Univ., Ekaterinburg, 1983 80 pp. (Russian).

[259] P. Zecca. Eliminabilita delle singolarita compatte per le soluzioni dei sistemi sopradeterminati ellittici omogenei a coefficienti constanti. *Ric. mat.*, 30(1): 171–175, 1981.

[260] N. Zeylon. Das Fundamentalintegral der allgemeinen partiellen linearen differentialgleichungen mit konstanten koeffizienten. *Ark. Mat. Astr. Fys.*, 6(38): 1–32, 1911.

[261] V. V. Zharinov. Green formula for non-linear systems of partial differential equations. *Dokl. AN SSSR*, 296(4): 785–787, 1987.

Name Index

Subject Index

Index of Notation

Other *Mathematics and Its Applications* titles of interest:

B.A. Plamenevskii: *Algebras of Pseudodifferential Operators*. 1989, 304 pp. ISBN 0-7923-0231-1

Ya.I. Belopolskaya and Yu.L. Dalecky: *Stochastic Equations and Differential Geometry*. 1990, 288 pp. ISBN 90-277-2807-0

V. Goldshtein and Yu. Reshetnyak: *Quasiconformal Mappings and Sobolev Spaces*. 1990, 392 pp. ISBN 0-7923-0543-4

A.T. Fomenko: *Variational Principles in Topology. Multidimensional Minimal Surface Theory*. 1990, 388 pp. ISBN 0-7923-0230-3

S.P. Novikov and A.T. Fomenko: *Basic Elements of Differential Geometry and Topology*. 1990, 500 pp. ISBN 0-7923-1009-8

B.N. Apanasov: *The Geometry of Discrete Groups in Space and Uniformization Problems*. 1991, 500 pp. ISBN 0-7923-0216-8

C. Bartocci, U. Bruzzo and D. Hernandez-Ruiperez: *The Geometry of Supermanifolds*. 1991, 242 pp. ISBN 0-7923-1440-9

N.J. Vilenkin and A.U. Klimyk: *Representation of Lie Groups and Special Functions. Volume 1: Simplest Lie Groups, Special Functions, and Integral Transforms*. 1991, 608 pp. ISBN 0-7923-1466-2

A.V. Arkhangelskii: *Topological Function Spaces*. 1992, 206 pp. ISBN 0-7923-1531-6

Kichoon Yang: *Exterior Differential Systems and Equivalence Problems*. 1992, 196 pp. ISBN 0-7923-1593-6

M.A. Akivis and A.M. Shelekhov: *Geometry and Algebra of Multidimensional Three-Webs*. 1992, 358 pp. ISBN 0-7923-1684-3

A. Tempelman: *Ergodic Theorems for Group Actions*. 1992, 400 pp. ISBN 0-7923-1717-3

N.Ja. Vilenkin and A.U. Klimyk: *Representation of Lie Groups and Special Functions, Volume 3. Classical and Quantum Groups and Special Functions*. 1992, 630 pp. ISBN 0-7923-1493-X

N.Ja. Vilenkin and A.U. Klimyk: *Representation of Lie Groups and Special Functions, Volume 2. Class I Representations, Special Functions, and Integral Transforms*. 1993, 612 pp. ISBN 0-7923-1492-1

I.A. Faradzev, A.A. Ivanov, M.M. Klin and A.J. Woldar: *Investigations in Algebraic Theory of Combinatorial Objects*. 1993, 516 pp. ISBN 0-7923-1927-3

M. Puta: *Hamiltonian Mechanical Systems and Geometric Quantization*. 1993, 286 pp. ISBN 0-7923-2306-8

V.V. Trofimov: *Introduction to Geometry of Manifolds with Symmetry*. 1994, 326 pp. ISBN 0-7923-2561-3

Other *Mathematics and Its Applications* titles of interest:

J.-F. Pommaret: *Partial Differential Equations and Group Theory. New Perspectives for Applications*. 1994, 473 pp. ISBN 0-7923-2966-X

Kichoon Yang: *Complete Minimal Surfaces of Finite Total Curvature*. 1994, 157 pp. ISBN 0-7923-3012-9

N.N. Tarkhanov: *Complexes of Differential Operators*. 1995, 414 pp. ISBN 0-7923-3706-9

GPSR Compliance

The European Union's (EU) General Product Safety Regulation (GPSR) is a set of rules that requires consumer products to be safe and our obligations to ensure this.

If you have any concerns about our products, you can contact us on ProductSafety@springernature.com

In case Publisher is established outside the EU, the EU authorized representative is:

Springer Nature Customer Service Center GmbH
Europaplatz 3
69115 Heidelberg, Germany

Batch number: 10607320

Printed by Printforce, the Netherlands